The Unity of the Fundamental Interactions

THE SUBNUCLEAR SERIES

Series Editor: **ANTONINO ZICHICHI**
European Physical Society
Geneva, Switzerland

1. 1963 STRONG, ELECTROMAGNETIC, AND WEAK INTERACTIONS
2. 1964 SYMMETRIES IN ELEMENTARY PARTICLE PHYSICS
3. 1965 RECENT DEVELOPMENTS IN PARTICLE SYMMETRIES
4. 1966 STRONG AND WEAK INTERACTIONS
5. 1967 HADRONS AND THEIR INTERACTIONS
6. 1968 THEORY AND PHENOMENOLOGY IN PARTICLE PHYSICS
7. 1969 SUBNUCLEAR PHENOMENA
8. 1970 ELEMENTARY PROCESSES AT HIGH ENERGY
9. 1971 PROPERTIES OF THE FUNDAMENTAL INTERACTIONS
10. 1972 HIGHLIGHTS IN PARTICLE PHYSICS
11. 1973 LAWS OF HADRONIC MATTER
12. 1974 LEPTON AND HADRON STRUCTURE
13. 1975 NEW PHENOMENA IN SUBNUCLEAR PHYSICS
14. 1976 UNDERSTANDING THE FUNDAMENTAL CONSTITUENTS OF MATTER
15. 1977 THE WHYS OF SUBNUCLEAR PHYSICS
16. 1978 THE NEW ASPECTS OF SUBNUCLEAR PHYSICS
17. 1979 POINTLIKE STRUCTURES INSIDE AND OUTSIDE HADRONS
18. 1980 THE HIGH-ENERGY LIMIT
19. 1981 THE UNITY OF THE FUNDAMENTAL INTERACTIONS

Volume 1 was published by W. A. Benjamin, Inc., New York; 2–8 and 11–12 by Academic Press, New York and London; 9–10 by Editrice Compositori, Bologna; 13–19 by Plenum Press, New York and London.

The Unity of the Fundamental Interactions

Edited by
Antonino Zichichi

European Physical Society
Geneva, Switzerland

PLENUM PRESS • NEW YORK AND LONDON

Library of Congress Cataloging in Publication Data

International School of Subnuclear Physics (19th: 1981: Erice, Italy)
  The unity of the fundamental interactions.

(The Subnuclear series; 19)
  "Proceedings of the Nineteenth Course of the International School of Subnuclear Physics, held July 31–August 11, 1981, in Erice, Trapani, Sicily"—Verso t.p.
  Includes bibliographical references and index.
  1. Nuclear reactions—Congresses. I. Zichichi, Antonino. II. Title. III. Series.
QC794.8.H5I577 1981   539.7′54   82-22466
ISBN 0-306-41242-X

Proceedings of the Nineteenth Course of the International School of Subnuclear
Physics, held July 31–August 11, 1981, in Erice, Trapani, Sicily

© 1983 Plenum Press, New York
A Division of Plenum Publishing Corporation
233 Spring Street, New York, N.Y. 10013

All rights reserved

No part of this book may be reproduced, stored in a retrieval system, or transmitted,
in any form or by any means, electronic, mechanical, photocopying, microfilming,
recording, or otherwise, without written permission from the Publisher

Printed in the United States of America

P. A. M. Dirac

PREFACE

From 31 July to 11 August 1981, a group of 108 physicists from 75 laboratories in 27 countries met in Erice for the 19th Course of the International School of Subnuclear Physics. The countries represented were Argentina, Australia, Austria, Belgium, Brazil, Bulgaria, Canada, Denmark, the Federal Republic of Germany, Finland, France, Greece, Hungary, India, Israel, Italy, Japan, Korea, the Netherlands, Norway, Poland, Sweden, Turkey, the United Kingdom, the United States of America, Venezuela, and Yugoslavia. The School was sponsored by the Italian Ministry of Public Education (MPI), the Italian Ministry of Scientific and Technological Research (MRST), the Regional Sicilian Government (ERS), and the Weizmann Institute of Science.

The programme of the School was mainly devoted to a review of the most significant results, both in theory and experiment, obtained in the field of high-energy interactions. The outcome of the School was to present a clear picture of how far we are along the fascinating route towards understanding the deep meaning of the natural laws of hadronic and leptonic matter -- the final goal being the unity of all forces.

As stated by P.A.M. Dirac in a discussion, physicists should try to be more modest. Nowadays the trend is to understand everything: from the origin and the fate of the Universe to the substructure of a quark, the unity of all forces of nature. This volume contains a very significant sample of the present status of this ambitious game in which subnuclear physicists have become involved. It is certainly not their fault if the research in high-energy physics has now developed to a degree which was unthinkable a few years ago. This unexpected and spectacular development needs to be guided by intellectual modesty -- not by arrogance -- if we want that our scientific activity should develop along a continuing successful road.

I hope the reader will enjoy the book as much as the students enjoyed attending the lectures and the discussion sessions, which are one of the most attractive features of the School. Thanks to the work of the Scientific Secretaries the discussions have been reproduced as faithfully as possible. At various stages of my work I have enjoyed the collaboration of many friends whose contributions have been extremely important for the School and are highly appreciated. I thank them most warmly. A final acknowledgement to all those who, in Erice, Bologna, and Geneva, have helped me on so many occasions and to whom I feel very indebted.

                                                Antonino Zichichi
                                                October 1981
                                                Geneva

CONTENTS

OPENING LECTURE

The end of the High-Energy Frontier 1
    S.L. Glashow

THEORETICAL LECTURES

The Magnetic Monopole Fifty Years later 21
    S. Coleman

Numerical Studies of Gauge Field Theories 119
    M.J. Creutz

What can we learn from the next generation of experiments?. . 157
    G.G. Ross

Supersymmetric Unified Models 237
    S. Dimopoulos and F. Wilczek

Erice lectures on Cosmology 251
    F. Wilczek

Introduction to Supersymmetry 305
    E. Witten

SEMINARS ON SPECIALIZED TOPICS

Neutrino Physics at Fermilab 373
    D. Jovanovic

Heavy Flavour Production at the Highest Energy (pp)
Interactions 409
    M. Basile, C. Bonvicini, G. Cara Romeo, L. Cifarelli,
    A. Contin, G. D'Ali, B. Esposito, P. Giusti, T. Massam,
    R. Nania, F. Palmonari, A. Petrosino, G. Sartorelli,
    G. Valenti, and A. Zichichi

b-Quark Physics 471
      K. Berkelman

Hadron Production in e^+e^- Annihilation 507
      R. Cashmore

Search for New Particles and Electroweak Interference
Effects in e^+e^- Interactions 601
      P. Duinker

Update on CP-Violation 677
      V.L. Fitch

What can we learn from High-Energy, Soft (pp) Interactions. . 695
      M. Basile, G. Bonvicini, G. Cara Romeo, L. Cifarelli,
      A. Contin, M. Curatolo, G. D'Ali, B. Esposito,
      P. Giusti, T. Massam, R. Nania, F. Palmonari,
      A. Petrosino, V. Rossi, G. Sartorelli, M. Spinetti,
      G. Susinno, G. Valenti, L. Votano, and A. Zichichi

THE GLORIOUS DAYS OF PHYSICS

My Life as a Physicist 733
      P.A.M. Dirac

My Life as a Physicist 751
      E. Teller

The Glorious Days of Physics 765
      E.P. Wigner

CLOSING LECTURE

What have we learned? 775
      E. Teller

CLOSING CEREMONY 791

PARTICIPANTS 795

INDEX 807

THE END OF THE HIGH-ENERGY FRONTIER

Sheldon L. Glashow

Lyman Laboratory of Physics
Harvard University
Cambridge, MA 02138 U.S.A.

I. INTRODUCTION

In the past twenty years, the great accelerators have dominated the field of high-energy physics. Consider the 16 basic particles of contemporary physics: six leptons, photons, gluons, W's and Z's. Seven of them were "discovered" before 1960 without the use of artificial acceleration. Five were discovered since 1960 and with accelerators. Brookhaven may claim muon neutrinos and part of charm; SPEAR found the tau lepton and the rest of charm; Fermilab found beauty; and PETRA saw the spoor of gluons. Four fundamental particles remain: the tau neutrino, the top quark, and the two weak-interaction intermediaries. These particles will almost certainly turn up at accelerators during the eighties. It seems folly to speak now of the end of the high-energy frontier. So much is to be done, and so many machines are abuilding.

Since we are concerned with the highest available center-of-mass energies, we will concentrate on particle colliders; hadron-hadron, lepton-hadron and lepton-lepton.

ISR has made energies of 60 GeV available in hadron-hadron collisions. The CERN collider will soon make a giant step to

500 GeV. This machine, and its American variations at Fermilab (2000 GeV) and at Brookhaven (800 GeV), should be able to produce and detect W's, Z's and, possibly, a new flavor of quark or two. They should be regarded as prototypes for a truly large hadron collider with c.m. measured in tens of TeV. Remember: there were no surprises at ISR energies. The absence of low-energy indicators strongly suggests that there will be no surprises at one TeV. Nevertheless, we must push on to tens of TeV. Here is the last hope for the truly high-energy frontier.

Electron-positron machines now operate in the high thirties of GeV. Forty-five GeV is in sight at a rejuvenated PETRA, and sixty GeV will be available at KEK by 1986. Soon afterwards, LEP-I will attain 100 GeV, and eventually, 280 GeV. True, these energies are small compared to what will be available in hadron colliders. Yet, electron-positron machines are the instruments of choice for the study of heavy quarks, for precision measurements at the Z^0 mass, for the possible discovery of the Higgs boson, and for other worthy purposes.

Lepton-hadron collisions have been studied at fixed-target machines at center-of-mass energies up to 30 GeV. Electron-proton colliders have not as yet been built, but I am sure that they will be. It is an obligatory direction for study, and new exciting phenomena may show up in the TeV domain.

Imagine one central machine complex that will provide hadron-hadron collisions at $\sim$10 TeV, lepton-hadron collisions at $\sim$1 TeV, and electron-positron collisions at $\sim$0.1 TeV. The present state of society, with its financial and energy constraints make a machine of the kind seem "ultimate". It may well represent our last sortie into the high-energy frontier. CUPID, the CERN ultimate particle interaction device (or STUPID, if it is built at SLAC), may be our last great machine.

In the past, many great achievements which relate to high-energy physics were discovered without the artificial aid of particle accelerators. Muons, pions, strange particles, and positrons were discovered in cosmic rays. Neutrons were first seen in table-top experiments, and neutrinos were seen at nuclear reactors. The V-A form of the beta interaction was determined from the study of radioactive materials. It is only recently that large accelerators have dominated the new discoveries. Although it is clear that they will continue to play a very important role in the future, it is also clear that much remains to be done in the neglected field of non-accelerator fundamental physics. I will devote this lecture to a few of these possibilities.

II. NEW FORMS OF STABLE MATTER

The discovery of argon at the close of the nineteenth century was a spectacular surprise. The existence of a zero-valence column of the periodic table was quite unanticipated, and was in fact rejected by Mendeleev. Similar surprises may await us in the future. Of course, there are no new stable elements waiting to be discovered. Of course, the putative new stable particle cannot have a terrestrial abundance comparable with argon. What we may have missed is something new and different and very rare. Here are some possibilities:

(a) <u>Magnetic Monopoles</u>

I shall not review this subject. Sensitive but unsuccessful searches have been done. Many of these searches would not have been sensitive to the very heavy monopoles of grand unified theories. Further cosmic ray searches are possible. So are searches for monopoles within old magnetic ores or within meteorites. Better limits on the net magnetic charge of the moon or of other memebers of the solar system can be obtained.

(b) <u>Antimatter</u>

New evidence indicates an antimatter component in cosmic rays. Its energy-dependence and its Z-dependence must be studied. Although, this is not strictly a new and unexpected form of stable matter, the existence of a measurable flux of cosmic antimatter must be understood in turns of the ambitious big-bang cosmogeny.

(c) <u>Free Quarks</u>

Here, I am referring to the possible occurrence of stable, isolated systems with electric charges of $N \pm 1/3$ where N is an integer. Fairbanks and his collaborators have reported evidence for the existence of such systems. Many experiments are in planning to test this result. The importance of this endeavor is clear.

(d) <u>Heavy "Isotopes"</u>

New forms of stable matter may bind to atomic nuclei forming heavy isotopes of conventional chemical elements. A recent Rutherford Laboratory experiment puts very strong limits on the existence of such isotopes of hydrogen. The following parable shows that such experiments must be repeated with other chemical elements: Suppose that the weak interactions were strangeness-conserving. Of the strange particles, only $K^\pm$, Λ, Ξ^0 and Ω^- would be stable. Bound states of one Ξ or Ω with nuclei are unstable under the decay into two strange systems. A captured K^- yields the production of a hyperon or a hyper-nucleus. We are left with conventional hypernuclei or with isolated stable $K^+ e^-$ atoms. Among the hyper-nuclei, none with $Z = 1$ is stable under strong interactions. The $K^+ e^-$ will behave as a super-light hydrogen isotope. Thus, the Rutherford search for heavy hydrogen could not have discovered strange particles in our imitation world. On the other hand, a search for heavy isotopes of helium or lithium could have been fruitful. The point of the allegory is to insist that similar experiments be done on heavier chemical elements. In this connection, we are delighted to learn of a Los Alamos proposal to search for stable superheavy technetium isotopes.

What are the masses of the putative heavy isotopes? Some supersymmetry theories predict stable massless fermions which are color octets. These "gluinos" should result in the existence of isotopes which are only 1 - 2 GeV heavier than is expected from their quark content. Thus, these are not truly "superheavy" isotopes, but their masses are not expected to have the nearly integer values of conventional isotopes. Technicolor, on the other hand, suggests isotopes with TeV masses. The search for these particles has recently been described by R. Cahn and myself in Science and by C. K. Jorgensen in Nature. Finally, there may exist a few stable relic particles with masses associated with the grand-unified theory. These would give isotopes with masses of order 10^{15} GeV. Such particles may have fallen to the center of the earth, so that the search for ambient superheavies would be fruitless. Perhaps, they could be searched for in newly fallen meteorites or in the orbiting debris of the solar system. Perhaps a space probe to the stable Lagrange point of the earth-moon is in order.

III. PROTON DECAY

Here is the grand challenge of the low-energy frontier. In the simplest grand unified theories, there is little or nothing surprising awaiting us at higher but accessible energies: only the top quark, the Higgs boson, perhaps a fourth family, and perhaps an axion (visible or invisible). In this theory, at least, there is coincidence between the intrinsic impracticallity of ever larger machines and the predicted absence of new fundamental phenomena to discover. A projected 10-TeV collider can offer the best evidence for this theory--the negative (but significant) observation of absolutely nothing new. Mercifully, the theory offers us an alternative window into the world of higher and inaccessible energies. In these theories, the proton must decay at a perhaps observable rate. The proton lifetime is linked to the chromodynamic coupling constant and to the value of the weak mixing angle. It also depends upon the choice of unified model. For the plausible values of $\lambda_{\overline{MS}} = 0.1 - 0.3$ GeV, and for Georgi-Glashow SU(5), Marciano and Sirlin compute a value of $\sin^2\theta_W = 0.210$ and a value of the proton lifetime of $10^{30\pm2}$ years. It is already known that the proton lifetime is greater than 10^{30} years. The next generation of experiments will certainly be sensitive to lifetimes up to 10^{32} years. In the next year or two we will learn whether or not simple grand unification is a viable theory. In this essay, we assume that the result will be positive, and we ask what can be learned in future proton decay experiments.

We begin with some bizarre possibilities, corresponding to theories or notions more elaborate than conventional SU(5) unification.

(a) $\underline{\Delta B = 2 \text{ Processes}}$

This is, of necessity, a six-fermion process. It is not permitted in simple SU(5), which admits B-L as an exact symmetry. However, it may be allowed in other grand unified theories. If the characteristic mass scale is the unification mass M_G (where G is for Grand or Georgi), then B-L processes are quite unobservable. For $\Delta B = 2$ processes to compete with "conventional" proton decay, its characteristic mass M scale must satisfy. $M \sim (M_G)^{2/5}$ (in GeV) ~ 1000 TeV. It is essential to arrange for these new not-so-heavy intermediaries not to mediate ordinary $\Delta B = 1$ processes in by an induced four-fermion coupling. Models which produce $\Delta B = 2$ processes are discussed in the literature. It is, of course, essential to search for $\Delta B = 2$ decays. These could show up as 2 GeV hadronic events in the same apparatus which looks for proton decay. A $\Delta B = 2$ coupling can also induce neutron-antineutron mixing in a magnetically shielded environment. In 1979, I argued that the available limits on proton decay ensure that the lifetime for $n\bar{n}$ mixing must be at least 10^5 seconds. Searches for this process can be done with slow reactor neutrons—indeed, I have two or three proposals in hand. Limits on the $n\bar{n}$ transition time of $10^8 - 10^9$ seconds can and must be obtained.

This will correspond to a direct sensitivity for B = 2 decays of $10^{34} - 10^{36}$ years, well beyond the sensitivity of any now conceivable underground experiment. Perhaps we will be lucky and find such an effect. Perhaps the desert between M_F and M_G is not barren after all.

It is also possible to imagine a scenario (within SU(5)) that leads $\Delta S = 2$ processes <u>without</u> the possibility of $n\bar{n}$ transitions. Suppose there exists an SU(5) decimet of scalar mesons which is coupled to baryons. The decimet contains $(3,2) + (\bar{3},1) + (1,1)$ under SU(3) x SU(2). In order not to run into phenomenological difficulty, we must assume that the mass of the (3,2) is of order M_G. The remaining states are assumed to be light. (Thus, the 10 of mesons is a split multiplet just like the familiar 5). Through some loop diagram, a fifth-order self-coupling of the 10's is induced:

$$\sim \frac{1}{m} \phi_{ia} \phi_{jb} \phi_{kc} \phi_{\ell d} \phi_{me} \, \varepsilon^{abcde} \, \varepsilon^{ijk\ell m}$$

where m is the mass of the light members of the multiplet. A $\Delta B = 2$, $\Delta L = -4$ ten-fermion interaction is induced. It can lead to $\Delta B = 2$ decays which violate strangeness by three:

$$N + N \rightarrow K^+ + K^+ + \ell^- + \ell^- + \bar{\nu} + \bar{\nu},$$

where ℓ^- denotes either e^- or μ^-. Clearly, such an interaction cannot contribute to $n\bar{n}$ mixing. I can identify no unacceptable consequences of this <u>ansatz</u>. If the light members of the decimet have masses comparable with M_F, then these $\Delta B = 2$ processes can be observed. This model may well seem contrived and implausible. Nonetheless, the search for these bizarre decay modes can be done for free with a second-generation proton decay experiment.

(b) $\underline{\Delta B = -\Delta L \text{ Decay}}$

Here is another variation on SU(5) with a decimet of scalar mesons. This time, assume that the $(\bar{3},2)$ and the $(3,1)$ have masses less than M_G. Suppose as well that there is a coupling of the form $\phi_{10}\phi_{10}\phi_5\phi_{24}$ among the various meson multiplets. An effective SU(2)-violating mass term of order $M_G M_F \phi_{10} \phi_{10}$ is induced which will lead to mixing between the (3,1) and the Q = 2/3 member of the (3,2). As a consequence, the $\Delta B = -\Delta L$ strangeness violating decays

$$N \to K^+ \mu^-$$
$$P \to K^0 \pi^+ \mu^-$$

are induced. These decay modes can be significant (or even dominant) providing that

$$M \sim (M_G^3 M_F)^{1/4} \sim 10^{11} \text{ GeV},$$

where M is the mass of the light members of the decimet. If these are the dominant decay modes of nucleons, we expect protons to be considerably longer lived than bound neutrons. Limits on the $K^+ \mu^-$ decay mode are an obligatory duty for a second-generation proton decay experiment.

(c) <u>Higgs Dominant Proton Decay</u>

Suppose that there exists but a single Higgs 5-plet, and that its colored members dominate proton decay. In this case, we can immediately conclude that the important modes are

$$P \to K^0 \mu^+$$
$$N \to K^0 \pi^- \mu^+$$

In this situation, it is the proton that is the shorter lived nucleon. The search for $K^0 \mu^+$ is mandatory, for most models make it a significant decay mode. It is imperative to determine whether or not it is the dominant decay mode.

(d) <u>The Conventional Scenario</u>

We have discussed a number of unconventional (and, I believe, unlikely) perversions of proton decay. In this section, we discuss the decay of nucleons as mediated by the gauge bosons of SU(5) or its simple extension to O(10). Thus, $\Delta B = \Delta L = -1$ is assumed to be a valid law. Moreover, the final state must contain

THE HIGH-ENERGY FRONTIER

no more than one lepton. Decay modes such as
$$P \to e^+ e^- \mu^-,$$
while $\Delta B = \Delta L = -1$, are necessarily six-fermion couplings, not gauge-mediated four-fermion couplings.

In SO(10), there are two distinct (3,2) multiplets of gauge bosons which contribute to proton decay. To a sufficient approximation, each multiplet is degenerate, and they do not mix. The electrical charges of one multiplet are (-1/3, -4/3), of the other (2/3, -1/3). The SU(5) limit is obtained when the first multiplet is much lighter than the second. The opposite case, which is also physically plausible, we refer to as anti-SU(5). Here are some ways in which the two limits differ:

Decay Mode	$(S=0)e^+$	$(S=1)\mu^+$	ν
SU(5)	67%	13%	20%
Anti-SU(5)	50%	0	50%

These are, of course, crude estimates of quantities subject to severe strong-interaction corrections. We should emphasize the fact that as much as ~13% of proton decay should produce muons. Those muons associated with $S=1$ final states should be unpolarized. The second generation of proton decay detector should have the capability of measuring muon polarization. Measurement of the K/π ratio in proton decay is evidently essential to unravel the structure of the effective four-fermion coupling. Note the predicted suppression of the $K\mu$ mode in anti-SU(5).

At least as important is the study of the Cabibbo suppressed modes, $K^0 e^+$ and $\pi^0 \mu^+$. For the SU(5) model, we compute two values for these depending upon two extreme models of fermion masses. In one case (F-masses), the mass matrices of down quarks and charged leptons are assumed to commute. For the case of S-masses, the up quark and lepton matrices commute. Of course, nature may not admit such simplicities, and indeed, the Cabibbo suppressed amplitudes may not be suppressed at all. In any case, here are our expectations for SU(5):

	F-MASSES	S-MASSES
$\left[\dfrac{K^0 e^+}{K^0 \mu^+}\right]$	3%	14%
$\left[\dfrac{\pi^0 \mu^+}{\pi^0 e^+}\right]$	1%	2%

For anti-SU(5), the first ratio is ill-defined since the numerator is Cabibbo suppressed while the denominator is absent. The πμ/πe ratio is predicted to be 5%. Clearly, the next generation of proton decay experiment must be sensitive to small branching ratios.

A simple discrimination between SU(5) and anti-SU(5) is possible should the ω decay modes be prevalent, as is suggested in recent calculations. In SU(5) we expect

$$\Gamma(P \to \omega e^+) = 5\Gamma(N \to \omega\nu),$$

while in anti-SU(5) the two rates are equal. Furthermore, the ων decay mode is one of the few neutrino decay modes which can be identified relatively easily.

(e) <u>The Total Decay Rate</u>

The proton decays in many ways. Only a few of these decay modes are detectable in the first generation of proton decay experiment. It is important to consider a decay-mode independent measure of the nucleon lifetime. Such an experiment has been suggested by E. Fireman. The key to this experiment is the decay of a nucleon in the potassium nucleus. He argues that ^{39}K will end up as radioactive ^{37}A in some 20% of the decays. A large and very deep underground tank of KOH solution is periodically examined for traces of ^{37}A. The experiment has much in common with the Davis search for solar neutrinos, with chlorine replaced by potassium. Fireman finds it likely that nucleon total lifetimes of up to 10^{31} years are detectable in this fashion.

IV. NEUTRINO PHYSICS

Neutrino masses, and the oscillations that come with them, have become fashionable once again. There are two reasons for this. One is theoretical. Years ago, it was regarded as natural and beautiful that neutrinos be massless. Today, a new philosophy has arisen to the effect that "the only good symmetry is a gauge symmetry", or equivalently, "any phenomenon that is not explicitly forbidden is allowed". Thus, it is that the proton __must__ decay, the neutron __must__ have an electric dipole moment, the $\mu \to e\gamma$ decay mode __is__ allowed, etc. The question is merely one of size or rate. By the same token, there is no established principle saying that neutrinos cannot have masses, and hence, they do. Moreover, within the framework of grand-unified theories, both the existence of, and the smallness of, neutrino masses can be understood. The general magnitude that is indicated is $m_F^2/m_G \simeq 10^{-2}$ eV. Small, but perhaps not too small. The second reason is experimental. Some results suggest neutrino oscillations: The Reines-Sobel reactor experiment and the Davis solar neutrino experiment are two examples. Other results suggest neutrino masses: The Soviet experiment on the tritium endpoint and the Doi, et al. reanalysis of tellurium double beta decay are another two. All this has made the subject a hot one. The community has embarked upon a series of new and more sensitive experiments:

(a) __Solar Neutrinos__

There certainly is a deficit in the solar neutrino flux which requires explanation. As is well-known, the Davis chlorine experiment is sensitive mostly to the high-energy neutrinos of the exotic ^{8}B side reaction. It is clearly necessary to perform an experiment which is sensitive to a wide range of neutrino energies. The most advanced plan involves fifty tons of expensive Gallium. It seems clear that this experiment must be performed. If the neutrino deficit is energy independent, its explanation in terms of neutrino oscillations becomes more likely. It is also possible that the lack of ^{8}B neutrinos is due to a time dependence of the central solar temperature. It has been pointed out that this hypothesis can be tested by an experiment measuring the integrated energetic neutrino flux over the past several million years. Cowan and Haxton suggest a search for neutrino-produced technetium in molybdenum ores which could reveal a time dependence of the ^{8}B neutrino flux.

(b) <u>Terrestrial Oscillation Experiments</u>

These may employ high-energy neutrinos from large accelerators, atmospheric neutrinos produced by cosmic rays, meson factories, nuclear reactors, or megacurie artificial sources. Many experiments have been done, are being done, and even more will be done. There is no firm evidence yet for oscillations, but future experiments may tell another story.

(c) <u>Double Beta-Decay</u>

The question is whether there is such a process as no-neutrino double beta decay, such as would be induced by a Majorana neutrino mass of order tens of eV. Recent reanalyses of geological data on the double beta decay of Tellurium isotopes suggest a neutrino mass of order 30 eV. Other renalyses of Selenium and Germanium data yield an upper limit of 15 eV to the Majorana neutrino mass. Clearly, there is room for much more experimental work. The radiochemical techniques that have been developed for the detection of solar neutrinos (chlorine and gallium), and for proton decay (potassium) strongly suggest the development of similar techniques for the study of double beta decay. A ton of Tellurium should produce 10^6 Xenon atoms per year by double beta decay. These can be separated and detected to give a reliable measurement of the double beta decay rates.

(d) <u>Endpoint Experiments</u>

A decade ago, Bergkvist obtained an upper limit of 60 eV to the mass of the electron neutrino by the study of the endpoint of tritium beta decay. Today, a Soviet experiment claims a measurement of a non-vanishing mass. This has stimulated several groups to try harder. Within a few years, we should know with some certainty whether or not the electron-neutrino has a mass greater than 10 eV. A particularly interesting development is the study of the photon endpoint in radiative electron capture. This can also provide a measurement of the neutrino mass. Experiments on appropriate isotopes of Platinum and Holmium are now in progress.

V. CONCLUSION

The drive towards higher energies has slowed. New machines will come, but not for many years. Some of us may live to see 10 TeV in the center of mass. It is indeed fortunate that there are other ways to reveal Nature's secrets. We have discussed a few, but there are many others. Much remains to be done in what was once the high-energy frontier. We must learn more about particles with charm and beauty. We need precise measurements to verify QCD and the electroweak theory. A more precise measurement of the CP-violation parameters in kaon decay is necessary, as is a precise determination of hyperon magnetic moments. The electric dipole moment of the neutron. The study of glueballs and baryonium. Better limits on decays such as $\mu \to e\gamma$ or $K \to \pi \mu^{\pm} e^{\mp}$. There is a long, long shopping list. I am convinced that there are very exciting experimental discoveries waiting to be made, and I am sure that they will be made.

DISCUSSION

CHAIRMAN: S.L. GLASHOW

Scientific Secretary: I. Cohen

DISCUSSION

- *GOLLIN:*

What is really the physical information from galaxies? How is it that massive neutrinos might help in the formation of galaxies?

- *GLASHOW:*

From studying the 21 cm radiation coming from atomic hydrogen we obtain galactic rotation curves which extend far outside the visible galaxies. This atomic hydrogen is in orbit around the galaxies. The analogous rotation curve for the planets of the solar system shows a velocity profile which falls off as $1/\sqrt{R}$ as one goes further and further away from the gravitating mass. The same analysis for a galaxy shows a rotation curve which actually increases with R, namely, $R^{0.2}$, well outside the galaxy itself. It extends outwards until there is simply no more atomic hydrogen around to measure. It might extend all the way from a galaxy to its nearest neighbour, 100 Kpc away. That tells you that the mass of the thing is 5 to 10 times more massive than the luminous galaxy. It does not quite close the universe, but it brings us significantly towards that. Whether neutrinos help or hinder galaxy formation or galaxy stability is quite another question. First of all most of the analyses that have been done on galaxies become irrelevant because the galaxy is now a small amount of baryonic mass contaminating a basically neutrino structure.

It was neutrino fluctuation not baryonic fluctuation, that gave rise to galaxies. Secondly, there is a problem of stability for the arms of galaxies, which tend to bar. This instability is cured, some say, if the galaxy is placed in a uniform background of mass. This background mass could be the relic neutrinos which form the galactic halo.

- TELLER:

Could you say what the required neutrino mass distribution is ?

- GLASHOW:

Newton's formula is

$$\frac{V^2}{R} = \frac{M(R)}{R^2}$$

where M(R) is the amount of mass in a sphere of radius R. What has been observed is that V^2 is roughly constant, and that means that M(R) is growing with R, that is the mass inside the distance R is proportional to R. This is associated with a density that looks like $1/R^2$, the alleged neutrino density, or perhaps the mass halo may be made of Jupiter-like planets or black holes. We don't know. One possibility is that this invisible matter is made of a distribution of gravitationally bound neutrinos with velocities of 1000 km/sec. There is a model in which these neutrinos are radially oscillating back and forward through the center of the galaxy, suggested to me by Shmuel Nussinov.

- TELLER:

There is a suspicion that inside each galaxy there may be a giant blackhole. The formation of that blackhole may help in capturing neutrinos and precisely in a radial mode.

- GLASHOW:

That is interesting.

- TELLER:

The trouble with these neutrinos is that while they are cooling down they still are moving very fast and they are hard to capture except by a gravitational mass distribution which changes very rapidly, like a blackhole formation.

- GLASHOW:

Incidentally, there are galaxies that do not have rotation curves, 3 out of 50, roughly speaking. All of these galaxies are irregular, suggesting that indeed the neutrinos are important for conventional galaxy formation or that irregular galaxies are those that suffered collisions and might have lost their neutrino clouds.

- WIGNER:

Are our present known laws of nature accurate enough to be applicable to the extraordinary conditions that existed at the time of the big bang ?

- GLASHOW:

We may extrapolate as much as we want but we will be judged on the basis of what is successfully predicted and whether what is deduced is in agreement with what we have seen. It is certainly plausible to extrapolate to temperatures of MeV where one gets truly solid predictions like the blackbody radiation and the nucleosynthesis of helium.

Going back much further we will allegedly explain the origin of baryons. But all we have till now is an indication of how the asymmetry between baryons and antibaryons could have been established at that very early time. That is not enough to make us believe in the physics of the big bang at the time of 10^{-38} sec. It shows that we may be on the right track, but it is not a proof. Many serious problems intervene both in particle physics and cosmology.

- KARLINER:

Suppose there are quarks with integer charges in addition to quarks with fractional charges. Would this mean that Fairbank is seeing fractionally charged hadrons ?

- GLASHOW:

That is very radical. Any of the normal attempts at grand unification demand that fractionally charged particles have colour in a rather natural way. If Fairbank is right, we will have taken a giant step backwards. Some of my colleagues have already done just this. I would advise a "wait-and-see" attitude.

- WITTEN:

You have drawn attention to the logical completeness of the $SU(3) \otimes SU(2) \otimes U(1)$ model as a framework for particle physics at ordinary energies. Is not our lack of understanding of the Higgs boson and specially its lightness, compared to the "large" mass scales of physics, an internal indication that the present framework is incomplete ?

- GLASHOW:

I sincerely hope you are right. With grand unification the

THE HIGH-ENERGY FRONTIER

desert extends from 100 GeV to 10^{14} GeV. Models that seriously address the hierarchy problem (e.g., supersymmetry and extended technicolour) tend to richly populate the nearer reaches of the desert. We shall soon see, if and when really big machines are built.

- SEIBERG:

You mentioned several exotic phenomena. One such phenomenon which you did not describe is the possible existence of axions. Can you say anything about this ?

- GLASHOW:

Peccei and his group have looked at the experimental data on conventional axions and have come to the conclusion that such axions cannot exist. There is also a positive experimental indication from the Aachen group that should be taken into account. The conventional view is that the axion has in fact been ruled out. It is for this reason that a number of people have suggested the possibility that the axion is invisible. It is invisible because the Peccei-Quinn symmetry is broken at a very high mass scale. Recently, Wise, Georgi and I have constructed a model in which (in a very natural way) the breaking of the Peccei-Quinn symmetry takes place at the unification mass. It becomes an undetectable relic of grand unification which plays no role except to solve the strong CP problem. "Tant pis", for the axion.

- HERTEN:

Do all present grand unified theories require a finite proton lifetime ?

- GLASHOW:

No. You can grand unify with $SU(3) \otimes (SU(3) \otimes SU(3)$ plus a discrete symmetry that interchanges the various $SU(3)$ factors. One $SU(3)$ is colour, one is the left-handed weak interaction, the other is the right-handed. It gives a very ugly theory but it can be done and it does not have proton decay. In other models (see lectures of Dimopoulos) the proton lifetime can be exceedingly long, and therefore undetectable.

- D'HOKER:

Could you explain why the lifetime of the proton is so strongly dependent on the strong coupling renormalization scale ?

- GLASHOW:

The first thing you realize when you grand unify is that the electromagnetic coupling constant is much weaker than the strong interaction coupling constant. If they were equal then you would not have to go to very high a mass scale to perform grand unification. The further they are apart the higher the unification mass. With lower values of Λ, the unification mass decreases, as does the proton lifetime, which depends upon the fourth power of the unification mass.

- COHEN:

What should we as physicists do to prevent a nuclear war ?

- GLASHOW:

I cannot answer that question. I think that each of us, in our own way, should do what we each think is best to prevent a nuclear war. I invite you to attend the next seminar at the Majorana Institute where we shall discuss the mechanisms by which such a calamity should be prevented.

- GOTTLIEB:

In your lecture you did not address the question of why generations are repeated. Do you consider that to be an uninteresting question, or one that can't be approached by either high energy or low energy experiments ?

- GLASHOW:

I consider that question to be one of the most interesting questions of all times. If there were just one family of quarks and leptons it would fit to SU(5) like a hand in a glove and you could not but believe that the theory was right. But that is not the way it works. There is the second family and there now is the third. It is true that it is convenient to parametrise CP violation Since we don't particularly understand why CP is violated in the first place it is hard to theologically argue, as some people do, that the third family exists for the sole purpose of violating CP and allowing baryons to be produced at some temperature of which neither Wigner nor I are willing to conceive. That is why I like these little toy theories which force upon you a fixed number of families. I can give you an SU(5) ⊗ SU(5) model with 5 families rather quickly, or an SU(6) ⊗ SU(6) model with 3 families. The trouble is that I can give you some kind of theory with any number of families. None of these theories has a special ring of truth.

The problem of superfluous replication first arose 40 years ago when I.I. Rabi, contemplating the muon, said "who ordered that". There has been little progress since. The problem remains both important and unsolved.

THE MAGNETIC MONOPOLE FIFTY YEARS LATER

Sidney Coleman

Lyman Laboratory of Physics, Harvard University

Cambridge, Massachusetts   02138

1. INTRODUCTION

This is a jubilee year. In 1931, P. A. M. Dirac[1] founded the theory of magnetic monopoles. In the fifty years since, no one has observed a monopole; nevertheless, interest in the subject has never been higher than it is now.

There is good reason for this. For more than forty years, the magnetic monopole was an optional accessory; Dirac had shown how to build theories with monopoles, but you didn't have to use them if you didn't want to. Seven years ago, things changed; 't Hooft and Polyakov[2] showed that magnetic monopoles inevitably occur in certain gauge field theories. In particular, all grand unified theories necessarily contain monopoles. (In this context, a grand unified theory is one in which a semi-simple internal symmetry group spontaneously breaks down to electromagnetic U(1).) Many of us believe that grand unified theories describe nature, at least down to the Planck length. So where are the monopoles?

As we shall see, grand unified monopoles are very heavy; a typical mass is roughly a hundred times greater than the grand unified scale. Thus they are not likely to be made by contemporary accelerators or supernovae. However, energy was more abundant shortly after the big bang. As Preskill[3] pointed out, naive estimates would lead one to believe that monopoles would have been produced so copiously in the very early universe and annihilated so inefficiently subsequently that they would at the current time form the dominant contribution to the mass of the universe.

Thus the absence of monopoles is significant. It tells us something about the extreme early universe or about extreme microphysics or about both. Since evidence on both these subjects is in short supply, monopoles are important.

This is the last I will say in these lectures about cosmology, or indeed about any reason for studying monopoles. We will have enough to occupy us just developing the basics of monopole theory.

I've tried in these lectures to go from the simple to the complex, from monopoles as seen at large distances to monopole internal structure, from classical physics to quantum mechanics. Of course, I've not been able to keep rigorously to this program; for example, elementary quantum mechanics will be with us from the very beginning, although we will not deal with the full complexities of quantum field theory until the very last lecture.

The organization of these lectures is as follows:

In Section 2 I shall discuss the theory of a classical magnetic monopole as seen from the outside. That is to say, I shall investigate only those questions that can be answered without looking at the monopole interior, without asking, for example, whether there is a real singularity at its core or only some complicated excitation of degrees of freedom that do not propagate out to large distances, like massive gauge fields.

In Section 3 I shall extend the analysis to a classical non-Abelian monopole. I emphasize that by this I do not mean an object that has massive non-Abelian fields in its core, but one which is surrounded by massless non-Abelian gauge fields that extend out to large distances. Since no such fields exist in nature, this may seem a silly exercise, but there are two good reasons for doing it.

Firstly, there is a pedagogical reason. In investigating monopoles in unbroken non-Abelian gauge theories, we will encounter mathematical structures that will be useful to us later, when we deal with the more complicated case of spontaneous symmetry breakdown. Secondly, there is a physical reason. Some of the monopoles that arise in grand unified theories have colored gauge fields surrounding them; they are color magnetic monopoles as well as ordinary electromagnetic monopoles. It is true that these colored fields are damped by confinement effects at distances greater than 10^{-13} cm. However, the cores of these monopoles are on the order of the grand unification scale, something like 10^{-28} cm. Thus we have fifteen orders of magnitude in which they look like non-Abelian monopoles, plenty of room for interesting physics.

In Section 4 we shall plunge into the belly of the beast, and see under what conditions the structures we have discovered at large distances can be continued to small distances without encountering singularities. This is a subject I discussed in my 1975 Erice lectures[4] (coming to it from a different starting point), and although I'll try to keep it to the minimum, there will be some unavoidable recycling of those lectures here.

Finally, in Section 5, I'll turn to quantum mechanics and discuss things like dyonic excitations and the effects of confinement.

There's a lot that's not in these lectures. I won't deal with cosmology, as I've said, nor will I have anything to say about the interactions of monopoles with ordinary matter, how they make tracks in emulsions and damp galactic magnetic fields. On a more theoretical level, I won't talk about exact solutions, index theorems, or fermion fractionalization. Finally, I've made no attempt to compose a full and fair bibliography. If I don't refer to a paper, please assume this is a result of ignorance and laziness, not of informed critical judgment.[5]

Most of what I know about this subject I've learned from conversations with Curtis Callan, Murray Gell-Mann, Jeffrey Goldstone,

Roman Jackiw, Ken Johnson, David Olive, Gerard 't Hooft, and Erick Weinberg. It is a pleasure to acknowledge my debt.

<u>Notational conventions</u>: As usual, Greek indices run from 0 to 3, latin indices from the middle of the alphabet from 1 to 3. When doing ordinary vector analysis (as in all of Sec. 2), the signature of the three-dimensional metric is (+++); when working in four dimensions, the signature is (+---). I will usually set $\hbar$ and c equal to one, although occasionally I will restore explicit $\hbar$'s when discussing the approach to the classical limit. Rationalized units are used for electromagnetism; thus, the electromagnetic Lagrange density is $\frac{1}{2}(E^2-B^2)$ and there are no π's in Maxwell's equations.

2. ABELIAN MONOPOLES FROM AFAR

2.1 The Monopole Hoax and a First Look at Dirac's Quantization Condition

Consider some arrangement of stationary charges and currents restricted to a bounded region. Outside this region there are only stationary electromagnetic fields, vanishing at spatial infinity. Without detailed knowledge of the distribution of charges and currents, what can we say about the exterior fields?

The answer is known by anyone who has taken a course in electromagnetic theory. The exterior fields are a sum of electric monopole, magnetic dipole, electric quadrupole, etc. The successive terms in this series fall off faster and faster at infinity; the leading term at large distances is the electric monopole,

$$\vec{E} = \frac{e\,\vec{r}}{4\pi r^3} \, , \qquad \vec{B} = 0 \, , \tag{2.1}$$

where e is a real number, called the electric charge (of the system inside the region).

This analysis depends on the assumption that the only things inside the region are charges and currents, that is to say, that the empty-space Maxwell's equations,

$$\vec{\nabla} \cdot \vec{B} = 0 = \vec{\nabla} \times \vec{E} + \partial \vec{B}/\partial t \, , \tag{2.2}$$

remain true inside the region. If we drop this assumption, we must add to the series above a dual series, consisting of magnetic monopole, electric dipole, etc. The leading term at large distances is the magnetic monopole,

$$\vec{E} = 0, \qquad \vec{B} = \frac{g\vec{r}}{r^3}, \qquad (2.3)$$

where g is a real number, called the magnetic charge (of the system within the region).[6] Systems carrying non-zero magnetic charge are somewhat confusingly also called magnetic monopoles.

As I stated in my introductory remarks, no one has ever observed a magnetic monopole. Let us imagine, though, that we attempt to hoax a gullible experimenter into believing that he has discovered a monopole. For this purpose we obtain a very long, very thin solenoid; it is best if it is many miles long and considerably thinner than a fermi. (This is very much a *gedanken* hoax.) We put one end of the solenoid in the experimenter's laboratory in Geneva and the other one here in Erice. We then turn on the current. The experimenter sees $4\pi g$ worth of magnetic flux emanating from his lab bench; he can not detect the fact that it is being fed in along the solenoid; he thinks he has a monopole.

Is there any way he can tell he has been hoaxed, that his monopole is a fake? Certainly not, if all he has access to are classical charged particles, for all these see are $\vec{E}$ and $\vec{B}$, and $\vec{E}$ and $\vec{B}$ are the same as they would be for a real monopole. (Except within the solenoid, but this is by assumption so thin as to be undetectable.)

However, the situation is very different if he has access to quantum charged particles. With these, a cunning experimenter can search for the solenoid via the Bohm-Aharonov effect.[7]

This is a variant of the famous two-slit diffraction experiment shown in Fig. 1. Charged particles emitted by the source A pass through the two slits in the screen B and are detected at the screen C. As anyone who has got six pages into a quantum mechanics text knows, the amplitudes for passage through the individual slits combine coherently; the probability density at C is

$$|\psi_1 + \psi_2|^2, \qquad (2.4)$$

where ψ_1 is the amplitude for passage through the first slit and ψ_2

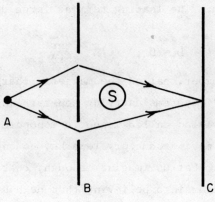

Figure 1

that for passage through the second. If a solenoid, S, shown end on in the figure, is placed between the slits, the probability density at C is changed; it becomes

$$|\psi_1 + e^{ie\Phi} \psi_2|^2 \qquad (2.5)$$

where e is the charge of the particle and Φ is the flux through the solenoid (in our case, $4\pi g$.)

Thus, by moving this apparatus about and observing the change in the interference pattern, our experimenter can detect the solenoid, unless eg is a half-integer,

$$eg = 0, \pm\tfrac{1}{2}, \pm 1, \ldots \quad . \qquad (2.6)$$

In this case, the expressions (2.4) and (2.5) are identical. The solenoid is undetected and our hoax has succeeded.

Equation (2.6) is the famous Dirac quantization condition.[1] As we shall see, if it is obeyed, the solenoid is not only undetectable by the Bohm-Aharonov effect, but undetectable by any conceivable method. It can be abstracted away to nothing, becoming a mathematical singularity of no physical significance, the Dirac string; all that is left behind is a genuine monopole. To demonstrate these assertions requires a closer analysis of the physics of a charged particle moving in a monopole field, to which I now turn.

2.2 Gauge Invariance and a Second Look at the Quantization Condition

Consider a spinless non-relativistic particle moving in some time-independent magnetic field (described in the usual way, by a vector potential, $\vec{A}$), and perhaps also in some scalar potential, V. The Schrödinger equation for this system is

$$i\frac{\partial \psi}{\partial t} = -\frac{1}{2m}(\vec{\nabla} - ie\vec{A})^2 \psi + V\psi . \qquad (2.7)$$

This system has a famous invariance, gauge invariance. For our purposes, we need only consider time-independent gauge transformations:

$$\psi \to e^{-ie\chi} \psi , \qquad (2.8a)$$

$$\vec{A} \to \vec{A} - \vec{\nabla}\chi = \vec{A} - \frac{i}{e} e^{ie\chi} \vec{\nabla} e^{-ie\chi} . \qquad (2.8b)$$

I have written the second of these equations in a somewhat unconventional form to emphasize that the only relevant quantity is $\exp(-ie\chi)$. Thus, for example, χ and $\chi + 2\pi/e$ are different functions, but they define the same gauge transformation.[8]

After these generalities, let us turn to the case of interest, a point monopole at the origin of coordinates. We can think of this as the field of a genuine point object, or as the field of some extended object as viewed from very large distances; it doesn't matter. Of course, it is impossible to find a vector potential whose curl is the monopole field, because the monopole field is not divergenceless. However, it is possible to find a potential that does the job except for a line extending from the monopole to infinity, "the Dirac string". Such a potential is simply the potential of our hoax, in the limit that the solenoid becomes infinitely long and infinitely thin.

For example, if I arrange the string along the negative z-axis, $\vec{A}$ is given by

$$\vec{A} \cdot d\vec{x} = g(1 - \cos\theta)d\phi . \qquad (2.9)$$

As promised, $\vec{A}$ is ill-defined on the string, $\theta = \pi$. Let's check that it has the right curl away from the string. The computation is done most simply using the methods of tensor analysis. The only non-vanishing covariant component of A is

$$A_\phi = g(1 - \cos\theta) . \qquad (2.10)$$

Thus, the only nonvanishing component of the field-strength tensor is

$$F_{\theta\phi} \equiv \partial_\theta A_\phi - \partial_\phi A_\theta = g \sin\theta \ . \tag{2.11}$$

This indeed corresponds to a radial magnetic field. To check that it has the right magnitude, we compute the flux through an infinitesimal element of solid angle:

$$F_{\theta\phi} \, d\theta \, d\phi = \frac{g}{r^2} \, r^2 \sin\theta \, d\theta \, d\phi \ . \tag{2.12}$$

This is the desired result, g/r^2 times the infinitesimal element of area.

I emphasize that I have made no attempt to define $\vec{A}$ or compute $\vec{B}$ on the string. As we shall see, there is no need to inquire into these matters.

Of course, the choice of the negative z-axis is totally arbitrary. For example, I could just as well have put the string along the positive z-axis:

$$\vec{A}' \cdot d\vec{x} = - g(1 + \cos\theta) d\phi \ . \tag{2.13}$$

It will be useful to us later to note that the two potentials we have introduced are simply related:

$$\vec{A}'(\vec{x}) = \vec{A}(-\vec{x}) \ . \tag{2.14}$$

In his classic monopole paper, Dirac showed that if the quantization condition was obeyed, the string was unobservable. The most straightforward way to demonstrate this is by showing that there is a gauge transformation that moves the string from the negative z-axis to any other desired location, that transforms the potential $\vec{A}$ into, for example, the potential $\vec{A}'$. However there are unpleasant features to this argument: the gauge transformation is necessarily singular at both the old and new locations of the string, and singular gauge transformations make people uneasy.

I will now give a refinement of this argument, due to Wu and Yang,[9] that avoids these difficulties. In the Wu-Yang construction, we never have to deal with singular gauge transformations, nor indeed with singular potentials (except, of course, at the origin, where $\vec{B}$

THE MAGNETIC MONOPOLE

blows up and there is a real singularity). The price we pay for this is the necessity of using different vector potentials in different regions of space.

Let me define the upper region of space as the open set $\theta < 3\pi/4$ and the lower region as the open set $\theta > \pi/4$. The union of these two regions is all of space (except for the origin of coordinates, which is a singular point anyway). Let me define the monopole field by using the potential $\vec{A}$ in the upper region and the potential $\vec{A}'$ in the lower region. Thus neither potential has any singularity in its region of validity; in each case the string lies outside the region.

This is an adaptation of a mapmaker's stratagem. There is no way of mapping the spherical surface of the earth onto a portion of a plane without introducing singularities. For example, in the United Nations flag, a north polar projection, the south pole, a single point, is mapped into the circumference of the map. However, we can easily do the job with two maps. For example, we could use a north polar projection for the region north of the Tropic of Capricorn, and a south polar projection for that south of the Tropic of Cancer. The two maps together would form a singularity-free representation of the globe. However, we must be sure that the two maps fit together properly, that we have not by error received a map of the northern part of the earth and the southern part of Mars. That is to say, in their equatorial region of overlap, the two maps must describe the same geography.

Transposing this criterion from cartography to field theory, we must be sure that the two vector potentials describe the same physics in the overlap of the upper and lower regions, that is to say, that $\vec{A}'$ is a gauge transform of $\vec{A}$. Thus, in the overlap region, we must find χ such that

$$\vec{A} - \vec{A}' = \vec{\nabla}\chi. \tag{2.15}$$

This is easy:

$$(\vec{A} - \vec{A}') \cdot d\vec{x} = 2g\, d\phi. \tag{2.16}$$

Hence,

$$\chi = 2g\phi. \tag{2.17}$$

This is not a well-defined function in the overlap region; it is

multiple-valued. Fortunately, as I explained at the beginning of
this section, the relevant object is not χ but $\exp(-ie\chi)$. This is
single-valued if
$$eg = 0, \pm\tfrac{1}{2}, \pm 1 \ldots \ . \tag{2.6}$$
This is the Dirac quantization condition. This completes the
argument.

In principle, the singularity-free Wu-Yang description of the
monopole field is much cleaner than the Dirac description with its
singular string. However, in practice, it's an awkward business to
be continually changing gauges as one moves about. Thus, I'll mostly
work with the Dirac description in the remainder of these lectures,
and use due care with the string, just as, for example, one uses due
care with the coordinate singularities when working with polar coordinates.

2.3 Remarks on the Quantization Condition

(1) All observed electric charges are integral multiples of a
common unit; this is called charge quantization. There is no explanation of this striking phenomenon in either classical or quantum
electrodynamics; nothing would go wrong if the proton charge were
π times that of the electron, for example. One of the most attract-
tive features of Dirac's monopole theory when it was first proposed
was that it explained charge quantization. If there were monopoles
in the universe (indeed, if there was just one monopole in the universe), all electric charges would be forced to be integral multiples
of a common unit, the inverse of twice the minimum magnetic charge.

Of course, nowadays no one believes in pure electrodynamics.
We believe the electrodynamic $U(1)$ group is part of a larger group
that suffers spontaneous symmetry breakdown. However, this explains
charge quantization only if the larger group is semi-simple. (Actually, this is overstating the case. The larger group can be the
product of a semi-simple group and a bunch of $U(1)$ factors, as long
as the electrodynamic group lies completely in the semi-simple
factor.) As I said in the Introduction, and as I will demonstrate

THE MAGNETIC MONOPOLE

later, in this case the theory inevitably contains monopoles.

Thus the connection between monopoles and charge quantization is firmer than ever, although the details are quite different from what was originally envisaged. Both are consequences of a common cause, spontaneous breakdown of a semi-simple symmetry.

(2) Although I won't have much to say about it in these lectures, considerable effort has been devoted over the years to developing quantum electrodynamics with monopoles, a relativistic field theory with both electrically and magnetically charged fundamental particles. A peculiar feature of this theory is that one can not investigate it using diagrammatic perturbation theory, the method that is so successful for ordinary QED. This is because perturbation theory of any kind is nonsense. The two coupling constants in the theory can not simultaneously be made arbitrarily small; because of the quantization condition, as e becomes small, g becomes large.

This does not mean that the effects of virtual monopoles are necessarily large, even for very small e. Large couplings enhance the effects of virtual particles, but large masses diminish them, both through large energy denominators and through the fall-off of form factors at large momentum transfers. Thus, even very strongly coupled particles can have small effects at low energies, if they are sufficiently massive. As we shall see, this is what happens for monopoles in gauge field theories. The masses of these particles are proportional to $1/e^2$; as e goes to zero, their effects are negligible at any fixed energy.

(3) Dyons are defined to be objects that carry both electric and magnetic charge. We can imagine constructing dyons by binding together electrically and magnetically charged particles, or we can imagine them as fundamental entities, not composed of more primitive objects. In either case, we can quickly work out the properties of dyons by exploiting the invariance of Maxwell's equations under duality rotations.

A duality rotation is defined by:

$$\vec{E} \to \vec{E}\cos\alpha + \vec{B}\sin\alpha ,$$
$$\vec{B} \to -\vec{E}\sin\alpha + \vec{B}\cos\alpha , \qquad (2.18)$$

where α is a real number. It is easy to check that this is an invariance of the empty-space Maxwell equations. We describe Eq. (2.18) by saying that $(\vec{E},\vec{B})$ is a duality vector. From Eqs. (2.3) and (2.4), $(e, 4\pi g)$ is also a duality vector.

Given two dyons, with electric and magnetic charges e_i and g_i, ($i = 1, 2$), we can form two duality invariants bilinear in the charges: the inner product,
$$e_1 e_2 + 16\pi^2 g_1 g_2 , \qquad (2.19a)$$
and the outer product,
$$4\pi(e_1 g_2 - g_1 e_2) . \qquad (2.19b)$$

Any observable property of the two-dyon system must be expressible in terms of these invariants.

For example, the force a dyon fixed at the origin exerts on another, moving, dyon must be a linear function of these invariants. Thus, in the nonrelativistic limit, for example,
$$\vec{F} = \left(\frac{e_1 e_2}{4\pi} + 4\pi g_1 g_2\right)\vec{r}/r^3 + (e_1 g_2 - g_1 e_2)\vec{v}\times\vec{r}/r^3 , \qquad (2.20)$$
where "1" is the moving dyon and "2" the fixed one. The reason is that this is the only expression that agrees with the known results for both $g_1 = g_2 = 0$ and $g_1 = e_2 = 0$.

By the same reasoning, the Dirac quantization condition for dyons is
$$e_1 g_2 - g_1 e_2 = 0, \pm\tfrac{1}{2}, \pm 1 \ldots . \qquad (2.21)$$

This equation has some bizarre solutions. For example, it is satisfied by
$$e_i = n_i e + m_i f ,$$
$$g_i = m_i/2e , \qquad (2.22)$$

where n_i and m_i are integers, e is an arbitrary real number, and so is f. That is to say, it is perfectly consistent with the quantization condition for all magnetically charged particles to have fractional electric charge. (We shall see eventually that not only is it consistent, there are circumstances in which it is inevitable.)

(4) When discussing the monopole hoax, I said that the solenoid

was invisible to classical charged particles but detectable (unless the quantization condition was satisfied) by quantum ones. However, the solenoid could also be detected by another kind of physical entity, a classical charged field. We have essentially already seen this. I have been talking about the Schrödinger equation as if it were the equation for a quantum probability amplitude (as it is), but all of our arguments would have been just as valid if ψ were just a classical field like any other (as Schrödinger briefly thought it was).

Thus, Dirac's quantization condition can be a consequence of classical physics, if classical physics contains charged fields. I make this point now because we will shortly be dealing with classical field theories that do contain charged fields, Yang-Mills field theories. There we will find the quantization condition again, and I don't want you to be disoriented by worrying about what a quantum effect is doing in a purely classical context.

(There is one difference between the classical-field and quantum-particle versions of the quantization condition that has been obscured by my use of units in which $\hbar$ is one. The electric charge of a classical field, the quantity that governs the strength of its electromagnetic interactions, is very different from the electric charge of a classical particle; the two even have different dimensions. They are linked together only by the quantum wave/particle duality; e_{field} is $e_{particle}/\hbar$. Thus, for fields, the condition says that ge is a half integer; for particles, that ge is a half-integral multiple of $\hbar$.)

2.4 Funny Business with Angular Momentum

Rotational invariance is a great simplifier of dynamical problems. For example, for a non-relativistic spinless particle moving in a central potential,

$$H = -\frac{1}{2m}\vec{\nabla}^2 + V(r) \ . \tag{2.23}$$

This operator commutes with angular momentum. On a subspace of states of given total angular momentum, H simplifies to:

$$H_\ell = -\frac{1}{2m}\left(\frac{\partial^2}{\partial r^2} + \frac{2}{r}\frac{\partial}{\partial r}\right) + \frac{\ell(\ell+1)}{2mr^2} + V \ , \tag{2.24}$$

where $\ell = 0, 1, 2 \ldots$.

In this subsection I will extend this result to a particle moving in the field of a monopole,[10]

$$H = -\frac{1}{2m}(\vec{\nabla} - ie\vec{A})^2 + V(r) , \quad (2.25)$$

where $\vec{A}$ is the monopole vector potential, Eq. (2.9). The first obstacle to the analysis is that the monopole potential spoils manifest rotational invariance. We will avoid this problem by introducing

$$\vec{D} \equiv \vec{\nabla} - ie\vec{A} . \quad (2.26)$$

This is a gauge-invariant operator. Further, H is expressible in terms of $\vec{D}$ and the position operator, $\vec{r}$,

$$H = -\frac{1}{2m}\vec{D}^2 + V(r) , \quad (2.27)$$

and these operators obey a rotationally-invariant set of commutation relations,[11]

$$[r_i, r_j] = 0 \quad (2.28)$$

$$[D_i, r_j] = \delta_{ij} \quad (2.29)$$

and

$$[D_i, D_j] = -ieg\,\varepsilon_{ijk} r_k / r^3 . \quad (2.30)$$

Our method will be to work as much as possible with Eqs. (2.27) to (2.30) and avoid invoking the explicit form of $\vec{A}$.

Our first step is to construct an angular momentum operator, $\vec{L}$, a vector function of $\vec{D}$ and $\vec{r}$ obeying

$$[L_i, D_j] = i\,\varepsilon_{ijk} D_k \quad (2.31)$$

$$[L_i, r_j] = i\,\varepsilon_{ijk} r_k . \quad (2.32)$$

As a consequence of these, $\vec{L}$ will automatically obey

$$[L_i, L_j] = i\,\varepsilon_{ijk} L_k \quad (2.33)$$

and

$$[L_i, H] = 0 . \quad (2.34)$$

The natural guess is

$$\vec{L} \stackrel{?}{=} -i\vec{r} \times \vec{D} . \quad (2.35)$$

However, this doesn't work; it has the right commutators with $\vec{r}$, but not with $\vec{D}$. The right answer turns out to be

THE MAGNETIC MONOPOLE

$$\vec{L} = -i\vec{r} \times \vec{D} - eg\vec{r}/r . \qquad (2.36)$$

The second term looks very strange indeed; in Rabi's immortal words about something else altogether, "Who ordered that?"

I know of three ways to answer this question. I will sketch out two of them (leaving the details as an exercise), and give the third in full.

(1) I have already given the first answer: Eq. (2.36) gives the right commutation relations. I leave the verification of this statement as an exercise.

(2) From the Heisenberg equations of motion,

$$\dot{\vec{r}} = -i\vec{D}/m . \qquad (2.37)$$

Hence
$$\vec{L} = m\vec{r} \times \dot{\vec{r}} - eg\vec{r}/r . \qquad (2.38)$$

Thus, there is angular momentum in the system even if the particle is at rest! The only possible source of this angular momentum is the angular momentum of the electromagnetic field,

$$\vec{L}_{em} = \int d^3x\, \vec{x} \times (\vec{E} \times \vec{B}) . \qquad 2.39)$$

It takes little labor to go far towards evaluating this integral. Firstly, it must be proportional to eg. Secondly, by dimensional analysis, it must be a homogeneous function of $\vec{r}$ of order zero. Thirdly, by rotational invariance, it must be proportional to $\vec{r}$, the only vector in the problem. Thus

$$\vec{L}_{em} = \beta eg\vec{r}/r , \qquad (2.40)$$

where β is a numerical constant. The evaluation of β is left as an exercise.

(3) In my first physics course, I held a spinning bicycle wheel by its axis and attempted to rotate the axis. To my surprise, I felt a force orthogonal to the applied force; later I learned that this was because the system had angular momentum.

This is just what happens if we attempt to move a charged particle at rest in a monopole field. The Lorentz force moves the particle in a direction orthogonal to the initial impulse. This suggests that this system also has angular momentum.

Let us attempt to compute this angular momentum by identifying its time rate of change with the external torque applied to the system. Inspired by the previous argument, let us make the *Ansatz*

$$\vec{L} = m\vec{r} \times \dot{\vec{r}} + \beta eg\, \vec{r}/r \, , \qquad (2.41)$$

where β is a constant we will fix in the course of the computation.

The equation of motion is

$$m\ddot{\vec{r}} = \vec{F}^{ext} + eg\,\dot{\vec{r}} \times \vec{r}/r^3 \, , \qquad (2.42)$$

where $\vec{F}^{ext}$ is the external force. Using this and the identity

$$\frac{d}{dt}(\vec{r}/r) = \dot{\vec{r}}/r - (\vec{r}\cdot\dot{\vec{r}})\vec{r}/r^3 \, , \qquad (2.43)$$

we find

$$\frac{d\vec{L}}{dt} = \vec{r} \times \vec{F}^{ext} + \vec{r} \times (eg\,\dot{\vec{r}} \times \vec{r})/r^3 + eg\beta[r^2 \dot{\vec{r}} - (\vec{r}\cdot\dot{\vec{r}})\vec{r}]/r^3. \qquad (2.44)$$

Thus, if $\beta = -1$,

$$d\vec{L}/dt = \vec{r} \times \vec{F}^{ext} \, . \qquad (2.45)$$

This completes the discussion of the extra term in the angular momentum. Now let us return to the analysis of the Hamiltonian (2.27).

We begin with the identity

$$\vec{D} \cdot \vec{D} = \vec{D} \cdot \vec{r}\,\frac{1}{r^2}\,\vec{r} \cdot \vec{D} - \vec{D} \times \vec{r} \cdot \frac{1}{r^2}\,\vec{r} \times \vec{D} \, . \qquad (2.46)$$

Here I have ordered the terms such that the identity is true whatever the commutators of $\vec{D}$ and $\vec{r}$. We will analyze the two terms in this expression separately.

Because $A_r = 0$,

$$\vec{r} \cdot \vec{D} = \vec{r} \cdot \vec{\nabla} = r\,\frac{\partial}{\partial r} \, . \qquad (2.47)$$

Also,

$$\vec{D} \cdot \vec{r} = \vec{r} \cdot \vec{D} + 3 \, . \qquad (2.48)$$

Thus,

$$\vec{D} \cdot \vec{r}\,\frac{1}{r^2}\,\vec{r} \cdot \vec{D} = (r\,\frac{\partial}{\partial r} + 3)\,\frac{1}{r}\,\frac{\partial}{\partial r} = \frac{\partial^2}{\partial r^2} + \frac{2}{r}\,\frac{\partial}{\partial r} \, . \qquad (2.49)$$

From the commutators of $\vec{r}$ and $\vec{D}$, $\vec{r} \times \vec{D} = -\vec{D} \times \vec{r}$ commutes with r^2. Thus,

$$-\vec{D} \times \vec{r}\,\frac{1}{r^2}\,\vec{r} \times \vec{D} = \frac{1}{r^2}\,(\vec{r} \times \vec{D})^2 \, . \qquad (2.50)$$

If we square the expression for the angular momentum, Eq. (2.36), we

find
$$\vec{L} \cdot \vec{L} = -(\vec{r} \times \vec{D})^2 + e^2 g^2. \tag{2.51}$$

Putting all this together, we find that on a subspace of states of given total angular momentum,

$$H_\ell = -\frac{1}{2m}\left(\frac{\partial^2}{\partial r^2} + \frac{2}{r}\frac{\partial}{\partial r}\right) + \frac{\ell(\ell+1) - e^2 g^2}{2m\, r^2} + V, \tag{2.52}$$

where, as usual, $\ell(\ell+1)$ is the eigenvalue of $\vec{L} \cdot \vec{L}$. Because $\vec{L}$ obeys the angular-momentum algebra, we know ℓ must be an integer or a half odd integer, but further analysis is required to determine which values of ℓ actually occur.

We all know the solution to this problem when $eg = 0$. Representations occur with $\ell = 0, 1, 2 \ldots$, and, at fixed r, each of them occurs only once. In elementary texts, this result is established by studying the solutions of the angular part of the wave equation. This method can certainly be extended to the case $eg \neq 0$, but it gets a bit sticky; one has to worry about singularities at the string, or, alternatively, about patching together solutions in the two regions. Therefore, I will solve the problem here by a siightly more abstract method that avoids these difficulties.

The general problem is this: Given a space of states that transform in some specified way under rotations, to find a set of basis vectors that transform according to the irreducible representations of the rotation group. Let me remind you of some standard formulas from the theory of rotations. If we label a general rotation in the standard way, with three Euler angles, α, β, and γ, then the states we are searching for obey

$$e^{-iL_z\alpha} e^{-iL_y\beta} e^{-iL_z\gamma} |\ell,m\rangle = \sum_{m'} D^{(\ell)}_{m'm}(\alpha,\beta,\gamma) |\ell,m'\rangle, \tag{2.53}$$

where
$$D^{(\ell)}_{m'm}(\alpha,\beta,\gamma) = e^{im'\alpha} d^{(\ell)}_{m'm}(\beta) e^{-im\gamma}, \tag{2.54}$$

and $d^{(\ell)}$ is a matrix that can be found in any quantum mechanics text. (We shall not need its explicit form here.)

To warm up, let's do a case where we already know the answer, $eg = 0$. We are working at fixed r, so a complete set of basis vectors are the eigenvectors of the angular position of the particle, which we describe in the usual way, by the two angles θ and ϕ. Any of these states can be obtained by applying an appropriate rotation to the state where the particle is at the north pole:

$$|\theta,\phi\rangle = e^{-iL_z\phi} e^{-iL_y\theta} |\theta = 0\rangle . \qquad (2.55)$$

(At $\theta = 0$, we don't need to specify ϕ.) We know the states we are searching for if we know their position-space wave-functions, $\langle\theta,\phi|\ell,m\rangle$. By the previous equations,

$$\langle\theta,\phi|\ell,m\rangle = \langle\theta = 0|e^{iL_y\theta} e^{iL_z\phi} |\ell,m\rangle$$

$$= \sum_{m'} e^{im\phi} d^{(\ell)}_{m'm}(\theta) \langle\theta = 0|\ell,m'\rangle . \qquad (2.56)$$

Thus we know everything if we know $\langle\theta = 0|\ell,m'\rangle$. These coefficients obey an important consistency condition that is a consequence of

$$e^{-iL_z\alpha} |\theta = 0\rangle = |\theta = 0\rangle . \qquad (2.57)$$

This implies
$$\langle\theta = 0|\ell,m'\rangle = 0, \ m' \neq 0 . \qquad (2.58)$$

Thus we can construct $\langle\theta,\phi|\ell,m\rangle$ only for $\ell = 0, 1, 2, \ldots$, and for each of these values of ℓ, the solution is unique, save for an irrelevant normalization. Once we have constructed these functions, it is easy to check, using the multiplication rules for the D-matrices, that they do indeed transform in the desired way.

Now let us extend this analysis to the case $eg \neq 0$. The commutators of $\vec{L}$ and $\vec{r}$ are the same as before, so

$$|\theta,\phi\rangle = e^{-iL_z\phi} e^{-iL_y\theta} |\theta = 0\rangle \times \text{(phase factor)} . \qquad (2.59)$$

The phase factor depends on what gauge we are working in. Fortunately, we don't need to know its explicit form; whatever it is, the main conclusion of the previous analysis is unchanged: we know everything if we know $\langle\theta = 0|\ell,m'\rangle$. The consistency condition, Eq. (2.58), is changed, though:

$$L_z|\theta = 0\rangle = [-i\,\vec{r}\times\vec{D} - eg\,\vec{r}/r]_z |\theta = 0\rangle$$

$$= -eg|\theta = 0\rangle . \qquad (2.60)$$

Thus,
$$\langle \theta = 0 | \ell, m' \rangle = 0, \quad m' \neq -eg. \quad (2.61)$$

and the allowed values of the total angular momentum are

$$\ell = |eg|, |eg| + 1, |eg| + 2 \ldots. \quad (2.62)$$

As before, each of these occurs only once.

This completes the analysis.

Remarks: (1) The effect of the monopole is surprisingly simple. It merely changes slightly the centrifugal potential in the radial Schrödinger equation. If we can solve the Schrödinger equation for a given central potential without a monopole, we can solve it with a monopole. (2) Because ℓ is always greater than or equal to $|eg|$, the centrifugal potential is always positive, and the monopole by itself does not bind charged spinless particles. As one would expect, and as we shall see in a special case, the situation is very different when the particle has spin. (3) Dirac's quantization condition allows eg to be a half integer. Thus it is possible for two spinless particles, one carrying electric charge and the other carrying magnetic charge, to bind together to make a dyon with half-odd-integral angular momentum. This is puzzling from the viewpoint of the spin-statistics theorem. I will now explain the solution of this puzzle.

2.5 The Solution to the Spin-Statistics Puzzle

One might think that there is nothing to be said about the connection between spin and statistics within the framework of non-relativistic quantum mechanics. This is not so; although relativistic field theory is indeed needed to show that spinless particles are bosons, nonrelativistic theory is then sufficient to deduce the statistics of composites made of these bosons. I will now show that a dyon made of a spinless electrically charged particle ("electron") and a spinless magnetically charged particle ("monopole") obeys Bose-Einstein statistics if eg is integral and Fermi-Dirac statistics if eg is half-odd-integral.[12]

We already know the Hamiltonian for an electron in the field of a monopole,

$$H = \frac{(\vec{p}_e - eg\vec{A}_D(\vec{r}_e - \vec{r}_m))^2}{2m_e} + \ldots \quad (2.63)$$

Here the triple dots represent possible non-electromagnetic interactions, and A_D is the standard Dirac string potential,

$$\vec{A}_D(\vec{x}) \cdot d\vec{x} = (1 - \cos\theta)d\phi . \quad (2.64)$$

By a duality rotation, we thus know the Hamiltonian for a monopole in the field of an electron,

$$H = \frac{(\vec{p}_m + eg\vec{A}'_D(\vec{r}_m - \vec{r}_e))^2}{2m_m} . \quad (2.65)$$

Here the sign has changed because the duality rotation that takes e into g takes g into minus e, and $\vec{A}'_D$ is some potential that is gauge-equivalent to $\vec{A}_D$.

To fix $\vec{A}'_D$, we consider a system made of one monopole and one electron. For any choice of $\vec{A}'_D$, $\vec{p}_e + \vec{p}_m$ is a constant of the motion, because H is translationally invariant. However, because

$$m_e \vec{v}_e = \vec{p}_e + eg\vec{A}_D(\vec{r}_e - \vec{r}_m) , \quad (2.66a)$$

and

$$m_m \vec{v}_m = \vec{p}_m - eg\vec{A}'_D(\vec{r}_m - \vec{r}_e) , \quad (2.66b)$$

$m_e\vec{v}_e + m_m\vec{v}_m$ is a constant of the motion (as the classical equations of motion say it should be) only if

$$\vec{A}'_D(\vec{x}) = \vec{A}_D(-\vec{x}) . \quad (2.67)$$

This is indeed gauge-equivalent to $\vec{A}_D$. (See Sec. 2.2.)

To summarize, the correct Hamiltonian for a monopole in the field of an electron is

$$H = \frac{(\vec{p}_m + eg\vec{A}_D(\vec{r}_e - \vec{r}_m))^2}{2m_m} + \ldots \quad (2.68)$$

In exchanging monopoles and electrons, one changes the sign in front of the vector potential, but not the order of the terms within the vector potential.

We can now go on to the system of interest, two dyons, each made of a spinless monopole and a spinless electron. Just as when dealing

THE MAGNETIC MONOPOLE

with two atoms, we describe the states of the system by a Schrödinger wave function,

$$\psi_{A_1 A_2}(\vec{r}_1, \vec{r}_2)$$

Here the $\vec{r}$'s are the positions of the dyons, and the A's are discrete variables that give the internal states of the dyons, spins or excitation energies or whatever. Because our electrons and monopoles are bosons, standard arguments lead to

$$\psi_{A_1 A_2}(\vec{r}_1, \vec{r}_2) = \psi_{A_2 A_1}(\vec{r}_2, \vec{r}_1) . \quad (2.69)$$

We would normally say this implies that dyons are bosons. However, let us look more closely at the Hamiltonian for the two-dyon system:

$$H = \frac{(\vec{p}_1 - eg\vec{A}_D(\vec{r}_1 - \vec{r}_2) + eg\vec{A}_D(\vec{r}_2 - \vec{r}_1))^2}{2m}$$

$$+ \frac{(\vec{p}_2 - eg\vec{A}_D(\vec{r}_2 - \vec{r}_1) + eg\vec{A}_D(\vec{r}_1 - \vec{r}_2))^2}{2m}$$

$$+ \ldots , \quad (2.70)$$

where the triple dots represent Coulomb interactions as well as possible non-electromagnetic interactions. I hope the origin of the terms in this equation is clear; the electron in the first dyon sees the monopole in the second dyon, the monopole in the first dyon sees the electron in the second dyon, etc.

Equation (2.70) looks as if it describes the most horrible velocity-dependent forces, but there can be no such forces between two identical dyons, for there exists a duality rotation that makes their magnetic charges simultaneously vanish. Indeed, from Eq. (2.16),

$$\vec{A}_D(\vec{x}) - \vec{A}_D(-\vec{x}) = 2\vec{\nabla}\phi . \quad (2.71)$$

Thus, if we make a gauge transformation,

$$\psi \to \psi' = e^{2ieg\phi_{12}} \psi , \quad (2.72)$$

the horrible interactions disappear,

$$H \to H' = \frac{\vec{p}_1^{\,2}}{2m} + \frac{\vec{p}_2^{\,2}}{2m} + \ldots . \quad (2.73)$$

But this can change the symmetry of the wave function, for when $\vec{r}_1$ and $\vec{r}_2$ are interchanged, ϕ_{12} goes into $\phi_{12} + \pi$. Hence,

$$\psi'_{A_1 A_2}(\vec{r}_1, \vec{r}_2) = e^{2\pi i e g} \psi'_{A_2 A_1}(\vec{r}_2, \vec{r}_1) \ .$$

That is to say, ψ' is symmetric under interchange of the dyons if eg is integral and antisymmetric if eg is half-odd-integral.

We thus have two descriptions of the same system. One, given by ψ and H, says that our dyons are bosons whatever their spin, but that there are extraordinary long-range forces between them. The other, given by ψ' and H', says that there are no extraordinary forces, but that dyons obey the statistics appropriate to their spin. These two descriptions are connected by a gauge transformation, and therefore make the same prediction for all observable quantities. Nevertheless, there is no ambiguity here; it would be madness to choose the first description when the second is available. (After all, we could use the same gauge transformation to make any pair of fermions look like bosons, whatever their electric or magnetic charges.) Dyons unambiguously obey the spin-statistics theorem.

3. NON-ABELIAN MONOPOLES FROM AFAR

3.1 Gauge Field Theory - A Lightning Review

This subsection is a collection of definitions and formulas from classical gauge field theory, with occasional comments. I have inserted it here to establish notation and to remind you of some features of the subject that will be important to us later. It is far too compressed to be a pedagogical exposition; if you don't know basic gauge field theory already, you won't learn it here.

Gauge Transformations

The dynamical variables in gauge field theories fall into two classes, gauge fields and matter fields. We begin with the matter fields, which we assemble into a big vector, Φ. A gauge transformation is labeled by a function, $g(x)$, from space-time into some compact connected Lie group, G. Under such a transformation, the matter fields transform according to some faithful unitary representation of G, D(g),

$$g(x) \in G: \Phi(x) \to D(g(x))\Phi(x) \ . \quad (3.1)$$

For our purposes, it will be convenient to identify the abstract group element g with the matrix D(g), and write this equation as

$$g(x) \in G: \Phi(x) \to g(x)\Phi(x) \ . \quad (3.2)$$

In the neighborhood of the identity of G, a group element can be expanded in a power series,

$$g = 1 + \sum_{a=1}^{\dim G} \varepsilon^a T^a + O(\varepsilon^2) \ . \quad (3.3)$$

Here the ε's are some coordinates for the group, and the T's are a set of matrices called the infinitesimal generators of the group. The generators span a linear space called the Lie algebra of G. Because g is unitary, the T's are anti-Hermitian

$$T^a = -T^{a+} \ . \quad (3.4)$$

(We are following mathematicians' conventions here; physicists frequently insert a factor of i in the definition of the generators, to make them Hermitian.)

For a general group element, g,

$$g T^a g^+ = D_{ba}^{(adj)}(g) T^b \ , \quad (3.5)$$

where $D^{(adj)}$ is a representation of the group, called the adjoint representation, and the sum on repeated indices is implied.

The commutator of any two generators is a linear combination of generators,

$$[T^a, T^b] = c^{abc} T^c \ , \quad (3.6)$$

where the c's are real coefficients called structure constants. They depend only on the abstract group G and not upon the particular representation we use to realize it. We will always choose the T's such that

$$\text{Tr } T^a T^b = -N \delta^{ab} \ , \quad (3.7)$$

where N is a normalization constant. Thus,

$$c^{abc} = -N^{-1} \text{Tr}[T^a, T^b] T^c \ . \quad (3.8)$$

from which it follows that c^{abc} is unchanged by even permutations of the indices and changes sign under odd permutations.

For example, if G is SU(2), the group of 2×2 unitary unimodular matrices, the standard choice is $T^a = -i\sigma^a/2$, where the σ's are the Pauli spin matrices. The adjoint representation is the vector representation, N is ½, and c^{abc} is ε^{abc}.

Gauge Fields

The gauge fields are a set of vector fields, A_μ^a, $a = 1 \ldots \dim G$. It will be convenient for us to assemble these into a single matrix-valued vector field,

$$A_\mu = A_\mu^a T^a . \tag{3.9}$$

The gauge-transformation properties of this field are defined to be

$$g(x): A_\mu \rightarrow g A_\mu g^{-1} + g \partial_\mu g^{-1} \tag{3.10}$$

If we define the covariant derivative of the matter field by

$$D_\mu \Phi \equiv (\partial_\mu + A_\mu) \Phi , \tag{3.11}$$

then, under a gauge transformation,

$$g(x): D_\mu \Phi \rightarrow g D_\mu \Phi . \tag{3.12}$$

The field-strength tensor, $F_{\mu\nu}$, is a matrix-valued field defined by

$$[D_\mu, D_\nu] \Phi = (\partial_\mu A_\nu - \partial_\nu A_\mu + [A_\mu, A_\nu]) \Phi$$
$$\equiv F_{\mu\nu} \Phi . \tag{3.13}$$

Under a gauge transformation, the field strength transforms according to the adjoint representation of G,

$$g(x): F_{\mu\nu} \rightarrow g F_{\mu\nu} g^{-1} . \tag{3.14}$$

The covariant derivative of the field strength is defined by

$$D_\lambda F_{\mu\nu} = \partial_\lambda F_{\mu\nu} + [A_\lambda, F_{\mu\nu}] . \tag{3.15}$$

This transforms in the same way as F itself under gauge transformations.

Parallel to Eq. (3.9),

$$F_{\mu\nu} = F_{\mu\nu}^a T^a , \tag{3.16}$$

THE MAGNETIC MONOPOLE

where

$$F^a_{\mu\nu} = \partial_\mu A^a_\nu - \partial_\nu A^a_\mu + c^{abc} A^b_\mu A^c_\nu . \tag{3.17}$$

Dynamics

The Lagrange density is

$$\mathscr{L} = \frac{1}{4Nf^2} \operatorname{Tr} F_{\mu\nu} F^{\mu\nu} + \mathscr{L}_m(\Phi, D_\mu \Phi) , \tag{3.18}$$

where $\mathscr{L}_m$ is some invariant function,

$$\mathscr{L}_m(g\Phi, g D_\mu \Phi) = \mathscr{L}_m(\Phi, D_\mu \Phi) , \tag{3.19}$$

and f is a real number, called the coupling constant. The gauge-field part of $\mathscr{L}$ can also be written as

$$\frac{1}{4Nf^2} \operatorname{Tr} F_{\mu\nu} F^{\mu\nu} = -\frac{1}{4f^2} F^a_{\mu\nu} F^{\mu\nu a} . \tag{3.20}$$

If there are no matter fields, then the gauge fields obey

$$D_\mu F^{\mu\nu} = 0 . \tag{3.21}$$

These are called the sourceless Yang-Mills equations.

If we define new fields, denoted by a prime, by

$$A^{a\prime}_\mu = f^{-1} A^a_\mu , \tag{3.22}$$

and

$$F^{\prime a}_{\mu\nu} = f^{-1} F^a_{\mu\nu} = \partial_\mu A^{\prime a}_\nu - \partial_\nu A^{\prime a}_\mu + f c^{abc} A^{b\prime}_\mu A^{c\prime}_\nu , \tag{3.23}$$

then

$$-\frac{1}{4f^2} F^a_{\mu\nu} F^{\mu\nu a} = -\frac{1}{4} F^{\prime a}_{\mu\nu} F^{\mu\nu \prime a} \tag{3.24}$$

and

$$D_\mu \Phi = \partial_\mu \Phi + f A^{\prime a}_\mu T^a \Phi . \tag{3.25}$$

From these we see that f is indeed a coupling constant, absent from the quadratic terms in the Lagrangian but present in the higher-order ones.

If the gauge group is a product of factors, simple groups and U(1) groups, then it is possible to have an independent coupling constant for each factor. The gauge-field part of the Lagrangian becomes

$$-\frac{1}{4} \sum_a \frac{1}{f_a^2} F^a_{\mu\nu} F^{\mu\nu a} , \tag{3.26}$$

where f_a is the same for fields associated with the same factor group. Likewise, Eq. (3.25) becomes

$$D_\mu \Phi = \partial_\mu \Phi + \sum_a f_a A_\mu^{a'} T^a \Phi .\tag{3.27}$$

Temporal Gauge

Temporal gauge is defined by $A_0 = 0$. To transform a gauge-field configuration into temporal gauge we need a function $g(x)$ such that

$$g A_0 g^{-1} + g \partial_0 g^{-1} = 0 .\tag{3.28}$$

A solution to this equation is

$$g^{-1}(x,t) = T \exp - \int_0^t dt' A_0(\vec{x},t') .\tag{3.29}$$

where T indicates the integral is time-ordered.

Temporal gauge is useful because the gauge-field part of the Lagrange density is $-\partial_0 A_i^a \partial^0 A^{ia} / 2f^2$, plus terms with no time derivatives. Thus the structure of the initial-value problem closely resembles that in a scalar field theory with non-derivative interactions, as does the form of Hamilton's equations.

The adaptation of temporal gauge does not destroy all gauge freedom; one can still make time-independent gauge transformations.

Group Elements Associated with Paths

A path in space-time is described by a function $x(s)$, where s goes from 0 to 1. With every such path, we associate a group element, g, defined by

$$g = P \exp - \int_0^1 ds\, A_\mu \frac{dx^\mu}{ds} ,\tag{3.30}$$

where P indicates the integral is path-ordered. g can also be identified with $g(1)$, where $g(s)$ is the solution to

$$\frac{Dg}{Ds} \equiv \frac{dg}{ds} + A_\mu \frac{dx^\mu}{ds} g = 0 ,\tag{3.31}$$

with the boundary condition $g(0) = 1$. (Note the similarity to Eqs. (3.28) and (3.29).)

Under a gauge transformation, $h(x)$,

$$h(x) \in G: g \to h(x(1)) g\, h(x(0))^{-1} .\tag{3.32}$$

This is even simpler if the path is closed, $x(0) = x(1)$,

$$h(x) \in G: g \to h(x(0)) g\, h(x(0))^{-1} .\tag{3.33}$$

In this case, despite the fact that it is a non-local integral, g transforms just like $F_{\mu\nu}$, a local field.

For a closed path, Tr g is the famous Wilson loop factor. It is gauge invariant, and independent of where along the closed loop one chooses to start and end the path. It does depend on the particular matrix representation one uses; in the language of group theory, different representations have different characters.

For ordinary electromagnetism, the group element associated with a closed path is
$$g = \exp(ie\Phi) . \tag{3.34}$$
where Φ is the magnetic flux passing through a surface bounded by the path. Thus, the concept of group element associated with a closed path is one possible generalization to non-Abelian theory of the concept of magnetic flux. (The obvious alternative, defining magnetic flux by integrating the magnetic field, doesn't work; one gets nowhere adding together quantities that have different gauge-transformation properties, as do non-Abelian magnetic fields at different points.)

3.2 The Nature of the Classical Limit

We are going to spend some time and effort investigating the properties of magnetic monopoles in classical non-Abelian gauge theories. Thus it is reasonable to begin by asking when we can expect classical physics to describe quantum reality.

Let $S(\Phi)$ be the action functional for any classical field theory, with Φ now all the fields in the theory, gauge fields or whatever. The Euclidean form of Feynman's path integral formula says that the quantum version of the theory is defined by the functional integral
$$\int (d\Phi) e^{-S(\Phi)/\hbar} . \tag{3.35}$$
If we define new fields, $\Phi' = \Phi/\sqrt{\hbar}$, then, aside from an irrelevant normalization constant, this can be written as
$$\int (d\Phi') e^{-S(\Phi'\sqrt{\hbar})/\hbar} . \tag{3.36}$$
This trivial transformation reveals $\sqrt{\hbar}$ to be a coupling constant; it

is absent from the quadratic terms in the argument of the exponential but present in the higher-order ones. The classical limit is a weak-coupling limit. Sending $\hbar$ to zero, with all coupling constants fixed, is the same as sending all coupling constants to zero (at appropriate rates), with $\hbar$ fixed.

Thus, classical gauge field theory (without symmetry breakdown), the subject of Sec. 3.1, should be a good guide to weakly-coupled quantum gauge field theory (again without symmetry breakdown). I know of two regimes in which such theories apply. One was mentioned in the introduction, quantum chromodynamics at distances less than the confinement length. Because the theory is asymptotically free, this is a regime of weak coupling. The other is the early universe. Weakly-coupled gauge fields which now have masses might well have been massless early on, when the structure of symmetry breakdown was different.

These then are the places where classical non-Abelian monopoles might be found: deep within hadrons or far in the past. Now let us see what they would look like if we found them.

3.3 Dynamical (GNO) Classification of Monopoles

At the beginning of Sec. 2, I described the standard series of stationary electromagnetic fields that could exist outside a black box of unknown content; the leading terms at large distances were the electric and magnetic monopole fields. The dynamical classification of monopoles, invented by Goddard, Nuyts, and Olive (GNO),[13] can be thought of as the extension of this analysis to non-Abelian gauge theories. I will now give a derivation of the GNO classification.

Just as in the electromagnetic case, the analysis will use only the equations that hold outside the black box, the sourceless field equations. Because these are difficult nonlinear equations, I will not attempt to construct a complete series of solutions, but just try to find the non-Abelian generalization of the magnetic monopole field.

I will exclude electric fields from the beginning by looking for solutions that are not just time-independent but time-reversal

THE MAGNETIC MONOPOLE

invariant (in some appropriate gauge). Time reversal is the operation

$$T: A_0(\vec{x},t) \to -A_0(\vec{x},-t)$$
$$\vec{A}(\vec{x},t) \to \vec{A}(\vec{x},-t) \qquad (3.37)$$

(The alternative definition allowed in the Abelian case, this operation times minus one, is not an invariance of the non-Abelian theory.) Thus,

$$A_0 = 0, \qquad \partial_0 A_i = 0, \qquad (3.38)$$

and F_{0i} vanishes.

Equation (3.38) still allows us the freedom to make time-independent gauge transformations. I will use this freedom to make

$$A_r = 0. \qquad (3.39)$$

Just as one integrates along time lines to construct the gauge transformation that takes one to temporal gauge, Eq. (3.29), one here constructs the required gauge transformation by integrating along radial lines, starting, for example, at the unit sphere. Of course, this procedure could lead to trouble at the origin, where radial lines intersect, but this is of no concern to us; the origin is inside the black box, where we don't want to use the field equations anyway. Note that we still have the freedom to make gauge transformations that depend only on θ and ϕ, a freedom we shall use shortly.

We now assume that for large r, $\vec{A}$ can be expanded in powers of $1/r$,

$$\vec{A} = \frac{\vec{a}(\theta,\phi)}{r} + O\left(\frac{1}{r^2}\right). \qquad (3.40)$$

Because the Yang-Mills equations are non-linear, they will involve, in general, cross terms between the leading term in this expansion and the higher-order terms. However, only the leading term enters into the part of the equations proportional to $1/r^3$. Thus, if we are only interested in the leading term, as we are, it is legitimate to ignore the higher-order terms, as we shall.

Just as in the discussion of the Abelian case, it is convenient to write $\vec{A}$ in terms of its covariant components,

$$\vec{A} \cdot d\vec{x} = A_\theta(\theta,\phi)d\theta + A_\phi(\theta,\phi)d\phi. \qquad (3.41)$$

If the field is to be nonsingular at the north and south poles, $A_\phi(0,\phi)$ and $A_\phi(\pi,\phi)$ must both vanish.[14]

I will now make one more gauge transformation to make

$$A_\theta = 0 \,. \tag{3.42}$$

This transformation can be constructed by integrating along lines of fixed ϕ, meridians, starting from the north pole. Of course, this choice of gauge can lead to an artificial singularity at the place where all the meridians intersect again, the south pole. In particular, it can lead to a non-zero (and ϕ dependent) $A_\phi(\pi,\phi)$, that is to say, to a Dirac-string singularity. Just as in the Abelian case, we'll live with this singularity for the time being. When we're done with solving the equations, we'll check that the singularity is indeed just a gauge artifact, that the string is undetectable.

We are now in a position to (finally) use the field equations. The field strength tensor has only one non-vanishing component,

$$F_{\theta\phi} = \partial_\theta A_\phi \,. \tag{3.43}$$

In curvilinear coordinates, the sourceless Yang-Mills equations take the form

$$\partial_\mu \sqrt{g}\, F^{\mu\nu} + [A_\mu, \sqrt{g}\, F^{\mu\nu}] = 0 \,. \tag{3.44}$$

In our case,

$$\sqrt{g}\, F^{\theta\phi} = \frac{1}{r^2 \sin\theta} \partial_\theta A_\phi \,. \tag{3.45}$$

Thus there are two non-trivial field equations. One is

$$\partial_\theta \frac{1}{\sin\theta} \partial_\theta A_\phi = 0 \,. \tag{3.46}$$

The general solution of this, consistent with the vanishing of $A_\phi(0,\phi)$, is

$$A_\phi = Q(1-\cos\theta) \,, \tag{3.47}$$

where Q is an arbitrary (matrix-valued) function of ϕ. However, the other field equation

$$\partial_\phi \sqrt{g}\, F^{\phi\theta} + [A_\phi, \sqrt{g}\, F^{\phi\theta}] = -\partial_\phi Q = 0 \,, \tag{3.48}$$

tells us Q must be a constant.

THE MAGNETIC MONOPOLE

Thus, up to a gauge transformation, the non-Abelian monopole field is constructed by multiplying the Abelian monopole field by a constant matrix. That this procedure leads to non-Abelian monopoles is trivial; what is non-trivial is that it leads to all of them.

Because the non-Abelian monopole is so simply related to the Abelian one, it is easy to work out the quantization condition. We just follow the arguments of Sec. 2.2, with trivial changes. The potential with the north-pointing string is defined by

$$A'_\phi = -Q(1 + \cos\theta) \ . \tag{3.49}$$

$\vec{A}$ and $\vec{A}'$ are transformed into each other by the gauge transformation,

$$g = \exp 2Q\phi \ , \tag{3.50}$$

which is single-valued if

$$\exp 4\pi Q = 1 \ . \tag{3.51}$$

This is the quantization condition.

As a small consistency check, let us verify that this reduces to the correct condition when we specialize to the Abelian theory of Sec. 2. There has been a slight notational change; what we now call $\vec{A}$ we then called $-ie\vec{A}$. (See the expression for the covariant derivative, Eq. (3.11).) Thus, Q becomes $-ieg$, and Eq. (3.51) becomes the Dirac quantization condition, as it should.

In high-energy theory, we tend to focus on the Lie algebra of a group and ignore its global structure; for example, we indiscriminately refer to the isospin group as SU(2) or SO(3). This is an especially bad habit in monopole theory, because the quantization condition is sensitive to the global structure of G; the allowed set of monopoles is different for SU(2) and SO(3). For example, let us suppose that Q is proportional to I_3, the generator of rotations about the third axis in isospin space. If G is SO(3), Q can be any half-integral multiple of I_3, because a rotation by 2π is the identity. If G is SU(2), though, only integral multiples are allowed, because a rotation by 4π is needed to get back to the identity.

There is nothing deep in this; it can all be understood in the elementary terms of Sec. 2.1. To say G is SO(3) is to say that all particles have integral isospin, and therefore integral values of I_3. To say G is SU(2) is to allow half-odd integral values. With a richer set of test particles one can do a richer set of Bohm-Aharonov experiments and detect solenoids formerly undetectable.

To have some examples to use later on, let me work out the explicit form of Q for some representative groups.

If G is SU(n), Q must be a traceless antihermitian n×n matrix. Such a matrix can always be diagonalized by a unitary transformation, that is to say, by a constant gauge transformation. Thus

$$Q = -\frac{i}{2} \text{diag}(q_1, q_2, \ldots q_n) . \qquad (3.52)$$

Tracelessness implies that the sum of the q's is zero, while the quantization condition implies that each q is an integer. Because we can arbitrarily permute the q's by a gauge transformation, only the set of q's is relevant, not their order.

This becomes expecially simple for SU(2). Here,

$$Q = -\frac{i}{2} \text{diag}(q,-q) = -\frac{i}{2} q \sigma_3 . \qquad (3.53)$$

We can use our freedom to permute the eigenvalues of Q to insure that q is always non-negative, $q = 0, 1, \ldots$.

Some gauge field theories based upon the Lie algebra of SU(n) have for their global group, G, not SU(n), but $SU(n)/Z_n$. (Z_n is the n-element finite group consisting of the nth roots of unity, that is to say, the integral powers of $\exp(2\pi i/n)$.) One example is the theory of gauge fields only, without matter fields, gluons without quarks. Gauge fields transform according to the adjoint representation of the group; two SU(n) matrices that differ only by a factor belonging to Z_n will be represented by the same matrix in the adjoint representation.

To treat this case without my equations growing too long, let me introduce some *ad hoc* notation. I will denote by Q_{fund} the matrix that represents a given abstract group generator in the n-dimensional fundamental representation of the group, and by Q_{adj} the matrix that represents the same generator in the (n^2-1)-dimensional adjoint representation. The quantization condition in the case at hand is $\exp(4\pi Q_{adj}) = 1$. This is true if and only if

$$\exp(4\pi Q_{fund}) \in Z_n \ . \tag{3.54}$$

Thus, as before,

$$Q_{fund} = -\frac{i}{2} \operatorname{diag}(q_1 \ldots q_n) \tag{3.55}$$

but now

$$q_m = \frac{r}{n} + \text{integer} \ , \tag{3.56}$$

where r is an integer, independent of m. As before, the q's must sum to zero, and only the set of q's is relevant, not their order.

This becomes especially simple for $SU(2)/Z_2 = SO(3)$. Just as in Eq. (3.53), Q_{fund} is $-\frac{1}{2} i q \sigma_3$, with $q \geq 0$. However, now q can be a half-integer.

3.4 Topological (Lubkin) Classification of Monopoles

The topological classification of monopole fields was developed almost twenty years ago by Elihu Lubkin in an important and unjustly ignored paper.[15] We begin by investigating a situation in which a black box of unknown content is surrounded by gauge fields, the same situation we studied in the dynamical classification. However, now we shall not assume the gauge fields are stationary solutions of the sourceless Yang-Mills equations, or indeed any kind of solutions to any dynamical equations. Instead, we shall associate with the gauge-field configuration a topological charge, a quantity that is unchanged by arbitrary continuous deformation of the configuration, and thus in particular is unchanged by time evolution in accordance with any sensible equation of motion.

I shall first describe Lubkin's construction and develop some of its important properties. This will involve a (very) short course in (very little) homotopy theory. No matter how low your tolerance for abstract mathematics, please pay close attention here; otherwise, you will understand nothing of what follows. After this, I shall investigate the connection of the topological classification with the dynamical classification.

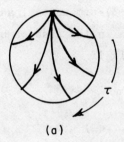 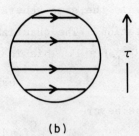

(a)   (b)

Figure 2

Figure 2a shows a family of closed paths lying on a sphere surrounding the black box. Each path begins and ends at the north pole; the return portions of the paths are on the back side of the sphere and can't be seen in the figure. These paths are labeled by a parameter τ that ranges from 0 to 1; the first and last paths in the family, $\tau = 0$ and 1, are trivial paths that never leave the pole. As τ goes through its range, the paths sweep around the black box, like a magician's hoop sweeping around a floating lady.

In case you have trouble visualizing this from Fig. 2a, I've drawn the identical structure in a different way in Fig. 2b. This is a south polar projection of the sphere; the north pole is represented by the circumference of the disc. The paths are the horizontal lines.

Along each path we may integrate the gauge field to obtain the group element associated with the path. In this way, from our family of paths, we obtain a path in group space, $g(\tau)$, beginning at the identity and ending at the identity. (We need not worry about encountering gauge-dependent singularities, like Dirac strings, as we sweep around the sphere. We can always make a gauge transformation to get the singularity out of our way; by Eq. (3.33), this will not affect $g(\tau)$, as long as we take care to insure that the gauge transformation is the identity at the north pole.)

The path $g(\tau)$ depends on practically everything in the problem: the sphere we choose, the details of how we construct the family of closed paths on the sphere, the time at which we do the computation (if our gauge field is changing with time), the gauge in which we are working. However, it depends continuously on all of these things, and therefore the associated element of the first homotopy group of $G, \pi_1(G)$, is unchanged.

To explain the preceding sentence requires the promised short course in homotopy theory.[16]

For any topological space, X, a path, x(t), is a continuous function from the interval [0,1] into X. Let x(t) and x'(t) be two paths with the same endpoints, that is to say, $x(0) = x'(0)$ and $x(1) = x'(1)$. Then we say these paths are "homotopic" or "in the same homotopy class" if one can be continuously distorted into the other, keeping the endpoints fixed. Phrased in equations, x(t) and x'(t) are homotopic if there exists a continuous function of two variables, $F(s,t)$, $0 \le s, t \le 1$, such that

$$F(0,t) = x(t),$$
$$F(1,t) = x'(t),$$

and
$$F(s,0) = x(0) = x'(0),$$
$$F(s,1) = x(1) = x'(1). \tag{3.57}$$

An example is shown in Fig. 3. The topological space, X, is a portion of the plane with a disc (the shaded region) removed. Four paths are shown; all have the same initial and final point, x_0. Path C is homotopic to path D and also to the trivial path, $x(t) = x_0$ for all t. Otherwise, no two paths shown are homotopic. (I hope these statements are obvious to you, because they are rather difficult to prove analytically.)

Given two paths such that one ends where the other begins, the product of the paths is defined by first going along the first path and then going along the second. In equations,

$$x \cdot x'(t) = x(2t), \quad 0 \le t \le \tfrac{1}{2},$$
$$= x'(2t-1), \quad \tfrac{1}{2} \le t \le 1. \tag{3.58}$$

The products of homotopic paths are clearly homotopic.

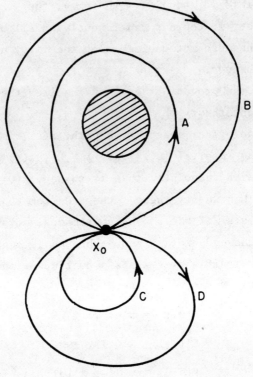

Figure 3

The inverse of a path is defined by going along the path in the opposite direction. In equations,

$$x(t)^{-1} = x(1-t) . \qquad (3.59)$$

The product of a path and its inverse (in either order) is clearly homotopic to a trivial (*i.e.*, constant) path. These concepts are also exemplified in Fig. 3; A is homotopic to the inverse of B.

If we consider all homotopy classes of closed paths, with fixed initial-final point, x_0, these operations of multiplication and inversion define a group structure. The group obtained in this way is called the first homotopy group of X, and is denoted by $\pi_1(X)$. We do not have to specify x_0, because $\pi_1(X)$ is independent of x_0, if X is connected. Proof: Let y be some path going from x_0 to some other point, x_1. Then the mapping, $x \to y \cdot x \cdot y^{-1}$, maps every closed path

THE MAGNETIC MONOPOLE

beginning and ending at x_1 into one beginning and ending at x_0 in such a way that the group operations are preserved.

For the example of Fig. 3, the homotopy classes are labeled by an integer, the winding number, the net number of times the path winds around the shaded disc. For example, path A has winding number 1, path B, -1, paths C and D, 0. The winding number of a product of paths is the sum of the winding numbers of the factors. Thus, π_1 is the additive group of the integers, sometimes called Z.

If X is connected and $\pi_1(X)$ is trivial, we say X is simply connected. It is easy to show that in this case any two paths connecting the same two points are homotopic.

This completes the first part of the short course on homotopy theory. There will be a second part to the course, on the computation of π_1 for compact Lie groups. However, even at this stage, we can understand some things about monopoles:

(1) You should now understand my cryptic statement of a few paragraphs back, that Lubkin's loop construction associates with the gauge fields outside the black box an element of $\pi_1(G)$. This element is the topological charge I referred to at the beginning of this section.

(2) We can see that the topological charge is gauge invariant. If we make a gauge transformation equal to h at the north pole, then $g(\tau)$ is transformed into $hg(\tau)h^{-1}$. Because G is connected, h can be continuously deformed to 1, and the transformed path is homotopic to the original one.

(3) Let us consider a world that contains two black boxes. We can compute the topological charge of each of the boxes, by surrounding it by a sphere that does not contain the other, or we can compute the topological charge of the total system, by surrounding both boxes by a large sphere. In the latter case, we can continuously distort the sphere into an hourglass, two spheres, each surrounding a box, connected by a tube pinched down to a point at the middle. We choose this midpoint to be the "north pole" of the distorted sphere.

Now, when we sweep loops about the distorted sphere, they pass first around one sphere and then around the other; the path in group space is the product of the paths for the individual spheres. Thus the topological charge of the combined system is the (group theoretical) product of the topological charges of the individual components. As a byproduct, we have produced a very indirect argument that $\pi_1(G)$ is Abelian, since it obviously doesn't matter in what order we do things. We'll shortly have a more direct demonstration of the same result.

(4) We see that the forbidden region, the black box, is essential if we are to obtain a non-trivial structure. For, if the box were not there, we could shrink the surrounding sphere to a point. In this limit, the associated element of π_1 would be the identity, because all paths on the sphere would be trivial. But, since it is an invariant under continuous transformations, if it is the identity in the limit, it must have been the identity to begin with.

If we are to obtain a non-trivial topological charge, there must be something inside the box other than just non-singular gauge fields. In Sec. 4 we shall see what that something is.

I now return to the course. As promised, I will tell you how to compute π_1 for any compact connected Lie group. The best approach to this problem is an indirect one; I'll begin by classifying all Lie groups with a given Lie algebra. Once this is done, the computation of π_1 will turn out to be trivial.

I assume you know that the Lie algebra of any compact Lie group is the direct sum of copies of certain fundamental Lie algebras. These are the Lie algebras of $U(1)$, of the three infinite families of classical groups, $SO(n)$, $SU(n)$, and $Sp(n)$, and of the five exceptional Lie groups.

For each of these, there exists a simply connected group with the given Lie algebra. For the algebra of $U(1)$, it is R, the additive group of real numbers. For the algebras of $SU(n)$ and $Sp(2n)$, it is these very groups. For the algebra of $SO(n)$, it is the double

THE MAGNETIC MONOPOLE

covering of SO(n), sometimes called Spin(n). I won't worry here about the five exceptional groups; we'll never use them in these lectures.

Thus, by taking direct products of these groups, we can construct a simply connected group whose Lie algebra is isomorphic to that of a given compact connected Lie group, G. I will denote this group by $\bar{G}$. $\bar{G}$ is called the covering group of G; the reason is that it covers G in the same way SU(2) covers SO(3) = $SU(2)/Z_2$. To be precise, it is possible to show that G is isomorphic to the quotient group $\bar{G}/K$, where K is some discrete subgroup of the center of $\bar{G}$. (I remind you that the center of a group is the subgroup consisting of all elements that commute with every element of the group.) This is a standard theorem of Lie-group theory and I ask you to take it on trust.

Thus, to classify all groups with a given algebra, we have to find all the discrete subgroups of the center of $\bar{G}$. In all the cases we shall encounter, finding the subgroups will be trivial once we find the center.

The center of $\bar{G}$ is the product of the centers of its factors. All of these are easy to compute. R is Abelian, and thus all center. The center of SU(n) is Z_n. The center of Sp(n) consists of 1 and -1. The center of Spin(n), for even n, consists of the two elements that are mapped into 1 in SO(n) and the two elements that are mapped into -1. For odd n, -1 is not in SO(n), so the center of Spin(n) consists only of the two elements mapped into 1.

As an example, let me use this apparatus to work out all the groups with Lie algebras isomorphic to that of $\bar{G}$ = SU(2) ⊗ SU(2). To emphasize the differences between these groups, as I construct the groups I will also describe their irreducible representations. If I describe the elements of $\bar{G}$ in the standard way, by (g_1, g_2), the center consists of (1,1), (1,-1), (-1,1) and (-1,-1). This has five subgroups, and thus we have five groups with the given algebra.

(a) K is (1,1). G is SU(2) ⊗ SU(2). The representations of G are of the form $D^{(s_1)}(g_1) \otimes D^{(s_2)}(g_2)$, with the D's representations of SU(2) and s_1 and s_2 half-integers.

(b) K consists of (1,1) and (1,-1). G is SU(2) ⊗ SO(3). The representations are as before; s_1 is a half-integer and s_2 an integer.

(c) K consists of (1,1) and (-1,1). This is the same as the previous case with the two factors transposed.

(d) K consists of all four elements of the center. G is SO(3) ⊗ SO(3). Both s_1 and s_2 are integers.

(e) K consists of (1,1) and (-1,-1). This is in many ways the most interesting case. G is not a direct product. Either s_1 and s_2 are both integers or they are both half-odd-integers. This is reminiscent of the kind of structures we find in realistic theories, where the unbroken gauge group has the algebra of SU(3) (color) ⊗ U(1) (electromagnetism), but it is not a direct product, because only particles of non-zero triality have fractional charge.

Now that we have classified the Lie groups, it is easy to compute their first homotopy groups. We reason as follows. In general, the mapping of $\bar{G}$ into G is many-to-one; there is no unique element of $\bar{G}$ mapped into a given element of G. However, because K is discrete, for every continuous path in G beginning at the identity, there is a unique <u>continuous</u> path in $\bar{G}$ beginning at the identity. If the path in G is closed, that is to say, if it ends at the identity, the path in $\bar{G}$ must end at some group element that is mapped into the identity, that is to say, at some element of K. Let us consider two closed paths in G such that the corresponding paths in $\bar{G}$ end at the same element of K. Because $\bar{G}$ is simply connected, the paths in $\bar{G}$ can be continuously deformed into each other; thus, so can the corresponding paths in G. Hence, the homotopy classes of closed paths in G are in one-to-one correspondence with the elements of K. It is easy to show that the product of two closed paths corresponds to the products of the two group elements; that is to say,

THE MAGNETIC MONOPOLE 61

that $\pi_1(G)$ is isomorphic to K.

This concludes the short course in homotopy theory. Now let us return to magnetic monopoles.

The Lubkin construction is totally gauge invariant, but carrying it through can be tedious; we have to solve an independent differential equation (or, equivalently, evaluate an independent path-ordered integral) for each value of τ. Things can be simplified considerably by going to a special gauge, "string gauge".

This is very much like the gauge we used in the dynamical classification of monopoles. On each sphere of fixed r, we start at the north pole and gauge transform along meridians to make $A_\theta(\theta,\phi)$ vanish. This may induce a non-zero $A_\phi(\pi,\phi)$, a Dirac string along the south polar axis. When we did this in Sec. 2.3, we assumed the vector potential was proportional to $1/r$; thus it sufficed to make the gauge transformation on some one sphere. Here, we make no such assumption, so we might have to make an independent gauge transformation on each sphere. This can induce a non-zero A_r, but this is no problem; A_r never enters into the Lubkin construction.

We now choose our family of closed paths as shown in Fig. 4. This is a north polar projection of the sphere; the circumference of the disc is the south pole, $\theta = \pi$. The path labeled by τ goes from the north pole to the south pole along the meridian $\phi = 0$ and returns along $\phi = 2\pi\tau$.

Because A_θ vanishes,

$$g(\tau) = P \exp - \int_0^{2\pi\tau} A_\phi(\pi,\phi) d\phi . \qquad (3.60)$$

Thus we need to evaluate the integral for only one path, an infini-

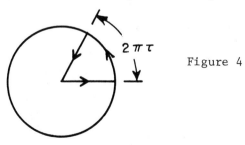

Figure 4

tesimal loop circling the Dirac string. This expression becomes especially simple for one of the Goddard-Nuyts-Olive solutions, Eq. (3.47),

$$g(\tau) = \exp(-4\pi\tau Q) \ . \tag{3.61}$$

This equation enables us to readily compute the topological class into which a given GNO monopole field falls. Let us begin with ordinary electromagnetism, for which G is U(1) and $Q = -ieg$. For $eg = n/2$,

$$g(\tau) = \exp(i2\pi n\tau) \ . \tag{3.62}$$

This winds n times around U(1) as τ goes from 0 to 1. Thus, for each element of π_1, there is one and only one monopole field. Topological charge is magnetic charge. The situation is quite different for other groups, though. For example, for SU(n), we found an infinite number of GNO monopole fields, but SU(n) is simply connected; π_1 has only a single element. Likewise, for $SU(n)/Z_n$ there are an infinite number of GNO monopole fields but only n elements in $\pi_1 = Z_n$. (As an exercise, you might want to work out which element of Z_n is associated with each of the fields.)

Thus, the topological classification is in general coarser than the dynamical one. This is what one might expect; the topological classification is based on fewer assumptions than the dynamical one, and therefore should contain less information.

I've been misleading you. Not because anything I've said in the preceding paragraph is a lie, but because I have been withholding an important piece of information: most GNO monopole fields are unstable.

3.5 The Collapse of the Dynamical Classification

Stability analysis is trivial for an Abelian monopole, because the field equations are linear. This is not so for a non-Abelian monopole. Even at large distances, the field equations do not linearize; if the gauge fields fall off like $1/r$, derivatives and commutators are of comparable magnitude, both $O(1/r^2)$.

I will now investigate small vibrations about an arbitrary SO(3) monopole field.[17] (Of course, this will also take care of SU(2) monopole fields, a subset of the SO(3) ones.) After the computation is

THE MAGNETIC MONOPOLE

done, I will discuss the extension to an arbitrary GNO monopole field for an arbitrary gauge group. The calculation is lengthy and I will present it in outline only; you should have no trouble filling in the missing steps, if you're interested.

We work in temporal gauge, $A_0 = 0$, and write the gauge field as an SO(3) monopole plus a small perturbation,

$$\vec{A} = -\tfrac{1}{2} i q \sigma_3 \vec{A}_D + \delta \vec{A} , \qquad (3.63)$$

where, as explained in Sec. 3.3, $q = 0, \tfrac{1}{2}, \ldots$, and A_D is the Dirac string potential, Eq. (2.64). $\pi_1(SO(3))$ is Z_2, so there are only two topological classes. It is easy to check that the monopoles with integer q's are all in one class and the ones with half-odd-integer q's are all in the other.

We write the perturbation as an explicit 2×2 matrix,

$$\delta \vec{A} = -\tfrac{1}{2} i \begin{pmatrix} \vec{\phi} & \vec{\psi} \\ \vec{\psi}^* & -\vec{\phi} \end{pmatrix} \qquad (3.64)$$

where $\vec{\phi}$ is a real vector field and $\vec{\psi}$ is a complex one. If we linearize the field equations in the perturbation, an easy computation shows that $\vec{\phi}$ obeys a free wave equation,

$$-\partial_0^2 \vec{\phi} = \vec{\nabla} \times (\vec{\nabla} \times \vec{\phi}) , \qquad (3.65)$$

while $\vec{\psi}$ obeys a more complicated equation

$$-\partial_0^2 \vec{\psi} = \vec{D} \times (\vec{D} \times \vec{\psi}) + \frac{i q \vec{r}}{r^3} \times \vec{\psi}$$

$$\equiv H \vec{\psi} , \qquad (3.66)$$

where

$$\vec{D} = \vec{\nabla} - i q \vec{A}_D . \qquad (3.67)$$

(Note that q here plays the role of eg in Sec. 2.)

The stability of the system under small perturbations is determined by the eigenvalue spectrum of the differential operator H. If H has a negative eigenvalue, the associated eigenmode has exponential time behavior, and the monopole field is unstable.

H has a large set of eigenmodes with eigenvalue zero. This is a consequence of the invariance of the temporal-gauge equations of motion under time-independent gauge transformations. We write such a transformation as

$$g(\vec{x}) = 1 - \tfrac{1}{2} i \begin{pmatrix} \lambda(\vec{x}) & \chi(\vec{x}) \\ \chi^*(\vec{x}) & -\lambda(\vec{x}) \end{pmatrix} + \ldots , \qquad (3.68)$$

where the triple dots indicate quadratic and higher-order terms in λ and χ. An easy computation shows that, under this transformation,

$$\vec{\psi} \to \vec{\psi} + \vec{D}\chi + \ldots . \qquad (3.69)$$

Since $\vec{\psi} = 0$ is certainly a solution of Eq. (3.66), so must be its gauge transform, $\vec{\psi} = \vec{D}\chi$. Thus,

$$H\vec{D}\chi = 0 \qquad (3.70)$$

for arbitrary χ.

We call these trivial eigenmodes gauge modes. All of the interesting physics is in the physical modes, the modes that are orthogonal to the gauge modes. For a physical mode,

$$\int d^3 x \, \vec{\psi}^* \cdot \vec{D}\chi = 0 , \qquad (3.71)$$

for any χ. Equivalently,

$$\vec{D} \cdot \vec{\psi} = 0 . \qquad (3.72)$$

As always, rotational invariance is a great simplifier. H commutes with

$$\vec{J} = \vec{L} + \vec{S} , \qquad (3.73)$$

where, just as in Sec. 3.4,

$$\vec{L} = -i\vec{r} \times \vec{D} - q\frac{\vec{r}}{r} , \qquad (3.74)$$

and $\vec{S}$ is the usual spin operator for vector fields, defined by

$$(\vec{a} \cdot \vec{S})\vec{b} = i \vec{a} \times \vec{b} , \qquad (3.75)$$

for any two vectors $\vec{a}$ and $\vec{b}$. As explained in Sec. 2.4, the orbital angular momentum takes on the values,

$$\ell = q, q+1, q+2 \ldots . \qquad (3.76)$$

THE MAGNETIC MONOPOLE

Thus, by the usual rules for adding angular momenta, the total angular momentum takes on the values

$$j = q-1, q, q+1 \ldots, \qquad q \geq 1,$$

$$= q, q+1 \ldots, \qquad q = 0 \text{ or } \tfrac{1}{2}. \qquad (3.77)$$

(I won't bother here to keep track of how often each value of j occurs.)

We can use Eq. (3.75) to purge H of cross products:

$$H = -(\vec{S}\cdot\vec{D})^2 + q\, \frac{\vec{S}\cdot\vec{r}}{r^2} . \qquad (3.78)$$

It is then straightforward to use the methods of Sec. 2.4 to show that, acting on functions of definite j,

$$H\vec{\psi} = \left[-\frac{\partial^2}{\partial r^2} - \frac{2}{r}\frac{\partial}{\partial r} + \frac{j(j+1)-q^2}{r^2}\right]\vec{\psi}$$

$$+ \vec{X}\vec{D}\cdot\vec{\psi} + \vec{D}(\vec{Y}\cdot\vec{\psi}) . \qquad (3.79)$$

Here $\vec{X}$ and $\vec{Y}$ are two ugly objects whose explicit form is of no interest to us, since the terms in which they occur make no contribution to the matrix element of H between physical modes. (The $\vec{X}$ term annihilates the mode on the right and the $\vec{Y}$ term the mode on the left.)

We are now home. For $q \geq 1$, j can be $q-1$, and

$$j(j+1) - q^2 = -q . \qquad (3.80)$$

That is to say, the centrifugal potential is attractive. This is bad news. Just how bad can be seen by computing the expectation value of H for the radial function

$$\psi = 0, \qquad\qquad r < R,$$

$$= \frac{1}{r}(\sqrt{r} - \sqrt{R})\, e^{-r/a}, \qquad r \geq R, \qquad (3.81)$$

where R and a are positive numbers. The expectation value is given by

$$\langle H \rangle = \int_0^\infty r^2\, dr\, \psi^* \left(-\frac{d^2}{dr^2} - \frac{2}{r}\frac{d}{dr} - \frac{q}{r^2}\right)\psi$$

$$= \int_0^\infty dr\, [r^2 (d\psi/dr)^2 - q\psi^2]$$

$$= (\tfrac{1}{4} - q)\ln a + \ldots, \qquad (3.82)$$

where the triple dots denote terms that have a finite limit as a goes
to infinity. For any fixed R, this expression becomes negative for
sufficiently large a.

Thus, not only does H have negative eigenvalues, the existence
of these eigenvalues is totally insensitive to the form of the gauge
field at short distances (or, indeed, at any finite distance). This
is good to know, because we don't trust our expressions at short distances, inside the black box. All the GNO monopole fields with $q \geq 1$
are unstable under arbitrarily small perturbations at arbitrarily
large distances; that is to say, they decay by the emission of non-Abelian radiation.

(In case you've been worrying about it, the modes we have been
studying are indeed physical modes. Taking the covariant divergence
is a rotationally invariant operation. Thus if $\vec{\psi}$ has $j = q-1$, so does
$\vec{D} \cdot \vec{\psi}$. But we showed in Sec. 2.4 that every non-zero scalar function
has $j \geq q$. Thus $\vec{D} \cdot \vec{\psi} = 0$.)

For SO(3), there are only two stable GNO monopole fields, $q = 0$
(no monopole at all) and $q = \frac{1}{2}$, one for each of the two topological
classes. It's reasonable that there should be at least one stable
monopole in each topological class: We can't radiate away topological
charge; the passage of radiation through a Lubkin sphere is just another continuous deformation. What's surprising is that there is
only one stable monopole for each topological class. For SU(2),
it's also true that there's only one stable monopole in each topological class. SU(2) is simply connected, so there's only one topological class, and only $q = 0$ is a SU(2) GNO solution.

Once we have gone through the hard work of the stability analysis for SO(3), there's no need to get involved with differential
equations again for more general groups. For example, let me do the
analysis for $SU(n)/Z_n$.

As in Eq. (3.55), we write the gauge field in terms of the $n \times n$
matrices of the fundamental representation. Writing these out in
components

$$\vec{A}_{ij} = -\tfrac{1}{2} i [q_i \delta_{ij} \vec{A}_D(\vec{x}) + \delta A_{ij}], \qquad (3.83)$$

where there is no sum on i. In this expression

$$q_i = \frac{r}{n} + \text{integer}, \qquad (3.56)$$

and the q's sum to zero. It's easy to check that $r = 0, 1 \ldots n-1$ labels the topological class of the monopole. It's also easy to check that δA_{ij} obeys a differential equation identical in form to that obeyed by $\vec{\psi}$, Eq. (3.66), with the substitution

$$q \to q_i - q_j. \qquad (3.84)$$

Thus, the condition for infinitesimal stability is

$$q_i - q_j = 0, \pm 1, \qquad \text{all } i, j. \qquad (3.85)$$

The only way this condition can hold is if the q's assume only two values. Because the q's sum to zero, one of these values must be non-negative. I will rearrange the q's such that the first s of them assume this value,

$$q_i = q_1 \geq 0, \qquad 1 \leq i \leq s, \qquad (3.86a)$$

Hence, by Eq. (3.86),

$$q_i = q_1 - 1, \qquad s+1 \leq i \leq n. \qquad (3.86b)$$

These sum to zero only if

$$q_1 = (n-s)/n. \qquad (3.87)$$

Thus, r is n-s; once again there is one and only one stable monopole in each topological class.

It's straightforward to extend this analysis to all of the classical groups, and, if you can remember their definitions, to the five exceptional Lie groups. Alternatively, if you're skilled at lifting weights and digging up roots, you can smash the general problem in one blow using the structure theory of Lie groups.[19] I won't do it either way here, but just tell you that no matter how you do it, the answer is the same: there is only one stable GNO monopole field for each topological class; the only stability is topological stability.

You may find it helpful in thinking about this to consider a simple mechanical problem that has the same property, an elastic loop constrained to lie on the surface of a sphere. For purposes of this problem, an elastic loop is a system whose potential energy is proportional to its length. Thus to find the time-independent solutions of the equations of motion is to find the closed geodesics on a sphere. These are the trivial geodesic (the loop bunched up at a single point), one transit of a great circle, two transits of a great circle, etc. Because a sphere is simply connected, there is only one topological class. Here it is easy to see that the only stability is topological stability. If the loop is wound around the sphere, we have but to move it ever so slightly, and it will spring away, scrunching itself up into a single point.

3.6 An Application

Topology is power. If we understand a dynamical system in topological terms, we can often deduce its qualitative features without messing around with detailed quantitative computations. As an example, I will here discuss the force between distantly separated non-Abelian monopoles. ("Distantly separated" only because we don't yet know what monopoles look like at short distances.)

For distantly separated Abelian monopoles, the force can be either repulsive or attractive, depending on whether the magnetic charges are of like or unlike sign. As we shall see, the situation is different for non-Abelian monopoles; if G is semi-simple, the force is always attractive. ("Semi-simple" means without U(1) factors; if U(1) factors are present in G, they can produce an Abelian repulsion that can overwhelm the attraction from the other factors.)

I will begin with SO(3) and afterwards generalize the result. As we have just seen, for SO(3) there is only one stable monopole,

$$\vec{A} = -\tfrac{1}{4} i \sigma_3 \vec{A}_D(\vec{x}) \ . \tag{3.88}$$

Of course, there are many gauge-equivalent ways of writing this; in particular, minus this expression is just the same monopole in another gauge.

Because of this, there are two ways of writing the field of two

THE MAGNETIC MONOPOLE

monopoles
$$\vec{A} = -\tfrac{1}{4} i \sigma_3 [\vec{A}_D(\vec{x}-\vec{r}_1) \pm \vec{A}_D(\vec{x}-\vec{r}_2)] . \tag{3.89}$$

This superposition of two solutions is a solution because everything lies in a single Abelian subgroup; this the nonlinear commutator terms in the Yang-Mills equations never appear. If we had attempted to add a monopole pointing in the 3-direction to its gauge transform pointing in the 2-direction, for example, the nonlinear terms would have made trouble. There may be other ways, less trivial then these, of putting together two monopoles, but I have been unable to find them; for purpose of this discussion, I will assume these two are the only ones.

In the Abelian case, an expression like Eq. (3.89) would be interpreted as either the superposition of two monopoles (plus sign) or of a monopole and an antimonopole (minus sign). At the risk of being repetitious, I emphasize that this is not the case here. The minus sign is simply a gauge transform of the plus sign; the two signs correspond to two different ways of putting together the same two monopoles, just as spin one and spin zero are two different ways of putting together the same pair of spin-$\tfrac{1}{2}$ objects.

Despite the difference in interpretation, the computation of the interaction energy stored in the magnetic field is the same as in Abelian gauge theory,
$$E_{int} \propto \pm 1/|\vec{r}_1-\vec{r}_2| , \tag{3.90}$$
repulsive in the plus case, attractive in the minus. The difference from the Abelian case appears when we study the field at large distances,
$$\vec{A} = -\tfrac{1}{4} i \sigma_3 \vec{A}_D(\vec{x}) [1 \pm 1] + O(1/r^2) . \tag{3.91}$$

Both of these are GNO monopole fields; they have to be, because they are time-independent solutions of the Yang-Mills equations. Both are in the same topological class; they have to be, because the topological charge of a two-monopole system is always the product of the topological charges of the individual monopoles, no matter how we weave the fields together. But only one of them is stable, because there is only one stable GNO field in each topological class, and it is the minus field.

Thus, even if we were able to put the two monopoles together in the repulsive plus configuration, they wouldn't stay there for a minute; they would emit non-Abelian radiation and settle down to the attractive minus configuration. This is much like the situation for two bar magnets, each free to pivot about its center of mass. It doesn't matter what the initial orientation of the magnets is; they will realign themselves until they are in the maximally attractive (anti-aligned) configuration. Here the realignment takes place in an internal rather than a geometrical space, but the physics is much the same.

Now let us go on to $SU(n)/Z_n$. Here we have n stable monopole fields, so the two monopoles may be gauge-inequivalent. Nevertheless, as we shall see, the force is still attractive. We write the field of the two monopoles as

$$\vec{A} = Q_1 \vec{A}_D(\vec{x}-\vec{r}_1) + Q_2^P \vec{A}_D(\vec{x}-\vec{r}_2) \ . \tag{3.92}$$

My notation here needs some explanation. Q_1 and Q_2 are the two $n \times n$ matrices that appear in the single-monopole fields. As before, we wish to avoid the nonlinear terms in the Yang-Mills equations, so we choose Q_1 and Q_2 to commute; this implies that they can be simultaneously diagonalized. This still leaves us the freedom to permute the eigenvalues of Q_2, say while keeping those of Q_1 fixed. We pick one arrangement of eigenvalues as the standard one, and denote the others by Q_2^P, where P is one of the n! permutations on n objects. Different ways of putting together the two monopoles correspond to different choices of the permutation P, but, just as before, the total topological charge is independent of P.

Because there is only one stable GNO field for a given topological charge, most choices of P will lead to unstable fields at large distances. We wish to find the energy in the unique stable configuration. For any P, just as before,

$$E_{int} \propto -\text{Tr}\, Q_1 Q_2^P / |\vec{r}_1 - \vec{r}_2| \ . \tag{3.93}$$

THE MAGNETIC MONOPOLE

(The minus sign is there because the Q's are anti-Hermitian.) If we sum over permutations,

$$\sum_P Q_2^P = (n-1)! \, \text{Tr} \, Q_2 = 0 \,. \tag{3.94}$$

Thus the energy averaged over all configurations vanishes, and therefore the energy in the unique stable configuration, the configuration of minimum energy, must be negative. Once again, the force is attractive.

If you know anything about the general theory of Lie groups, you can see that it is trivial to generalize this argument. The only property of the Q's we needed was their tracelessness, and this holds for an arbitrary semi-simple group. (For group mavens, the precise statement required is that only the zero element of the Cartan subalgebra is invariant under the Weyl group.) Thus, even in this case, any two monopoles attract each other.

4. INSIDE THE MONOPOLE

4.1 Spontaneous Symmetry Breakdown - A Lightning Review

Up to now, we have kept our distance from the monopole. We are now going to open the black box, to see whether the structures we have found at large distances can be continued down to $r = 0$.

We will work in the context of gauge field theories with spontaneous symmetry breakdown. These theories are the daily bread of contemporary high-energy physics; nevertheless, I'll give a lightning review of them here, much like the review of Sec. 3.1, but even more compressed. My purposes are the same as before, to establish common notation and emphasize salient points.

In Sec. 3.1, we divided the fields in our theory into gauge fields and matter fields; it will now be convenient to divide the matter fields into scalar fields (assumed, for convenience, to be all real) and other (typically, spinor) fields. We will change notation slightly and use Φ to denote a big vector made up of the scalar fields only.

We will restrict ourselves to theories for which the matter Lagrange density is of the form

$$\mathscr{L}_m = \tfrac{1}{2} D_\mu \Phi \cdot D^u \Phi - U(\Phi) + \dots , \tag{4.1}$$

where U is some G-invariant function and the triple dots indicate terms involving the non-scalar matter fields.

Up to gauge transformations, the ground states of the theory are states for which all non-scalar fields vanish, and the scalar fields are space-time independent and at an absolute minimum of U. We will use quantum language for this classical situation, and refer to the ground states as "vacua", and the ground-state value of Φ as "the vacuum expectation value of Φ". We will always assume a constant has been added to U such that the ground-state energy is zero.

Let $\langle\Phi\rangle$ be some (arbitrarily selected) minimum of U; then $g\langle\Phi\rangle$ is also a minimum, for any g in G. We will assume all minima of U are of this form. Then all the ground states are physically equivalent, and with no loss of generality we can restrict ourselves to the case in which the vacuum expectation value of Φ is $\langle\Phi\rangle$. (This assumption excludes interesting phenomena such as accidental degeneracy, in which no symmetry connects the ground states, and Goldstone bosons, in which a symmetry connects them but not a gauge symmetry. We make the assumption here just to keep our arguments as simple as possible; it's not difficult to extend the analysis to the more general case.)

We define H to be the subgroup of G that leaves $\langle\Phi\rangle$ invariant,

$$h \in H \text{ iff } \quad h\langle\Phi\rangle = \langle\Phi\rangle . \tag{4.2}$$

We say the symmetry group G has spontaneously broken down to H. The gauge fields associated with H remain massless; the others combine with some of the scalar fields to form massive vector fields. (This is the famous Higgs mechanism.) The vector mass matrix is given by

$$\mu^2_{ab} = f_a T_a \langle\Phi\rangle \cdot f_b T_b \langle\Phi\rangle , \tag{4.3}$$

where there is no sum on repeated indices. The scalar fields absorbed into the massive vector fields are those that correspond to perturbations of the vacuum of the form $\delta\phi = T_a\langle\Phi\rangle$.

Had we made a different arbitrary choice of $\langle\Phi\rangle$, H would have been a different subgroup of G, although one isomorphic to our original H. Occasionally we will find ourselves working in a gauge such that the vacuum expectation value of Φ varies in space; it will be important then to remember that H varies with it.

As an example of these ideas, let G be $SO(n)$, Φ an n-vector, and

$$U = \frac{\lambda}{4}(\Phi \cdot \Phi - c^2)^2, \qquad (4.4)$$

where λ and c are positive numbers. The possible choices for $\langle\Phi\rangle$ are all vectors of length c. If we make the choice

$$\langle\Phi^a\rangle = c\,\delta^{an}, \qquad a = 1\ldots n \qquad (4.5)$$

then H is the subgroup of rotations on the first n-1 coordinates, $SO(n-1)$. Of the original $\frac{1}{2}n(n-1)$ gauge fields, $\frac{1}{2}(n-1)(n-2)$ remain massless; the other n-1 combine with n-1 scalar fields to become massive; one scalar field remains untouched.

In what follows, we shall make much use of this theory with n = 3. This case faintly resembles reality in that there is only one massless gauge meson, which we can identify with the photon; for this reason it was incorporated by Georgi and Glashow into an ingenious but erroneous alternative to the Weinberg-Salam model, and is sometimes called the Georgi-Glashow model.

4.2 Making Monopoles [19]

I will now explain how magnetic monopoles arise in spontaneously broken gauge field theories. This phenomenon was first discovered by 't Hooft and Polyakov,[2] working in the SO(3) model I've just described. It is one of the most dazzling effects in field theory; magnetic monopoles miraculously appear in theories that contain no fundamental magnetically charged fields at all.

It turns out that the non-scalar matter fields play no role in building monopoles, so, for simplicity, I will assume that we are working in a theory of gauge fields and scalar fields only. In this theory, we will study finite-energy nonsingular field configurations at some fixed time. Later on, we'll worry about how these configurations evolve in time.

The energy density is

$$\Theta^{\infty} = \tfrac{1}{2} \vec{D}\Phi \cdot \vec{D}\Phi + U(\Phi) + \text{other positive terms} . \qquad (4.6)$$

For the energy integral to converge, each of the two terms displayed must vanish at large r. To keep the argument simple, I will assume here that they are strictly zero outside some radius R,

$$U = \vec{D}\Phi = 0 , \qquad r \geq R . \qquad (4.7)$$

I stress that this is a totally bananas assumption. It is not a consequence of finiteness of the energy and it is not even true of any of the known finite-energy solutions to the field equations. I make it here for pedagogical reasons only, so I can construct an argument in which the underlying structure is not buried under analysis of how rapidly and uniformly limits are attained. I invite those of you skilled at real analysis to generalize the argument to weaker and more sensible assumptions.

Equation (4.7) implies that Φ must be at a minimum of U for $r \geq R$, but it may be at different minima at different points. However, we can always gauge transform such that

$$\Phi = <\Phi> , \qquad r \geq R . \qquad (4.8)$$

We do this in two steps. First we make a gauge transformation depending only on r such that Φ is transformed to $<\Phi>$ along the north polar axis, for $r \geq R$. Then, for each fixed r, we start at the north pole and transform along meridians to make $\Phi = <\Phi>$ everywhere. This second step may introduce a Dirac string singularity along the south polar axis, where all meridians intersect again.

Equations (4.7) and (4.8) imply that

$$\vec{D}\Phi = \vec{A}<\Phi> = 0 , \qquad r \geq R . \qquad (4.9)$$

Thus, after we have made our gauge transformations, the only gauge fields that exist for $r \geq R$ are those associated with the unbroken subgroup H. Of course, this is just what we would have expected;

only massless fields can extend to large distances.

Thus we have the same structure as in Sec. 3; outside of a black box, the sphere of radius R, there is nothing but massless gauge fields. The only change is a mild notational one; the group we called G in Sec. 3 we now call H. Thus, outside the sphere, we have the usual classification of field configurations by their topological charges, elements of $\pi_1(H)$.

If all there were inside the sphere were H gauge fields, then a nonsingular field would always have trivial topological charge. I gave the argument for this in Sec. 3.4, but let me give it again here: With every sphere, we associate a path in group space. If we take a sphere around a monopole and shrink it to a point, the associated path must become the constant path. Thus there are only two possibilities. Either the path changes continuously, in which case it was in the trivial homotopy class to begin with, or it changes discontinuously, in which case we have encountered a singularity. (Note that since the topological charge is gauge-invariant, singularities that are mere gauge artifacts, like Dirac strings, will not do the job; a genuine gauge-invariant singularity is needed.)

We now see what is different in the case at hand. Inside the black box, we have G gauge fields, not just H ones, and the path may move out of H and into the larger group G. It is quite possible for a path that is homotopically nontrivial in H to be homotopically trivial in G; a topological knot that can not be disentangled in the smaller space may fall apart in the larger. In this way we can have a nontrivial topological charge without a singularity.

This can be phrased in somewhat more abstract language. Because H is a subgroup of G, every path in H is a path in G. This induces a mapping of $\pi_1(H)$ into $\pi_1(G)$. The kernel of this mapping is defined, as always, as the subgroup that is mapped into the identity. Thus our result can be stated as follows:

> The condition for a nonsingular monopole is that the topological charge is in the kernel of the mapping $\pi_1(H) \to \pi_1(G)$.

We have seen that this condition is necessary. I will now demonstrate that it is sufficient by constructing a nonsingular finite-energy field configuration for every topological charge in the kernel.

For any topological charge, there is a field at large distances with that charge, the appropriate GNO field,

$$\vec{A}_{GNO} \cdot d\vec{x} = Q(1 - \cos\theta)d\phi . \quad (4.10)$$

Here, Q is the Lie algebra of H,

$$Q\langle\Phi\rangle = 0 , \quad (4.11)$$

and the path

$$g(\tau) = e^{4\pi Q\tau} , \quad 0 \leq \tau \leq 1 , \quad (4.12)$$

is in the specified homotopy class.

In G, $g(\tau)$ is homotopic to the trivial path. Thus, there exists a continuous function of two variables, $g(\theta,\phi) \in G$, $\theta \in [0,\pi]$, $\phi \in [0,2\pi]$, such that

$$g(0,\phi) = g(\theta,0) = g(\theta,2\pi) = 1 \quad (4.13a)$$

and

$$g(\pi,\phi) = e^{2Q\phi} . \quad (4.13b)$$

We will use this to define our vector and scalar fields for $r \geq R$,

$$\Phi = g\langle\Phi\rangle , \quad (4.14a)$$

$$\vec{A} = g\vec{A}_{GNO}g^{-1} + g\vec{\nabla}g^{-1} . \quad (4.14b)$$

This is simply a gauge transform of the GNO field, so it is still a finite-energy field configuration. However, the Dirac string has completely disappeared; all our fields are now manifestly nonsingular at all angles. The price we have paid is that we have made the vacuum expectation value of Φ angle-dependent.

It is now trivial to continue this to $r \leq R$,

$$\Phi(r,\theta,\phi) = \frac{r^2}{R^2} \Phi(R,\theta,\phi) , \quad (4.15a)$$

$$\vec{A}(r,\theta,\phi) = \frac{r^2}{R^2} \vec{A}(R,\theta,\phi) . \quad (4.15b)$$

This is manifestly nonsingular and of finite energy all the way down to $r = 0$. Note that we could not have continued things into the interior in this way if we had not first removed the string; if we had attempted to simply scale down the GNO field, we would have violated the quantization condition.

This completes the proof of sufficiency.

The fact that the condition is both necessary and sufficient makes it the jewel of monopole theory, well worthy of being honored with a box. It enables us to tell instantly, without solving a single differential equation, whether a given theory admits non-singular monopoles, and what kinds it admits. I will give three examples:

(1) G is SO(3) and H is SO(2). This is the theory described at the end of Sec. 4.1, the Georgi-Glashow model. If we identify the unbroken subgroup with the electromagnetic group, then large distance analysis allows eg to be 0, $\pm\frac{1}{2}$, ± 1, etc. Topologically, these correspond to paths that go 2eg times around SO(2). Only paths that go an even number of times around SO(2) can be deformed into the trivial path in SO(3); thus the allowed values of eg are the alternate terms in the series, eg = 0, ± 1, etc.

(2) G is U(2) and H is U(1), embedded as the subgroup of U(2) that leaves the first basis vector in the two-dimensional unitary space invariant. This is the Weinberg-Salam model. G is locally isomorphic to U(1) $\otimes$ SU(2), and any path that winds around H also winds around the U(1) factor in G. Thus only trivial topological charge is allowed, eg = 0.

(3) G is any semisimple group and H is any group with a U(1) factor. These are the grand unified models referred to in Sec. 1. $\pi_1(G)$ is finite and $\pi_1(H)$ is infinite, so the kernel of the mapping must be infinite and the theory must contain monopoles, as I asserted in Sec. 1. Of course, we can't tell precisely what magnetic charges are allowed until we know the details of the theory.

We have concerned ourselves here with finite-energy nonsingular field configurations at fixed time. If we have a well-posed initial-value problem, any such configuration will evolve into some solution of the equations of motion, but it need not be time-independent. Thus, none of our analysis demonstrates the possibility of building time-independent monopoles. This is not important. We theorists like time-independent solutions, but that is because they are easy to analyze and we are lazy. If an experimenter finds a black box surrounded by a monopole field, this is of interest whether the box contains time-independent fields or oscillating, rotating, ergodically quivering fields.

We do know one thing about the time evolution of configurations with nontrivial topological charge. Whatever they do, they can not dissipate utterly, simply leak out of the box in the form of ordinary radiation fields, massive or massless. This is because radiation does not carry topological charge, and topological charge is conserved. Topology is power.

4.3 The 't Hooft-Polyakov Object

Despite these sour words about time-independent solutions, I will devote some time here to discussing the famous time-independent monopole found in the Georgi-Glashow model, Eq. (4.4), by 't Hooft and Polyakov.[2] Searching for general time-independent solutions is difficult because one has to deal with many functions of three variables; the trick is to simplify things by looking for solutions that are symmetric under some astutely chosen subgroup of the symmetry group of the theory.

It turns out that a fruitful subgroup for this purpose is the SO(3) subgroup consisting of simultaneous spatial and internal rotations. Scalar fields invariant under this are necessarily of the form

$$\phi^a = f(r^2) \frac{r^a}{r}, \qquad (4.16)$$

where the internal index a runs from 1 to 3. To keep the energy finite at infinity, $f(\infty)$ must equal c. To keep the field non-singular at $r = 0$, $f(0)$ must vanish.

THE MAGNETIC MONOPOLE

This is not only invariant under the stated SO(3) group, it is also invariant under parity, if we define Φ to be pseudoscalar. The only gauge fields invariant under both these symmetries are of the form

$$A^{ia} = h(r^2)\epsilon^{iak} r_k \ . \tag{4.17}$$

Finiteness of the energy puts restrictions on the large-r behavior of h, but I won't bother to work them out here.

Thus finding time-independent solutions becomes a simple problem in the calculus of variations; we must minimize the energy as a functional of two functions of a single variable. This problem is not beyond the reach of either functional or numerical analysis; I hope you will find it plausible when I tell you that it is possible both to prove a solution exists and to compute this solution with good accuracy on a pocket calculator.

We have a time-independent solution, but is it a monopole? The easiest way to answer this question is to transform the solution into string gauge. From Eq. (4.16),

$$\Phi(r,\theta,\phi) = g(\theta,\phi)\Phi(r,\theta=0) \ , \tag{4.18}$$

where

$$g(\theta,\phi) = e^{T_3\phi} e^{T_2\theta} e^{-T_3\phi} \ , \tag{4.19}$$

and the T's are the generators of SO(3). Equation (4.18) would be true even if I left the last factor out of Eq. (4.19); I put it in so g would be well defined at $\theta = 0$.

Let us now make a gauge transformation, using g^{-1}:

$$\Phi(r,\theta,\phi) \to g^{-1}\Phi = \Phi(r,\theta=0) \ . \tag{4.20}$$

This aligns the vacuum expectation value throughout space; in all directions, the unbroken group H is the SO(2) subgroup generated by T_3. Under the same gauge transformation,

$$\vec{A} \to g^{-1}\vec{A}g + g^{-1}\vec{\nabla}g \ . \tag{4.21}$$

Only the second term in this expression produces a Dirac-string singularity on the south polar axis. Here,

$$g(\pi,\phi) = e^{T_2\pi} e^{-2T_3\phi} \ , \tag{4.22}$$

and
$$A_\phi(\pi,\phi) \to -2T_3 \ . \tag{4.23}$$

Integrating this, we see that we circle the SO(2) subgroup twice when we go once around the string. This is a monopole with eg = 1, the minimal value allowed by the topological considerations of Sec. 4.2.

4.4 Why Monopoles are Heavy

I said in the introduction that monopoles are typically very massive. Now that we understand their topological structure we can see why this is so.

Let me begin by considering a theory in which all the heavy gauge fields have similar masses, on the order of some typical mass, μ, and all of the gauge couplings are on the order of some typical coupling, e. In any non-radiant configuration, the heavy gauge fields fall off with distance like $\exp(-\mu r)$. Thus we expect the core of the monopole, the region in which the heavy fields are signigicant, to have a size on the order of $1/\mu$. Outside the core, only massless fields should be significant, and the monopole should look like a magnetic Coulomb field (in the Abelian case) or a GNO field (in the general case).

It is easy to estimate the energy stored outside the core. Just to be definite, let me do the computation for an Abelian monopole of magnetic charge g,

$$\begin{aligned}
E_{magnetic} &= \tfrac{1}{2} \int_{r > O(1/\mu)} d^3\vec{x} |\vec{B}|^2 \\
&= 2\pi g^2 \int_{O(1/\mu)}^{\infty} dr/r^2 \\
&= 2\pi g^2 O(\mu) \ . \tag{4.24}
\end{aligned}$$

Because the energy density of the theory is positive, the energy inside the core can only add to this. Thus,

$$m \geq O(\mu/e^2) \ . \tag{4.25}$$

In this order-of-magnitude form, the bound is clearly also valid

THE MAGNETIC MONOPOLE

for non-Abelian monopoles. Monopoles are heavy because the integral for the electromagnetic energy of a Coulomb field is divergent at short distances.

If we're willing to make some plausible guesses, we can replace this inequality with an equality. If we imagine increasing the core radius, the magnetic energy decreases. Since the monopole is in equilibrium, this must be compensated for by an increase of the internal energy. (If the monopole is time-dependent, this is still true, if we average over time.) Thus, it is plausible to assume that the core energy is of the same magnitude as the magnetic energy.

Now let us go on to a theory which has many mass scales. This typically occurs when there is a hierarchy of symmetry breakdown,

$$H \subset G_1 \subset G_2 \subset G_3 \ldots , \qquad (4.26)$$

with an associated hierarchy of masses

$$\mu_1 \ll \mu_2 \ll \mu_3 \ldots , \qquad (4.27)$$

where μ_i is the typical mass of a gauge field in G_i (but not G_{i-1}). For example, in the grand unification model of Georgi and Glashow,[20] SU(5) breaks down to SU(3) (color) $\otimes$ U(2) (electroweak); the gauge fields associated with this get masses on the order of 10^{14} GeV. U(2) in turn breaks down to U(1), as in the Weinberg-Salam model; the gauge fields here get masses on the order of 10^2 GeV.

Associated with the sequence (4.26), there is a sequence of mappings

$$\pi_1(H) \to \pi_1(G_1) \to \pi_1(G_2) \ldots . \qquad (4.28)$$

We can build a monopole with any topological charge that is eventually mapped into the identity. Let us suppose this first happens at the group G_i. Then the monopole field remains Coulombic down to distances of the order of $1/\mu_i$, and the monopole mass obeys

$$m \gtrsim O(\mu_i/e^2) . \qquad (4.29)$$

As before, if we're willing to make some plausible guesses, this inequality can be replaced by an equality.

As an example, in the Georgi-Glashow grand unified theory, we have to go to SU(5), as we saw in Sec. 4.3; thus the monopole mass is on the order of 10^{16} GeV, or 10^{-8} grams.

Of course, it's possible for different elements of $\pi_1(H)$ to be mapped into the identity at different stages of the hierarchy; this gives the possibility of a wide variety of monopole mass scales. For example, consider a theory in which SU(3) breaks down to its real subgroup, SO(3), at some large mass scale; at a much smaller scale, SO(3) breaks down to SO(2) in the familiar way. This has monopoles with both integral and half-odd-integral values of eg, the latter much heavier than the former.

4.5 The Bogomol'nyi Bound and The Prasad-Sommerfield Limit

We have derived rough estimates of monopole masses. For certain theories, it is possible to derive a rigorous lower bound, the Bogomol'nyi bound.[21] I will give the derivation here for the Georgi-Glashow model; it can readily be extended to any theory in which the scalar fields transform according to the adjoint representation of the gauge group.

We write the energy of the theory as

$$E = \tfrac{1}{2} \int d^3 \vec{x} \left[\vec{E}^a \cdot \vec{E}^a + \vec{B}^a \cdot \vec{B}^a + (D_0 \phi^a)^2 + \vec{D}\phi^a \cdot \vec{D}\phi^a + \frac{\lambda}{2}(\phi^a \phi^a - c^2)^2 \right]. \tag{4.30}$$

Here the vectors indicate spatial transformation properties and the indices internal symmetry ones. Also, E^a and B^a are defined as in electromagnetism,

$$E_i^a = F_{0i}^a, \qquad B_i^a = \tfrac{1}{2} \varepsilon_{ijk} F_{jk}^a, \tag{4.31}$$

and we have rescaled the fields as in Eq. (3.23), so there is no coupling constant in the expression for the energy.

To derive the bound, we shall need two properties of the magnetic field. One is

$$\vec{D} \cdot \vec{B}^a = 0. \tag{4.32}$$

This is a consequence of the Jacobi identity for covariant differentiation. The other is

$$\lim_{r \to \infty} \int d^2 S\, \vec{n} \cdot \vec{B}^a\, \phi^a = 4\pi g c, \tag{4.33}$$

where d^2S is the usual element of surface area and $\vec{n}$ is the outward-pointing normal. This is most easily seen to be true by going to the gauge introduced in Sec. 4.2, where, at large distances,

$$\phi^a = \langle \phi^a \rangle = \delta^{a3} c \ , \qquad (4.34)$$

and the only surviving component of $\vec{B}^a$ is $\vec{B}^3$, a magnetic monopole field.

We are now ready to go.

$$\begin{aligned} E &\geq \tfrac{1}{2} \int d^3\vec{x} (\vec{B}^a \cdot \vec{B}^a + \vec{D}\phi^a \cdot \vec{D}\phi^a) \\ &= \tfrac{1}{2} \int d^3\vec{x} (\vec{B}^a \pm \vec{D}\phi^a)^2 \\ &\mp 4\pi g c \ , \end{aligned} \qquad (4.35)$$

by integration by parts and Eqs. (4.32) and (4.33). Thus

$$E \geq |4\pi g c| \ , \qquad (4.36)$$

the desired result.

Because g is $O(1/e)$ and μ is $O(ce)$, the right-hand side of this equation is $O(\mu/e^2)$, consistent with the arguments of Sec. 4.4. Actually, this is evidence for nothing but dimensional analysis; once we have discarded the scalar coupling constant, λ, the only quantity we can build with the dimensions of energy is μ/e^2.

It is actually possible to saturate the bound in an appropriate limit, first studied by Prasad and Sommerfield.[22] In deriving the bound, we discarded three of the five terms in Eq. (4.30). Two of them, the E^2 term and the $(D_0\phi)^2$ term, automatically vanish if we assume time-reversal invariance. We make the third one, the scalar potential, vanish by going to the limit $\lambda \to 0^+$. Phrased less formally, we eliminate all λ-dependent terms from the equations of motion, but retain the boundary condition that $\phi^a \phi^a = c^2$ at large distances.

In this limit, Eq. (4.35) is an equality, not an inequality, and we can saturate the bound if we can find solutions of

$$\vec{B}^a \pm \vec{D}\phi^a = 0 \ , \qquad (4.37)$$

where the sign in this equation depends on the sign of g. As a bonus, any solution of Eq. (4.36) is also a time-independent solution of the equations of motion; a minimum of the energy functional is *a fortiori* a stationary point.

Equation (4.37) is a first-order differential equation, and considerably easier to analyze than the second-order equations of motion. I don't have time to give the analysis here, or even to describe its main results in any detail. However, it turns out that the equation does have solutions, and, delightfully, not just single-monopole solutions but also many-monopole ones. This is not so incredible as it may seem. In the Prasad-Sommerfield limit, the scalar field becomes massless, and therefore it is possible for many monopoles to exist in static equilibrium, their scalar attraction balancing their magnetic repulsion.

5. QUANTUM THEORY

5.1 Quantum Monopoles and Isorotational Excitations[19]

In Sec. 3.2, I argued that a quantum field theory should most closely resemble the corresponding classical theory in the limit of weak coupling. In this section, I will build on this observation to develop a quantitative approximation scheme for theories whose classical limits possess time-independent monopole solutions.[23]

For definiteness, we'll work with the Georgi-Glashow model,

$$\mathcal{L} = -\frac{1}{4f^2} F^a_{\mu\nu} F^{\mu\nu a} + \frac{1}{2} D_\mu \phi^a \cdot D^\mu \phi^a$$
$$- \frac{\lambda}{4} (\phi^a \cdot \phi^a - c^2)^2 \ . \tag{5.1}$$

We define new variables by $\Phi' = \Phi f$, $c' = cf$, and $\lambda' = \lambda/f^2$. In terms of these,

$$\mathcal{L} = \frac{1}{f^2} \left[-\frac{1}{4} F^a_{\mu\nu} F^{\mu\nu a} + \frac{1}{2} D_\mu \phi'^a D^\mu \phi'^a \right.$$
$$\left. - \frac{\lambda'}{4} (\phi'^a \cdot \phi'^a - c'^2) \right] \ . \tag{5.2}$$

I propose to study this theory in the limit of small f, with e' and λ' held fixed. This is the classical limit, as discussed in

THE MAGNETIC MONOPOLE

Sec. 3.2; the relevant quantity in the quantum theory is $\mathscr{L}/\hbar$, so the limit of small f with fixed $\hbar$ is the same as the limit of small $\hbar$ with fixed f. Also, when things are written in this way, the classical monopole solution is independent of f; a factor that multiplies the total Lagrangian divides out of the equations of motion. (We will use these redefined parameters for the rest of this discussion, so I'll drop the primes from now on.)

The theory is especially easy to work with in temporal gauge, $A_0 = 0$,

$$\mathscr{L} = \frac{1}{f^2}\left[\frac{1}{2}\partial_0 \vec{A}^a \cdot \partial_0 \vec{A}^a + \frac{1}{2}\partial_0 \Phi^a \partial_0 \Phi^a + \text{terms without time derivatives}\right]. \quad (5.3)$$

Thus the canonical momentum density conjugate to Φ^a is

$$\pi^a = f^{-2} \partial_0 \Phi^a, \quad (5.4)$$

that conjugate to A^a is

$$\vec{\pi}^a = f^{-2} \partial_0 \vec{A}^a, \quad (5.5)$$

and the Hamiltonian is given by

$$f^2 H = \frac{1}{2} f^4 \int [\vec{\pi}^a \cdot \vec{\pi}^a + \pi^a \pi^a] d^3\vec{x} + V, \quad (5.6)$$

where V is minus the integral of the terms without time derivatives in Eq. (5.3) The classical monopole solution is a stationary point of the functional V; indeed, if the monopole is stable (as we shall assume it is), it is a local minimum of V. I emphasize that all powers of f are explicitly displayed in Eq. (5.6); V is independent of f.

Equation (5.6) defines a peculiar Hamiltonian from the viewpoint of ordinary perturbation theory. Firstly, there is an explicit f^2 on the left-hand side of the equation. Of course, this is a trivial peculiarity; if we can find an expansion for the energy eigenfunctions and eigenvalues of $f^2 H$, we can find one for those of H. Secondly, the small parameter multiplies the kinetic energy, the term quadratic in canonical momenta, rather than the potential energy, the term independent of canonical momenta. This is very strange; have we ever encountered such a system before?

Yes, we have. For this is the situation for a diatomic molecule:

$$H = \frac{\vec{p}^2}{2M} + V(r) , \quad (5.7)$$

where M is the reduced nuclear mass. The standard expansion in the study of the spectra of diatomic molecules uses as the small parameter 1/M, the coefficient of the kinetic energy. Of course, our system is not exactly a diatomic molecule. What it is exactly is a polyatomic molecule,

$$H = \sum_{i=1}^{N} \frac{\vec{P}_i^2}{2M_i} + \vec{V}(\vec{r}_1 \ldots \vec{r}_N) . \quad (5.8)$$

To be precise, it is an infinitely polyatomic molecule, where all the nuclei have mass $1/f^4$. Thus the problem of constructing quantum monopoles is one that was solved completely fifty years ago.

I will now explain this solution, first by reminding you of the familiar results for a diatomic molecule, then by telling you the trivial extension to a polyatomic molecule, and, finally, by making the even more trivial transcription of this extension into the language of field theory.

For the diatomic molecule, we assume the interatomic potential is as shown in Fig. 5. The minimum of V is at $r = r_0$, and $V(r) = E_0$. The first three approximations to the low-lying energy eigenstates and their eigenvalues are shown in Table 1.

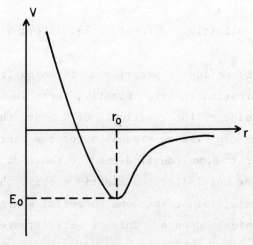

Figure 5

Table 1

The Diatomic Molecule

Order of Approximation	Energy Eigenstate	Energy Eigenvalue
0	$\|r_0, \theta, \phi\rangle$	E_0
1	$\|n, \theta, \phi\rangle$	$+(n+\tfrac{1}{2})\sqrt{V''(r_0)/M}$
2	$\|n, \ell, m\rangle$	$+\ell(\ell+1)/2M r_0^2 + \ldots$

As we see from the table, the proper expansion parameter for energy eigenvalues is $1/\sqrt{M}$; this will become f^2 in the field-theory problem. (The right-hand side of the table is cumulative; that is to say, the energy in first order is the sum of the first two entries, etc.) I will now explain the origin of the table.

In zeroth order, we neglect the kinetic energy altogether. The particle sits at the bottom of the potential well, in an eigenstate of the position operator, $\vec{r}$. The magnitude of the position is fixed at r_0, but the angular position is arbitrary. This is not much like the real spectrum revealed by molecular spectroscopy; in particular, there is a totally spurious infinite angular degeneracy. As we shall see, this degeneracy is lifted only in second order.

In first order, we begin to see the effects of the vibration of the particle about its equilibrium position. Since M is very large, the particle does not vibrate very far, and, to first order, we can replace the potential near equilibrium by a harmonic potential,

$$V(r) = E_0 + \tfrac{1}{2} V''(r_0)(r-r_0)^2 . \tag{5.9}$$

The energy eigenfunctions are now harmonic-oscillator wave functions in r, but still angular delta-functions. They are labeled by the usual oscillator excitation number, n, and have the usual oscillator energy. These are the famous vibrational levels of molecular spectroscopy.

It is only in second order that we begin to see the effects of rotation; this is because the zeroth-order moment of inertia of the molecule is Mr_0^2. The degeneracy in angle is removed; angular eigenstates are replaced by angular-momentum eigenstates,

$$|n,\ell,m\rangle = \int d\Omega\, Y_{\ell m}(\theta,\phi)|n,\theta,\phi\rangle\ , \qquad (5.10)$$

and a rigid-rotator term is added to the energy. These are the famous rotational levels of molecular spectroscopy. Note that the rotational structure involves no properties of V that have not entered earlier approximations. In addition, we begin to see the effects of departures from the harmonic approximation, Eq. (5.9). I have indicated these terms (vibrational-vibrational coupling) in the table by triple dots. They depend on the detailed form of V (in particular, on its third and fourth derivatives at r_0). Unlike the rotational term, they do not affect the qualitative features of the problem, nor do the higher terms in the expansion.

The extension of all this to a polyatomic molecule is trivial. Unless the equilibrium configuration of the molecule is one in which all the nuclei are aligned, the equilibrium configurations are labeled, like the positions of a rigid body, by three Euler angles rather than two polar angles. As a consequence of this, the rotational spectrum, once it appears in second order, will be that of a rigid body, rather than a rigid rotator. Also, there are many ways to vibrate about equilibrium, and the single integer n is replaced by a string of integers n_i, one for each normal mode.

It is straightforward to transcribe this to field theory; the only question is what replaces the rotational spectrum. The zeroth-order molecular energy eigenstates were rotationally degenerate because the classical equilibrium state was not invariant under rotations. We will have similar phenomena in field theory whenever the classical equilibrium state, the monopole, is not invariant under the unbroken symmetry group of the theory. How rich a "rotational spectrum" we have depends on how assymetric the monopole is, how many solutions we can generate by application of the symmetry group.

At the very minimum, the monopole solution is not translationally invariant, so we have at least a three-parameter family of time-

Table 2

The Quantum Monopole

Order of Approximation	Energy Eigenstate	Energy Eigenvalue
0	$\|\vec{r}\rangle$	V_0/f^2
1	$\|n_1,n_2\ldots\vec{r}\rangle$	$+\sum_i \omega_i(n_i+\tfrac{1}{2})$
2	$\|n_1,n_2\ldots\vec{P}\rangle$	$+\dfrac{f^2\vec{P}^2}{2V_0}+\ldots$

independent solutions, labeled by the position of the center of the monopole, $\vec{r}$. I have constructed Table 2 on the assumption this is the only degeneracy.

This is very much a transcription of Table 1. The small parameter $1/M$, the coefficient of the kinetic energy in Eq. (5.7), has been replaced by the small parameter f^4, the coefficient of the kinetic energy in Eq. (5.6), and all the eigenvalues have been divided by f^2, because of the f^2 on the left in Eq. (5.6), but otherwise the right-hand columns in the two tables are almost identical.

Now let's go through the table in detail.

To zeroth order, the energy eigenstates are eigenstates of the field operators, with eigenvalues given by the classical solutions to the field equations. In equations,

$$\phi^a_{op}(\vec{x})|\vec{r}\rangle = \phi^a_{cl}(\vec{x}-\vec{r})|\vec{r}\rangle \,, \qquad (5.11)$$

where "op" indicates an operator and "cl" the classical solution. Of course, a similar equation holds for $\vec{A}^a$. V_0 is the value of the functional V at the monopole solution. We see that to leading order the monopole mass is proportional to $1/f^2$, something we already knew on other grounds.

In first order, we have a sum over normal modes, the eigenmodes of classical small vibrations about the monopole solution. Of course, we have a system with an infinite number of degrees of freedom, so we can have continuum eigenmodes as well as discrete ones; for these the sum should be replaced by an integral. As usual when passing from particle mechanics to field theory, we reinterpret harmonic-oscillator excitation numbers as meson normal-mode occupation numbers. The state where all n's vanish is an isolated monopole; states with nonvanishing n's correspond to one or more mesons bound to the monopole (discrete eigenmode) or passing by the monopole (continuum eigenmode). To this order, there is no sign of meson-meson interactions in the energy because meson-meson interactions are of order f^2; Their effects are analogous to the vibrational-vibrational coupling in the molecule, and, like it, they are lurking behind the three dots in the second-order energy.

There is a first-order correction to the mass of the monopole, $\Sigma \tfrac{1}{2}\omega_i$. This sum is divergent, but, at least in a renormalizable theory, the difference between it and the corresponding sum for the vacuum state is finite and is a genuine correction to the classical monopole mass.

In second order, the degeneracy is lifted. Because the degeneracy is caused by translational invariance, not rotational invariance, the energy eigenstates are not angular-momentum eigenstates, as in Eq. (5.10), but linear-momentum eigenstates,

$$|n_1 \ldots \vec{P}\rangle = \int \frac{d^3 \vec{r}}{(2\pi)^{3/2}} e^{i\vec{P}\cdot\vec{r}} |n_1 \ldots \vec{r}\rangle . \qquad (5.12)$$

By a fluke, we know the form of the second-order energy without having to do any computations; the expression in the table just comes from

$$\sqrt{\vec{P}^2 + M^2} = M + \frac{f^2 \vec{P}^2}{2V_0} + O(f^4) . \qquad (5.13)$$

Now let us relax our assumption that the only degeneracy is translational. There can be either further geometrical degeneracy or internal-symmetry degeneracy.

Geometrical degeneracy leads to a spectrum resembling that of molecular physics. If the monopole solution is not spherically symmetric, but does have an axis of rotational symmetry, the classical solutions are labeled by two polar angles, and in second order the

THE MAGNETIC MONOPOLE

system develops a rigid-rotator spectrum, like the diatomic molecule. If the system has no axis of symmetry at all, three Euler angles are required, and the spectrum is a rigid-body spectrum, like that of the polyatomic molecule. Interesting variations on these phenomena can appear if the classical solution is invariant under some discrete subgroup of O(3). For example, if there is both an axis of rotational symmetry and an orthogonal plane of reflection symmetry, the odd angular momenta do not appear in the rotator spectrum.

Internal-symmetry degeneracy occurs if the classical solution is not invariant under H, the unbroken subgroup of G. As always, the second-order energy eigenstates are linear combinations of the eigenstates of the field operators. However, now the linear combinations transform as irreducible representations of H, rather than the translation or rotation group. The "rotational spectrum" lies in internal-symmetry space, and "rotational levels" carry H quantum numbers. I will call these states "isorotational levels". (Spin is to isospin as rotation is to isorotation.)

To compute the energies of the isorotational levels requires knowing the form of the kinetic energy operator acting on wavefunctions restricted to the surface of minima of V. This is a generalization of the most straightforward way of solving the rigid rotator, by analyzing the angular part of the Laplace operator. We don't need to do the computation to see that the H-singlet state is always the lowest level; this state has a constant wave-function, and is always annihilated by any generalized Laplace operator, whatever its detailed form.[24]

As an example, let us consider the 't Hooft-Polyakov monopole. To avoid confusion caused by a spatially-dependent H, we will work in the string gauge defined in Eqs. (4.19-21); Φ points everywhere in the 3-direction, and H is everywhere the SO(2) subgroup of rotations about the 3-axis. As usual, we identify this with the group of electromagnetism. The fields in the theory consist of a neutral scalar, a neutral massless vector (the photon), and positively and negatively charged massive vectors. The monopole solution is invariant under H only if the charged fields vanish everywhere. But they can not; if all charged fields vanish, the electromagnetic field obeys the free Maxwell equations, and a monopole field at large distances implies a gauge-invariant singularity at the origin. Note that this argument is independent of the details of the model; it is true for any magnetic monopole in any spontaneously broken gauge field theory.[25]

Thus the zeroth order eigenstates are labeled not just by the position of the center, r, but also by a complex number of modulus one, $\exp(i\alpha)$, the value of the charged field at some standard point. Under SO(2) rotations,

$$e^{-iQ_{op}\lambda} |e^{i\alpha},\vec{r}\rangle = |e^{i(\alpha+e\lambda)},\vec{r}\rangle , \quad (5.14)$$

where I have normalized the electric charge operator, Q_{op}, such that the charged field has charge e. The isorotational levels are

$$|m,\vec{P}\rangle = \int_0^{2\pi} \frac{d\alpha}{\sqrt{2\pi}} e^{-im\alpha} \int \frac{d^3\vec{r}}{(2\pi)^{3/2}} e^{-i\vec{P}\cdot\vec{r}} |e^{i\alpha},\vec{r}\rangle , \quad (5.15)$$

where m is an integer and I have suppressed the vibrational quantum numbers. These are charge eigenstates as well as momentum eigenstates,

$$Q_{op}|m,\vec{P}\rangle = m e |m,\vec{P}\rangle . \quad (5.16)$$

The monopole has brought forth dyons.[26]

I emphasize that these are not fundamental dyons; they are inevitable, not optional, and their properties are computable, not adjustable. Neither are they bound states of a monopole and a charged meson; their excitation energies are proportional to f^2, while the meson mass is $O(1)$. For the same reason, the dyons are stable, at least until very high m; they can not de-excite by emitting a charged meson.

Only the last of these statements is not true in general. In a theory with a hierarchy of symmetry breakdown and mass scales, f^2 times a large mass scale may still be much greater than the masses

THE MAGNETIC MONOPOLE

of the light particles in the theory. If these light particles carry the proper quantum numbers, the dyons may all decay into the ground-state monopole.

5.2 The Witten Effect

There is a famous CP-violating term that can be added to the Lagrangian of a gauge field theory, the θ-term,

$$\mathcal{L}' = \frac{\theta}{32\pi^2} \epsilon^{\mu\nu\lambda\sigma} F^a_{\mu\nu} F^a_{\lambda\sigma} , \qquad (5.17)$$

where θ is a real number, and the fields are normalized as in Sec. 3.1. This term is a total derivative, so it has no effect on the equations of motion; nevertheless, it has profound effects on the physics of the theory. How this comes about is part of the theory of instantons, which I have no intention of reviewing here. However, I will need one result of this theory: θ is an angle, that is to say, all physical phenomena are periodic functions of θ with period 2π.

Witten showed that for Abelian monopoles the θ-term has a striking influence on the dyon spectrum.[27] I will give here a derivation of the Witten effect which shows that it is independent of the dynamics of the heavy fields inside the monopole. The heavy fields are important only in that they insure that a monopole exists in the first place; everything that happens afterwards involves only the fields outside the monopole, the ordinary electromagnetic fields. Had I the courage, I could have discussed the Witten effect in Sec. 2.

Let me write out the θ-term, totally ignoring the heavy fields. That is to say, we replace $F^a_{\mu\nu}$ by a single field $F_{\mu\nu}$, where

$$F_{0i} = e E_i , \qquad F_{ij} = e \epsilon_{ijk} B_k , \qquad (5.18)$$

and $\vec{E}$ and $\vec{B}$ are the conventionally normalized electric and magnetic fields. In terms of these,

$$\mathcal{L}' = \frac{\theta e^2}{4\pi^2} \vec{E} \cdot \vec{B} . \qquad (5.19)$$

We now write these fields as ordinary electromagnetic fields (for simplicity assumed static) plus a monopole background:

$$\vec{E} = \vec{\nabla}\phi \;,$$

$$\vec{B} = \vec{\nabla}\times\vec{A} + g\vec{r}/r^3 \;. \qquad (5.20)$$

where $\vec{A}$ and ϕ are the conventional vector and scalar potentials. (We will generalize to dyon backgrounds shortly.) We find

$$L' = \int d^3\vec{r}\,\mathscr{L}'$$

$$= \frac{e^2 g\theta}{\pi}\int d^3\vec{r}\,\phi(\vec{r})\delta^{(3)}(\vec{r}) \;, \qquad (5.21)$$

by integration by parts.

This is the standard coupling of the scalar potential to an electric charge of magnitude $e^2 g\theta/\pi$ localized at the monopole. In the presence of the θ-term, magnetic charge induces electric charge.

We can understand this in a rough way. If we had added an $\vec{E}\cdot\vec{E}$ term to the Lagrange density, it would have represented a dielectric constant, a term that would make a given electric charge induce a (typically shielding) additional electric charge. A $\vec{B}\cdot\vec{B}$ term would have a similar effect on magnetic charge. Thus it is not surprising that an $\vec{E}\cdot\vec{B}$ term causes magnetic charge to induce electric charge.

The problem with this argument is that the effect is not reciprocal; if we add an electric monopole background, its contribution to L' vanishes upon integration by parts. However, this makes it easy to extend things to dyons. In the presence of the θ-term, the dyon charges are given by

$$Q = e(m + eg\theta/\pi) \;, \qquad (5.22)$$

with m an integer. This violates CP, but not the quantization condition. (See the discussion after Eq. (2.22).)

Because eg is always a half-integer, when θ increases by 2π, the original series of charges is recreated, with each term moving 2eg places forward. With the methods used here, we can't say anything about the θ dependence of dyon energies, but Witten has looked into the interior of the monopole, and he reports that by the time a dyon has replaced another, its energy has become that appropriate to its new charge. In a word, θ is an angle; physics is a periodic function of θ with period 2π. This is a charming result, considering that instantons never entered the argument.

5.3 A Little More About SU(5) Monopoles

In the earlier parts of these lectures, I've made some comments about the monopoles that occur in the SU(5) theory of Georgi and Glashow. We'll now look a little more closely at these objects.[28] There are two reasons for doing this. Firstly, the SU(5) theory is in itself interesting. It is the simplest of a family of grand unified theories, one of which might well describe reality, at least at energies below the Planck mass. Secondly, the theory is sufficiently rich in its structure to serve as a good example of the interplay of many of the fundamental ideas of monopole theory.

I'll begin by summarizing the relevant features of the theory. The theory is a gauge field theory with gauge group SU(5). This is a simple group, so there is only one coupling constant, f; f is on the order of e, the ordinary electromagnetic coupling. (It is not precisely e both because of Clebsch-Gordon coefficients and because of renormalization effects in going from large distances, where e is defined, to 10^{-28} cm, where monopoles live.)

All we will need to know about the scalar fields in the theory is that their interactions are such that there are two mass scales, or, equivalently, two scales of scalar vacuum expectation value.

The larger expectation value breaks SU(5) down to [SU(3) ⊗ SU(2) ⊗ U(1)]/Z_6. If we realize SU(5) as 5 × 5 matrices, SU(3) consists of transformations on the first three coordinates, SU(2) on the last two, and U(1) of diagonal matrices of the form

$$\text{diag}(e^{2i\theta}, e^{2i\theta}, e^{2i\theta}, e^{-3i\theta}, e^{-3i\theta}) \qquad (5.23)$$

If $e^{6i\theta} = 1$, this is in SU(3) ⊗ SU(2); this is why we must take the quotient of the direct product by Z_6. The SU(3) subgroup is identified with color; the SU(2) ⊗ U(1) subgroup with the group of the Weinberg-Salam model. As a result of this symmetry breakdown, twelve of the original twenty-four gauge fields acquire masses on the order of 10^{14} GeV. (It is only at these high mass scales that the color gauge coupling is comparable in strength to the electroweak ones.) These superheavy mesons transform as the $(3,\bar{2}) \oplus (\bar{3},2)$ representation of $SU(3) \otimes SU(2)$.

The smaller vacuum expectation value breaks the group down further, to $SU(3)$ (color) $\otimes$ $U(1)$ (electromagnetism); as a result of this symmetry breakdown, three gauge fields acquire masses on the order of 10^2 GeV. The electromagnetic group is generated by

$$Q_{em} = -i\,\text{diag}(\tfrac{1}{3}, \tfrac{1}{3}, \tfrac{1}{3}, -1, 0)\,, \qquad (5.24)$$

where the generator has been normalized such that the proton has unit charge. The factors of $\tfrac{1}{3}$ are a sign that the theory contains fractionally charged particles: quarks, of course, but also the superheavy vector mesons. Because of these factors,

$$\exp(2\pi Q_{em}) \neq 1\,. \qquad (5.25)$$

We have to go three times as far,

$$\exp(6\pi Q_{em}) = 1\,. \qquad (5.26)$$

$\exp(2\pi Q_{em})$ is an $SU(3)$ matrix though, so really the group H is not $SU(3) \otimes U(1)$ but $SU(3) \otimes U(1)/Z_3$. This is just a fancy way of saying that only particles of non-zero triality have fractional charge. (*cf.* example (e) at the end of Sec. 3.4.)

In addition to these gauge symmetries, the theory possesses a continuous global internal symmetry which, when all the dust of symmetry breakdown has settled, leads to the conservation of the difference of baryon and lepton number, $B - L$. (B and L are not separately conserved; the theory famously predicts proton decay.)

We already know some things about this theory. We know that it has monopoles (Sec. 4.2), that the masses of these monopoles are of order 10^{16} GeV and their cores of order 10^{-28} cm in size (Sec. 4.4), and that they have a spectrum of isorotational excitations, dyons, with a level spacing on the order of 10^{12} GeV (Sec. 5.1).

Now let us look at the monopoles in more detail. At large

THE MAGNETIC MONOPOLE

distances, the monopole field must be of the standard form,

$$\vec{A} = Q\vec{A}_D , \qquad (5.27)$$

where Q is a generator of $SU(3) \otimes U(1)$. One's first thought is that because the theory contains fractionally charged particles, $g = \frac{1}{2} e$ is not allowed; $g = \frac{3}{2} e$ is needed. Indeed, $Q = Q_{em}/2$ does not satisfy the quantization condition;

$$Q = 3Q_{em} / 2 \qquad (5.28a)$$

is needed, as we see from Eqs. (5.25) and (5.26).

(Let me raise and lay a spectre. All fractionally charged particles are confined; for example, we can not separate the components of quark-antiquark pair by more than roughly 10^{-13} cm. In using confined particles to derive the quantization condition, are we not making an error? No, we are not. The Dirac string is infinitely thin and infinitely long. We can imagine bringing the pair to within 10^{-14} cm of the string, a thousand light years from the monopole, and diffracting the quark around the string while holding the antiquark fixed. Confinement is irrelevant to this point.)

However, there is a second possibility,[29]

$$Q = (Q_{em} + Q_Y) / 2 \qquad (5.28b)$$

where Q_Y is the generator of color hypercharge,

$$Q_Y = -i \, \text{diag}(\tfrac{2}{3}, -\tfrac{1}{3}, -\tfrac{1}{3}, 0, 0) . \qquad (5.29)$$

Thus,

$$Q = -i \, \text{diag}(1, 0, 0, -1, 0) / 2 , \qquad (5.30)$$

and the quantization condition is satisfied.

This is a combination of an electromagnetic monopole and a chromomagnetic monopole. Colorless particles, like leptons and hadrons, see only the electromagnetic field, and find $g = \frac{1}{2} e$. However, because all fractionally charged particles are colored, a fractionally charged particle sees both the electromagnetic and the chromomagnetic field, and the combined effect of these two fields renders the string undetectable.

Of course (5.28a) and (5.28b) are not the only monopole fields there are. Up to a gauge transformation, the solution of the quantization condition and the non-Abelian stability condition is

$$Q = -i\,\mathrm{diag}(r, s, s, -r-2s, 0)/2 , \qquad (5.31)$$

where r and s are integers such that

$$r-s = 0, \pm 1 . \qquad (5.32)$$

Because SU(3) is simply connected, the only topological charge is the Abelian magnetic charge, r+2s.

We can estimate the energies of these configurations by the method of Sec. 4.4. The energy stored in the field outside the core is proportional to

$$-\mathrm{Tr}\,Q^2 = s^2 + \tfrac{1}{2}(r+s)^2 . \qquad (5.33)$$

The nontrivial monopole of lowest energy is $r = 1$, $s = 0$, the combination monopole, Eq. (5.28b). If we take Eq. (5.33) dead seriously as an estimate of energy, it is easy to show that every other nontrivial monopole has more than enough energy to decay into an appropriate number of combination monopoles. For example, the pure electromagnetic monopole, Eq. (5.28a), with $r = s = 1$, has six times the energy of a combination monopole, and is thus unstable to decay into three such objects, the minimum number needed to conserve topological charge.

(Another spectre: In our energy estimate we neglected factors of coupling constants. By treating the color coupling constant as if it were the same as the electromagnetic coupling constant, are we not making an error? No, we are not. The two couplings are indeed very different at large distances, but they are the same at 10^{-28} cm, near the monopole core, and this is the region that dominates the energy integral.)

Dokos and Tomaras[28] have constructed a combination monopole that is a time-independent solution of the field equations; their method is a transposition of the 't Hooft-Polyakov construction to an appropriate SU(2) subgroup of SU(5).

For our purposes, all we need to know about this solution is its degeneracy, for this suffices to determine the spectrum of isorotational levels. It turns out that one of the superheavy vector fields, X^a (a = 1, 2, 3) is nonzero in the Dokos-Tomares solution. This field is a color 3, with electric charge $-4/3$ and $B - L = -2/3$. Further, once the value of this field at some standard point is given, the solution is uniquely determined.

Since any complex 3-vector can be turned into any other of the same magnitude by an SU(3) matrix, the manifold of solutions is in one-to-one correspondence with the set of all unit complex three-vectors,
$$X^a \bar{X}_a = 1 , \qquad (5.34)$$
and the problem of constructing the isorotational levels is the problem of constructing the functions of these vectors that transform as irreducible representations of the symmetry group.

This is easily done. A complete set of functions on the manifold consists of all monomials in X and $\bar{X}$. We can write these as tensors,
$$X^{a_1 \cdots a_n}_{b_1 \cdots b_m} = X^{a_1} \cdots X^{a_n} \bar{X}_{b_1} \cdots \bar{X}_{b_n} . \qquad (5.35)$$

These tensors are almost the objects that form the basis space for the irreducible representation of SU(3) called (n,m). The only difference is that the irreducible tensors are traceless. However, it is straightforward to subtract the trace from these expressions. Because of Eq. (5.34), the tensors of lower rank we obtain by taking the traces are not new functions, just objects already constructed as monomials of lower degree. To compute the electric charge and (B-L) assignments of these tensors is trivial; we just get n-m times the contribution of a single X.

Thus: The isorotational levels transform as the representation (n,m) of color SU(3), with each representation occurring once and only once. These levels have an electric charge of $4(m-n)/3$ and a B-L of $2(m-n)/3$. Note the coupling of charge excitation to color excitation. These are chromodyons.

Let us look more closely at the physics of chromodyons. To

begin with, I will ignore the existence of fermions.

Given two levels with the same value of m-n, the higher can always decay into the lower by the emission of massless color gluons. (The fact that the gluons are confined is irrelevant. They are confined at 10^{-13} cm and the decay takes place at 10^{-28} cm; worrying about the effects of gluon confinement on the decay of a chromodyon is like worrying about the effects of the walls of the laboratory on the decay of a radioactive nucleus.) Thus we expect only one stable level for a given m-n. These levels can not decay by the emission of superheavy vector mesons, because their excitation energy is too low; they can not decay by the emission of color gluons because these do not carry charge; they can not decay by the emission of Weinberg-Salam gauge mesons because these do not carry color; they are stable.

Although they are stable, they are colored (except for the ground state monopole). Thus an (n,m) chromodyon and an (m,n) chromodyon will form a colorless pair, bound together by the confining force. The fate of this pair depends on the interaction between its components at short distances. (Although not so short that the cores overlap.)

This is dominated by the magnetic interaction. If the components have opposite Abelian magnetic charge, that is to say, if they are excitations of monopole and antimonopole, the magnetic force is attractive and the components will annihilate each other.

A short computation is needed if the components have the same Abelian charge; we have to worry about the competition between Abelian repulsion and non-Abelian attraction, as explained in Sec. 3.6. If we represent the field of one monopole by a matrix Q_1 and the other by a matrix Q_2, the magnetic interaction energy is proportional to $-\operatorname{Tr} Q_1 Q_2 / r_{12}$. We can always choose our gauge so

$$Q_1 = -i \operatorname{diag}(1,\ 0,\ 0,\ -1,\ 0)\ /\ 2\ . \tag{5.36}$$

Q_2 can be any matrix obtained from this by permutation on the first three entries (the color entries). We minimize the interaction energy by choosing

$$Q_2 = -i\,\text{diag}(0, 1, 0, -1, 0)/2 , \qquad (5.37)$$

but even at the minimum the energy is still positive. Abelian repulsion wins over non-Abelian attraction.

Thus we have a system composed of two particles with an attractive interaction between them at large distances and a repulsive interaction at short distances. This is the diatomic molecule again, not as an analogy this time but as the real thing, with the full panoply of vibrational and rotational levels, though on a somewhat larger energy scale than is usual in molecular physics.

But all of this is mere fantasy. In the real world there are light fermions, quarks and leptons, and all the chromodyons can decay to the ground-state monopole by emitting quark-lepton pairs. Pity.

5.4 Renormalization of Abelian Magnetic Charge

Some years ago, in a famous series of papers, Julian Schwinger generalized quantum electrodynamics to include fundamental magnetic monopoles.[30] In the course of this work, he investigated the renormalization of magnetic charge. Even though fundamental monopoles are not our primary interest, I'll review Schwinger's results here for later comparison with ours.

Schwinger found that a necessary condition for consistent quantization of the theory was

$$e_0 g_0 = n_0/2 , \qquad (5.38)$$

where e_0 and g_0 are the bare coupling constants, the parameters that appear in the Lagrangian of the theory, and n_0 is an integer. (In fact, at first Schwinger argued that n_0 had to be even, but this point is irrelevant to the issues at hand, and I'll ignore it here.) The standard arguments for the Dirac quantization condition have nothing to do with the fundamental or composite character of the monopoles; they tell us that

$$eg = n/2 , \qquad (5.39)$$

where e and g are the physical coupling constants and n is an integer.

The analysis of electric charge renormalization proceeds much as in ordinary electrodynamics. Because there are magnetic currents as well as electric ones, the propagator for the unrenormalized electromagnetic field is expressed as an integral over three spectral weight functions rather than one; however, there is still only one photon, and only one single-photon pole term. Thus there is no problem in following the usual path, defining Z_3 to be the coefficient of this term and showing that

$$e = Z_3^{\frac{1}{2}} e_0 . \qquad (5.40)$$

As usual, the spectral representation implies that if the theory is nontrivial, Z_3 is strictly less than one.

The analysis of magnetic charge renormalization is trivially obtained by performing the duality rotation that exchanges E and B. The spectral weight functions change places but the photon remains the photon. Thus Z_3 is unchanged and

$$g = Z_3^{\frac{1}{2}} g_0 . \qquad (5.41)$$

Hence,
$$eg = Z_3 e_0 g_0 . \qquad (5.42)$$

This is the origin of Schwinger's statement that Z_3 is a rational number.

This is for fundamental monopoles. Our interest, though, is in composite monopoles, like the 't Hooft-Polyakov object. Is Eq. (5.42) still true for these? Clearly not, but neither is it false; it is meaningless. In a theory of composite monopoles, e_0 still appears as a parameter in the Lagrangian, but g_0 is nowhere to be found.

Nevertheless, all is not lost. If we define e_λ as the quantity that gives the strength of the interaction between an electron and a tiny test charge at distance λ, then e_0 is the limit of e_λ as λ goes to zero. This suggests that we define g_λ in a similar way, as the parameter that gives the strength of the interaction between a monopole and a tiny test current loop at distance λ. In this case, though, we can not send λ to zero; once λ is smaller than the size of the monopole core there is no unambiguous way to separate electromagnetism from the larger non-Abelian structure in which it is embedded, no unambiguous way to define the current loop. However, until we get to the core there is no problem. Thus we can ask how e_λ and g_λ, for λ larger than the core radius, are related to e and g, their large-distance limits. This is not the same question as that addressed by Schwinger, but it is an interesting one; there is plenty of room for large renormalization effects between infinite distance and the monopole core.

However, there is no room for renormalization effects caused by virtual monopoles. For weak coupling, monopole Compton wavelengths are much smaller than monopole geometrical sizes. (That is to say, monopoles are much heavier than massive vector mesons.) Thus, at the distance scales we are working at, between λ and infinity, the effects of virtual monopoles are negligible. (I emphasize that "weak coupling" here does not mean merely "to lowest order in e". If we think in terms of an expansion in powers of $\hbar$, the effects of virtual monopoles are barrier-penetration effects and are exponentially suppressed.) Thus, for our purposes, it is completely legitimate to replace the full theory by a simplified one in which the only dynamical variables are the electromagnetic field and the fields of ordinary charged particles.

For notational simplicity, we will restrict ourselves to the case in which the only charged field is a Dirac electron; the generalization is trivial. The theory we will study is defined by

$$\mathcal{L} = -\frac{1}{4}(\partial_\mu A_\nu - \partial_\nu A_\mu)^2 + \frac{1}{2\alpha}(\partial_\lambda A^\lambda)^2$$
$$+ \bar{\psi}(i\slashed{\partial} - e_\lambda \slashed{A} - e_\lambda g_\lambda \slashed{A}_D - m_0)\psi$$
$$+ e_\lambda J_\mu (A^\mu + g_\lambda A_D^\mu) \:. \quad (5.43)$$

This expression requires some explanation. Firstly, as to notation: All fields are unrenormalized, α is the usual gauge-fixing parameter, $A_D^\mu = (0, \vec{A}_D)$ is the standard Dirac monopole potential, and J^μ is an external c-number conserved current,

$$\partial_\mu J^\mu = 0 \:, \quad (5.44)$$

which we will use to construct test charges and current loops. Secondly, as to physics: This is supposed to contain only the degrees of freedom which are relevant at distances larger than λ. Thus, although it is not indicated explicitly, the theory is supposed to be cut off in some gauge-invariant way (say, with regulator fields) at λ. It is for this reason that the places of the bare coupling constants are taken by e_λ and g_λ, and also why the monopole is replaced by an external Dirac potential.

I emphasize that Eq. (5.43) is not intended to give an exact description of the theory at distances greater than λ. It is certainly possible to construct an effective Lagrangian that would give such a description, but it would be much more complicated, full of nonlocal and nonpolynomial interactions. Equation (5.43) is intended only to give a simple model of the relevant physics.

Arguments which by now should be excessively familiar show that the string singularity in the monopole potential is undetectable only if $e_\lambda g_\lambda$ is a half-integer. I will therefore assume that this is the case. It is not at all clear at this stage that this implies that eg is a half-integer, but we have a well-defined Lagrangian, so let's just go ahead and calculate.

To warm up, let us make sure that our test apparatus is properly calibrated by computing the effects of the external current far from the monopole, or, equivalently, with g_λ set equal to zero. In this case, the vacuum-to-vacuum transition matrix element is

$$<0|S|0> = 1 - \frac{e_\lambda^2}{2} \int d^4 x \, d^4 y \, J^\mu(x) J^\nu(y) T<0|A_\mu(x) A_\nu(y)|0> + O(J^4) \,. \tag{5.45}$$

I emphasize that this is not an (illegitimate) expansion in powers of e_λ; it is an expansion in powers of J^μ. There is nothing wrong with this; the interactions with J^μ are gauge-invariant and thus incapable of detecting the Dirac string whatever the magnitude of J^μ, so long as it is conserved. (This is just another way of saying that there is no quantization condition for classical charged particles.)

The relevant object in Eq. (5.45) is the propagator for the un-

THE MAGNETIC MONOPOLE

renormalized field,

$$T\langle 0|A_\mu(x)A_\nu(y)|0\rangle = Z_3^\lambda D^0_{\mu\nu}(x-y) + \ldots, \tag{5.46}$$

where $D^0_{\mu\nu}$ is the free propagator, the triple dots indicate terms that fall off at large distances more rapidly than the term displayed, and the superscript is on Z_3 to remind you of its cutoff-dependence. Thus, two distantly separated current elements interact as they would in free electromagnetism, with interaction strength given by

$$e_\lambda^2 Z_3^\lambda = e^2. \tag{5.47}$$

This is famously the right result.

Let us now return to nonzero g_λ. There is now a linear term in the transition matrix element,

$$\langle 0|S|0\rangle = 1 - ie_\lambda \int d^4x\, J^\mu(x)[g_\lambda A^D_\mu(x) + \langle 0|A_\mu(x)|0\rangle] + O(J^2). \tag{5.48}$$

I emphasize that $|0\rangle$ is now the ground state of the theory computed to zeroth order in J^μ but to all orders in the external monopole field. We wish to calculate eg in the same way we calculated e^2 in the preceding paragraph. This can be done if the total integral in Eq. (5.48) has the same form as its first term when the support of the current is far from the monopole; we can then identify the coefficient of the total expression with eg. In equations,

$$\langle 0|S|0\rangle - 1 \xrightarrow[\text{large distance}]{?} - ieg \int d^4x\, J^\mu(x) A^D_\mu(x) + O(J^2). \tag{5.49}$$

I've put a question mark here because we don't yet know that things will have the right form in this limit. (Let's hope they do; if they don't, we're in real trouble.)

The equation of motion for A_μ implies that

$$\langle 0|A_\mu(x)|0\rangle = -ie_\lambda \int d^4y\, D^0_{\mu\nu}(x-y)\langle 0|J^\nu(y)|0\rangle. \tag{5.50}$$

The object on the right is the matrix element of a gauge-invariant operator; thus, despite the Dirac string, it is rotationally invariant and CP invariant. (Note that this would not be true if we had foolishly attempted to expand things in powers of the monopole field.)

By rotational invariance,
$$\langle 0|\bar{\psi}\gamma^0\psi|0\rangle = f(r^2) ,\tag{5.51}$$
and
$$\langle 0|\bar{\psi}\gamma^i\psi|0\rangle = r^i g(r^2) ,\tag{5.52}$$
for some functions f and g. CP invariance implies f vanishes. The conservation equation
$$\partial_\mu \langle 0|\bar{\psi}\gamma^\mu\psi|0\rangle = 0 ,\tag{5.53}$$
implies g vanishes. Thus Eq. (5.49) is satisfied trivially, and
$$eg = e_\lambda g_\lambda .\tag{5.54}$$
This is in striking contrast to Eq. (5.42), but, as I have explained, the contradiction is only apparent; despite the appearance of similar symbols, the two equations are answers to different questions in different theories.

Equation (5.54) is very satisfying. Monopole theory says that eg must be a half integer, but a modern field theorist would ask, "eg defined at what distance scale?" The answer is, "At any distance scale, from infinity down to the monopole core. It doesn't matter."

5.5 The Effects of Confinement on Non-Abelian Magnetic Charge

We have analyzed the renormalization of Abelian magnetic charge, computed the electromagnetic monopole field at large distances in terms of the electromagnetic monopole field just outside the monopole core. To complete the analysis, we should now extend the argument to non-Abelian magnetic charge, and compute, for example, the chromomagnetic monopole field at large distances from the grand unified monopoles of Sec. 5.3. But such a computation would be folly. We know very well that there is no chromomagnetic monopole field at large distances; there is no chromodynamic field of any kind proportional to any inverse power of distance. Because of confinement, all chromodynamic forces fall off exponentially with distance, with a coefficient given by the lightest hadron mass. (If we ignore fermions, as we have been doing, this would be the glueball mass.)

This leads to an apparent paradox. For grand unified monopoles,

THE MAGNETIC MONOPOLE

the simultaneous existence of chromomagnetic and electromagnetic monopoles was necessary to keep the Dirac string undetectable; when we diffracted a fractionally charged particle around the string, the chromomagnetic phase factor was needed to cancel the electromagnetic one. (I remind you that the fact that fractionally charged particles are themselves confined is irrelevant to this point, as we showed in Sec. 5.3.) But at large distances the chromomagnetic monopole field disappears, while the electromagnetic one does not. Does this mean that at large distances the Dirac string is observable?

To ask this question is to be carried away by the momentum of my own rhetoric. Of course, the string can not become observable; confinement does not spoil gauge invariance. Nevertheless, we would like to understand how effects that fall off exponentially with distance are nevertheless able to produce a distance-independent phase factor. The remainder of this section is devoted to this problem.

Although the relevant group for grand unified monopoles is $SU(3) \otimes U(1)/Z_3$, to keep things simple I will work here with $SO(3)$, the smallest non-Abelian group. The extension is trivial. I shall first give general arguments and then back them up with explicit computations in a highly simplified but exactly soluble two-dimensional model.

The Lubkin construction associates a path in group space, $g(\tau)$, with every sphere in ordinary space; the homotopy class of this path tells us whether the sphere contains a monopole. Although the homotopy class is gauge-invariant, the path itself is not. When investigating tricky questions in gauge field theory, it's wise to work as much as possible with gauge-invariant entities. Thus I will work with the Wilson loop factor,

$$W(\tau) = \tfrac{1}{2} \operatorname{Tr} g(\tau) \,. \tag{5.55}$$

Although the loop factor is independent of the gauge, it does depend on the particular matrix representation we use to realize the group. For the problem at hand, it will turn out to be advantageous to choose the spin-$\tfrac{1}{2}$ representation of $SO(3)$, that is to say, to choose g to be an element of SU(2). This is a double-valued representation, so there is a binary ambiguity in computing the loop factor. We will eliminate this ambiguity by adopting the convention that $g(0) = 1$, so $W(0) = 1$. Of course, we could have chosen the opposite convention, $g(0) = -1$, in which case $W(\tau)$ would everywhere have been replaced by $-W(\tau)$.

The behavior of W tells us whether our sphere contains a monopole. If $g(\tau)$ is in the trivial homotopy class, $W(1) = W(0) = 1$. Figure 6a shows this behavior for a typical field of this class, the zero field. If $g(\tau)$ is in the nontrivial homotopy class, $W(1) = -W(0) = -1$. Figure 6b shows this behavior for a typical field of this class, the GNO monopole field. Phrased in terms of W, our problem is to understand how effects that fall off exponentially with distance are nevertheless able to produce a distance-independent change in W.

But this formulation is not quite right. Confinement is a quantum effect, and in quantum theory, W is an operator. Except in the leading semi-classical approximation (which does not display confinement) the one-monopole state is not an eigenstate of W. The object we must study is <W>, the expectation value of W.

Of course, for a small sphere, one whose radius is much less than the confinement length, the semi-classical approximation is valid, and Fig. 6 gives accurate plots of <W> for the vacuum and the one-monopole state. But for a large sphere, one whose radius is much greater than the confinement length, even the graph for the vacuum, Fig. 6a, is wrong. As our loop sweeps around a large sphere, it necessarily becomes, at the half-way point, a loop of large area, and there-

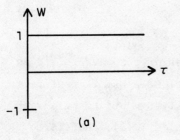
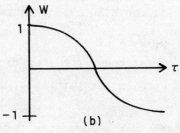

Figure 6

THE MAGNETIC MONOPOLE

fore <W> becomes very small, because of Wilson's area law. This is sketched in Fig. 7a. The sketch is very much not to scale. <W(½)> is $O(\exp{-\pi K r^2})$, where K is the string tension. This is so small that if the graph were drawn to scale you couldn't see that <W(½)> wasn't zero.

Our problem is solved. The monopole does not need to produce a distance-independent change in <W>. The area law reduces <W> from 1 to a small positive value. All the monopole field has to do is make a small additional change to bring <W> to a small negative value. The area law then brings <W> back to -1. This is sketched in Fig. 7b. I emphasize that the right side of this graph represents the same physics as the left side; the evolution of negative values of <W> can be deduced from that of positive values by a change of convention.

It's amusing to note that if we had defined W using an integer-spin representation of SO(3), we would not have had the area law, but neither would we have had a test for monopoles.

This concludes the general arguments. Now, as promised, I'll do some explicit computations in a two-dimensional Euclidean gauge theory.

Two-dimensional gauge theories have two big advantages. One is that they are easy to work with. This is basically because they have very few degrees of freedom; in Minkowski space, there is no radiation field, just a Coulomb field. The other is that they trivially display confinement. In one spatial dimension, even ordinary Abelian electrodynamics yields a linear potential. For our purposes, they have one big disadvantage; they don't possess monopoles. I will take

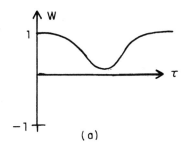 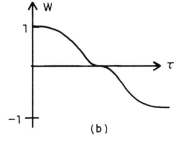

Figure 7

care of this by studying SO(3) gauge field theory not on flat two-space but on a two-dimensional sphere. Ordinary two-dimensional Euclidean field theory can be thought of as obtained from four-dimensional theory by holding z and t fixed, leaving x and y to vary. Likewise, the spherical two-dimensional theory can be obtained by holding t and r fixed, leaving θ and φ to vary. The monopole is in the part of space we have discarded, near r = 0, but its effects are still felt on the sphere we have retained.

Now let's make this more precise. We are dealing with an SO(3) gauge field theory on a sphere. Just as in Sec. 3, we will go to string gauge, so A_θ vanishes everywhere. In general, this gauge introduces a Dirac string singularity at the south pole, a nonzero $A_\phi(\pi,\phi)$. We define $g(\phi)$, an element of SU(2), by

$$\frac{dg}{d\phi} = - A_\phi(\pi,\phi) g(\phi) , \qquad (5.56)$$

and

$$g(0) = 1 . \qquad (5.57)$$

Then, for the string singularity to be just a gauge artifact,

$$g(2\pi) = \pm 1 . \qquad (5.58)$$

The plus sign corresponds to no monopole (inside the sphere), the minus sign to a monopole. If we choose our Lubkin loops as in Fig. 4,

$$W(\tau) = \tfrac{1}{2} \operatorname{Tr} g(2\pi\tau) . \qquad (5.59)$$

The Euclidean action, S_E, is the conventional one, restricted to the sphere,

$$S_E = - \frac{1}{f^2} \int \operatorname{Tr} F_{\theta\phi} F^{\theta\phi} \sqrt{g}\, d\theta\, d\phi$$

$$= - \frac{1}{r^2 f^2} \int \operatorname{Tr} (\partial_\theta A_\phi)^2 \frac{d\theta\, d\phi}{\sin\theta} , \qquad (5.60)$$

by Eqs. (3.43) and (3.45).

As always, the expectation value of a gauge-invariant observable, $\mathcal{O}$, is given by the ratio of two functional integrals,

$$\langle \mathcal{O} \rangle = \frac{\int (d A_\phi) \mathcal{O} e^{-S_E}}{\int (d A_\phi) e^{-S_E}} . \qquad (5.61)$$

We will compute $\langle \mathcal{O} \rangle$ in the absence (presence) of a monopole by integrating over all gauge field configurations with $g(2\pi) = +1\ (-1)$. This guarantees from the very beginning that $\langle W \rangle$ will have the appropriate behavior, $\langle W(1) \rangle = \pm 1$.

We will begin with the denominator in Eq. (5.61); the extension to the numerator will be straightforward. We write A_ϕ as a sum,

$$A_\phi = \hat{A}_\phi + \tfrac{1}{2}(1-\cos\theta)A_\phi(\pi,\phi) \ . \tag{5.62}$$

Thus,

$$\hat{A}_\phi(0,\theta) = \hat{A}_\phi(\pi,\theta) = 0 \ . \tag{5.63}$$

This separation diagonalizes S_E:

$$S_E = \hat{S} + S_1 \ , \tag{5.64}$$

where

$$\hat{S} = \frac{\mathrm{Tr}}{r^2 f^2} \int d\theta\, d\phi\, \frac{(\partial_\theta \hat{A}_\phi)^2}{\sin\theta} \ , \tag{5.65}$$

and

$$S_1 = -\frac{\mathrm{Tr}}{2r^2 f^2} \int_0^{2\pi} d\phi\, [A_\phi(\pi,\phi)]^2$$

$$= -\frac{\mathrm{Tr}}{2r^2 f^2} \int_0^{2\pi} d\phi\, [g^{-1}\, dg/d\phi]^2 \ . \tag{5.66}$$

Our theory has been revealed as the sum of two independent systems. One, described by $\hat{S}$, is a two-dimensional field theory, but it is a trivial one; the dynamical variables obey linear boundary conditions, and the action is a quadratic functional of these variables. The other, described by S_1, is a nontrivial system, but it is a one-dimensional one. That is to say, it is not a Euclidean field theory at all, but simply the imaginary-time version of an ordinary mechanical system, with ϕ the imaginary time.

In fact, it is a very well-studied mechanical system. The states of the system are labeled by elements of the rotation group, that is to say, by sets of three Euler angles. The system is a rigid body with one point fixed, or more properly, since g is an element of SU(2) and not SO(3), it is the double covering of a rigid body; half-odd-integral angular momenta are allowed as well as integral ones.

If we consider motions near $g = 1$,

$$g = 1 - i\varepsilon^a \sigma^a / 2 + O(\varepsilon^2) \ , \tag{5.67}$$

then

$$S_1 = \frac{1}{4r^2 f^2} \int \frac{d\varepsilon^a}{d\phi} \frac{d\varepsilon^a}{d\phi} d\phi + O(\varepsilon^3) \ , \tag{5.68}$$

from which we see that the rigid body is isotropic; it has equal principal moments of inertia,

$$I_1 = I_2 = I_3 \equiv I = 1/(2r^2 f^2) \ . \tag{5.69}$$

This enables us to analyze the system totally for all values of r. However, the answer is especially simple for very small r and very large r, so I will restrict myself to these two cases here.

For very small r, $f^2 r^2 \ll 1$, the functional integral is dominated by the stationary points of the action, the solutions of the classical equations of motion. For a rigid body, these are steady rotations about a fixed axis, for example, the positive 3-axis,

$$g = \exp(-\frac{i}{2}\omega \sigma_3 \phi) \ , \tag{5.70}$$

with ω an arbitrary non-negative constant, the angular velocity. (It's a good thing the moments of inertia turned out to be all equal; otherwise we'd be worrying about force-free precession.) Of these motions, the dominant one is the one of minimum action, that is to say, minimum ω, consistent with the boundary conditions. In the absence of a monopole, the boundary conditions are $g(0) = g(2\pi) = 1$. The dominant motion is $\omega = 0$, and the corresponding vector potential is $A_\phi = 0$. In the presence of a monopole, the boundary conditions are $g(0) = 1$, $g(2\pi) = -1$. The dominant motion is $\omega = 1$, and the corresponding vector potential is

$$A_\phi = -\frac{i}{4} \sigma_3 (1 - \cos\theta) \ , \tag{5.71}$$

the GNO monopole field. At small distances, quantum field theory looks like classical physics, just as it should.

To study large r, we need the energy eigenstates and eigenvalues for the (doubly covered) isotropic rigid body. These are discussed in standard texts,[31] so I will merely give the results here. The eigenstates are of the form $|j,m,m'\rangle$, where $j = 0, \frac{1}{2}, 1 \ldots$, and m and m' individually range from $-(2j+1)$ to $2j+1$ by unit steps. The eigenvalues are given by

$$H|j,m,m'\rangle = E_j|j,m,m'\rangle, \quad (5.72)$$

where

$$E_j = j(j+1)/2I,$$

and I is the moment of inertia. In any one of these states, the amplitude for finding the system in the configuration labeled by the group element g is

$$\langle g|j,m,m'\rangle = (2j+1)^{\frac{1}{2}} D^{(j)}_{mm'}(g), \quad (5.73)$$

where D is the usual representation matrix.

We are now ready to go. The contribution of the rigid body to the functional integral is

$$\int (dg) e^{-S_1} = \langle g = \pm 1| e^{-2\pi H} |g = 1\rangle, \quad (5.74)$$

by Feynman's path-integral formula. Inserting a complete set of energy eigenstates,

$$\langle g = \pm 1| e^{-2\pi H} |g = 1\rangle = \sum_{mm'j} (2j+1) D^{(j)}_{mm'}(\pm 1) D^{(j)*}_{mm'}(1) e^{-2\pi E_j}$$

$$= \sum_j (2j+1)^2 (\pm 1)^{2j} e^{-\pi j(j+1)/I}. \quad (5.75)$$

This is exact. For large r, $f^2 r^2 \gg 1$, this expression is dominated by the first term in the series, $j = 0$, which is totally insensitive to the presence or absence of the monopole. To see the monopole at all, we have to retain the second term, $j = \frac{1}{2}$. We find

$$\int (dg) e^{-S_1} = 1 \pm 4 e^{-3\pi f^2 r^2/2}. \quad (5.76)$$

The effect is miniscule.

Of course, this is just the denominator of the equation for the expectation value of an operator, Eq. (5.61), but similar reasoning applies to the total expression. A particularly easy case to work out is the expectation value of the action itself,

$$\langle S_E \rangle = \langle \hat{S} \rangle + \langle S_1 \rangle$$

$$= \langle \hat{S} \rangle + f^2 \frac{d}{df^2} \ln \int (dg) e^{-S_1}$$

$$= \langle \hat{S} \rangle \mp 6\pi f^2 r^2 \, e^{-3\pi f^2 r^2/2} \, , \qquad (5.77)$$

where, as before, I've dropped terms exponentially small compared to the terms retained. Another easy computation is

$$\langle W(\tau) \rangle = e^{-3\pi f^2 r^2 \tau/2} \pm e^{-3\pi f^2 r^2 (1-\tau)/2} \, , \qquad (5.78)$$

where again I've retained only the leading terms. (Here I've left the details of the calculation as an exercise.) This expression has a very simple physical interpretation; it is the sum of the area factors for the two areas into which the loop divides the sphere.

All of this is in perfect agreement with our general reasoning. At small distances, everything looks like classical physics; at large distances, all effects of the monopole are miniscule, but nevertheless the monopole makes $\langle W \rangle$ go from $+1$ to -1.

Gravity theorists say, "A black hole has no hair." What this means is that a black hole has limited hair; the only things that stick out of a black hole are massless gauge fields associated with strictly conserved quantities, the gravitational fields associated with total energy and angular momentum, and the electromagnetic fields associated with total electric and magnetic charge.

Chromodynamic forces are short-range forces, because of confinement. Nevertheless, topological charge is strictly conserved, and can be computed from chromodynamic effects at arbitrarily large distances. These two statements would be in contradiction in classical physics, but are perfectly compatible in quantum mechanics, as we have seen. Problem for the student: Can a black hole have colored hair?

Footnotes and References

1. P. A. M. Dirac, Proc. Roy. Soc. (London) Ser. A, **133**, 60 (1931).
2. G. 't Hooft, Nucl. Phys. **B79**, 276 (1974). A. M. Polyakov, JETP

Lett. **20**, 194 (1974).

3. J. Preskill, Phys. Rev. Lett. **43**, 1365 (1979).
4. S. Coleman, "Classical Lumps and Their Quantum Descendants", in *New Phenomena in Subnuclear Physics*, edited by A. Zichichi (Plenum, 1977).
5. Ref. 4 has a more extensive bibliography. See also the reviews of P. Goddard and D. Olive, Rep. Prog. Phys. **41**, 1357 (1978), and E. Amaldi and N. Cabibbo, in *Aspects of Quantum Theory*, edited by A. Salam and E. Wigner (Cambridge Univ. Press, 1972).
6. Note that there is no 4π in the definition of magnetic charge.
7. Y. Aharonov and D. Bohm, Phys. Rev. **115**, 485 (1959).
8. This is a truism for non-Abelian gauge theories, where everyone thinks of gauge transformations as labeled by functions from space-time into the gauge group. For the Abelian theory at hand, the gauge group is U(1), and the function into the group is $\exp(-ie\chi)$.
9. T. T. Wu and C. N. Yang, Nucl. Phys. **B107**, 365 (1976).
10. This problem was solved very early on by I. Tamm, Z. Phys. **71**, 141 (1931), and my results are the same as his, although my method is somewhat different. To my knowledge, the treatment in the literature closest to that given here is that of H. J. Lipkin, W. I. Weisberger, and M. Peshkin, Ann. of Phys. **53**, 203 (1969).
11. These commutators are not so innocuous as they seem. If we attempt to verify the Jacobi identity for three D's, we obtain, instead of zero, a term proportional to $\delta^{(3)}(r)$. This is no real problem, for two reasons. Firstly, we don't believe the monopole field all the way down to the origin; it's just the long-range part of something that is more complicated at short distances. (In Sec. 4 we'll see what that something is.) Secondly, even if we believe the field all the way down to the origin, there is, as we shall see, a centrifugal barrier in *all* partial waves that keeps the particle away from the origin.
12. The analysis given here follows that of A. Goldhaber, Phys. Rev. Lett. **36**, 1122 (1976). Goldhaber's work was stimulated by

investigations by R. Jackiw and C. Rebbi (*ibid.*, 1116) and by P. Hassenfratz and G. 't Hooft (*ibid.*, 1119).
13. P. Goddard, J. Nuyts, and D. Olive, Nucl. Phys. B125, 1 (1977).
14. I'm being very sloppy here about singularities. Here's a more careful argument: We define a gauge-field configuration to be locally non-singular in some region if the region is the union of a family of open sets, such that in each set the gauge field is non-singular and such that in the intersection of any two sets the gauge fields in the two sets are connected by a non-singular gauge transformation. It is possible to show that a gauge-field configuration that is locally non-singular in all of space except for the origin is gauge-equivalent to one that is non-singular (in the ordinary sense) everywhere except for the south polar axis. (This theorem is proved in Ref. 4.) That is to say, if all singularities, other than ones at the origin, are gauge artifacts, then they can all be shoved onto the Dirac string by an appropriate choice of gauge.
15. E. Lubkin, Ann. Phys. (N.Y.) 23, 233 (1963). See especially Sec. XV.
16. A somewhat longer (though still hopelessly vulgar) course can be found in Ref. 4, together with references to the mathematical literature.
17. This stability analysis was first done by R. Brandt and F. Neri, Nucl. Phys. B161, 253 (1979).
18. W. Nahm and D. Olive (private communication, summer 1979).
19. The work described here is drawn from many sources; for references, see Ref. 4.
20. H. Georgi, and S. L. Glashow, Phys. Rev. Lett. 32, 438 (1974).
21. E. Bogomol'nyi, Sov. J. Nucl. Phys. 24, 449 (1976). S. Coleman, S. Parke, A. Neveu, and C. Sommerfield, Phys. Rev. D15, 544 (1977).
22. M. Prasad and C. Sommerfield, Phys. Rev. Lett. 35, 760 (1975).
23. Much of this section is blatant plagiarism from Ref. 4. (Copyright holder take note!)

24. Actually, one can avoid this computational horror by group-theoretic tricks, just as we did for the Laplace operator in Sec. 2.
25. This note is for experts only. The discussion in the text leaves the impression that the detailed computation of the isorotational spectrum is easier than it is. There are technical complications associated with gauge invariance. These can all be dealt with, but they make the calculation lengthier than it would be if they weren't around. For example, in temporal gauge, there are many invariances of the Hamiltonian that do not leave the monopole solution unchanged, to wit, time-independent gauge transformations. No one in his right mind expects these to lead to isorotational levels. However, to demonstrate this, and to disentangle the spurious excitations from the genuine ones, requires fiddling around with the subsidiary condition that is the bane of temporal-gauge quantization. Subsidiary conditions can be avoided by working in Coulomb gauge, for example, where none are needed, but then the form of the Hamiltonian is more complicated, and this makes things messy. (For a careful Coulomb-gauge treatment of the dyons discussed immediately below, see E. Tomboulis and G. Woo, Nucl. Phys. B107, 221 (1976).
26. These dyons were first discovered by B. Julia and A. Zee, Phys. Rev. D11, 2227 (1975), using quite different methods from these.
27. E. Witten, Phys. Lett. 86B, 283 (1979).
28. C. Dokos and T. Tomaras, Phys. Rev. D21, 2940 (1980).
29. This trick was discovered by G. 't Hooft, Nucl. Phys. B105, 538 (1976) and by E. Corrigan, D. Olive, D. Fairlie, and J. Nuyts, Nucl. Phys. B106, 475 (1976).
30. J. Schwinger, Phys. Rev. 144, 1087; 151, 1048; 151, 1055 (1966).
31. For example, see L. Landau and E. Lifshitz, *Quantum Mechanics* (3rd ed.)(Pergamon, 1977) p. 410.

NUMERICAL STUDIES OF GAUGE FIELD THEORIES*

Michael J. Creutz

Physics Department
Brookhaven National Laboratory
Upton, NY 11973

INTRODUCTION

Monte Carlo simulation of statistical systems is a well established technique of the condensed matter physicist. In the last few years, particle theorists have rediscovered this method and are having a marvelous time applying it to quantized gauge field theories. The main result has been strong numerical evidence that the standard SU(3) non-Abelian gauge theory of the strong interaction is capable of simultaneously confining quarks into the physical hadrons and exhibiting asymptotic freedom, the phenomenon of quark interactions being small at short distances.

In four dimensions, confinement is a non-perturbative phenomenon. Essentially all models of confinement tie widely separated quarks together with "strings" of gauge field flux. This gives rise to a linear potential at long distances

$$E(r) \underset{r \to \infty}{\sim} Kr \tag{1.1}$$

where r is the quark-antiquark separation and the constant K is referred to as the string tension. As K is physical, it must

---

*The submitted manuscript has been authored under contract DE-AC02-76CH00016 with the U. S. Department of Energy.

satisfy the renormalization group equation. This implies the form

$$K \sim \frac{1}{a^2} \exp\left(-\frac{1}{\gamma_0 g_0^2(a)}\right) (g_0^2)^{-\gamma_1/\gamma_0^2} (1 + O(g_0^2)) \qquad (1.2)$$

where g_0 is the bare coupling when an ultraviolet cutoff of length a is introduced, and the parameters γ_0 and γ_1 are the first terms in a perturbative expansion of the Gell-Mann Low function[1]

$$a \frac{\partial}{\partial a} g_0 = \gamma(g_0) = \gamma_0 g_0^3 + \gamma_1 g_0^5 + O(g_0^7) \qquad (1.3)$$

The important observation is that eq. (2) precludes any perturbative expansion of K in terms of g_0.

A non-perturbative treatment requires a non-perturbative regulator to control the ultraviolet divergences so rampant in field theory. Wilson's elegant lattice formulation provides this needed cutoff.[2] Once formulated on a lattice, the gauge theory becomes a statistical mechanics problem in which temperature corresponds to the square of the field theoretical coupling constant. It is this analogy which permits us to borrow Monte Carlo technology from the solid state physicist.

A Monte Carlo program generates a sequence of field configurations by a series of random changes of the fields. The algorithm is so constructed that ultimately the probability density for finding any given configuration is proportional to the Boltzmann weighting. We bring our lattices into "thermal equilibrium" with a heat bath at a temperature specified by the coupling constant. Thus we do computer "experiments" with four-dimensional "crystals" stored in a computer memory. As the entire field configuration is stored, we have access to any correlation function desired.

In the remainder of these lectures I will describe the kinds of experiments we have been doing and the implications of these results for strong interaction physics.

THE MODEL

We work with Wilson's formulation of a gauge field on a lattice.[2] A link variable U_{ij}, which is an element of the gauge group, is

GAUGE FIELD THEORIES

associated with every nearest neighbor pair of sites i and j on a four-dimensional hypercubic lattice. The reversed link is associated with the inverse element

$$U_{ij} = (U_{ji})^{-1} \tag{2.1}$$

The path integral

$$Z = \int \left(\prod_{\{i,j\}} dU_{ij} \right) e^{-\beta S(U)} \tag{2.2}$$

defines the quantum theory. Here we integrate over all independent link variables with the invariant group measure. The action S is a sum over all elementary squares or "plaquettes" in the lattice

$$S(U) = \sum_\square S_\square \tag{2.3}$$

where for SU(N) we normalize

$$S_\square = (1 - \frac{1}{N} \text{Tr}\, (U_{ij} U_{jk} U_{kl} U_{li})) \tag{2.4}$$

and for U(1)

$$S_\square = (1 - \text{Re}(U_{ij} U_{jk} U_{kl} U_{li})) \tag{2.5}$$

Here i, j, k, and l label the sites circulating about the square $\square$.

In a classical continuum limit we identify

$$U_{ij} = e^{ig_0 a A_\mu} \tag{2.6}$$

where a is the lattice spacing, and A_μ is the gauge potential in the direction μ which points from i to j. The potential is regarded as an element of the Lie algebra for the gauge group. In a naive continuum limit for SU(N) the action reduces to an integral over space-time of the conventional Yang-Mills Legrangian[3]

$$S = \frac{g_0^2}{2N} \int \frac{1}{4} F_{\mu\nu}^\alpha F_{\mu\nu}^\alpha d^4 x \tag{2.7}$$

We refer to this limit as "naive" because for the full quantum theory the bare coupling constant must be renormalized.

Equation (2.2) is the partition function for a statistical system at temperature $T = 1/\beta = g_0^2/2N$. The most intuitive Monte Carlo algorithm consists of successively touching a heat bath at this temperature to each link of the lattice. By this I mean to take each U_{ij} in turn and replace it with a new group element U'_{ij}, selected randomly from the entire group manifold but with a weighting proportional to the Boltzmann factor

$$dP(U') \sim \exp[-\beta S(U')] \, dU' \qquad (2.8)$$

where $S(U')$ is the action evaluated with the given link having the value U' and all other links fixed at their current values. One Monte Carlo iteration refers to the application of this procedure to each link in the entire lattice.

I have used precisely this algorithm for the SU(2) theory[4] and smaller groups. For larger groups I have found it computationally simpler to use less intuitive but standard algorithms from statistical mechanics. These are discussed in the next section and can be competitive with or better than the heat bath algorithm for groups with sufficiently complicated manifolds.

MONTE CARLO ALGORITHMS

The goal of a Monte Carlo program is to generate a sequence of field configurations in a stochastic manner so that the ultimate probability density of encountering any given configuration C is proportional to the Boltzmann weighting

$$p(c) \sim e^{-\beta S(c)} \qquad (3.1)$$

where $S(c)$ is the action of the given configuration. We thus use the computer as a "heat bath" at inverse temperature β. Each state in the Monte Carlo sequence is obtained in a Markovian process from the previous configuration. Thus we have a probability distribution $P(c',c)$ of taking any configuration c into configuration c'. The choice of $P(c',c)$ is by no means unique.[5] Most algorithms in practice change one statistical variable at a time in a manner satisfying a condition of detailed balance

$$P(c',c)e^{-\beta S(c)} = P(c,c')e^{-\beta S(c')} \qquad (3.2)$$

Indeed, this condition plus an eventual access to any configuration will ultimately give the Boltzman distribution of eq. (3.1).

I will now show that an algorithm satisfying eq. (3.2) brings an ensemble of configurations closer to equilibrium. To do this,

GAUGE FIELD THEORIES

I need a definition of "distance" between ensembles. Suppose we have two ensembles E and E', each of many configurations. Suppose also that the probability density of configuration c in E or E' is p(c) or p'(c), respectively. Then I define the distance between E and E' as

$$||E-E'|| = \sum_c |p(c) - p'(c)| \tag{3.3}$$

where I sum over all possible configurations. Now suppose that E' is obtained from E by the Monte Carlo algorithm defined by a P(c',c) satisfying eq. (3.2). This means that

$$p'(c) = \sum_{c'} P(c,c') p(c') \tag{3.4}$$

As P(c',c) is a probability, it satisfies

$$P(c',c) \geq 0 \tag{3.5}$$

$$\sum_{c'} P(c',c) = 1 \tag{3.6}$$

Note that if we sum eq. (3.2) over c' and use eq. (3.6), we obtain

$$e^{-\beta S(c)} = \sum_{c'} P(c,c') e^{-\beta S(c')} \tag{3.7}$$

This means that the equilibrium ensemble E_{eq}, defined by eq. (3.1), is an eigenvector of the algorithm. Using this, we can now compare the distance of E' from E_{eq} to the distance of E from equilibrium

$$||E'-E_{eq}|| = \sum_c |\sum_{c'} P(c,c') (p(c') - p_{eq}(c'))|$$
$$\leq \sum_{c,c'} P(c,c') |p(c') - p_{eq}(c')| \tag{3.8}$$
$$= ||E - E_{eq}||$$

This is just the result I set out to prove; the algorithm reduces the distance of an ensemble from equilibrium.

The detailed balance condition of eq. (3.2), which is sufficient but not necessary to approach equilibrium, still does not uniquely determine P(c',c). The intuitive heat bath algorithm was discussed in the last section. In its turn, each link variable U is replaced with a new group element U' selected randomly from the gauge group with a weighting given by the Boltzmann factor

$$P(U') \sim e^{-\beta S(U')} \tag{3.9}$$

Here the action is calculated with all other links fixed, at their current values. A detailed discussion on how to implement this technique for SU(2) is given in ref. (4).

For complicated group manifolds, the heat bath generation of new elements may be too tedious to carry out efficiently, either for computational or human reasons. Several less intuitive algorithms based on the detailed balance condition have shown their value through their simplicity.[5] A popular procedure begins with the selection of a trial U' as a tentative replacement for U. This test variable is selected with a distribution $P_T(U,U')$ depending on U and symmetric in U and U'

$$P_T(U,U') = P_T(U',U) \qquad (3.10)$$

Beyond this constraint, P_T is arbitrary and can be selected empirically to optimize convergence. Once U' is selected, the new action S(U') is calculated. If the action is lowered by the change U→U', then this change is accepted. The detailed balance condition then determines the remainder of the algorithm; if the action is raised, the change is accepted with conditional probability $\exp\{-\beta(S(U')-S(U))\}$.

To implement eq. (3.10) in practice, I usually obtain U' by multiplying U with a random group element from a table, where this table is itself of random elements with a convenient weighting towards the identity. The table contains the inverse of each of its elements. I revise this table frequently and adjust its distribution as a function of the temperature to improve convergence.

In a gauge theory, the interaction is rather complicated and involves considerable arithmetic to evaluate. Therefore it can be extremely beneficial to do as good a job as possible in selecting the stochastic changes. In terms of computer time to reach equilibrium, it is usually of value to test several trial group elements before proceeding to the next link. In this way I typically use on the order of 10 to 20 tries.

SOME "EXPERIMENTS"

I will now display the results of some simple experiments. Figure 1 shows several Monte Carlo runs with the gauge group SU(2)

GAUGE FIELD THEORIES

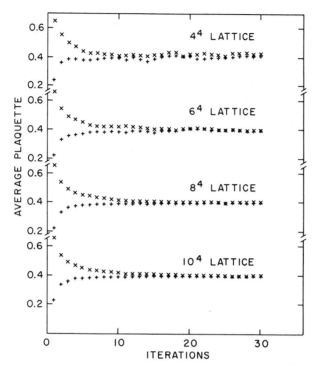

Fig. 1 The average plaquette for SU(2) gauge theory at $\beta = 2.3$ as a function of number of Monte Carlo iterations.

at the particular coupling

$$\beta = \frac{4}{g_o^2} = 2.3 \tag{4.1}$$

This value was selected as representative of the slowest convergence in this model. Runs are shown on lattices of 4^4 to 10^4 sites. I have plotted the "average plaquette" which is just the expectation value of $S_\Box$ defined in eq. (2.4). This is shown as a function of the number of Monte Carlo iterations. For each size lattice, two different initial configurations were studied, one totally ordered with each U_{ij} set to the identity and one with each U_{ij} selected randomly from the group. Thus we approach equilibrium from opposite extremes, zero and infinite temperature. Note that for all lattice sizes convergence is essentially complete after only 20-30 iterations. Thermal fluctuations are apparent on the smallest systems but are relatively small on the 10^4 site crystal.

The situation can be much worse if a phase transition is nearby. In Fig. 2 I show the convergence of the U(1) lattice theory on a 6^4 lattice near the known critical temperature for this model. In addition to the slow convergence compared to SU(2), note the large critical fluctuations. Thus we conclude that convergence is rapid away from a phase transition and slow near one.

In Fig. 3 I show a different type of experiment. Here I have performed rapid thermal cycles on the SU(2) theory in 4 and 5 space-time dimensions and the SO(2) = U(1) theory in four dimensions.[6] Each point was obtained by running on the order of 20 iterations from an either hotter or cooler state. Phase transitions are to be suspected in regions where the heating and cooling points do not agree. Such "hysteresis" phenomena are clear for the 5 dimensional SU(2) and the four dimensional U(1) models. More detailed analysis has indicated that the U(1) transition is second order[7] and the 5 dimensional SU(2) transition is first order. Note that the 4

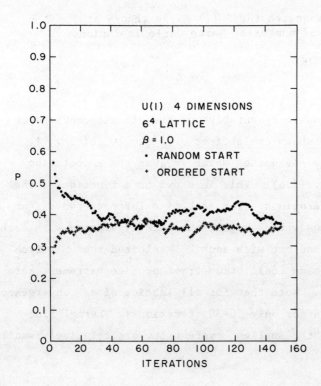

Fig. 2 The convergence of U(1) lattice gauge theory at $\beta = 1.0$.

GAUGE FIELD THEORIES

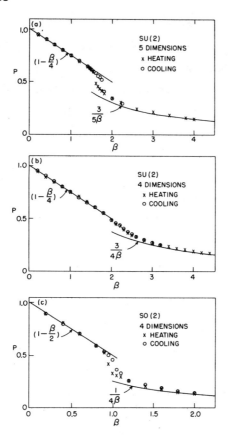

Fig. 3 Thermal cycles on several models.

dimensional SU(2) model is in sharp contrast to the others. The lack of any clear hysteresis shows the critical nature of four dimensions.

THE STRING TENSION

As mentioned earlier, because the entire lattice is accessible, any desired correlation function can be obtained. As we are interested in the interquark potential, I can insert sources with quark quantum numbers into the lattice and measure the response. In particular, I wish to extract the coefficient K of the hypothetical long distance linear potential of eq. (1.1). Measuring distances in units of the lattice spacing, one actually measures the dimensionless combination $a^2 K$ as a function of the bare coupling. If the linear potential survives a continuum limit, the weak coupling dependence of a^2K follows from eq. (1.2). Verification of that behavior is essential to understanding confinement.

The extraction of K is made using Wilson loops. For a closed contour C of links, the Wilson loop is defined

$$W(C) = \langle \frac{1}{2} \text{Tr} \, (\Pi_C U)_{P.O.} \rangle \qquad (5.1)$$

Here P.O. represents "path ordering"; that is, the U_{ij} are ordered and oriented as they are encountered in circulating around the contour. If, for large separations, the interaction energy of two static sources in the fundamental representation of the gauge group increases linearly with distance, then, for large contours, one expects

$$\ln W(C) = -K\,A(C) + 0\,(p(C)) \qquad (5.2)$$

where $A(C)$ is the minimal area enclosed by C and $p(C)$ is the perimeter of C. The constant K is precisely the desired string tension.

Thus motivated, I measured the expectation values of rectangular loops $W(I,J)$ when I and J are the dimensions of the loop in lattice units.[8] From these loops I construct the quantities

$$\chi(I,J) = -\ln \left(\frac{W(I,J)W(I-1,J-1)}{W(I,J-1)W(I-1,J)} \right) \qquad (5.3)$$

In this combination overall constant factors and perimeter behavior cancel out. Whenever the loops are dominated by an area law, $\chi(I,J)$ directly measures the string tension

$$\chi \to a^2 K \qquad (5.4)$$

This happens when I and J are large and also when the bare coupling is large. However, in the weak coupling limit with I and J held fixed, χ should have a perturbation expansion

$$\chi(I,J) = C_1 g_o^2 + O(g_o^4) \qquad (5.5)$$

For example,

GAUGE FIELD THEORIES

$$\chi(1,1) \underset{g_o^2 \to 0}{\sim} \begin{cases} \frac{3}{16} g_o^2 & SU(2) \\ \frac{1}{3} g_o^2 & SU(3) \end{cases} \tag{5.6}$$

This power behavior is radically different from the essential singularity expected for the right hand side of eq. (5.4)

$$a^2 K \underset{g_o^2 \to 0}{\sim} \frac{K}{\Lambda_o^2} (\gamma_o g_o^2)^{(-\gamma_1/\gamma_o^2)} \exp(-1/(\gamma_o g_o^2)) \tag{5.7}$$

This defines the asymptotic freedom scale Λ_o and is just a rewriting of eq. (1.2). In summary, for strong coupling we expect all $\chi(I,J)$ to equal the coefficient of the area law but, as g_o^2 is reduced, small I and J should give a χ deviating from the desired value. Thus the envelope of curves of $\chi(I,J)$ plotted versus the coupling should give the true value of $a^2 K$.

In Fig. 4 I plot $\chi(I,I)$ for I = 1-4 versus $1/g_o^2$ for the gauge group SU(2). At strong coupling the large loops have large relative errors but are consistent with χ approaching the values from smaller loops. On this graph I also plot the strong coupling limit for all χ

$$\chi(I,J) = \ell n(g_o^2) + O(g_o^{-4}) \tag{5.8}$$

The weak coupling behavior of eq. (4.7) is shown as a band corresponding to

$$\Lambda_o = (1.3 \pm .2) \times 10^{-2} \sqrt{K} \quad (SU(2)) \tag{5.9}$$

Figure 5 shows the same analysis for SU(3). Here the strong coupling limit is

$$\chi(I,J) = \ell n(3g_o^2) + O(g_o^{-2}) \tag{5.10}$$

and the band for Λ_o is

$$\Lambda_o = (5.0 \pm 1.5) \times 10^{-3} \sqrt{K} \quad (SU(3)) \tag{5.11}$$

At first sight the small numbers in eqs. (4.9) and (4.11) were rather surprising, as the pure gauge theories have no small dimension-

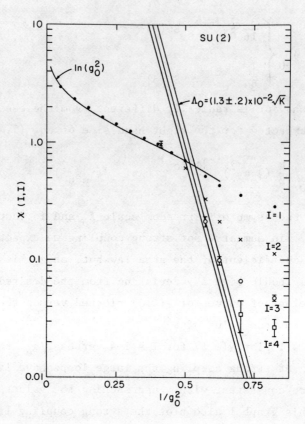

Fig. 4 The quantities $\chi(I,I)$ for SU(2) gauge theory as a function of g_o^{-2}.

less parameters. However, the value of Λ_o is strongly dependent on renormalization scheme. Hasenfratz and Hasenfratz[9] have done a perturbative calculation relating this Λ_o to the more conventional Λ^{MOM} defined by the three-point vertex momentum subtracted in Feynman gauge. They find

$$\Lambda^{MOM} = 57.5 \, \Lambda_o \quad SU(2) \tag{5.12}$$

$$\Lambda^{MOM} = 83.5 \, \Lambda_o \quad SU(3) \tag{5.13}$$

These factors largely compensate for the small numbers for Λ_o. If we accept the string model correction[10] between K and the Regge slope α'

$$K = 1/(2\pi\alpha') \tag{5.14}$$

GAUGE FIELD THEORIES

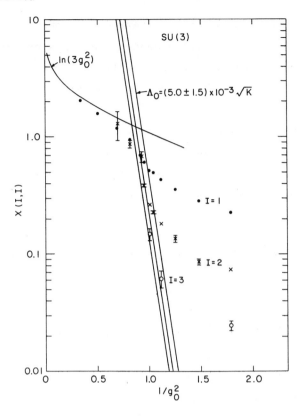

Fig. 5   The quantities $\chi(I,J)$ for the SU(3) theory.

and use $\alpha' = 1.0$ $(\text{GeV})^{-2}$, then we conclude for SU(3)

$$\Lambda^{MOM} = 170 \pm 50 \text{ MeV} \tag{5.15}$$

Phenomenological interpretation of this value requires an understanding of the neglected effects of virtual quark loops.

THE CONTINUUM LIMIT AND THE RENORMALIZATION GROUP

One of the marvelous features of Monte Carlo simulation is that the entire lattice is stored in the computer memory and therefore one can in principle measure any desired function of the fields. Indeed, the most difficult part of this technique is deciding just what to measure. Of course, we are ultimately interested in taking the continuum limit of our lattice theory. Renormalization group techniques tell us how to adjust the coupling constant for this limit. Non-Abelian gauge theories are asymptotically free, which for our purposes means that the bare charge must be taken to zero.[11] This should be done in such a manner that physical observables remain finite. In this section I will review the renormalization group prediction, and then present some Monte Carlo measurements verifying asymptotic freedom.

In a conventional perturbative treatment, one defines a renormalized coupling g_R in terms of a physical observable at a scale of mass μ. The precise definition is merely a convention, but to lowest order it should agree with the bare charge

$$g_R(g_o,\mu,a) = g_o + O(g_o^3) \tag{6.1}$$

where a is the lattice spacing or cutoff scale. The variation of g_R with the scale of definition gives rise to the Gell-Mann Low function

$$\gamma(g_R) = -\mu \frac{\partial}{\partial \mu} g_R(g_o,\mu,a) \tag{6.2}$$

If the continuum limit is physically sensible, then $\gamma(g_R)$ should remain a finite function as a is taken to zero. For SU(N) a perturbative evaluation of $\gamma(g_R)$ gives

$$\gamma(g_R) = \gamma_o g_R^3 + \gamma_1 g_R^5 + O(g_o^7) \tag{6.3}$$

where γ_o and γ_1 are independent of renormalization scheme and have the values[12]

$$\gamma_o = \frac{11}{3} (N/16 \pi^2) \tag{6.4}$$

$$\gamma_1 = \frac{34}{3} (N/16 \pi^2)^2 \tag{6.5}$$

Remarkably, if a perturbative analysis is ever valid, then eq. (6.3) tells us how g_o must be varied as a function of a for a continuum limit. In this limit, the renormalized g_R should not vary; thus, we conclude

$$0 = a \frac{d}{da} g_R(g_o,\mu,a) = a \frac{\partial g_R}{\partial a} + \frac{\partial g_R}{\partial g_o} a \frac{\partial g_o}{\partial a} \tag{6.6}$$

Simple dimensional analysis tells us

GAUGE FIELD THEORIES

$$a \frac{\partial g_R}{\partial a} = \mu \frac{\partial g_R}{\partial \mu} = \gamma(g_R) \tag{6.7}$$

Combining the previous equations gives

$$a \frac{\partial g_o}{\partial a} = \gamma(g_o) + O(g_o^7) \tag{6.8}$$

A little algebra shows that the $O(g_o^5)$ corrections cancel in this equation. Indeed, dropping the corrections in eq. (6.8) gives rise to a definition of an alternative Gell-Mann Low function.

If $g_o(a)$ is ever small enough that $O(g_o^7)$ terms can be neglected in eq. (6.8), then we can integrate to obtain

$$g_o^2(a) = (\gamma_o \ln(1/(\Lambda_o^2 a^2))) + (\gamma_1/\gamma_o)\ln(\ln(1/(\Lambda_o^2 a^2))) + O(g_o^2))^{-1} \tag{6.9}$$

Here Λ_o is an integration constant which determines the scale of a logarithmic decrease of g_o^2 as the lattice spacing is reduced.

We would like to check this logarithmic decrease with our Monte Carlo simulation. In particular, if we measure some general physical observable P as a function of g_o, a, and the scale $r = \mu^{-1}$,

$$P = P(r, a, g_o(\mu)) \tag{6.10}$$

then P should not change as we vary a and g_o in the way indicated in eq. 6.9. Thus, for a factor of two change in cutoff we expect

$$P(r, \frac{a}{2}, g_o(\frac{a}{2})) = P(r, a, g_o(a)) + O(a^2) \tag{6.11}$$

In general, there are two classes of dimensional parameters which set the scale for the finite cutoff corrections in this equation. First is the scale r used to define P. As our lattices in practice are rather small, these corrections must be optimistically ignored. Second, there are the dimensional parameters characterizing the continuum theory. Regardless of the value of r, we expect corrections of order $a^2 m^2$ where m is a typical hadronic mass. One should not trust the lattice theory phenomenologically when the lattice spacing is larger than a proton.

Assuming P is dimensionless, we can scale a factor of two from both r and a to obtain

$$P(2r,a,g_o(\tfrac{a}{2})) = P(r,a,g_o(a)) \tag{6.12}$$

Thus a measurement at two different physical scales relates the bare coupling at two values of cutoff.

The most studied "order parameter" in lattice gauge theory is the Wilson loop.[2] We would like to use the loops to define a physical observable. Unfortunately, the bare loop at finite fixed size cannot be used because of ultraviolet divergences. These are of a rather trivial nature, arising from the infinitely thin contour. They represent the self energy of pointlike sources circumnavigating the loop. To proceed, we assume that removing divergences proportional to the loop perimeter and divergences from sharp corners, inevitable in our lattice formulation, as well as appropriately renormalizing the bare charge, will leave the finite physical part of the Wilson loop. This immediately implies that ratios of loops with the same perimeter and number of corners remain finite in the continuum limit. Thus motivated, we define the two functions of bare coupling[13]

$$F(g_o) = 1 - \frac{W(2,2)\,W(1,1)}{(W(1,2))^2} \tag{6.13}$$

and

$$G(g_o) = 1 - \frac{W(4,4)\,W(2,2)}{(W(2,4))^2} \tag{6.14}$$

Here $W(I,J)$ is a rectangular Wilson loop of dimensions I by J in lattice units.

In eq. (6.13) we have effectively taken $r = 2a$ in the more general ratio

$$P(r,a,g_o) = 1 - \frac{W(\tfrac{r}{a},\tfrac{r}{a})\,W(\tfrac{r}{2a},\tfrac{r}{2a})}{(W(\tfrac{r}{a},\tfrac{r}{2a}))^2} \tag{6.15}$$

For small g_o this can be expanded perturbatively

$$P(r,a,g_o) = p_1 g_o^2 + O(g_o^4) + O(\tfrac{a^2}{r^2} g_o^2) \tag{6.16}$$

where for SU(2)

GAUGE FIELD THEORIES

$$P_1 = \frac{3}{16\pi^2} [8 \arctan 2 + 2 \arctan \tfrac{1}{2} - 2\pi - 4 \ln(\tfrac{5}{4})] \tag{6.17}$$

$$= 0.049559841$$

Conversely, in the strong coupling limit

$$F(g_o) = 1 - \frac{1}{g_o^2} + O(g_o^{-6}) \tag{6.18}$$

$$G(g_o) = 1 - \frac{1}{g_o^8} + O(g_o^{-12}) \tag{6.19}$$

In Fig. 6 I show the Monte Carlo results for the functions F and G of eqs. (6.13) and (6.14). Note that the function G always appears to lie above F and that both functions are monotonic increasing in g_o^2. Thus we conclude from eq. (6.12) that the bare charge is a monotonic function of the cutoff, decreasing to zero as the cutoff is removed. In this figure I also show the weak coupling limit of eq. (6.16) and the strong coupling limits in eqs. (6.18) and (6.19).

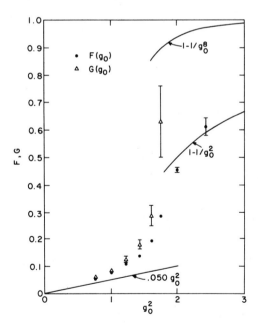

Fig. 6 The ratios F and G as functions of the SU(2) coupling.

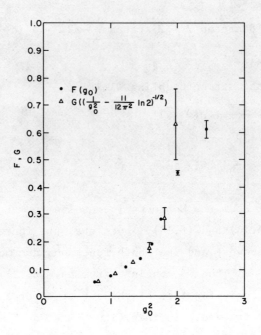

Fig. 7 Testing asymptotic freedom.

Asymptotic freedom predicts a logarithmic decrease of g_o^2 with cutoff when we approach the continuum theory. Using eq. (6.9), we thus expect

$$F(g_o) = G((\frac{1}{g_o^2} - \frac{11}{12\pi^2} \ln 2)^{-1/2} + O(g_o^3)) \tag{6.20}$$

In Fig. 7 I show the function F and G plotted again versus g_o^2 but with G shifted by the amount indicated in eq. (6.20). Note the excellent agreement with the asymptotic freedom prediction. This is rather astonishing in the light of neglected finite cutoff corrections.

The function P in eq. (6.15) should have a finite continuum limit. Therefore we can use it to define a renormalized charge at scale r. Thus absorbing the higher order terms in eq. (6.16), we define

$$g^2(r) = \lim_{a \to o} P(r,a,g_o(a))/p_1 \tag{6.21}$$

Optimistically assuming a is small enough in our function F, we calculate

$$g^2(r=2a) = F(g_o(a))/p_1 \qquad (6.22)$$

In Fig. 8 I plot $g^{-2}(2a)$ versus g_o^{-2}. At large inverse charge these points approach a straight line, for which the unit slope is a test of our neglect of finite cutoff corrections. The intercept of this line measures the Λ parameter associated with this definition of g

$$\frac{1}{g_o^2(a)} - \frac{1}{g^2(2a)} = 2\beta_o \ln(\frac{2\Lambda}{\Lambda_o}) + O(g_o^2) \qquad (6.23)$$

From the graph we obtain

$$\Lambda = 22 \Lambda_o \qquad (6.24)$$

This number is in principle calculable in perturbation theory.

PHASE TRANSITIONS AND VARIANTS ON THE WILSON ACTION

In Fig. 9 I show the results of some Monte Carlo experiments with the gauge group SO(3). Figure 9a is a thermal cycle of this model around $\beta_A = 2.5$. Note the apparent hysteresis effect. Figure 9b shows the results of 100 iteration runs at $\beta_A = 2.5$ from ordered and disordered initial states. Note the appearance of two distinct stable phases. The action per plaquette is normalized

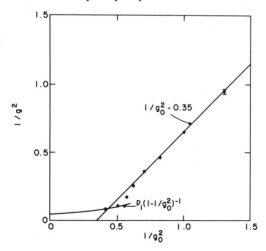

Fig. 8   The inverse renormalized charge squared at r = 2a versus the inverse bare charge squared.

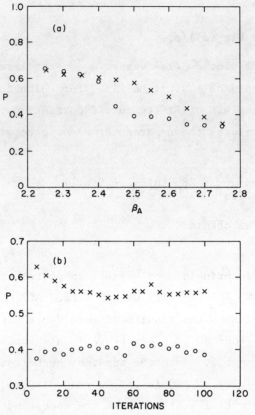

Fig. 9 a. A thermal cycle on the SO(3) model on a 5^4 lattice.
       b. Evolution of the SO(3) model at $\beta_A = 2.5$ from ordered and disordered initial states.

$$S_{\square} = \beta_A(1 - \frac{1}{3} \text{Tr} U_{\square}) \tag{7.1}$$

where $U_{\square}$ is the product of the SO(3) matrices around the plaquette in question. The quantity P plotted in Fig. 9 is the expectation value of $S_{\square}$. Thus SO(3) lattice gauge theory has a first order phase transition.[14] This is rather peculiar in that the continuum SO(3) and SU(2) gauge theories are identical.

In Fig. 10 I show the results of four Monte Carlo runs with the gauge group SU(5) at $g_0^{-2} = 1.67$. The upper and lower runs are from random and ordered starts, respectively. The intermediate runs are from superheated and supercooled states. The appearance of two distinct asymptotic values for P is indicative of a first order

GAUGE FIELD THEORIES

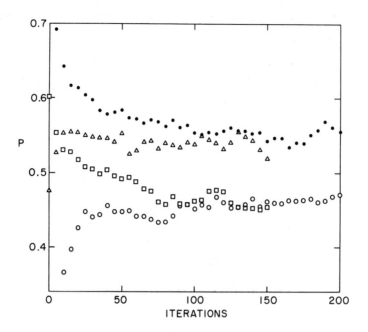

Fig. 10 Four Monte Carlo runs with the gauge group SU(5) at $\beta = 1.67$.

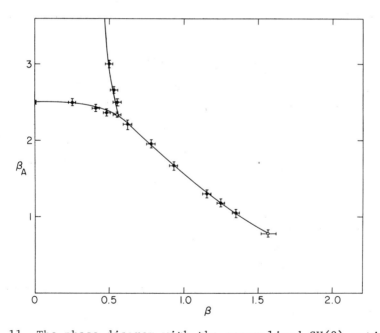

Fig. 11 The phase diagram with the generalized SU(2) action.

phase transition in this system as well.[15] The critical coupling is $g_o^{-2} = 1.66 \pm 0.03$.

As these groups are non-Abelian, these transitions are presumably not simple deconfinement. One possibility is that they represent a dynamical symmetry breakdown into smaller gauge groups. More mundanely, they might all be artifacts of the lattice action. To see that a change in formulation can modify the phase structure of a gauge theory, Bhanot and I have studied the SU(2) theory with the more general action.[16]

$$S_\Box = \beta(1-1/2\ \text{Tr}U_\Box) + \beta_A(1-1/3\ \text{Tr}_A U_\Box) \qquad (7.2)$$

Here the first term is the usual Wilson action and in the second the trace Tr_A is taken in the adjoint representation of SU(2). This model has three simple limits: (1) $\beta_A = 0$ is the usual SU(2) theory, (2) $\beta = 0$ is the SO(3) model, and (3) $\beta_A \to \infty$ reduces to Z_2 lattice gauge theory at inverse temperature β. Both limits (2) and (3) have nontrivial phase structure.

Using Monte Carlo techniques, we have obtained the phase diagram shown in Fig. 11. Both the Z_2 and SO(3) transition are stable and meet at a triple point located at

$$(\beta,\beta_A) = (0.55 \pm 0.03,\ 2.34 \pm 0.03) \qquad (7.3)$$

A third first order line emerges from this point and ends at a critical point at

$$(\beta,\beta_A) = (1.57 \pm 0.05,\ 0.78 \pm 0.05) \qquad (7.4)$$

The conventional SU(2) theory exhibits a narrow but smooth peak in its specific heat[17] at $\beta = 2.2$. This is directly in line with a naive extrapolation of the above first-order transition to the β axis. Thus this peak is a remnant of that transition and a shadow of the nearby critical point.

The continuum SU(2) theory should be unique for all physical observables. The connection between the bare field theoretical coupling and the parameter (β,β_A) is

$$g_o^{-2} = \beta/4 + 2\beta_A/3 \qquad (7.5)$$

GAUGE FIELD THEORIES

A continuum limit requires $g_o^2 \to 0$, however, this can be done along many paths in the (β, β_A) plane. Previously we concentrated on the trajectory $\beta_A = 0$, $\beta \to \infty$. Along that line no singularities are encountered and thus confinement, present in strong coupling, should persist into the weak coupling domain. However, an equally justified path would be, for example, $\beta = \beta_A \to \infty$. In this case we cross a first order transition. Because one can continue around it in our larger coupling constant space, the transition is not deconfining and is simply an artifact of the lattice action.

To test whether physical observables are indeed independent of direction in the (β, β_A) plane, we measured Wilson loops in the weak coupling regime for several values of β_A. The loop by itself is not an observable, for reasons discussed previously. As then, we constructed ratios of loops with the same perimeters and numbers of corners. Thus we define

$$R(I,J,K,L) = \frac{W(I,J)\,W(K,L)}{W(I,L)\,W(J,K)} \qquad (7.6)$$

Wishing to compare points which give similar physics, we searched at each value of β_A for the values of β for which $R(2,2,3,3)$ had the values 0.87 and 0.93. This gave the points in the (β, β_A) plane shown in Fig. 12. The dashed lines in this figure represent constant bare charge from eq. (7.5). If physics is indeed similar at the corresponding points, then all ratios R of eq. (7.6) should match. In Fig. 13 I show several such ratios as functions of β_A and at the $R(2,2,3,3) = 0.87$ points from above. The comparison is quite good considering that finite cutoff corrections are ignored.

Note that in this comparison the bare charge is far from being a constant. In Fig. 12 we see that g_o^2 varied from less than unity to nearly 4 while holding $R(2,2,3,3)$ fixed at 0.87. Such variation is permissible and perhaps even expected since the bare charge is unobservable and should depend on the cutoff prescription. The dependence can be characterized by a β_A dependent renormalization scale defined as in eq. (3.9). Using this relationship, I show

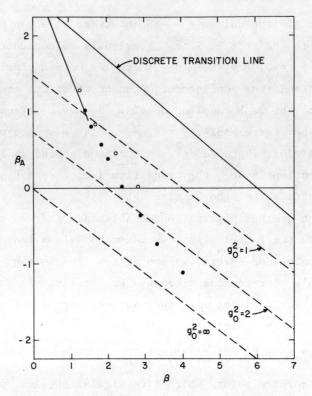

Fig. 12 Points of "constant physics". The solid points give
R(2,2,3,3) = 0.87 and the open circles give
R(2,2,3,3) = 0.93.

in Fig. 14 Λ_o as a function of β_A. Note the consistency of the
R(2,2,3,3) = 0.93 and 0.87 results. Remarkably, the addition of
β_A can change Λ_o by several orders of magnitude. This dependence
can in principle be checked with a perturbative calculation.

CONCLUDING REMARKS

I hope to have conveyed to you that lattice gauge theory is
both an exciting and a rapidly evolving technique for the particle
physicist. We are coming frighteningly close to calculating some
real numbers for the strong interactions. The main stumbling block
at this stage is the inclusion of light quarks. Courageous attempts
at such calculations are being made, however, at present these are
severely demanding on computer time and only practical on the most
modest lattices. Technical breakthroughs are likely in this area.

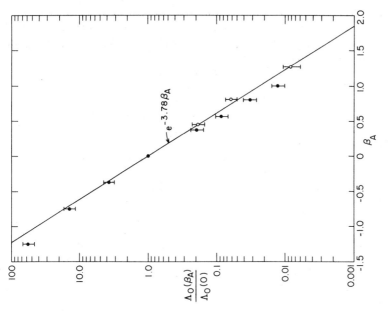

Fig. 14 The β_A dependence of the renormalization scale. The solid circles and open circles are from R(2,2,3,3)=0.87 and 0.93, respectively.

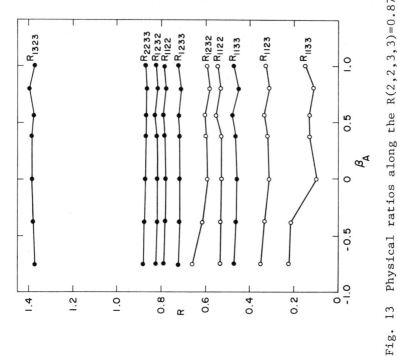

Fig. 13 Physical ratios along the R(2,2,3,3)=0.87 contour. Solid circles are from loops in the fundamental representation, open circles from the adjoint.

REFERENCES

1. M. Gell-Mann and F. E. Low, Phys. Rev. 95:1300 (1954);
   W. E. Caswell, Phys. Rev. Lett. 33:244 (1974); D. R. T. Jones, Nucl. Phys. B75:531 (1974).
2. K. Wilson, Phys. Rev. D19:2445 (1974).
3. C. N. Yang and R. Mills, Phys. Rev. 96:191 (1954).
4. M. Creutz, Phys. Rev. D21:2308 (1980).
5. K. Binder, in "Phase Transitions and Critical Phenomena", C. Domb and M. S. Green, eds., Academic Press, New York (1976), Vol. 5B.
6. M. Creutz, Phys. Rev. Lett. 43:553 (1979).
7. B. Lautrup and M. Nauenberg, Phys. Lett. 95B:63 (1980);
   T. A. Degrand and D. Toussaint, Preprint UCSB-TH-26 (1981);
   G. Bhanot, Preprint BNL 29087 (1981).
8. M. Creutz, Phys. Rev. Lett. 45:313 (1980).
9. A. Hasenfratz and P. Hasenfratz, Phys. Lett. 93B:165 (1980).
10. P. Goddard, J. Goldstone, C. Rebbi, and C. B. Thorn, Phys. Rev. D20:2096 (1979).
11. H. D. Politzer, Phys. Rev. Lett. 30:1343 (1973); D. Gross and F. Wilczek, Phys. Rev. Lett. 30:1346 (1973) and Phys. Rev. D8:3633 (1973).
12. W. E. Caswell, Phys. Rev. Lett. 33:244 (1974); D. R. T. Jones, Nucl. Phys. B75:531 (1974).
13. M. Creutz, Phys. Rev. D23:1815 (1980).
14. I. G. Halliday and A. Schwimmer, Preprint ICTP/80/81-15, (1981);
    J. Greensite and B. Lautrup, Preprint NBI-HE-81-4, (1981).
15. M. Creutz, Phys. Rev. Lett. 46:1441 (1981); for further analysis see H. Bohr and K. Moriarty, ICTP/80/81-19 (1981).
16. G. Bhanot and M. Creutz, preprint (1981).
17. B. Lautrup and M. Nauenberg, Phys. Rev. Lett. 45:1755 (1980).

DISCUSSIONS

CHAIRMAN: M.J. CREUTZ

Scientific Secretary: M. Karliner

DISCUSSION 1

- MORGAN:

Considering the calibration of the string tension K by the formula $K = (2\pi\alpha')^{-1}$, is it possible to understand this relationship in a simple way ? Is the formula reliable ?

- CREUTZ:

The connection is provided by the string model, in which one is tempted to believe because of the very successful prediction of linearly rising Regge trajectories. Failing this, one would be driven to draw conclusions from analyses of potential models for the heavy $Q\bar{Q}$ ('onium') compounds.

- KARLINER:

Could you tell us what is the present status of the Migdal conjecture in view of the Monte-Carlo calculations ?

- CREUTZ:

There are several places where it doesn't work very well, first of all in four dimensional Z(2) lattice gauge theory. Balian, Drouffe and Itzykson stated many years ago that it probably had a first order phase transition, whereas from the Migdal analogy with the two dimensional Ising model, one would expect a second order phase transition. Monte-Carlo and perturbation expansions confirm the existence of first order transition.

The nature of the phase transition in four dimensional U(1) model is also different from that in the two dimensional XY model.

The latter has an essential singularity (Kosterlitz-Thouless) whereas the U(1) model seems to have a conventional second order transition.

- KARLINER:

What about the number of phases ?

- CREUTZ:

The number of phases in every case has been just right. There are two phases in both U(1) in D=4 and XY in D=2. The Z(N) models in both 4 and 2 dimensions develop three phases for N large enough, although I am not sure if it happens for the same N in both cases.

- D'HOKER:

It is generally believed that in the limit of vanishing bare coupling, the Euclidean O(4) symmetry is restored. Is there any direct evidence from Monte-Carlo that this is indeed true ?

- CREUTZ:

In principle you just have to take your Wilson loops and turn them at an angle. In practice it turns out to be rather messy and I have never tried to do it, mainly because of a strong faith that the symmetry is going to be restored. There is evidence from strong coupling expansions and Hamiltonian theories that the symmetry does begin to be restored. Also, there are lots of solvable field theories on the lattice, which are initially terribly Lorentz non-invariant, but when you take the continuum limit the symmetry automatically comes out.

- ETIM:

Would you really say that potential models have confirmed the value of the string tension ?

- CREUTZ:

The evidence is rather weak, since potential models primarily probe the area where the potential is essentially logarithmic, so you would be skeptical. I think it is nice that there is no indication that the number is any different but if you don't like the Regge model that's the only other thing you have to turn to.

GAUGE FIELD THEORIES

- ETIM:

Are you familiar with the work of J.S. Bell who recovered the qq potential from data and found it to be an r^4 power law instead of the usual linear dependence on distance r ?

- CREUTZ:

No, I am not familiar with it.

- TELLER:

Is the gluon mass zero in your calculation ? If yes, how can there be forks ?

- CREUTZ:

You start with a Lagrangian which has no gluon mass term, and therefore perturbatively there are massless gluons. In the resulting theory, however, if the confinement really occurs, there are no massless excitations. Forks will never appear as one real gluon coupling to two real ones. A virtual gluon can however give rise to one. This is part of the non-Abelian nature of the theory.

- COLEMAN:

I think it is somewhat misleading to describe what you have done as assuming the existence of gluons. That's like saying that Dirac assumed the existence of photons. Dirac wrote down and solved a quantum field theory, namely electromagnetism, from which he deduced the existence of photons. He was very happy about this because photons were known to exist. Mike, in an approximate way, has written down, and partially solved, a different field theory, quantum chromo-dynamics. He has not found gluons, and that's good because gluons have not been seen, at least not at large distances.

- SEIBERG:

The continuum SU(2) in five dimensions is not renormalizable. Is the phase structure you found on the lattice influenced in any way by this fact ?

- CREUTZ:

The phase transition one finds in five dimensions is a first order one. The standard picture is that you can only take the continuum limit at a second order transition. I believe, but cannot prove that you could take the continuum limit of that theory anywhere below the phase transition, but you would end up with a free field theory.

- SEIBERG:

You can approximate $U(1)$ by $Z(N)$ by taking N as large as you want. That is not so for $SU(2)$, where you have only a finite number of discrete non-Abelian subgroups. How does that affect the accuracy of your results ?

- CREUTZ:

It turns out that the largest discrete subgroup of $SU(2)$ gives an extremely good approximation to $SU(2)$ calculations in the region of interesting physics, where the lattice is comparable to hadronic size. It is not until the lattice gets quite small, that this discreteness shows up, and shows up drastically: there is in fact a first order phase transition which has nothing to do with particle physics. The value of the approximation lies in the fact that it saves, by almost an order of magnitude, in computer time.

- YAMAGISHI:

There is a recent result by mathematical physicists that the lattice ϕ^4 theory becomes trivial in the continuum limit for spacetime dimension greater or equal 4. Can this be verified by Monte-Carlo calculation ?

- CREUTZ:

It is an interesting question which is currently being investigated with Monte-Carlo by Freedman and Weingarten and their preliminary results support the conjecture. The idea is that the renormalized coupling constant may be driven to zero as the cut-off is removed, regardless of what one does to the bare charge.

- BANKS:

Quite apart from the technical problems of including fermions in the theory, what would you expect to find if you succeeded in doing so ? In particular, what would be the criterion for confinement ?

— *CREUTZ:*

With fermions included, you are dealing with what we believe is the real theory. It is both exciting and frightening: we might for instance calculate the $\Delta_{3,3}$ mass and get it wrong. As to defining confinement there is a complication: the Wilson loop no longer obeys an area law, because when you pull two quarks far apart enough, they create a pair out of the vacuum and form two mesons, which only have a short range interaction. You might verify absence of free gluons by calculating the mass gap, but this would be obscured by the probable presence of massless pions. The only real test would be adjusting the mass parameters in order to get the right pion mass and the Λ parameter correct and see if everything else agrees with experiment. If it does, you don't care whether you can prove confinement or not.

— *BANKS:*

Has anyone succeeded in getting the massless pion, through chiral symmetry ?

— *CREUTZ:*

Because of the axial anomaly, chiral symmetry in lattice models is a complicated subject. Most treatments so mutilate the γ_5 symmetries that to see a massless pion will be very difficult.

— *MOTTOLA:*

What is known analytically about the convergence of the heat bath method ?

— *CREUTZ:*

It is very probable that such an analysis exists in the solid state literature, but I do not know any details.

— *MOTTOLA:*

Why is $\Lambda_0/K \sim 10^{-2}$ or 5×10^{-3} so small compared to one ?

— *CREUTZ:*

I think that this is an artifact of the definition of the Wilson action. The ratio certainly depends on the cut-off scheme and in the next lecture I will show how to get almost any desired value for this ratio.

- GIPSON:

If you introduce Higgs fields, does it change any of your results qualitatively ?

- CREUTZ:

If the Higgs field is in the fundamental representation of the gauge group, it has been shown by Fradkin and Shenker that the Higgs phase and confinement phase are analytically connected. In both cases there are no massless excitations.

If the Higgs field is invariant under some subgroup of the full gauge group there will be a limit where you get a pure gauge theory for the quotient group. These phenomena have been verified with Monte-Carlo in a few simple cases.

- GIPSON:

Do you have any intuitive explanation of how changing the dimension induces changes in the phase structure ?

- CREUTZ:

No, I think a simple intuitive explanation of why four dimensions are critical would be quite valuable.

- SPIEGELGLAS:

In the harmonic oscillator calculation, how did you manage to obtain the ground state wave function, without admixtures from excited states ?

- CREUTZ:

Since the contribution of each state to the path integral falls exponentially with time and with the appropriate energy eigenvalue, working in a large time box, gives the coordinate distributed according to the ground state wave function.

- SPIEGELGLAS:

Can you use parity to isolate the first excited state ?

- CREUTZ:

Yes, it can be done for a symmetric potential, where the ground state is even and the first excited state is odd under parity.

GAUGE FIELD THEORIES   151

For the first excited state, study the potential $V(x) + \infty \times \theta(-x)$ to obtain half the wave function.

- SPIEGELGLAS:

Can you identify non-local objects, especially in self-dual theories ?

- CREUTZ:

Vortices have been seen in two dimensional XY model. In four dimensions I haven't figured out how to display the relevant configurations.

- FORGACS:

It seems that any gauge theory would confine in the strong coupling limit. Do you know of any counter example ?

- CREUTZ:

In the conventional Wilson formulation of the lattice action all gauge groups confine in strong coupling.

- COLEMAN:

Even in Wilson's formulation of the action, if instead of the usual trace, which is the character in the triplet representation, (for SU(3)), you would use the character in the octet representation, of course you wouldn't have confinement. By changing the details of the action you can avoid confinement in the strong coupling phase, writing an action which looks much like Wilson's.

- CREUTZ:

With the action in the adjoint representation all the loops in the fundamental representation will vanish and the ones in the adjoint do not have an area law. However, you could still say you have confinement because at least in strong coupling the theory does have a mass gap, and there are no free gluons. Defining confinement by saying there is a mass gap, that theory would still confine.

DISCUSSION 2   (Scientific Secretary: M. Karliner)

- ETIM:

My question is about your attempts to compute path integrals using Trotter's finite product formula. Have you checked the dependence of the integral on the cut-off exponent N_o and how large N_o has to be in order to have reliable results ?

- CREUTZ:

N_o is essentially the size of the lattice. We have performed computations in SU(2) and SU(3) theories with different lattice sizes and the dependence on N_o is much like the dependence of numerically evaluated Riemanian integrals on the cut-offs.

- KARLINER:

You have mentioned the possibility that SU(5) is self-breaking. Could it be some kind of a tumbling phenomenon ?

- CREUTZ:

SU(5) self-breaking is still to be regarded as a speculation rather than an established result.

- COLEMAN:

The original tumbling mechanism of Susskind and collaborators, of course, depended on the presence of fermions in the theory, but you could have gluonic tumbling.

- BANKS:

The original tumbling scenario depended not only on having fermions, but on having them in a complex representation, so I don't quite see how you could make that up with gluons.

- GOLLIN:

Mike Chanowitz suggested that glueballs might already have been seen in experiments but not recognized as glueballs. Are his predictions at odds with lattice gauge theory predictions ?

- CREUTZ:

Not at all. Based on SLAC data from the decay $J/\psi \to \gamma E$, Chanowitz concludes that glueballs might exist in the E region, around 1.4 GeV. In lattice gauge theory, we are getting numbers around and of the order of 1 GeV, which I think is very nice.

- *FORGACS*:

If there's a deconfining phase transition in a lattice theory, the string tension goes to zero. So why not measure the string tension as a simple criterion for a phase transition ?

- *CREUTZ*:

It is a correct approach in principle, but in practice it turns out (e.g. in SU(5)) to be extremely demanding in terms of computer time.

- *GIPSON*:

All your results are based on at most 10^4 lattice. This seems awfully small at first sight, so how seriously should we take your results ?

- *CREUTZ*:

Empirically it seems that what matters is the total number of sites, not the overall linear dimension.

- *COLEMAN*:

It looks like an awfully small lattice. A 10 by 10 lattice in 4 dimensions has, after all, more surface points than interior points, since it has 8^4 interior points and 10^4 points altogether. Of course, we have periodic boundary conditions, but still it makes you nervous, until you think...

You have to ask two questions: how small a lattice spacing in terms of physical length units do you need, so that a lattice theory is a good approximation, and how large a lattice, again in physical units, you need so that you get interesting results from doing a lattice computation. The first question is, of course, imprecisely stated because if I take a lattice of any size, I cannot provide any answers for length scales smaller than the lattice size.

If I only ask questions about things larger than the lattice scale, I can always get a lattice theory that is arbitrarily accurate. What I really want to ask is, what lattice scale size gives a good lattice theory with only nearest neighbour interactions?

This is the scale size at which the coupling of the continuum theory is weak. We know, experimentally, from Bjørken scaling that it is around 2 or 3 GeV. The lattice becomes big enough to be interesting when we reach the confinement length, which corresponds to energy between 1/2 to 1/3 GeV. Therefore the size of the lattice we should expect to get away with is determined by the ratio of those two energy scales, that is to say something of the order of 6 to 9 lattice spacings on each side.

- D'HOKER:

Consider the phase diagram of a theory based on an action which contains plaquette links' product in both the fundamental and the adjoint representations. One can follow different paths to the continuum limit by letting $\beta \to \infty$ and $\beta_A \to \infty$, keeping $\alpha = \frac{\beta_A}{\beta}$ constant. Do you get the same continuum limit for various values of α?

- CREUTZ:

There is hopefully one renormalizable SU(2) gauge theory in the continuum limit, so this has to give you the unique theory, no matter how you do it. If there is more than one theory in the continuum limit, the whole idea of renormalizability is questionable, since renormalizability means your results must not depend on the details of the cut-off procedure.

- D'HOKER:

How about the Z(2) limit you obtain when $\beta_A \to \infty$?

- CREUTZ:

The Z(2) fields decouple from the system and are irrelevant to the physics of the Yang-Mills fields.

- SPIEGELGLAS:

To what extent can the discrete subgroups be useful in simplifying future calculations ?

- CREUTZ:

Bhanot and Rebbi have tried Monte-Carlo with discrete subgroups of SU(3) and they have found that with the simplest SU(3) action there is a discrete phase transition before you enter the region of interesting physics. Maybe, if the action is modified, one will be able to use discrete subgroups for SU(3). That would really save a tremendous amount of computer time.

- WOLFF:

Why not use some dense subset of group elements and make up an approximate multiplication table ?

- CREUTZ:

I tried that with a table of 50 randomly chosen elements of the group and the result was rather poor; it deviated from strong coupling very early.

- MOTTOLA:

Do all those different parametrizations of the action really give the same continuum limit ?

- CREUTZ:

In all cases the action reduces to $1/4g^2$ $Tr(F^2)$ in the continuum limit.

- MOTTOLA:

In order to prove confinement, is it really sufficient to show that the strong coupling limit and the continuum limit are smoothly connected by a contour in some enlarged parameter space ?

- CREUTZ:

Technically, not quite. It could be that the string tension goes to zero faster than the asymptotic freedom prediction. In this way, without having a phase transition, you could lose confinement.

- COLEMAN:

In principle, there is absolutely no reason why we should do strong coupling on a lattice: we choose the lattice size to be 10^{-14} cm and we choose the appropriate coupling constant. And then, if we find confinement at 10^{-13} cm, we have demonstrated that the theory confines and that's the end of the matter. The reason why we look into strong coupling is that in the strong coupling regime we have an intuitive idea of what confinement is like. If we find no phase transition passing to the weak coupling limit, then we have reasonable hope that that intuitive picture is probably qualitatively, if not quantitatively correct.

- CREUTZ:

That is true about Monte-Carlo in strong coupling, but not about strong coupling expansions in perturbation theory. These can yield meaningful quantitative results when continued to the weak coupling regime.

- COLEMAN:

Yes, of course.

- ZICHICHI:

Can you comment on the fantastic discrepancy between Λ_o lines and the effective results you get in $1/g_o^2$ below 0.5 to 0.0 ?

- CREUTZ:

In that region you must obtain the asymptotic behaviour $-\ln g_o^2$. Here the lattice spacing has become bigger than a hadron and therefore the lattice should not be used for hadronic phenomenology.

- MOTTOLA:

In theories which are known to have a phase transition, is it possible to avoid that transition by enlarging parameter space ?

- CREUTZ:

Not if some parameter, such as the mass gap, is identically zero on only one side of the transition.

- GATHERAL:

Have you measured the glueball mass in U(1) theory ?

- CREUTZ:

I haven't, though it might be interesting to compare a Monte-Carlo result with known results from strong coupling expansion.

- GATHERAL:

Has anyone found a non-zero result for the glueball mass in U(1) ?

- CREUTZ:

You can do that, but the strong coupling regime of U(1) is just a toy model and has nothing to do with real physics. On the weak coupling side there should be a massless photon and we have some evidence from renormalization calculations, that the potential indeed falls like 1/R.

WHAT CAN WE LEARN FROM THE NEXT GENERATION OF EXPERIMENTS?

G.G. Ross

University of Oxford
Theoretical Physics Department
1 Keble Road, Oxford

and Rutherford Appleton Laboratory
Chilton, Didcot
Oxfordshire OX11 OQX

1.1 INTRODUCTION

Particle physics is in an unusual period. For once theory leads experiment; there is a standard model for the electromagnetic, weak and strong interactions based on the gauge group SU(3) x SU(2) x U(1) which is consistent with all present experimental data. Moreover, these separate gauge theories have been elegantly and persuasively combined in a single Grand Unified Theory (GUT) which purports to describe particle interactions over the range 0 to 10^{15} GeV!

Faced with this impressive theoretical edifice it sometimes seems experiment has merely to confirm the standard structure. Indeed, many of the next generation of machines and experiment have been designed solely with the view to produce the most accessible and best understood of the new particles predicted by the standard model, namely, the intermediate vectors bosons $W^{\pm}$, Z which mediate the weak interactions. However there is a growing awareness that the evidence for the standard model is at best circumstantial and there are many possible alternative models which could give the observed low energy phenomena. Moreover, the beautiful edifice of GUT's has, many believe, a fatal flaw, the 'so called' hierarchy problem, whose solution requires a much richer structure than in the minimal GUT's. It is now clear that the next generation of experiments will be crucial in clarifying the theoretical picture and it will be surprising indeed if the new energy scales which will be probed by forthcoming experiments do not produce totally unexpected phenomena.

In these lectures, I would like to answer the question posed by the title of these talks by asking what is expected in the standard $SU(3) \times SU(2) \times U(1)$ model (the $(3,2,1)$ model) and what are the crucial tests of the theory. I will also try to discuss alternatives to the standard model where they suggest different things we should be looking for. In the second part of these lectures, I will extend the question beyond the $(3,2,1)$ model and discuss GUT's and the alternatives/modifications necessary to avoid the hierarchy problem. This will lead to a discussion of supersymmetric models, technicolour models and constituent models. Finally, I discuss some of the exotica predicted in these models and the experimental searches for them.

1.2 EXPERIMENTAL PROSPECTS

Over the past decade, high-energy accelerators have provided the main source of information on particle physics. Available are lepton-lepton, lepton-hadron and hadron-hadron machines.

Electron-positron collisions have provided a vast amount of new information[1] and is currently a favourite process for new machines. Existing machines have probed to centre of mass energies of 40 Gev and shortly will get to 45 Gev. They have proved particularly useful in producing new quarks and leptons and have given rise to the detailed $c\bar{c}$ and $b\bar{b}$ spectroscopy studies, the study of charmed hadrons and the tests of the parton model and QCD via R and multijet studies[2]. There are several proposed machines, some under construction which will ultimately take us to 200 Gev centre of mass energy. In Table 1a we list these machines together with their proposed completion dates.

Lepton-hadron machines have run at energies up to 30 Gev and have been central in our understanding of the strong force through the observed (near) scaling behaviour. They have also shed much light on weak interactions, discovering neutral weak currents and much more[3]. Somewhat surprisingly, this line of investigation is not very popular(c.f. Table 1b) in the new machines although many have the possibility of adding proton or electron beams at a later date.

Hadron-hadron collisions give the greatest potential for centre of mass energy although this may not be all "useful" energy. At the ISR pp collisions up to 60 Gev have been studied giving much information in hadronic processes and providing further QCD tests. The CERN $p\bar{p}$ collides will be first to provide energies at which the Z and W may be produced. Further machines, Isabelle and the Tevatron, should provide higher intensity and higher energies (see Table 1C). These should be good tools for discovering strongly interacting particles, heavy quarks, techni-hadrons, gluons, subquarks etc.[4]

Table 1a

E^+E^-

	E_B (GeV)	$\mathcal{L}$ (cm^{-2}s^{-1})	E	Year
SPEAR				
DORIS		10^{30-31}	10 GeV	
PETRA			38 GeV	
TRISTAN (Japan)	25 x 25	10^{30}	50 GeV	'84
S.L.C.	50 x 50	10^{30}	100 GeV	'85
L.E.P.	50 x 50	10^{31}	100 GeV	'87
	(120 x 120 S.C.)		200 GeV	

Table 1b

EP

	E_B (GeV)	$\mathcal{L}$ (cm^{-2}s^{-1})	$E_{\gamma P}$	$E_{\gamma Q}$	Year
SLAC		10^{32}			
SPS.					
TRISTAN	30 x 300	10^{32}	180 GeV	35 GeV	
HERA	30 x 800		310 GeV	60 GeV	'90
SLED			500 GeV	100 GeV	
ISABELLE	200 x 300				

Table 1c

PP P$\bar{P}$

	E_B (GeV)	$\mathcal{L}$ (cm^{-2}s^{-1})	$\sqrt{S_{P\bar{P}}}$	$\sqrt{S_{Q\bar{Q}}}$	Year
CERN$_2$ ISR PP	30 x 30	10^{31}	.06 TeV	10 GeV	
CERN$_1$ SPS P$\bar{P}$	270 x 270	10^{30}	.54 TeV	100 GeV	'81
FNAL$_1$ FIXED TARGET TEVATRON	1000		1 TEV	200 GeV	'83
COLLINEAR P$\bar{P}$	500 x 500	10^{29-30}	2 TeV	400 GeV	'84
	1000 x 1000				
ISABELLE$_2$ PP	300 x 300	10^{33}	.6	100 GeV	?

Machines rapidly become impracticable for energies above 1 TeV and unless there is a breakthrough in machine technology we must look elsewhere for a probe of physics at higher energies. This is possible if a process proceeds only through new interactions characteristic of a new energy scale. For example, in GUT's proton decay proceeds only through the exchange of new X bosons with a very large mass $\simeq 10^{15}$ GeV. Proton decay experiments are sensitive to this scale and such experiments will play an important role in our quest to understand the physics at high energies. In this category we will consider $n\bar{n}$ mixing, rare Kaon processes including CP violation, muon, electron and tau number violating processes and neutrino mass experiments; all sensitive to scales beyond the standard (3,2,1) model.

1.3 THEORETICAL STATUS - THE STANDARD (3,2,1) MODEL

The standard model is based on the gauge groups SU(3) x SU(2) x U(1) describing the strong force (QCD)[5] and the weak and electromagnetic interactions.[6] The vector boson spectrum (Table 2) consists of eight strongly interacting gauge fields, the gluons, three weakly interacting bosons the $W^{\pm}$ and Z^0 and the photon.

Table 2
The vector boson spectrum in the (3,2,1) model

Name		Mass	Coupling
$A_\mu^{a=1..8}$	Gluons	0	$\alpha_s = g_s^2/4\pi$
A_μ^γ	Photon	0	α_{EM}
$W_\mu^{\pm}$	W-Boson	$\frac{37.3}{\sin\theta_Q} \simeq 78$ GeV	g, g'
Z_μ	Z-Boson	$\frac{M_W}{\cos\theta_W} \simeq 89$ GeV	$\frac{g^2}{8M_W^2} = \frac{G_F}{\sqrt{2}}$ $\sin W = \frac{g'^2}{g^2 + g'^2}$

The interaction of fermions is specified once their transformation under (3,2,1) is specified, see Fig. 1(a). It seems fermions fall into 3 families whose representations are identical. If this is the case, we still have to find the top quark to complete the 3rd family. In the standard model no explanation is given for the multiplet structure favoured, nor for the number of generations. The spectrum of the (3,2,1) model also includes scalar particles, necessary to give mass to the W bosons in a manner compatible with gauge invariance and renormalisability. The minimal set necessary is one doublet (Fig.1b)

The theory has 18 parameters describing the interaction strengths, the masses and mixing angles. If the theory is to be checked, it is necessary to verify that the particles of Table 2 and Fig. 1 actually exist and that they interact according to the (3,2,1) gauge principle. It will also be important to measure the 18 parameters which are the fundamental parameters of the theory and may give clues as to the structure beyond the (3,2,1) scheme.

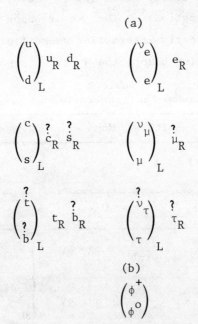

Fig. 1
(a) Fermion and (b) scalar SU(2) X O(1) assignments in the standard model

NEXT GENERATION OF EXPERIMENTS

We now consider the (3,2,1) model in detail starting with $SU(3)_C$ the strong interaction gauge group.

(2.0) Strong Interaction Tests - Is QCD Correct?

The fundamental mediators of the strong force are the 8 vector gluons. These are supposed to be massless, and bind quarks (and gluons) strongly at large distances giving rise to colour confinement; only colour singlets are expected to be asymptotic states, as only they can escape the long range strong colour force. At short distances (large momenta) the effective strong coupling becomes small, the theory is asymptotically free and perturbative calculations are justifiable. In order to test QCD we must answer the following questions

Do gluons exist? With spin 1?

Do they have non-Abelian gauge couplings?

Do quarks exist? Are they colour triplets?

What is the strong coupling? (i.e. what is Λ?)

At present, due to the difficulty in making predictions about the strong coupling regime, it is mainly through the perturbative predictions that the theory may be quantitatively verified. This necessitates large momentum transfer and the new machines should provide significant improvements in sensitivity to QCD effects. The fundamental parameter in QCD is the strength of the strong coupling constant, which may be re-interpreted in terms of a scale Λ^2 at which the perturbative calculations give a pole in the effective strong coupling.[7]

In a process such as $R(Q^2) = \frac{\sigma_{e^+e^-\to hadrons}(Q^2)}{\sigma_{e^+e^- \to \mu^+\mu^-}(Q^2)}$, QCD makes unambiguous predictions[8] of the form

$$R(Q^2) \underset{Q^2 \to \alpha}{=} \Sigma e_i^2 (1 + \frac{\alpha_s(Q^2)}{\pi} + C \frac{\alpha_s^2(Q^2)}{\pi^2}) \qquad (2.1)$$

where

$$\frac{\alpha_s(Q^2)}{4\pi} = \frac{1}{\beta_0 \ln(\frac{Q^2}{\Lambda^2})} - \frac{\beta_1}{\beta_0^3} \frac{\ln\ln(Q^2/\Lambda^2)}{\ln^2(\frac{Q^2}{\Lambda^2})} \qquad (2.2)$$

and C is a constant which depends on the renormalisation scheme
(the dependence occurs only in a truncated perturbation expansion).
In minimal and momentum subtraction C is 5.7 and -1.7 respectively.
(In Table 3 these subtractions are defined[9]). In addition, there
are small corrections suppressed by inverse powers of Q^2.

Eq.(2.1) illustrates the ideal QCD type prediction. The
perturbative corrections are small and the leading, parton model
term measures the quark charge, and is in good agreement with the
fractional quark model. However, the determination of the strong
coupling constant, or equivalently Λ^2, requires measurement of
the $0(\alpha_s)$ term, which is small. There are further complications
to do with the choice of subtraction scheme[9] (we should use one
in which perturbation theory is rapidly converging) and the fact
that (c.f. eq.(2.2) $\alpha(Q^2)$ for timelike Q^2 develops an imaginary
part[10] (eq.(2.1) should be reinterpreted to depend on $|\alpha_s(Q^2)|$).

Table 3

Definition of some commonly used regularisation schemes

Subtraction Scheme	Definition
Minimal Subtraction (MS)	Subtraction of $\frac{1}{\varepsilon}$ piece when computing integrals in $(4-2\varepsilon)$ dimensions.
$\overline{MS}$	Subtraction of $\frac{1}{\varepsilon} + \ln 4\pi - \gamma_E$
Momentum Subtraction (MOM) $(\alpha_{MOM} = \alpha_{\overline{MS}}(1+3.22 \frac{\alpha_{\overline{MS}}}{\pi} + ...))$	Vertex defined at symmetric point in Landau gauge

These latter complications become less severe at large Q^2 but there
remains the difficulty of determining $\alpha_s(Q^2)$ in the presence of a
large constant term (c.f. eq. (2.1)). At present, the analysis of
R gives[11] $\alpha_s(Q^2=10^3 GeV) = 0.23 \pm 0.26$.

QCD predictions for other processes usually involve the separation of small and large momentum transfer effects via the factorisation theorem[12]. (Two examples are given in Fig.2). In electroproduction, for example, the singlet moments are predicted to have Q^2 dependence given by

$$M_n(Q^2) = M_n(Q_o^2) \exp \int_{Q_o^2}^{Q^2} - \frac{\alpha(Q'^2)}{8\pi} (\gamma_o^n + \tilde{\gamma}_1^n \frac{\alpha_s(Q'^2)}{4\pi} + ...) \frac{dQ'^2}{Q'^2} \quad (2.3)$$

The Q^2 variation is given by the short distance (perturbative part) but the normalisation is not predicted perturbatively. However, different processes may depend on the same $M_n(Q_o^2)$ (or equivalently the same distribution or decay functions) and so the relative normalisation of different reactions may be predicted. An example of this is illustrated in Fig.2.

QCD tests will therefore involve looking for the predicted hard scattering processes described by $d\hat{\sigma}$ (scattering of quarks and gluons), the measurement of the predicted Q^2 behaviour and the measurement of the relative normalisation of different processes,

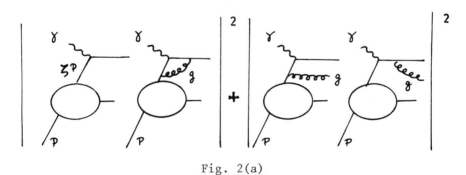

Fig. 2(a)

Feynman graphs showing factorisation in electroproduction into large momentum transfer parts (blobs) and small momentum parts calculated to $O(\alpha_s)$

$$\sigma^{eh\to ex}(Q^2,X) = \sum_j \int_x^1 \frac{d\zeta}{\zeta} \hat{\sigma}^j (\frac{x}{\zeta},Q^2) \zeta f_j^P(\zeta,Q^2) + O(\frac{1}{Q^2})$$

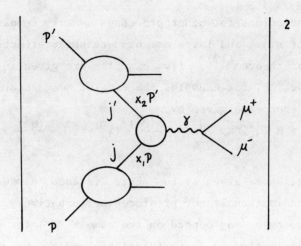

Fig. 2b
Leading order graphs showing Drell-Yan cross section in terms of the same large momentum transfer parts (blobs) found in electroproduction

$$d\sigma^{h_1 h_2 \to \mu^+ \mu^- x}(P,P') = \frac{1}{3} \frac{4\pi\alpha^2}{3Q^2} \int \frac{dx_1}{x_1} \int \frac{dx_2}{x_2}$$

$$\sum_{j,j'} \left\{ x_1 f_j^{h_1}(x,Q^2) \right\} \left\{ x_2 f_{j'}^{h_2}(x_2,Q^2) \right\} d\hat{\sigma}(x_1,x_2,Q^2)$$

which tests factorisation[†].

In Table 4 we summarise the present status of such tests. It is important to check that the perturbative predictions make sense i.e. that higher order corrections are not too large. In many cases the QCD predictions have been computed to second order and we tabulate the quantity B which is a measure of the relative size of the next to leading to leading order predictions. This parameter is sensitive to the renormalization scheme and to the choice

---

[†] Doubt has recently been cast on the validity of the latter predictions due to initial state interactions not covered by the factorisation proofs[13].

Table 4. Summary of QCD Perturbative Tests

Process	Leading Order (LO) Predictions×$(1+B\frac{\alpha_s}{\pi})$ $B(\overline{MS})$	Results and Comments $\Lambda(\overline{MS})$ (MeV), $\alpha_s(\overline{MS})$
$e^+e^- \to$ hadrons (15)	$(1 + \frac{1.52}{\pi}\alpha_s)_{\overline{MS}}$ $(1 - \frac{1.7}{\pi}\alpha_s)_{MOM}$	$\alpha_s = 0.26 \pm 0.25$
LEPTOPRODUCTION (16) $\frac{d \ln Mn(Q^2)}{d \ln Q^2}$ Shape of moment e.g. $\ln M_n$ v/s $\ln M_m$	3 (n = 2) 4.5 (n = 8) 1 (n = 2) 0.5 (n = 8)	$\alpha_s(13) = 0.29^{+0.08}_{-0.12}$ (17) $\Lambda = 690^{+300}_{-400}$ with new data (18) $\Lambda = 400$ Recent fit to structure function $\Lambda = 200$ (19)
Moments of Fragmentation Functions (31)	~ 5	$\Lambda = 400$
Drell-Yan moments (20)	~ π^2 + (n) f(2) = 0 ~ (n) (after Sukakov extrapolation) f(8) = 2.5	
$\gamma\gamma$ moments (21)	$\left.\begin{array}{l}-4.3 \ (n=4)\\-5.1 \ (n=8)\end{array}\right\|_{\overline{MS}}$ $\left.\begin{array}{l}-1 \ (n=4)\\-1.85 \ (n=8)\end{array}\right\|_{MOM}$	

(Continued)

Table 4. Continued

Process	Leading Order (LO) Predictions $\times (1 + B \frac{\alpha_s}{\pi})_{\overline{MS}}$	Results and Comments $\Lambda(\overline{MS})$ (MeV), $\alpha_s(\overline{MS})$
$e^+e^- \xrightarrow{T} 3$ jets (22,23) $\frac{1}{\sigma}\int_{0.5}^{T} \frac{d\sigma}{dT} = 1.156 \frac{\alpha_s(s)}{\pi}\left[1 + B\frac{\alpha_s(s)}{\pi}\right]$	18 $\overline{MS}$ (22) 15 $\overline{MS}$ $\rightarrow 10^{13}$ $\alpha(Q^2/9)$ $\dot{=}$ E gluon $\rightarrow 5$? After Sudakov Extrapolation	$\alpha_s = 0.122 \pm 0.01$ $\Lambda = 110$ But of doubtful validity due to the poor convergence
$\eta_c \rightarrow gg$ (24)	14 $\overline{MS}$ $\alpha(2m_{\eta_c}^2)$ 8 $\overline{MS}$ $\alpha(m_{\eta_c}^2)$ 1.8 MOM $(m_{\eta_c}^2)$	
Branching Ratios $\rightarrow 2\gamma$ (25) EG $\frac{B(0^{-+})}{B(0^{++})}$	1 ~ 6 0.9	
$\psi, \Upsilon \rightarrow 3g$ (26)	12 $\overline{MS}$ $\alpha(m_\psi^2)$ 0 MOM $\alpha(m_\psi^2)$ $\rightarrow$ 0 $\overline{MS}$ $\alpha(m_\psi^2/4)$ 12 MOM $\alpha(m_\psi^2/4)$	ψ $\alpha_s = 0.19 \pm 0.02 \rightarrow 0.34$ (with Glueball mass corrections)$\equiv \Lambda = 400$ η_c $\alpha_s = 0.054 {+0.04 \atop -0.024}$ cf. 0.37 with $\Lambda = 400$

Process	Leading Order (LO) Predictions $\times (1+B \cdot \frac{\alpha_s}{\pi})$ $B_{\overline{MS}}$	Results and Comments $\Lambda_{(\overline{MS})}$ (MeV), $\alpha_s(\overline{MS})$
$B(\psi)_{LO} \equiv \frac{\Gamma(\psi \to e^+e^-)}{\Gamma(\psi \to \text{hadrons})}$ $= \frac{81\pi}{10(\pi^2-9)} \frac{Q^2 \alpha^2}{\alpha_s^2}$		Υ $\alpha_s = 0.16^{+0.04}_{-0.2}$ $\Lambda = 138$
High $P_\perp$ Hadron-Hadron $qq \to qX$ (27)	$\begin{array}{ll} 25 & \overline{MS} \\ 19 & \text{MOM} \end{array} \alpha_s(s)$ $\begin{array}{ll} 16 & \overline{MS} \\ 12 & \text{MOM} \end{array} \alpha_s(p_\perp^2) \to$ $\begin{array}{ll} 3 & \overline{MS} \end{array} \alpha_s(\frac{p_\perp^2}{4}) \to$	
Pion Form Factor (28,29) $F_\pi(Q^2)_{LO} = \frac{16\pi\alpha_s(Q^2)}{Q^2} f_\pi^2$	$\begin{array}{ll} 6.5 & \overline{MS} \\ 2.1 & \text{MOM} \end{array}$ $\to 0.3 \; \alpha_s \left(\frac{Q^2}{16}\right) \overline{MS}$ $- 0.9 \; \alpha_s \left(\frac{Q^2}{4}\right) \text{MOM}$	

(Continued)

Table 4. Continued

Process	Leading Order (LO) Predictions $(1+B\frac{\alpha_s}{\pi})$ $B_{\overline{MS}}$	Results and Comments $\Lambda_{(\overline{MS})}$ (MeV), $\alpha_s(\overline{MS})$
Baryon Form Factor (28) $G_M(Q^2) \sim \frac{1}{Q^4} \alpha_s(Q^2)^{2+\frac{4}{3\beta_0}} [1+0(\alpha_s)]$		$\Lambda_{L.O.} = 100 - 300$ MeV
$\frac{d\sigma}{dt}$ (pp → pp) $\frac{[\alpha_s(t)]^6}{t^8}$ (28) $\times \exp\left[\frac{\alpha_s}{2\pi}(-18\ln s \ln t + 10\ln^2 t)\right]$ (Sudakov Form factor)		No evidence for $[\alpha_s(t)]^6$ Term or for Sudakov Form Factor(30) but momentum transfer/quark is small $\approx 1 - 2$ GeV^2.

of argument for α_s and we show the variation of B for various choices. Where this variation is large it is unclear whether the perturbative calculations make sense at presently accessible energies. We also show in Table 4 the value for $\Lambda_{\overline{MS}}$ extracted from the fits; if QCD is correct a single value for Λ should fit all experiments for which there are reliable perturbative predictions.

Broadly there is agreement between theory and experiment with a value of Λ in the range 100-400 MeV. There are however discrepancies in the value for Λ extracted from various processes which may be significant. Clearly future experiments will be invaluable in improving these tests and in particular high Q^2 is needed to distinguish logs from higher twist ($\sim \frac{1}{Q^2}$). Also it will be important to complete the calculation of next to leading corrections and hopefully understand whether the perturbative calculations are reliable.

Ultimately QCD will be tested by comparisons in many different processes for which QCD makes definie predictions. The vector nature of the gluon is tested by the scaling violations in quark operators but to measure the triple gluon coupling it is necessary to study the scaling violation in gluon operators, e.g. by looking at gluon jets. $\alpha_s(Q^2)$ is sensitive to both the vector and non-abelian nature of the gluons. The study of different processes will provide a test of factorisation and of the validity of the resummation of the large perturbative corrections. Measurement of the hard scattering cross sections tests the nature of the interacting components. For example in $e^+e^- \to q\bar{q}g$ or $\Upsilon \to ggg$ measurement of the angular distribution tests for the spin of the gluon and already favours spin 1[2]

3.0 Tests of the SU(2) x U(1) Standard Model - The Fermion Sector

At present the standard model has only been tested at low energies ($<< M_{W,Z}$) and so the only vector boson in Table 2 to have been seen is the photon. Of the fermions in Fig. 1 we have still to find the t quark and there is as yet no experimental evidence in support of those assignments of the entries with question marks. Most of our knowledge of the structure of the weak interactions rests on the first two generations where experiment is consistent[32] with an effective Lagrangian

$$\mathcal{L}^{eff} = \frac{4G_F}{\sqrt{2}} \left[(2 J_\mu^+ J^{\mu-} + \rho(J_\mu^3 - J_\mu^{em}\sin^2\theta Q)^2 + c\, J_\mu^{em} J^{\mu em} \right]$$
(3.1)

where $J_\mu^{\pm 3}$ are SU(2) currents.

The leptonic multiplets are as given in Fig. 2 and the quark multiplets are the Cabibbo rotated ones

$$\begin{pmatrix} u \\ d\cos\theta_c + s\sin\theta_c \end{pmatrix}_L \qquad \begin{pmatrix} c \\ s\cos\theta_c - d\sin\theta_c \end{pmatrix}_L \qquad (3.2)$$

A fit to all available data gives[33]

$$\rho = 1.004 \pm 0.019$$
$$\sin^2\theta_W = 0.235 \pm 0.016 \qquad (3.3)$$
$$c < 0.04$$

In the standard model $\rho = 1$ and $c = 0$. For a full discussion of the status of the standard model see the review given by Sakurai[34] in last year's Erice lectures. Here I would like to look forward and ask whether we will be able to establish the missing pieces.

(3.1) Must the ν_t exist?

The answer is a qualified yes, although no direct measurement has yet been made. In τ decays, $\tau \rightarrow x + (e\bar{\nu}_e, \mu\bar{\nu}_\mu, d\bar{u})$, x cannot be $\bar{\nu}_e$ (or $\bar{\nu}_\mu$) otherwise the rates for $\tau \rightarrow x\bar{\nu}_e e, x\bar{\nu}_\mu \mu$ would differ, in conflict with experiment. In the standard model x cannot be ν_e (or ν_μ) for then the (LH) multiplets must be mixed with τ

eg. $\begin{pmatrix} \nu_e \\ e\cos\phi + \tau\sin\phi \end{pmatrix}_L$ and $(-e\sin\phi + \tau\cos\phi)_L$. This violates the GIM mechanism and gives lepton number violating processes $\tau \to 3e$, $e\gamma$ at unacceptable rates[35]. Experiment is consistent with a new neutrino in a left handed doublet; the Michel parameter in τ decay is $\rho = 0.72 \pm 0.10$[36] consistent with the (V-A) value of 0.75 ((V+A) would have $\rho = 0$).

(3.2) **Must the t quark exist?**

Probably. Without the t the b quark is mainly a SU(2) singlet and decays through mixing with the d or s doublets; similar arguments to that for the τ discussed above give

$$\frac{\Gamma(b \to X e^+ e^-)}{\Gamma(b \to X e\nu)} > 0.125 \qquad (3.4)$$

compared with the result from CLEO[37] of

$$\frac{\Gamma(b \to X e^+ e^-)}{\Gamma(b \to X e\nu)} < 0.13 \qquad (3.5)$$

Can the b decay through a new current? For example with a SU(3) gauge group we might try the assignments

$$\begin{pmatrix} u \\ d \\ B_1 \end{pmatrix} \qquad \begin{pmatrix} c \\ s \\ B_2 \end{pmatrix}$$

But then $B \to q$ + leptons in conflict with the fact that at CESR the B decays nonleptonically and the number of kaons produced per b is $2.5^{+0.5}_{-0.6}$. To construct an alternative we need to allow B to decay via a neutral current. For example[38] with the assignments

$$\begin{pmatrix} \cos\theta_c u - \sin\theta_c c \\ a \\ b' \end{pmatrix}_L \qquad \begin{pmatrix} \sin\theta_c u + \cos\theta_c c \\ \cos\beta s + \sin\beta b \\ -\sin\beta s + \cos\beta b \end{pmatrix}_L$$

the principal decays are b → cX both nonleptonically and semi-
leptonically. Such a pattern can arise naturally[39]. However,
the model has characteristic differences in the decay patterns of
the b.

For example b → s$\nu\bar{\nu}$, B_S^0 → e^+e^- through the neutral current.
Also decay ratios are characteristically different

$$\sigma(e^+e^-X) : \sigma(\mu^+\mu^-X) : \sigma(\mu^\pm e^\mp X) \qquad (3.6)$$

$$1 \quad : \quad 1 \quad : \quad 2 \qquad \text{Standard Model}$$

$$2.5 \quad : \quad 2.5 \quad : \quad 2 \qquad \text{Topless Model}$$

Experimental information of b decays is already almost good
enough to rule out such a model[39]. The b decays are consistent
with the assignment $\begin{pmatrix} c \\ b \end{pmatrix}_R$ but theoretically this is unsatisfactory
(unnatural) since there is no understanding of why the b should
not mix with the singlet d_R and s_R states violating the GIM
mechanism. The most satisfactory assignment is that of the
standard model requiring the existence of a t quark. Of course,
the question will only be finally settled by producing the t.

(3.3) <u>What is the mass of toponium</u>

The t quark will be found in e^+e^- collisions provided the
toponium mass is within the range of energy available.

In Fig.(3) I give a plot of the various theoretical predictions
for $M_{t\bar{t}}$. Below 32 GeV of course, the predictions are already ruled
out but theorists ingenuity is such that as the lower bound moves
up so does the theoretical prediction! Although the plot serves to
show how poorly understood are fermion masses let me comment on
several of the predictions. Radiative corrections (see section 6)
are sensitive to heavy quarks and using the results on the K_L-K_S
mass difference and K_L → 2μ rate Buras[40] finds

$$m_{t\bar{t}} = 52 \pm 14 \text{ GeV} \qquad (3.7)$$

Unfortunately his result relies sensitively on a bag model estimate for an operator matrix element and he would get no constraint on m_t if a free quark estimate were used instead.

Perhaps the most promising prediction comes from the general mass formula

$$M_{ij}(X) = \sum_a M_{ij}^{(a)} G^{(a)}(X) \qquad (3.8)$$

where X labels quark or lepton species and i,j (= 1,2,3) are family indices. If there are three terms (a = 1,2,3) in the sum the masses are all arbitrary. One term only gives the unacceptable result $e/\mu/\tau = u/c/t = d/s/b$, and vanishing Kobayashi-Maskawaa mixing angles. Two terms give a relation, which with QCD corrections gives[41] a mass tantalisingly close to the present limits

$$m_{tt} \lesssim 40 \text{ GeV} \qquad (3.9)$$

This form of the mass matrix does arise in specific GUT^S, for example in $O(10)$[42].

Finally, if the Yukawa coupling h_t giving the top quark a mass should initially (at a GUT scale) be large, radiative corrections drive this coupling (as measured at low energies) to an infra-red fixed point so that the resultant value is independent of the starting value and thus the top quark is uniquely determined[43]. Consider the evolution equation for h_t

$$\frac{dh_t}{dt} \simeq h_t(ah_t^2 - bg_s^2) \qquad (3.10)$$

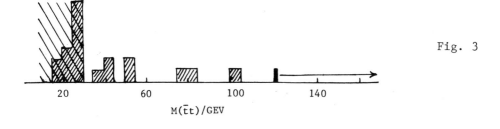

Fig. 3

M($\bar{t}t$)/GEV

The derivative vanishes for $\dfrac{h_t^2}{g_s^2} = \dfrac{b}{a}$ and there in an infra-red fixed point nearby. Thus h_t is given in terms of the QCD coupling g_s giving the favoured value for $m_t = 180 \pm 60$ GeV, the spread coming from the fact that the fixed point is never reached. Such a large value for m_t is inevitable if h_t is of the order of a gauge coupling and it is interesting to compare the result to the bound[44] coming from calculating radiative corrections to ρ, defined in eq.(3.1) and eq.(3.3) arising from the graph of Fig.(4).

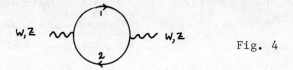

Fig. 4

This gives

$$\rho = 1 + \dfrac{G_F}{8\pi^2}\left\{m_1^2 + m_2^2 - \dfrac{2m_1^2 m_2^2}{(m_2^2 - m_1^2)}\ln\dfrac{m_2^2}{m_1^2}\right\} \times \begin{cases} 3 \text{ (quarks)} \\ 1 \text{ (leptons)} \end{cases} \qquad (3.11)$$

which imply

$$m_t \lesssim \begin{array}{ll} 420 \text{ GeV} & \rho = 1.018 \\ 260 \text{ GeV} & \rho = 1.002 \end{array} \qquad (3.12)$$

This formula is interesting because it also can be used to limit the number of extra generations[44].

(3.4) The parameters of the 6 quark model

It seems likely that the 6 quark model is necessary and that we will be able to produce the top quark with the next generation of experiment. If the standard model assignments of Fig.(1) are correct this means we must generalise the Cabibbo mixing of eq.(3.2) to include the third generation. This gives rise to the Kobayashi-Maskawa model[45] with 3 mixing angles θ_i and one phase δ (which gives rise to CP violation[46]). In terms of these the charged weak current is

$$J_\mu^+ = (u\ c\ t)_L\ \gamma_\mu \begin{bmatrix} c_1 & c_2 s_1 & s_1 s_3 \\ -c_2 s_1 & c_1 c_2 c_3 - s_2 s_3 e^{i\delta} & c_1 c_2 s_3 + c_3 s_2 e^{i\delta} \\ s_1 s_2 & -c_1 c_3 s_3 - c_2 s_3 e^{i\delta} & -c_1 s_2 s_3 + c_2 c_3 e^{i\delta} \end{bmatrix}$$

$$X \begin{pmatrix} d \\ s \\ b \end{pmatrix}_L \quad (3.13)$$

The neutral current does not mix flavours in this model as it has been carefully constructed to preserve the GIM mechanism[7]. In order to test the model it is necessary to check whether eq.(3.13) does describe the charged weak currents and measure the parameters. In this b (and t) decays will be crucial. A start to this program has already been made and review of the progress may be found in the Erice lectures of Berkleman[39]. Toponium decays promise to be a very useful laboratory for studying weak interactions[48] for, at a mass > 60 GeV, t decay competes with the annihilation process and the modes

$$V(t\bar{t}) \to ggg,\ q\bar{q}$$
$$t\bar{b}X,\ \bar{t}bX \propto (-c_1 s_2 s_3 + c_2 c_3 e^{i\delta})$$

are approximately equal. Thus one will be able to test QCD predictions and measure new combinations of the K-M angles. One can do even better for 10% of the time the resultant $t\bar{b}$ and $\bar{t}b$ will be the vector mesons $v(t\bar{b})$, $v(\bar{t}b)$ which subsequently decay producing $c\bar{s}$, $u\bar{d}$, $e\bar{\nu}$, $\mu\bar{\nu}$, $\tau\bar{\nu}$ measuring further combinations of the K-M angles {The decay $V \to P\pi$ is relatively disfavoured since its amplitude $\alpha\ \frac{1}{m_t}$ whereas the above decays grow with increasing m_t}

Toponium will also provide a good means to measure the number of neutrinos N_ν for

$$\frac{(v(t\bar{t}) \xrightarrow{Z} \sum_i \nu_i \bar{\nu}_i)}{(v(t\bar{t}) \xrightarrow{\gamma} e^+ e^-)} \simeq .2 \times 10^{-8}\ m_V^4\ N_\nu \quad (3.14)$$

Applied to the J/ψ this gives a bound $N_\nu < 5 \times 10^5$. With $m_{t\bar{t}} > $ 65 GeV this should be improved by a factor $> 10^5$. Higgs bosons should also be produced in toponium decay as we will see.

(4.0) TESTS OF THE STANDARD MODEL - THE BOSON SECTOR

We have not yet discussed possible tests for the W,Z and their Yang-Mills coupling. At present the neutral currents are consistent with a low energy Lagrangian of the form of eqs.(3.1) and (3.3) but this does not imply the standard model is correct. Here I want to discuss several variants of the standard scheme, consistent with eqs.(3.1) and (3.3), as a guide to what tests will be needed to establish the standard model.

(4.1) <u>Is there a new neutral current?</u>

Much work has been done with mild variants of the standard model in which the gauge group is SU(2) x U(1) x U(1) or SU(2) x U(1) x G. It has been shown[49] that the standard model low energy neutral current structure is recovered for any fermion neutral under G although the Z^S may be a mixture of the U(1) and U(1)' gauge bosons. However in this case we find that

$$M_{Z_{min}} < M_{Z_{standard}} = \frac{M_W}{\cos\theta_W} \qquad (4.1)$$

i.e. at least one Z must be lighter than that found in the standard model.

It is possible to relax the above condition and to duplicate the standard model with $M_{Z_{min}} > M_{Z_{standard}}$ [50], (for example in SU(2) x SU(2) x U(1)) but then it follows that $M_{W_{min}} > M_{W_{standard}}$. Moreover these models require an unnatural tuning of parameters to reproduce the form of eqs.(3.1) and (3.3).

It seems reasonable to hope that there will be a neutral

boson with a mass < 90 GeV but there is no support at present for the single Z of the standard model and tests at higher energies[51] are necessary to establish this.

(4.2) Are right hand currents absent?

In the standard model parity is violated by assigning different SU(2) x U(1) quantum numbers to LH and RH fermion components (cf. Fig. 1), and destroying left right symmetry. This symmetry cannot be restored by assigning, for example, e_R to a right handed doublet with a massive partner because the observed neutral current processes are consistent with e_R or SU(2) singlet[32]. However L-R symmetry can be restored by extending the gauge group to $SU(2)_L \times SU(2)_R \times U(1)$ with $e_{L(R)}$ transforming as (2,1) ((1,2)) under the $SU(2)_L \times SU(2)_R$. Parity will be violated (spontaneously) if the RH gauge bosons are heavier than the LH gauge bosons At present, RH charged currents are known to contribute < 10 % of the LH currents[34] which gives a limit on the right handed boson mass $M_R > 0$ (225 GeV). However it has been noticed[52] that there are solutions with $M_{W_R} \sim 100$ GeV and a $\sin^2\theta_W' = 0.28$ (this is the angle that enters GUT predictions - see section 5 - not the angle of eq.(3.1)) which agree with experiment and which can be included in simple GUT^s. It is therefore of considerable interest to look for such RH currents and again the new experiments at high energies should be able to distinguish these L-R symmetric models from the standard one.

(4.3) Must there be gauge bosons?

A more heretical variant of the standard model suggests that $W^\pm$ and Z may be composite fields bound at a scale > 100 GeV. At first sight this seems extremely implausible for how can the universal couplings and vector nature of the gauge fields be naturally explained. To show that this may indeed happen I will briefly discuss a model of Abbott & Farhi[53] which realises an idea of Bjorken[54]. It provides a simple illustration of the properties of many composite theories.

The model is based on a SU(2) x U(1) gauge group with the same multiplet structure as in the standard model. However it is supposed that there is no spontaneous breakdown, that the SU(2) gauge bosons remain massless, but that, at a scale $\Lambda \approx G_F^{-\frac{1}{2}}$, $SU(2)_L$ is confining so that at energies $\ll \Lambda$ physical states are $SU(2)_L$ singlets. This means that the U_L and D_L states must be composite operators

$$U_L \alpha \phi_i^* \psi_L^i$$
$$D_L \alpha \varepsilon_{ij} \phi^i \psi_L^j \qquad (4.2)$$

Here we have formed U and D from the lowest dimension $SU(2)_L$ singlet combination of elementary fields. ϕ_i is the Higgs doublet and ψ_{iL} the fermion doublet. U_R and D_R may be identified with the elementary u_R and d_R fermion fields as they are already $SU(2)_L$ singlets.

The usual Yukawa couplings can generate masses for these objects

$$\lambda_1 \bar{U}_R \phi_i^* \psi_L^i = \lambda_1 \Lambda \bar{U}_R U_L$$
$$\lambda_2 \bar{D}_R \varepsilon_{ij} \phi^i \psi_L^j = \lambda_2 \Lambda \bar{D}_R D_L \qquad (4.3)$$

Here we have chosen the (dimensionful) constant of proportionality in eq.(4.2) to be the confinement scale Λ and, as in the standard model, fermion masses are small because the Yukawa couplings λ are small. (There is a chiral symmetry which prevents the fermions from acquiring a mass if $\lambda_i = 0$).

The photon in this model must be chosen to be the U(1) boson as this remains massless. This means that the associated coupling $g' = e$ and not $\frac{e}{\cos\theta_W}$ as in the standard model. This is perfectly consistent with the usual $SU(2)_L \times U(1)$ charge assignments for the weak hypercharge $Y' = Q - T_3 = Q$ when acting on $SU(2)_L$ singlet states.

The W bosons in the model must also be $SU(2)_L$ singlets and indeed one can form a triplet of W^S from the Higgs fields

$$\phi_i^* D_\mu \varepsilon^{ij} \phi_j^* \propto W^+$$

$$\phi^i \varepsilon_{ij} D_\mu \phi^j \propto W^- \qquad (4.4)$$

$$\phi_i^* D_\mu \phi^i \propto W_3$$

The W^S have no symmetry preventing them acquiring a mass and we expect this mass to be of order Λ. However there is a residual global $SU(2)$ symmetry of the Lagrangian which ensures $M_{W^\pm} = M_{W^0}$. The coupling of these W^S to hadrons must preserve this symmetry and it has the form

$$\frac{g}{2} (\bar{U}_L, \bar{D}_L) W_\mu^a \tau^a \gamma^M \begin{pmatrix} U_L \\ D_L \end{pmatrix} \qquad (4.5)$$

where g is a coupling constant which is related in a complicated (unknown) way to the $SU(2)_L$ coupling. The W^S couple only to LH quarks as only they feel the strong $SU(2)_L$ force.

This gives an effective low energy Lagrangian

$$\mathcal{L}^{eff} = \frac{4g^2}{8M_W^2} \left(2 J_{\mu L}^+ J_L^{\mu -} + J_{\mu L}^3 J_L^{\mu 3} \right) \qquad (4.6)$$

This gives the first two terms of eq.(3.1). However there is another contribution which arises since, due to strong $SU(2)_L$ interactions, the photon has an electric form factor in its coupling to the quarks. This is given by $K \frac{Q^2}{M^2} \bar{F}_L \gamma^\mu F_L$ where K is an unknown constant. (Note there is no magnetic form factor involving LH fields only). Now a Born diagram with a photon exchanged generates a new term[54].

$$\left\{ \frac{e^2}{Q^2} \cdot \frac{KQ^2}{M^2} \right\} J^{\mu^3} J_\mu^{em} \qquad (4.7)$$

If we set

$$\frac{e^2 K}{M^2} = \frac{4G_F}{\sqrt{2}} (\sin^2\theta_W) \qquad (4.8)$$

then it generates the correct interference term $\propto J^3_\mu J^{\mu em}$ in eq.(3.1). Amazingly the resulting model automatically generates the standard model effective Lagrangian up to terms $\propto (J_\mu^{em})^2$. Both fermions and vector bosons are composite but the compositeness will only show up at a scale $\approx \Lambda$.

However the model has two flaws which persist in many composite models. Firstly one can construct an isoscalar partner of the W^S in eq.(3.18) and there is no explanation why this should be suppressed[55]. Secondly there is now experimental evidence for the $(J_\mu^{em})^2$ term of eq.(3.1). To cancel it one must choose $C = -\sin^4\theta_W \approx 0.05$ which is disfavoured by the bounds of eq.(3.3). I will not discuss composite models further but I hope this example shows that surprising things may happen and that a true test of the standard model requires the production of the intermediate vector bosons.

(4.4) $W^\pm$, Z Physics

Of course the major discoveries we expect from the next generation of experiments are the Z and W bosons. Indeed the new machines are designed to allow production of the standard model W and Z. In this model[44,59]

$$\begin{aligned} M_{Z^0} &= \frac{37.3}{\sin\theta_W \cos\theta_W} + (3.3 \pm .3) \text{ GeV} \\ M_W &= \frac{37.3}{\sin\theta_W} + (3.0 \pm .3) \text{ GeV} \end{aligned} \qquad (4.9)$$

The additional terms in brackets are the radiative corrections coming from the graphs of Fig.(4). A measurement of M_{Z^0} to an accuracy of 0.3 GeV will allow determination of $\sin^2\theta_W$ to an accuracy of 0.001. (In practice this will be better done from a fit to the predicted Z resonance shape). This will be of great importance in testing GUT^S (see section 6) and in checking the calculations of radiative corrections which rely on the renormalisable nature of the interactions. The width and shape of the Z also depends on the number of leptons and quarks with mass $<< M_Z$. The relative branching ratios are

$$\bar{\nu}_e \nu_e \;:\; e^+ e^- \;:\; u\bar{u} \;:\; d\bar{d} \qquad (4.10)$$

$$= 2 \;:\; 1 + (-1 + 4\sin^2\theta_W)^2 \;:\; 3\left(1 + \left(1 - \frac{5}{3}\sin^2\theta_W\right)^2\right.$$

$$:\; 3\left[1 + \left(-1 + \frac{4}{3}\sin^2\theta_W\right)^2\right]$$

$$= 2 \;:\; 1.01 \;:\; 3.45 \;:\; 4.44$$

and the width to neutrinos is given by

$$\Gamma(Z^0 \rightarrow \nu_e \bar{\nu}_e) = \frac{G m_Z^3}{12\sqrt{2}} \qquad (4.11)$$

which gives a contribution of 2.5 GeV to the total width Γ_Z. So $\frac{\Delta\Gamma_Z}{\Gamma_Z} = 6\%$ for every extra ν species.

Similarly W^S decay unambigously with branching ratios

$$e\bar{\nu}_e \;:\; \mu\bar{\nu}_\mu \;:\; \tau\bar{\nu}_\tau \;:\; d\bar{u} \;:\; s\bar{c} \;:\; b\bar{t}$$

$$= 1 \;:\; 1 \;:\; 1 \;:\; 3 \;:\; 3 \;:\; 3$$

For example $e\bar{\nu}_e$ gives a contribution of 2.5 GeV to the total W width.

It is clear that W and Z physics will be very exciting - but will we be able to produce them? Present bounds on the mass of W^S and Z^S rely on their propagator effects modifying scaling. Present bounds are $M_W > 30$ GeV with $E_{lab} \sim 200$ GeV. At Hera we would be able to measure $M_W \lesssim 500$ GeV and indeed ep machines are the best places to look for heavy W^S.

The first production of W^S and Z should be at the $p\bar{p}$ collider in CERN[57]. The Drell-Yan production mechanism gives an expected rate of several 100 per day with a luminosity of 10^{30} cm^{-2} sec^{-1}. Because quark jets coming from W,Z decay are expected to be much smaller than QCD jets we must look at the leptonic modes to identify the W,Z events. From eq.(4.10) and (4.12) these modes have branching ratios

$$W \to e\nu, \mu\nu \qquad \text{B.R.} = 8\% \qquad (4.13)$$
$$Z \to e^+e^-, \mu^+\mu^- \qquad \text{B.R.} = 3\%$$

The integrated luminosity over 1000 hours gives $\simeq 300$ W events and 50 Z events. For Z the lepton pair should have cm energy of M_Z; for W^S one must look for the Jacobian peak. The energy resolution is ≈ 3 GeV and the expected electromagnetic background is a factor of 10^{-3} down.

Production of W^S and Z^S in ep machines is through the graphs of Fig.(5)

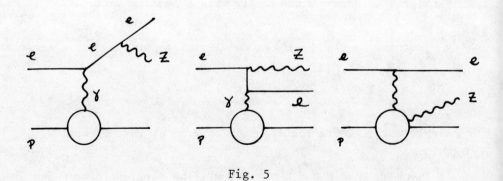

Fig. 5

This gives

$$\sigma^Z \sim 5 \times 10^{-7} \text{ cm}^2 \text{ for } M_Z = 89 \text{ GeV and } S = 10^5 \text{ GeV}$$
$$\sim 10^{-37} \text{ cm}^2 \text{ for } M_Z = 125 \text{ GeV} \quad (4.14)$$
$$\sigma^W \sim 2 \times 10^{-37} \text{ cm}^2 \text{ for } M_W = 78 \text{ GeV}$$

HERA with 30 GeV electrons on 800 GeV protons and a luminosity of 3×10^{32} cm^{-2} sec^{-1} will produce approximately 500 Z events and 200 W events in 100 hours.

The most attractive machines for producing and studying Z^S are the e^+e^- colliders. The relevant cross section is

$$\sigma(e^+e^- \to Z \to F) = \frac{12\pi}{S} \frac{\Gamma(Z \to e^+e^-)\Gamma(Z \to F) M_Z^2}{(S - M_Z^2)^2 + M_Z^2 \Gamma_{Tot}^2} \quad (4.15)$$

with $\Delta E_B \approx 100$ MeV and $\Gamma_{Tot} \approx 3$ GeV.

With a luminosity of 10^{31} cm^{-2} sec^{-1} this gives approximately 20,000 Z events per day. The beauty of the e^+e^- colliders, in addition to the high production rate, is that all of the final states of eq. (4.10) may be studied.

It is clear that if W^S and Z^S exist we will be able to find them. Their production and a measurement of the decay modes of eqs. (4.10) and (4.12) will be very strong tests of the standard model. In addition it will ultimately be possible to look for the non-abelian nature of the vector boson coupling. For example the process $e^+e^- \to W^+W^-$ proceeds via the three graphs of Fig.(8).

Each diagram has a rising cross section but with Yang-Mills couplings there is a strong cancellation between them necessary to restore perturbative unitarity[58]. So there is a big difference between cross sections with and without Yang-Mills couplings. For example at LEP in 50 days well above W^+W^- threshold we would expect 750 events in the standard model but 7500 events if the ZW^+W^- coupling is omitted.

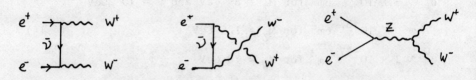

Fig. 6
Graphs contributing to $e^+e^- \to W^+W^-$

(4.5) The Scalar Bosons

The scalar sector is the least understood part of the standard model in the sense that there is no evidence yet for a Higgs scalar nor is there a definite prediction of its mass. In the standard model it is necessary to include a SU(2) doublet of (colourless) scalars $\equiv \begin{pmatrix} \phi^+ \\ \phi^o \end{pmatrix}$. Their self interaction is described by the potential

$$V(\phi) = -\mu^2 \phi^+\phi + \lambda(\phi^+\phi)^2 \qquad (4.16)$$

Minimising this potential leads to a nonzero vacuum expectation value for the neutral scalar

$$\langle \phi^o \rangle = \frac{\mu^2}{2\lambda} \equiv \frac{v}{\sqrt{2}} \qquad (4.17)$$

The charged Higgs $\phi^\pm$ and one component of the neutral Higgs provide the longtitudinal components for the $W^\pm$ and Z, leaving one physical neutral scalar H

$$H = \frac{(\phi^o + \phi^{o+})}{\sqrt{2}} - v \qquad (4.18)$$

with mass $m_H^2 = 2\mu^2$.

Since μ^2 is a parameter in the theory m_H^2 is not determined.

A lower bound for m_H^2 was found by Linde[60] and Weinberg[61], following from the condition that radiative corrections to eq.(4.16) should not remove the minimum leading to eq.(4.17). They obtained

$$m_H > 6.4 \text{ GeV} \qquad (4.19)$$

An interesting prediction for m_H^2 follows if we set $\mu^2 = 0$ in eq.(4.16) (what else could it be?). Coleman and Weinberg[62] showed that radiative corrections induce a minimum of $V(\phi)$ away from the origin; the position of the minimum is then determined by the gauge couplings and gives[62]

$$m_H = 10.5 \ (10.9) \text{ GeV} \qquad (4.20)$$

where the second value in brackets follows when higher order corrections are included[63].

This prediction must be modified if the top quark is very heavy for then radiative corrections due to Yakawa couplings become important. If m_t is determined by an infra-red fixed point (see section 3) then so too is m_H giving

$$m_H \approx 70 \text{ GeV} \qquad (4.21)$$

Finally an upper bound for m_H follows if we ask that (perturbative) partial wave unitarity should hold. This gives[76]

$$m_H < \frac{8\pi\sqrt{2}}{3G_F} \sim .3 \text{ TeV} \qquad (4.22)$$

Although the Higgs boson mass is not well determined its couplings are, simply because these couplings give rise to the particle masses. Thus

$$g_{f\bar{f}H} = \frac{m_f}{v}$$
$$g_{WWH} = \frac{2m_W^2}{v} \qquad (4.23)$$
$$g_{ZZH} = \frac{2m_Z^2}{v}$$

Such simple relations apply if there is only one Higgs doublet.

Since Higgs scalars are an essential feature of the standard model, it is important to look for them. We postpone the discussion of Higgs searches until we have discussed some alternatives to the standard model for, as we will see, the scalar sector promises to be a good proving ground for various theories.

(5.0) BEYOND THE STANDARD MODEL - I

So far we have discussed what we may hope to learn from the next generation of experiments with a view to establishing the standard $(3,2,1)$ model. Consequently we have concentrated on looking for effects predicted by the model. If however we want to look for a signal showing departure from the standard scheme it is sensible to look for effects forbidden by the model. In Table 5 we summarise the (possible approximate) symmetries of $SU(3)_c \times SU(2) \times U(1)$. Charge conservation and Lorentz invariance are, we believe, exact symmetries realised locally with the photon and the graviton as the associated massless gauge fields. In the "age of the gauge" it is conjectured that all continuous symmetries should be local gauge symmetries and consequently all exact continuous symmetries should be associated with zero mass gauge bosons. Since we have found no more it seems likely that all other continuous symmetries will be broken at some level and for this reason we have put a question mark against the baryon and lepton number entries which are exactly conserved in the standard model.

If any of the symmetries, exact in the standard model, are found to be broken it will be a clear signal of new physics, possibly due to processes occuring at energy scales $M_X \gg M_{W,Z}$.

Table 5

Symmetries of the Standard Model

Quantum Number	Extent of Conservation
Q	Exact
Fermion Number	Exact
B	Exact?
L	Exact?
d,s,b	Weakly Broken
e,μ,τ	Exact?
C	Weakly Broken
P	Weakly Broken
CP	< Weakly Broken
CPT	Exact

The low energy phenomenological consequences of a theory with new interactions occurring at a scale M_X may be conveniently parameterised[64] in terms of an effective Lagrangian involving SU(3) x SU(2) x U(1) symmetric operators of dimension d, O^d, expanded in inverse powers of M_X

$$\mathcal{L}^{eff} = \sum_d C_d \frac{1}{M_X^{d-4}} O^d \qquad (5.1)$$

For example for baryon number violation O must involve at least 3 quark fields (to form a colour singlet) and have an even number of fermions to be consistent with Lorentz invariance. Thus the minimal dimension operator is qqqℓ and

$$\mathcal{L}^{eff} = \frac{C}{M_X^2} (qqq\ell) + \ldots \qquad (5.2)$$

where C will be determined by the couplings of the new interaction. With $C \approx g^2$, and reasonable estimates for operator matrix elements we can calculate the possible proton decay rate in terms of M_X. For a proton lifetime in the potentially observable range of

$10^{30} - 10^{33}$ years M_X would be 3×10^{14} to 3×10^{15} GeV. Thus proton decay experiments are sensitive to incredible energy scales, and indeed studies of forbidden processes hold out the best hope for tests of the new grand unified theories which attempt to unify the strong, weak and electromagnetic interactions, and which involve such large energy scales.

The utility of the expansion of eq.(5.1) lies in the fact that the operators of a given dimension often have characteristic properties. For example the operator of eq.(5.2) conserves (B-L). If a (B-L) violating process if found it must arise from a higher dimension operator and must be due to a lower scale for new physics. In Table 6 we list[65] some of the interesting operators together with the typical mass scale probed if the process is found with a rate $10^{30} - 10^{33}$ years.

The construction of the allowed operators is made easy through the introduction of F parity[64]

$$F = +1 \quad q, \ell$$
$$F = -1 \quad \bar{q}, \bar{\ell}, \gamma, g, W, Z, \partial_\mu \phi \qquad (5.3)$$

Allowed operators have F parity even: for example $qq\bar{q}$ has $F = +1$ and is allowed but qqq has $F = -1$ and is forbidden. Then it is easy to see there are six operators with $d = 6$, violating baryon number. All conserve (B-L). Only two can arise due to (superheavy) vector boson exchange.

$$O_1 = \varepsilon_{\alpha\beta\gamma} \bar{u}_L^{c\gamma} \gamma^\mu u_L^\beta \bar{e}_R^{+} \gamma_\mu d_R^\alpha$$
$$- \varepsilon_{\alpha\beta\gamma} \bar{u}_L^{c\gamma} \gamma^\mu d_L^\beta \bar{\nu}_R^c \gamma_\mu d_R^\alpha \qquad (5.4)$$
$$O_2 = \varepsilon_{\alpha\beta\gamma} \bar{u}_L^{c\gamma} \gamma^\mu u_L^\beta \bar{e}_L^{+} \gamma_\mu d_L^\alpha$$

NEXT GENERATION OF EXPERIMENTS

Table 6

Operators giving forbidden processes
q, ℓ, φ and D represent quarks, leptons, Higgs
scalars and derivatives respectively

O	d	Process	ΔB, ΔL	M_X (GeV)
$qqq\ell$	6	$p \to e^+ \pi^0$	$\Delta B = \Delta L = -1$	$3 \times 10^{14} - 3 \times 10^{15}$
$qqq\ell^c\phi$	7	$n \to e^- K^+$	$\Delta B = -\Delta L = -1$	$2 \times 10^{10} - 10^{11}$
$qqq\ell^c D$	7	$n \to e^- \pi^+$	$\Delta B = -\Delta L = -1$	$4 \times 10^9 - 2 \times 10^{10}$
$qqq\ell^c\ell^c\ell^c\phi$	10	$n \to \nu\nu e^- \pi^+$	$\Delta B = -\frac{1}{3}\Delta L = -1$	$(3-7) \times 10^4$
$qqq\ell\ell\ell\phi^2$	11	$p \to e^+ \nu^c \nu^c$	$\Delta B = \frac{1}{3}\Delta L = -1$	
$qqq\;qqq$	9	$n \to \bar{n}$	$\Delta B = -2$	$4 \times 10^5 - 10^6$
		$pn \to \pi^+\pi^0$	$\Delta L = 0$	
		$nn \to 2\pi^0$		
$\ell\ell\phi\phi$	5	ν Majorana mass	$\Delta B = 0$, $\Delta L = \pm 2$	$10^{11} - 10^{15}$ for $m_\nu = 10^{-2} - 10^2$ ev.

Thus in a theory with vector bosons mediating proton decay the general form for $\int\mathcal{L}^{eff}$ is

$$\int\mathcal{L}^{eff} = \frac{4G_1}{\sqrt{2}} O_1 + \frac{4G_2}{\sqrt{2}} O_2 \tag{5.5}$$

Different models may have different values for G_1 and G_2 and their measurement will distinguish between such models. For example in SU_5, $G_2 = 2G_1$, and the proton decay modes are given in terms of the Kobayashi-Maskawa angles alone[66]. Then for example

$$\frac{N \to \mu^+ + \text{Nonstrange}}{N \to e^+ + \text{Nonstrange}} \approx \frac{\sin^2\theta_c \cos^2\theta_c}{(1 + \cos^2\theta_c)^2 + 1}$$

$$\frac{N \to e^+ + \text{Strange}}{N \to \mu^+ + \text{Strange}} \approx \frac{\sin^2\theta_c \cos^2\theta_c}{(1 + \sin^2\theta_c)^2 + 1} \tag{5.6}$$

If proton decay occurs at a reasonable rate second generation experiments should be able to test these fine details of the quark and lepton assignments.

The above discussion illustrates how non-accelerator experiments may be the best way to learn about physical processes beyond the standard model. Neutrino masses, $n-\bar{n}$ oscillation and lepton number violating processes are other examples of such windows on the high energy world which are currently being explored, and which we may hope will shed light on physics beyond the standard model.

Another type of test for grand unified theories follows from a comparison with experiment of GUT predictions for the low energy parameters of the standard model. These follow from the observation[67] that if $SU(3) \times SU(2) \times U(1) \subset G$ the couplings g_S, g and g' are all related to the single grand unified coupling g_{GU} of the simple group G. Below the threshold M_X for the full group G the couplings evolve as

$$\frac{1}{g_i^2(q^2)} = \frac{1}{g_i^2(\mu^2)} + \beta_i \ln \frac{\mu^2}{q^2} + O(g_i^2)$$

where

$$\beta_{g_S} = -\frac{1}{16\pi^2}\left(11 - \frac{2}{3} nf\right) \quad (5.7)$$

$$\beta_g = -\frac{1}{16\pi^2}\left(\frac{22}{3} - \frac{2}{3} nf\right)$$

$$\beta_{g'} = \frac{1}{24\pi^2} nf$$

where nf is the number of flavours.

The values for $g_i(\mu^2)$ for $\mu^2 \approx 1$ GeV^2 can be measured in the laboratory. The values of $g_i^2(M_X^2)$ are given in terms of g_{GU}. For the prototype GUT, SU(5), and indeed for any GUT in which a family of quarks and leptons fits into complete representations of G,

NEXT GENERATION OF EXPERIMENTS

$$g_S^2(M_X) = g_{Gu}^2$$

$$g^2(M_X) = g_{Gu}^2 \tag{5.8}$$

$$g'^2(M_X) = \frac{5}{3} g_{Gu}^2$$

Using eq.(5.7) and (5.8) we can immediately determine M_X^2 and $\sin^2\theta_W$ as [67,68]

$$M_X^2 = \mu^2 \exp\left\{\left(\frac{1}{\alpha_S(\mu^2)} - \frac{1}{\alpha_W(\mu^2)}\right) \frac{12\pi}{(11 + \frac{nf}{2})} + O(\alpha)\right\}$$

$$\sin^2\theta(\mu^2) = \frac{3}{8}\left\{1 - \frac{\alpha_{EM}}{4\pi} \frac{110 - nf}{9} \ln \frac{M_X^2}{\mu^2} + O(\alpha)\right\} \tag{5.9}$$

These predictions may be improved by including the $O(\alpha)$ terms giving[69] for the minimal SU(5) theory

$$M_X = 1.5 \times 10^{15} \Lambda_{\overline{MS}} \quad \pm 30\%$$

$$\sin^2\hat{\theta}_W(80) = .206 \quad {}^{+.016}_{-.004} \tag{5.10}$$

$$\tau_{P,N} = (0.6 - 2.5)\left[\frac{M_X}{5 \times 10^{14} \text{ GeV}}\right]^4$$

$$\approx 8 \times 10^{30} \text{ yrs } (\Lambda_{\overline{MS}} = 400 \text{ MeV})$$

Here $\sin^2\hat{\theta}_W(80)$ refers to $\sin^2\theta_W$ evaluated in a (modified) minimal subtraction scheme at a scale of 80 GeV[69].

To compare experiment with theory for $\sin^2\hat{\theta}(80)$ it is necessary to include also $O(\alpha)$ radiative corrections in the calculated cross section for the particular process considered. This has now been done[70,71] giving[71]

$$\sin^2\hat{\theta}^{Expt}(80) = 0.216 \pm .015 \quad \text{eD asymmetry}[72]$$

$$\sin^2\hat{\theta}^{Expt}(80) = 0.219 \pm 0.012 \quad \nu N \ [73]$$

These determinations will be considerably improved once the Z and W bosons are produced (see section 3) and will provide quite stringent tests of the minimal SU(5) theory.

(6.0) BEYOND THE STANDARD MODEL - II

(6.1) <u>Effective field theories</u>

We have discussed present and future tests of the standard SU(3) x SU(2) x U(1) model. We have also briefly discussed ways of looking beyond the standard model by measuring rare processes forbidden in the standard scheme, and testing relations amongst the parameters of the standard model. However one may ask whether we need to go beyond the standard model - could it not be the final theory?

Before attempting to answer this let me remind you of the situation before the standard model was invented. Low energy weak interactions were well described by 4-Fermi theory with a Lagrangian density given by

$$\mathcal{L} = \frac{4G_F}{\sqrt{2}} (J_\mu^\ell + J_\mu^h)^+ (J_\mu^\ell + J_\mu^h) \qquad (6.1)$$

where

$$J_\mu^\ell = \bar{e} \gamma_\mu \tfrac{1}{2}(1 - \gamma_5) V_e + \bar{\mu} \gamma_\mu \tfrac{1}{2}(1 - \gamma_5) V_\mu \qquad (6.2)$$

$$J_\mu^h = \cos\theta_c \bar{u} \gamma_\mu \tfrac{1}{2}((1 - \gamma_5)d + \sin\theta_c\, u \gamma_\mu \tfrac{1}{2}(1 - \gamma_5) S$$

However it was realised that such a theory could, at best, only be an effective low energy theory. The reason is that the theory has bad high energy behaviour. For example the process $\nu_e \mu \to \nu_\mu e$ is predicted to have a cross section $\propto G_F^2 E^2$ from the interaction of eq.(6.1) whereas partial wave unitarity says the cross section $\propto \frac{1}{E^2}$. For large E^2 these two statements are

NEXT GENERATION OF EXPERIMENTS

incompatible and putting in the constants of proportionality one finds the bound violated for $E \approx 300$ GeV. Something new must happen for $E < 300$ GeV or else one must give up the perturbative use of eq.(6.1). To avoid this it was suggested that the effective 4 Fermi Lagrangian of eq.(6.1) is generated by the exchange of a massive intermediate vector boson coupled to the currents of eq.(6.2). As a result eq.(6.1) is modified by the substitution

$$\frac{G_F}{\sqrt{2}} \rightarrow \frac{g^2}{8(M_W^2 + E^2)} \qquad (6.3)$$

where E is the centre of mass energy. For $E^2 \gg M_W^2$ things are improved and $\sigma_{\nu\mu \rightarrow \nu e} \propto \frac{g^4}{E^2}$ consistent with the unitarity bound.

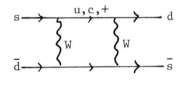

Fig. 7

Graph contributing to $K_L - K_S$ mass difference

The point is that one can sometimes predict the range of validity of an effective theory. This is seen even more dramatically when one considers higher order corrections in the 4 Fermi theory. The most severe limits are found when computing a $\Delta S = 2$, $\Delta Q = 0$ process. These may occur in higher order through the graph of Fig. 7 and gives an amplitude

$$\begin{aligned}A_{\substack{\Delta S = 2 \\ \Delta Q = 0}} &\propto (G_F \sin\theta_c \cos\theta_L)^2 \int \frac{d^4k}{k^2} \\ &\sim G_F^2 \sin^2\theta_c \int_0^{\Lambda^2} dk^2 = G_F^2 \sin^2\theta_c \Lambda^2.\end{aligned} \qquad (6.4)$$

The integral in eq.(6.4) is U.V. divergent signalling bad high energy behaviour in the 4 Fermi theory. Treating it as an effective field theory and truncating the momentum integral at a scale Λ gives the result in eq.(6.4). $A_{\substack{\Delta S = 0 \\ \Delta Q = 0}}$ contributes to the $K_L - K_S$ mass difference and, again putting in the constant factors, one finds a bound for Λ^2 if eq.(6.4) is not to generate too large a mass difference

$$\Lambda^2 < 1 \text{ GeV}^2 \qquad (6.5)$$

In this case calculating with intermediate vector bosons does not help as the bound on Λ^2 is so low that the vector boson theory is still an effective 4 Fermi theory. We are left with the conclusion that the effective theory breaks down at a scale < 1 GeV^2! The solution is to modify the theory to include the c quark and add new charged weak currents (cf. eq.(3.2)). There is then a cancellation between the u and c quark contributions (the GIM mechanism[47]) giving[74]

$$A_{\substack{\Delta S = 2 \\ \Delta Q = 0}} \sim G_F^2 \sin^2\theta_c (m_c^2 - m_u^2) \qquad (6.6)$$

This is consistent with the $K_L - K_S$ mass difference provided $m_c < 1$ GeV.

The point I wish to emphasise is that radiative corrections sample momentum scales beyond those presently accessible to experiment and can be used to limit the range of validity of an effective field theory. Indeed this type of analysis has been elegantly taken to its limit and it has been shown that the only field theories which do not violate unitarity bounds and have calculable radiative corrections, modulo a few renormalisation terms, are spontaneously broken Yang-Mills gauge field theories[75]. Thus one is led to the standard SU(2) x U(1) gauge theory with the GIM mechanism as the minimal viable candidate theory. Moreover it can be shown that the residual scalar field in the model must have a mass less than 10^3 GeV to preserve (perturbative) unitarity[76].

It seems then that there is no immediate need to go beyond the

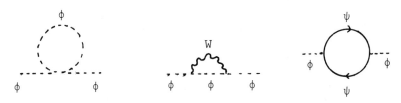

Fig. 8
Graphs contribution to scalar boson mass

standard model. Admittedly there are 18 parameters which one believes ultimately will be related as, for example, in a grand unified theory, but apparently there is no reason why SU(3) x SU(2) x U(1) should not be a good effective theory until a 10^{15} GeV. However there is a flaw in this picture - the flaw is known as the hierarchy problem.

Once we accept that ultimately the parameters in the standard model will be related we can treat the radiative corrections to them as meaningful and not just to be absorbed in a counter term. For example computing the graphs of Fig.(8) the Higgs scalar mass will obey an equation of the form

$$M^2(\mu) = M^2(\Lambda) + \sum c_i \alpha_i \int_{\mu^2}^{\Lambda^2} dk^2 + \ldots \qquad (6.7)$$

where $M^2(\mu)$ is the effective mass at low energies and $M^2(\Lambda)$ is the effective mass at the scale Λ at which "new physics" appears. There are other terms which may depend logarithmically on Λ. Barring an unnatural cancellation[†] this means $\sum c_i \alpha_i \Lambda^2 = O(M^2(\mu^2))$. In the standard model we need $M^2(\mu^2) < 1$ TeV^2 and the c_i are nonvanishing so $\Lambda^2 < \frac{1 \text{ TeV}^2}{\alpha}$ where α is a gauge coupling. This means that far from extending to 10^{15} GeV it seems likely the standard model will break down at a scale which will be probed by the next generation of experiments! What could this new physics be?

There are three obvious possibilities. One is that the perturbative analysis leading to eq.(6.7) fails due to an interaction becoming strong. If this happens only in the scalar sector the GUT predictions may still be valid for gauge and Yukawa couplings will not be greatly affected. In this case the departure from the standard model will be relatively difficult to see until the scalar sector is measured. For example $e^+e^- \to W_L^+ W_L^-$ will have strong interactions in the final state as longtitudinal W^S are generated by Higgs scalars, but clearly this is not a first generation experiment.

A second possibility is that there are no elementary scalars, as in technicolour models[79] or in other composite models. In this case the calculation leading to eq.(6.7) fails for $\Lambda <$ scalar binding energy. Provided this energy is $< \frac{1 \text{ TeV}}{\alpha}$ there will be no hierarchy problem.

Finally it may be that $(\sum c_i \alpha_i) = 0$. However it is not enough to arrange this only in leading order in perturbation theory for

---

[†] This would require that parameters on the microscopic scale Λ combine to give a massless scalar only at the macroscopic scale μ unnatural in the sense defined by 't Hooft[78].

then the next order will require $\Lambda < \frac{1 \text{ TeV}}{\alpha^2}$ and so on. If we are to arrange this cancellation involving both fermion and boson loops as in Fig.(8) we need a symmetry. The only known symmetry that can do this is supersymmetry[80]. I will discuss the phenomenological consequences of supersymmetry after a brief resumé of technicolour.

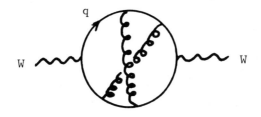

Fig. 9

Typical graph contributing to $\pi(k^2)$. The curly lines are gluons

(6.2) <u>Technicolour</u>

If we consider the standard SU(3) x SU(2) x U(1) model with no Higgs scalars we would naively think the W and Z bosons remain massless. This is not the case! Consider for example the inverse W boson propagator $P_{\mu\nu}(k)$. It has the form

$$P_{\mu\nu}(k) = \frac{\left(g_{\mu\nu} - \frac{k_\mu k_\nu}{k^2}\right)}{k^2 \left\{1 - g_W^2 \frac{\pi(k^2)}{2}\right\}} \qquad (6.8)$$

where the term involving $\pi(k^2)$ arises from radiative corrections of the type shown in Fig. (9).

Unless $\pi(k^2)$ is singular at $k^2 = 0$, $P_{\mu\nu}(k)$ will have a zero mass pole. However we know that, due to the strong interactions, $\pi(k^2)$ does have a zero mass pion pole. This results from the dynamical breaking of the chiral $SU(2)_L \times SU(2)_e$ symmetry and the pion in the resultant Goldstone boson. The contribution of this pole to $\pi(k^2)$ is

$$\pi(k^2) = \frac{f_\pi^2}{2} \frac{1}{k^2}$$

and

$$P_{\mu\nu}(k) = \frac{g_{\mu\nu} - \frac{k_\mu k_\nu}{k^2}}{k^2 - g_w^2 f_\pi^2/4} \quad (6.9)$$

with $f_\pi \simeq 100$ MeV this gives a mass of 30 MeV to M_W.

Of course this mass is far too small. However a reasonable theory may be built with realistic masses for the W bosons by postulating[79] a new strong gauge interaction called technicolour and a type of quark, the techniquark, carrying both flavour and technicolour quantum numbers. The techniquarks and technicolour gluons replace the quarks and gluons in Fig.(9) and, provided the technicolour interaction becomes strong at a scale ≈ 1 TeV the mass generated for the W boson due to dynamical breakdown of the techni-chiral symmetry is of the right magnitude as can be seen in the translation table (7). An extremely attractive feature of this idea is that as there are no elementary scalars with their associated couplings everything can, in principle, be determined in terms of the gauge couplings of the theory.

Technicolour theories are rich in new phenomena, which should occur at a scale < 1 TeV. The techniquarks will bind forming technihadrons with a mass $\gtrsim 1$ TeV. Moreover there are many pseudo-Goldstone bosons whose masses can be estimated in a specific model and some of which are $<< 1$ TeV. An example of the pseudo-Goldstone spectrum from an $SU(N)$ technicolour theory is, with $SU(2)$ doublets $\binom{U}{D}_{i=1...N}$ and $\binom{N}{E}_{i=1...N}$ transforming as quarks and

Table 7

Technicolour properties inferred from QCD

QCD	QT_cD
$\dfrac{g_c^2}{4\pi}$ (1 GeV) = 1	$\dfrac{g_{Tc}^2}{4\pi}$ (1 GeV) = 1
f_π = 100 MeV	$f_{\pi Tc}$ = 250 GeV
M_W = 30 MeV	M_W = 80 GeV
Hadrons ∿ 1 GeV	Technihadrons ∿ 1 TeV
Pseudo Goldstone Bosons π ∿ 100 MeV	Pseudo Goldstone Bosons 2 GeV Upwards

leptons respectively under $SU(3)_c$, given in Table (8)[81]. In the table we see that there are light charged scalars $P^\pm$ whose mass is 8-14 GeV and light neutral scalars P^0, P^3 whose mass is < 2.5 GeV.

Charged scalars with mass ≲ 13 GeV would already have been seen at PETRA[82] so this model is close to becoming disproved. There is another problem associated with technicolour theories which already rules out all models so far constructed. The problem is that the conventional quarks and leptons remain massless as the technicolour condensate does not break the chiral symmetries associated with the light fermions. (In the standard model these symmetries are broken by Yukawa couplings with elementary Higgs scalars). To give masses to the light fermions it is necessary to introduce another gauge interaction, "extended technicolour", which couples light quarks to techniquarks, and generates light fermion masses via the graph of Fig. (10).

Table 8

Pseudo-Goldstone Boson States

	Colour	Charge	Mass (GeV)
$P_8^0, P_8^3, P_8^\pm$	8_C	$0, 0, \pm 1$	245
$P_{\overline{EU}}, P_{\overline{ED}}, P_{\overline{NU}}, P_{\overline{ND}}$	3_C	$\frac{5}{3}, \frac{2}{3}, \frac{2}{3}, -\frac{1}{3}$	160
$P^\pm$	1_C	± 1	5 to 8
P^0, P^3	1_C	0	$0^{*)}$
$\Pi^{0,\pm}$ (eaten up by $Z^0, W^\pm$)	1_C	$0, \pm 1$	0

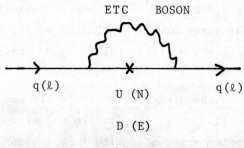

Fig. 10

Graph generating light fermion(q, ℓ) masses in extended technicolour

In order to generate the Cabibbo angle it is necessary for different generations of quarks to couple to the same generation of techniquark so the pattern of ETC interactions is as shown in Fig.(11).

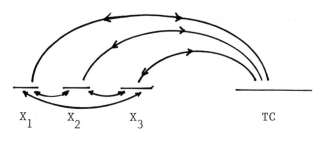

Fig. 11

Pattern of ETC couplings. X_i are the light fermion generations and TC are the technifermions

The SU(2) doublets $\begin{pmatrix} U \\ D \end{pmatrix}$ and $\begin{pmatrix} N \\ E \end{pmatrix}$ give masses to the quarks and leptons respectively via the graph of Fig.(10).

The (so far) insurmountable problem follows from the fact that the extended technicolour interaction couples quarks of different generations (cf. Fig.(11)) and generates large strangeness changing neutral currents. As usual the most dangerous is the $\Delta S = 2$ $\Delta Q = 0$ processes, the amplitude being inversely proportional to the mass squared of the ETC boson (m_{ETC}^2). However this mass is constrained by the fact that light quark and lepton masses are also inversely proportional to m_{ETC}^2. In order to get the required light quark mass m_{ETC} must be so small as to generate an unacceptably large $\Delta S = 2$, $\Delta Q = 0$ amplitude[79].

To date no acceptable way out of this problem has been proposed. Technicolour by itself seems an attractive idea but the extended technicolour models tried have serious problems. For the moment it seems sensible to keep an open mind and look for possible technicolour states and in particular pseudo-Goldstone bosons. They have characteristic differences from the usual Higgs bosons, being pseudoscalar and having different coupling patterns[80].

We will mention some of these tests in Section (6.4).

(6.3) Supersymmetry and supersymmetric zoology

As we discussed above a possible resolution of the hierarchy problem is to introduce a new symmetry which ensures scalar masses do not receive large perturbative corrections[85,86]. Supersymmetry achieves this by associating all scalars with fermion partners so that the scalar mass is equal to the fermion mass. Since fermion masses can be forbidden by chiral symmetries the associated scalar mass too may be forbidden by a chiral symmetry. As a result in a supersymmetric theory there are cancellations between fermion and boson graphs of fig. (8) and eq. (6.7) is replaced by

$$M^2(\mu^2) = M^2(\Lambda^2) \left(\frac{g^2(\Lambda^2)}{g^2(\mu^2)} \right)^{\gamma} \qquad (6.10)$$

so that $M^2(\mu^2)$ vanishes if $M^2(\Lambda^2)$ does. Thus there is no constraint on Λ^2, and the supersymmetric model may be valid up to the GUT scale. Supersymmetry breaking cannot occur above > 1 TeV, otherwise this evasion of the hierarchy problem will fail.

The simplest supersymmetric model which can be constructed is a direct product of the internal symmetry gauge group with a global (N = 1) supersymmetry[85,85,87]. The basic building blocks are massless supersymmetry multiplets[88] of the chiral or vector type as shown in Table (9).

Table 9
Fundamental massless supersymmetric multiplets in N = 1 supersymmetry

Chiral	$\begin{pmatrix} \psi \\ \phi \end{pmatrix}$	2 component majorana fermion 2 real scalar fields ($\equiv$ 1 complex)
Vector	$\begin{pmatrix} \nu_\mu \\ \psi \end{pmatrix}$	2 component massless vector 2 component majorana fermion

Table 10

Multiplet structure of the minimal
supersymmetric SU(3) x SU(2) x U(1) model

Vector Supermultiplets		Spin J
V_G	$G_\mu^{a=1\ldots 8}$ Gluons	1
	$\lambda_g^{a=1\ldots 8}$ Gluinos	$\frac{1}{2}$
V_W	$W_\mu^\pm, Z_\mu$ W,Z bosons	1
	$\lambda_{W^\pm,Z}$ Winos, Zinos	$\frac{1}{2}$
V_γ	A_μ Photon	1
	λ^γ Photino	$\frac{1}{2}$

Chiral supermultiplets		Spin J
$S_q, T_{\bar{q}}$	q_L, q_R Quarks	$\frac{1}{2}$
	ϕ_{q_L}, ϕ_{q_R} Scalar quarks	0
$S_\ell, S_{\bar\ell}$	ℓ_L, ℓ_R Leptons	$\frac{1}{2}$
	$\phi_{\ell_L}, \phi_{\ell_R}$ Scalar leptons	0
S, T	ψ_L', ψ_L'' Fermionic Higgs	$\frac{1}{2}$
	ϕ', ϕ'' Higgs doublets	0

Thus the supersymmetric model contains at least twice the number of particles needed in the nonsupersymmetric version. The minimal SU(3) x SU(2) x U(1) model can be made supersymmetric simply by assigning the usual states to supermultiplets of the type given in Table 9[85,86]. This leads to the multiplet structure of Table (10). In this table the new superpartners carry the same SU(3) x SU(2) x U(1) quantum numbers as their conventional partners. Note that there are two Higgs doublets (plus their supersymmetric partners) rather than the single one usually included in the standard model. This proves to be necessary if there is to be a reasonable mass spectrum[86].

Of course supersymmetry must be broken, for we would have seen the superpartners of the quarks and leptons if they were degenerate in mass with them. How supersymmetry breaks and what is the resulting mass spectrum is still a matter of theoretical uncertainty and discussion[89]. To avoid the hierarchy problem they must have masses < 1 TeV, but probably some will have masses considerably less. In addition there will be a massless goldstone fermion, the goldstino, the goldstone particle resulting from the breaking of global supersymmetry. It is evident therefore that if supersymmetric states exist they should be within the energy range of the next generation of experiments and we now turn to a discussion of the phenomenology to be expected for supersymmetric particles. Much of this is based on work by Fayet and Farrar[90] and for a recent review of this see last year's Erice lecture notes by Fayet[91].

The coupling of supersymmetric particles is simply related to that of their conventional partners. For example the photino coupling follows from the photon coupling to charged particles just by the replacement of the photon by the photino and one of the charged particles by its super partner - e.g. see Fig. (12)

Fig. 12

Photon and photino coupling to charged fermions

The other fermions in vector supermultiplets ('inos) have couplings similarly related to the gauge couplings of their partners. The scalar quarks and leptons have gauge couplings with the vector supermultiplets, as in Fig.(12), and also couplings to other chiral supermultiplets related to the usual Yukawa couplings necessary to give masses to the quarks and leptons.

The goldstino couples to the partners in a supermultiplet with a strength given by[88,91]

$$\frac{e}{g} = \pm \frac{\Delta m^2}{d^2} \qquad (6.11)$$

where Δm^2 is the mass difference of the superpartners and d is a dimensional quantity related to the scale of supersymmetry breaking (< 1 TeV), and the sign depends on the chirality of the fermion considered.

(6.4) <u>Supersymmetric phenomenology. The 'ino sector</u>

In most supersymmetry models there is a global "R" parity[88,91] which ensures the supersymmetric partners are not produced singly. The usual particles of the SU(3) x SU(2) x U(1) model have even R while their super partners have odd R. Thus superpartners will only be pair produced when starting with a conventional initial state.

For instance, in the decay of onium states a pair of 'inos may be produced via the graph of Fig.(13)

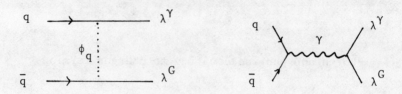

Fig. 13

Graphs contributing to 'ino production in onium decay

This would give for the goldstino/photino mode

$$r = \frac{\Gamma(\psi \to \text{goldstino} + \text{antiphotino}) + \Gamma(\psi \to \text{photino} + \text{antigoldstino})}{\Gamma(\psi \to e^+ e^-)}$$

$$= \left[\frac{m_\psi^2}{ed^2}\right]^2 \qquad (6.12)$$

Current limits on the rate $\Gamma(\psi \to \text{unobserved neutrals})$ imply[91]

$$d > 9.5 \text{ (GeV/c)} \qquad (6.13)$$

This is not yet a very stringent bound for we expect d < 1 TeV. However the rate eq.(6.12) is proportional to m_ψ^4 so, with a top quark mass 30 times the charm quark mass, toponium decay would be sensitive to the process, eq.(6.12) with d 0.3 TeV.

The gravitino remains massless in broken global supersymmetric theories. If supersymmetry is made local there is a supersymmetric analogue of the Higgs mechanism whereby the gravitino is eliminated while the corresponding gauge particle is the spin 3/2 gravitino, partner to the graviton. Supersymmetry contains the Poincaré group so local supersymmetry includes general relativity. The graviton remains massless but the gravitino acquires a mass

… where

$$m_{\frac{3}{2}} = \frac{kd}{\sqrt{6}} \qquad (6.14)$$

where

$$k = (8\pi G_{Newton})^{\frac{1}{2}} \simeq 4 \times 10^{-19} (GeV/c^2)^{-1}$$

Fayet has shown that[91] although the gravitino is eliminated by the Higgs mechanism, supplying the missing spin states for a spin 3/2 objects, in the limit $k \to 0$, the only states of the gravitino which would be produced and interact in a significant way would have $\pm \frac{1}{2}$ polarisations. The phenomenology would essentially be the same as for a massless goldstino. Thus tests for the gravitino may have profound implications for supergravity. For example eqs.(6.13) and (6.14) would give a lower limit on the gravitino mass

$$m_{\frac{3}{2}} > 1.5 \times 10^{-8} \, eV/c^2 \qquad (6.15)$$

Of the various 'inos of Table (10) the gluinos should be easiest to produce if their mass is low for they couple strongly. Gluinos may combine with quarks and gluons to give new colour singlet hadronic states, known as R-hadrons. These will decay into ordinary hadrons plus a photino or goldstino, with a lifetime typically somewhat shorter than nonleptonic weak decays. R hadrons will be pair produced in hadronic reactions and these R hadrons will subsequently decay. The characteristic signals for such events will be short tracks, reinteraction of the emitted photino or goldstino, or missing final state energy.

If the production cross section can be scaled from the charm production cross section

$$\frac{\sigma(hh \to \lambda^g + X)}{\sigma(hh \to c\bar{c} + X)} = O\left(\frac{m_c^2}{m_{\lambda_g}^2}\right) \qquad (6.16)$$

and if the reinteraction cross section, which depends on d and the scalar quark and lepton masses, is comparable to a weak interaction process then beam dump experiments$^{(92)}$ would give$^{(93)}$ $m_{\lambda_g} > 2$ GeV.

Another gluino signal, if gluinos are light, would be in deep inelastic scattering where gluinos would contribute to the nucleon "sea" and affect the scaling predictions and the sum rule measuring the fractional quark momentum$^{(93,94)}$.

Unfortunately it turns out that their effects are small and will be hard to measure. For example, the fractional quark momentum changes from 0.43 to 0.38$^{(95)}$ for four quark flavours.

One may devise many other good ways for producing gluinos, usually by looking for the best gluon production modes and changing the gluon to a gluino. For example $e^+e^- \to qqg$ is a good way to look for gluons equally $e^+e^- \to q\phi_q \lambda^g$ will be a good way to look for gluinos and scalar quarks once the threshold for producing them is passed. The latter process will have a characteristic angular and energy distribution$^{(93)}$. Another example$^{(93)}$ is onium $\to \lambda^g \bar{\lambda}^g g$ where again there should be characteristic signals in the gluino jets and should be sensitive to gluinos with $m_{\lambda}{}^g < m_q$ where m_q is the quark mass in the onium.

Finally we consider winos and zinos. These typically mix with fermionic partners of the Higgs scalars and acquire masses $\simeq M_{W,Z}$. They will decay to photinos or goldstinos plus quarks and/or leptons. If they are lighter than their partners the W and Z should decay into them and a goldstino (or photino for W^S). At LEP energies this should give a characteristic and clear signal. If they are lighter than an onium state then they will be produced via graphs analogous to Fig.(13) and should be visible in onium decays. As with gluinos there are many other possible production mechanisms analagous to those for the W^S and and Z^S.

(6.5) The Scalar Sector [96]

Supersymmetric theories have many new scalar states as compared to the standard model (cf. Table (10)). There are scalar quarks and leptons and in addition an extra Higgs multiplet which, after spontaneous symmetry breakdown, leaves two charged and three neutral scalar states. Their masses are not known except in particular models although they may be expected to have masses $\simeq$ the supersymmetry breaking scale due to the arguments which led to the hierarchy problem (cf. eq.(6.7)). Their coupling are however completely specified.

The best production mechanism for the charged scalars is in e^+e^- annihilation where charged scalars contribute at $\frac{1}{4}$ the rate of their fermion partners. Charged scalar leptons will decay to their lepton partners + a photino or goldstino giving the characteristic reaction

$$e^+e^- \to \text{Pair of scalar leptons}$$
$$\to \text{Noncoplanar pair } (e^+e^-, \mu^+\mu^-, \tau^+\tau^-) \qquad (6.17)$$
$$+ \text{ unobserved pair of photinos/goldstinos}$$

PETRA has already looked for such events and gives the limits[82]

$$m \text{ (scalar electron)} > 16 \text{ GeV}/c^2$$
$$m \text{ (scalar muon)} > 15 \text{ GeV}/c^2 \qquad (6.18)$$

Scalar quarks decay to the associated quark and photino or gluino. However they may also produce gluinos if energetically allowed, and these subsequently decay to hadrons plus a photino or goldstino, giving less missing energy than the direct photino or gluino emission.

Scalar quark production may be separated from spin $\frac{1}{2}$ production by its $\sin^2\theta$ angular distribution and different threshold dependence together with the characteristic difference of the scalar quark jet (large $p_\perp$, missing neutral energy).

At present the best limit on the scalar quarks mass comes from the beam dump experiments[92] in which they are pair produced in hh collisions and the photino or goldstino produced in their decay subsequently reinteracts. Model dependent estimates of this give[91,93]

$$m_{\phi_q} > 2 \text{ GeV}$$

There is also a limit coming from the non-observation of scalar onia. In this case the lowest state coupling to the virtual photon is $^1P_1$ and as a result the production rate, relative to the usual $^3S_1$ quarkonia state, is very small varying between (2-5)% for a scalar quark mass of 1 GeV to (.1-.3)% for a scalar quark mass of 17 GeV.[93,97] ADONE and SPEAR searches would not have seen such a term so the only limit comes from the very low energies which gives $m_{\phi_q} > 0.5$ GeV, a relatively model independent result.

We now turn to the Higgs system. As we mentioned above, compared to the standard model, supersymmetry requires two Higgs multiplets which leave, after spontaneous symmetry breaking, two charged and three neutral states. This is not unique to supersymmetric models for the standard model may easily be modified, including extra Higgs doublets without changing the usual weak current phenomenology. In these models one doublet gives up quarks a mass and the other is responsible for down quarks and lepton masses. As we discussed above other types of scalar states are expected in technicolour model and it is useful to devise tests to distinguish these from the usual Higgs states.

The charged scalar states will be best produced via a virtual photon and, both in technicolour and standard theories, these are expected to couple to heavy fermion pairs. As a result they should have been detected by JADE[80] if their mass is less than $\simeq$ 13 GeV. This is the expected mass range for the technicolour states (cf. Fig. 10) and suggests the simplest extended technicolour models are not correct.

If heavier charged (Higgs or technicolour) scalar states exist they will dominate the toponium final states because their coupling is expected to be strong to heavy quarks. Thus toponium may be an excellent factory for scalar states. Once produced $p^{\pm}$ will decay to the heaviest states available. For example in a technicolour model the couplings are found to be

$$\Gamma(p^{\pm} \to \tau\nu) : \Gamma(p^{\pm} \to c\bar{s}) : \Gamma(p^{\pm} \to c\bar{b}) \qquad (6.19)$$
$$\simeq 3 : |c_1 c_2 c_3 - s_2 s_3 e^{i\delta}|^2 \simeq 1 : 10|c_1 c_2 s_3 + s_2 c_3 e^{i\delta}|^2 < 1$$

while in the two Higgs doublet model such as found in the supersymmetric model discussed above

$$\Gamma(p^{\pm} \to \tau\nu) : \Gamma(p^{\pm} \to c\bar{s}) : \Gamma(p^{\pm} \to c\bar{b})$$
$$\simeq m_\tau^2 \quad : m_s^2 |c_1 c_2 c_3 - s_2 s_3 e^{i\delta}|^2 : m_s^2 |c_1 c_2 s_3 + s_2 s_3 e^{i\delta}|^2 \qquad (6.20)$$
$$\text{or} \simeq 0 \quad : m_c^2 |c_1 c_2 c_3 - s_2 s_3 e^{i\delta}|^2 : m_c^2 |c_1 c_2 s_3 + s_2 s_3 e^{i\delta}|^2$$

where the alternatives correspond to p principally coming from Higgs doublets responsible for down or up quark masses respectively. Thus the decay products of $p^{\pm}$, if found, should cast light on the type of scalar state.

The neutral Higgs and technicolour states are more difficult to see. Extensive reviews of Higgs boson phenomenology have been given in ref.{98}, here we concentrate on tests which may distinguish between Higgs and technicolour states. One possible signal is to look for the γ_5 coupling of the technicolour states[99] but in the two doublet model there is also a pseudo scalar Higgs boson which would have γ_5 coupling[100]. It has been shown that the technicolour states p^0 and p^3 of Table 8 should have significant flavour changing couplings either to up quarks or down quarks and this gives for these possibilities respectively

$$B(D \to p^{0,3} + X) = O(10^{-2}) \text{ or } O(10^{-10}) \; (m_{p^{0,3}} < 1.5 \text{ GeV})$$

$$B(B \to p^{0,3} + X) = O(10^{-3}) \text{ or } O(1) \; (m_{p^{0,3}} < 4 \text{ GeV}) \quad (6.21)$$

The p^3 branching ratio to $\mu^+\mu^-$ is sufficiently large for the CESR limit[39]

$$B(B \to \begin{smallmatrix} e^+e^- \\ \text{or} \\ \mu^+\mu^- \end{smallmatrix} + X) < 0.2\% \quad (6.22)$$

to already rule out the second possibility.

Unfortunately flavour changing couplings are not a unique signal for technicolour states. It has been shown by Wise[100] that radiative induced flavour changing decays may be significant for the two doublet Higgs model. He finds, for example,

$$\frac{\Gamma(b \to sH_2^0)}{\Gamma(b \to X)} = (4-100)\% \text{ for } m_{\phi_2} = 20 \text{ GeV and} \quad (6.23)$$
$$20 < m_5 < 50 \text{ GeV}.$$

One significant way of testing between Higgs and technicolour scalars is in their coupling to Z^0. The branching ratio to technicolour states

$$B(Z^0 \to p^{0,3} + \gamma) = O(10^{-7} \text{ to } 10^{-9}) \quad (6.24)$$

is much smaller than that to Higgs states

$$B(Z^0 \to H^0 + \gamma) = O(10^{-6}) \quad m_H \sim 45 \text{ GeV}[101] \quad (6.25)$$

$$B(Z^0 \to H^0 + e^+e^-) = O(10^{-4}) \quad m_H \sim 10 \text{ GeV}[102]$$

and correspondingly their electromagnetic production is smaller too[84].

$$\frac{\sigma(e^+e^- \to Z^0 + P^0)}{\sigma(e^+e^- \to Z^0 + H^0)} = O(10^{-4}) \qquad (6.26)$$

7. CONCLUSIONS

I hope that I have been able to convince you that theorists are far from the comfortable position of being able to predict with confidence what will happen in the next generation of experiments. The situation is confused and interesting. It seems something new must happen within the accessible energy range of the new experiments but, although theoretical ideas abound for what this may be, there is no overwhelmingly convincing model. Thus we rely, as usual, on experiment to show the way. It is clear what we can learn from the next generation of experiments is very exciting. Let us hope that what we do learn will be even more so.

REFERENCES

1. For a general review and further refs. see B.H. Wiik, Proceedings of the twenty-first Scottish Universities Summer School in Physics, ed. K. Bowler and D. Sutherland, SUSSP (1980).
2. R. Cashmore, these proceedings
3. For a recent review see F. Sciulli, XX International Conference, Madison, Wisconsin (1980), AIP Conference Proceedings No. 68. 12
4. M Jacob and R. Horgan, 1980 CERN/DESY School of Physics, CERN 81-04.
5. H. Fritzsch, M. Gell-Mann and H. Leutwyler, Phys. Lett. $\underline{47B}$ (1973) 365.
6. S.L. Glashow, Nucl. Phys. $\underline{22}$ (1961) 579;
   S. Weinberg, Phys. Rev. Lett. $\underline{19}$ (1967) 1264;
   A. Salam, Proc. 8th Nobel Symposium, ed. N. Svartholm, Almquist and Wicksells, Stockholm (1968), 367.
7. W. Caswell, Phys. Rev. Lett. $\underline{33}$ (1974) 244;
   D.R.T. Jones, Nucl. Phys. $\underline{B75}$ (1974) 531.
8. T. Appelquist and H. Georgi, Phys. Rev. $\underline{D8}$ (1973) 4000;
   A. Zee, Phys. Rev. $\underline{D8}$ (1973) 4038.
9. G. 't Hooft and M. Veltman, Nucl. Phys. $\underline{B44}$ (1972) 189;
   C.G. Bollini and J. Giambiagi, Phys. Lett. $\underline{40B}$ (1972) 566;
   W. Celmaster and R. Gonsalves, Phys. Rev. $\underline{D20}$ (1979) 1420;
   Phys. Rev. Lett. $\underline{42}$ (1980) 1435.
10. R.G. Moorhouse, M.R. Pennington and G.G. Ross, Nucl. Phys. $\underline{B124}$ (1977) 285.
11. See R. Cashmore, ref. 2 and A. Ali ref. 23 .
12. R.K. Ellis, H. Georgi, M. Machacek, H.D. Politzer and G.G. Ross, Nucl. Phys. $\underline{B152}$ (1979) 285. D. Amati, R. Petronzio and G. Veneziano, Nucl. Phys. $\underline{B140}$ (1978) 54, Nucl. Phys. $\underline{B146}$ (1978) 29. S. Gupta and A.H. Mueller, Phys. Rev. $\underline{D20}$ (1979) 118;
    J.C. Collins and D.E. Soper, Proceedings of the Moriond Workshop Les Arcs, France (1981).
13. G. Bodwin, S.J. Brodsky, and G.P. Lepage, SLAC-PUB-2787 (1981).

14. G.G. Ross, Proceedings of the twenty-first Scottish Universities Summer School in Physics, ed. by K. Bowler and D. Sutherland, SUSSP (1980); R.G. Roberts, ref. 18;; R.K. Ellis, Proceedings of the Coral Gables Conference (1981).
15. K.G. Chetyrkin, A.L. Kataev, F.V. Tkachov, Phys. Lett. $\underline{85B}$ (1979) 277; M. Dine and J. Sapirstein, Phys. Rev. Lett. $\underline{43}$ (1979) 668; W. Celmaster and R. Gonsalves, Phys. Rev. Lett. $\underline{44}$ (1979) 560; Phys. Rev. $\underline{D21}$ (1980) 3112
16. E.G. Floratos, D.A. Ross and C.T. Sachrajda, Nucl. Phys. $\underline{B129}$ (1977) 66 and Erratum, Nucl. Phys. $\underline{B139}$ (1978) 545; W.A. Bardeen, A.J. Buras, D.W. Duke and T. Muta, Phys. Rev. $\underline{D18}$ (1978) 3998.
17. M.R. Pennington and G.G. Ross, Phys. Lett. $\underline{102B}$ (1981) 167.
18. R.G. Roberts, Proceedings of the Sixteenth Rencontre de Moriond, Les Arcs, France (1981).
19. I. Gabathuler, these proceedings.
20. G. Altarelli, R.K. Ellis and G. Martinelli, Nucl. Phys. $\underline{B143}$, (1978) 521, Erratum Nucl. Phys. $\underline{B146}$ (1978) 544 and Nucl. Phys. $\underline{B157}$ (1979) 461.
21. W.A. Bardeen and A.J. Buras, Phys. Rev. $\underline{D20}$ (1979) 166; D.W. Duke and J.F. Owens, Phys. Rev. $\underline{D22}$ (1980) 2280.
22. J. Ellis, M.K. Gaillard and G.G. Ross, Nucl. Phys. $\underline{B111}$ (1976) 253; R.K. Ellis, D.A. Ross and A.E. Terrano, Phys. Rev. Lett. $\underline{45}$ (1980); Nucl. Phys. $\underline{B178}$ (1981) 421; A.M. Vermaseran, K. Gaemers and S. Oldham, CERN preprint TH-3002.
23. K. Fabricius, I. Schmitt, G. Schierholz and G. Kramer, Phys. Lett. $\underline{97B}$ (1980) 431. A. Ali, DESY preprint DESY 81-059.
24. R. Barbieri, G. Curci, E. d'Emilio and E. Remiddi, Nucl. Phys. $\underline{B154}$ (1979) 535.
25. R. Barbieri, M. Caffo, R. Gatto and E. Remiddi, Phys. Lett. 95B (1980) 93; G. Parisi and R. Petronzio, Phys. Lett. $\underline{94B}$ (1980) 51.
26. G. Lepage and P. Kackenzie, Cornell University preprint CLNS 81/498.

27. R.K. Ellis, M. Furman, I. Hinchliffe, Nucl. Phys. B173 (1980) 397
28. G.P. Lepage and S.J. Brodsky, Phys. Rev. D22 (1980) 2157.
29. R.D. Field, R. Gupta, S. Otto and L. Chang, University of Florida preprint, Gainesville (1980).
30. A. Donnachie and P.V. Landshoff, Z. Physik C2, 55 and 372 (1979).
31. G. Curci, W. Furmanski and R. Petronzio, Nucl. Phys. B175 (1980) 27; I. Antoniadis and C. Kounnas, Ecole Poly preprint A399.0580 (1980).
32. J.J. Sakurai, Proceedings of Hawaii Neutrino Conference (1981).
33. J.E. Kim, P. Langacker, M. Levine and H.H. Williams, Rev. Mod. Phys. 53 (1981) 211.
34. J.J. Sakurai, Proceedings of the Conference on the Unification of the Fundamental Particle Interactions, Erice (Italy) 1980 (Plenum Pub. Corp., N.Y.)
35. D. Horn and G.G. Ross, Phys. Lett. 67B (1977) 460.
    G. Altarelli, N. Cabibbo, L. Maiani and R. Petronzio, Phys. Lett. 67B (1977) 463.
36. W. Bacino et al., Phys. Rev. Lett. 42 (1979) 749.
37. A. Silverman, Bonn symposium on Lepton and Photon Interactions at High Energy (1981).
38. Y. Achiman and B. Stech, University of Heidelberg preprint (1981)
39. K. Berkelmann, these proceedings.
40. A.J. Buras, Fermilab preprint FERMILAB-PUB 81/022 THY (1981).
41. S.L. Glashow, Proceedings of the twenty-first Scottish Universities Summer School in Physics, ed. by K. Bowler and D. Sutherland SUSSP (1980).
42. S.L. Glashow, Phys. Rev. Lett. 45 (1980) 1914;
    H. Georgi and D.V. Nanopoulos, Nucl. Phys. B155 (1979) 52.
43. B. Pendleton and G.G. Ross, Phys. Lett. 98B (1981) 291, C.T. Hil Phys. Rev. D24 (1981) 691.
44. M. Veltman, Phys. Lett. 91B (1980) 95; M.S. Chanowitz, M.A. Furman and I. Hinchliffe, Nucl. Phys. B153 (1979) 402.
45. M. Kobayashi and T. Muskawa, Progr. Theoret. Phys. 49 (1973) 652
46. J. Ellis, M.K. Gaillard and D.V. Nanopoulos, Nucl. Phys. B109 (1976) 213; R.E. Shrock, S.B. Treiman and L. Wang, Phys. Rev. Lett. 42, (1979) 1589.

47. S.L. Glashow, J. Iliopoulos and L. Maiani, Phys. Rev. D2 (1970) 1285.
48. J. Ellis, Physica Scripta 23 (1980) 328.
49. H. Georgi and S. Weinberg, Phys. Rev. D17 (1978) 275.
    G.G. Ross and T. Wyler, J. Phys. G (1979) 733.
50. V.S. Berezinsky and A.Y. Smirnov, Phys. Lett. 94B (1980) 505.
    F. del Aguila and A. Mendez, Oxford preprint 13/81
51. A.H. de Groot, G.J. Gounaris and D. Schildknecht, Phys. Lett. 85B (1979) 399, Z. Phys. C5, (1980) 127, V. Barger, W.Y. Keung and E. Ma, Phys. Rev. Lett. 44 (1980) 1169, Phys. Rev. D22 (1980) 727, Phys. Lett. 94B (1980) 377.
52. T.G. Rizzo and G. Senjanovic, Phys. Rev. Lett. 46 (1981) 1315 and Brookhaven preprint (1981).
53. L. Abbott and E. Farhi, Phys. Lett. 101B (1981) 69 and CERN preprint TH.3057 (1981).
54. J.D. Bjorken, Phys. Rev. D19 (1979) 335.
55. N. Seiberg - private communication.
56. C.H. Llewellyn Smith, G.G. Ross and J.F. Wheater, Nucl. Phys. B177 (1981) 263.
57. C. Quigg, Rev. Mod. Phys. 49 (1977) 297; see also ref. 4.
58. K.J.F. Gaemers and G.J. Gounaris, Z. Phys. C1 (1979) 259.
59. A. Sirlin, Phys. Rev. D22 (1980) 971; A. Sirlin and W.J. Marciano, New York University reprint NYO/TR3/81.
60. A.D. Line, JETP Letters 23 (1976) 73.
61. S. Weinberg, Phys. Rev. Lett. 36 (1976) 294.
62. S. Coleman and E. Weinberg, Phys. Rev. D7 (1973) 1888.
63. K.T. Mahanthappa and M.A. Sher, University of Colorado preprint COLO-HEP 18 (1980).
64. S. Weinberg, Phys. Rev. Lett. 43 (1979) 1566;
    F. Wilczek and A. Zee, Phys. Rev. Letters 43 (1979) 1571;
    S. Weinberg, Harvard University preprint HUTP 80-A/023 (1980).
65. P. Langacker, Phys. Reports 72 (1981) 186.

66. A. De Rújula, H. Georgi and S.L. Glashow, Phys. Rev. Lett. 45 (1980) 413; J. Ellis, M.K. Gaillard and D.V. Nanopoulos, Phys. Lett. 88B (1980) 320.
67. H. Georgi, H. Quinn and S. Weinberg, Phys. Rev. Lett. 32 (1974) 451.
68. A.J. Buras, J. Ellis, M.K. Gaillard and D.V. Nanopoulos, Nucl. Phys. B135 (1978) 66.
69. C.H.Llewellyn Smith, G.G. Ross and J.F. Wheater, Nucl. Phys. B177 (1981) 263.
70. D. Yu Bardin, O.M. Fedorenko and N.M. Shumeiko, Yad. Fiz. 32 (1980) 782 and Dubna preprint E2-80-503.
71. C.H. Llewellyn Smith and J.F. Wheater, Phys. Lett. 105B (1981) 486.
72. For a comprehensive list of experimental references see the review by Kim et al. ref. 33 .
73. C.Y. Prescott et al., Phys. Lett. 84B (1979) 524.
74. M.K. Gaillard and B.W. Lee, Phys. Rev. D10 (1974) 897.
75. C.H. Llewellyn Smith, Phys. Lett. 46B (1973) 233.
76. B.W. Lee, C. Quigg and H.B. Thacker, Phys. Rev. Lett. 38 (1977) 883; Phys. Rev. D16 (1977) 1519; M. Veltman, Acta Phys. Pol. B8 (1977) 475; Phys. Lett. 70B (1977) 253.
77. E. Gildener, Phys. Rev. D14 (1976) 1667; Phys. Lett. 92B (1980) 111.
78. G. 't Hooft, Proceedings of Advanced Study Institute, Cargèse 1979, Eds. G. 't Hooft et al., (Plenum Press B.Y., 1980).
79. L. Susskind, Phys. Rev. D20 (1979) 2169; S. Weinberg, Phys. Rev. D19 (1979) 1277.
80. Yu. Gol'fand and E.P. Lekhtman, JETP Lett. 13 (1971) 323; P. Ramond, Phys. Rev. D3 (1971) 2415; A. Neveu and J. Schwartz, Nucl. Phys. B31 (1971) 86; J. Wess and B. Zumino, Nucl. Phys. B70 (1974) 39.

81. S. Dimopoulos, Nucl. Phys. B168 (1980) 69; M.E. Peskin, Nucl. Phys. B175 (1980) 197; S. Dimopoulos, S. Raby and G. Kane, Nucl. Phys. B182 (1981) 77; J. Ellis, D.V. Nanopoulos, and P. Sikivie, Phys. Lett. 101B (1981) 529; S. Chadha and M.E. Peskin, Nucl. Phys. B185 (1981) 61, and CERN preprint TH. 3098 (1981); P. Binetruy, S. Chadha and P. Sikivie, CERN preprint TH.3122((1981).

82. P. Duinker, these proceedings.

83. S. Dimopoulos and J. Ellis, Nucl. Phys. B182 (1981) 505.

84. J. Ellis, M.K. Gaillard, D.V. Nanopoulos and P. Sikivie, Nucl. Phys. B182 (1981) 529; A. Ali and M.A.B. Bég, Phys. Lett. 103B (1981) 376.
    A. Ali, DESY preprint, DESY 81/032.

85. S. Dimopoulos and S. Raby, ITP preprint (1981).
    S. Dimopoulos, S. Raby and F. Wilczek, ITP preprint 81-31 (1981).

86. L.E. Ibanez and G.G. Ross, Phys. Lett. 105B (1981) 439.

87. P. Fayet, Phys. Lett. 69B (1977) 489.

88. For a general review see P. Fayet and S. Ferrara, Phys. Reports 32C (1977) 250.

89. E. Witten, "Dynamical Breaking of Supersymmetry", Princeton preprint (1981). S. Dimopoulos and S. Raby, "Supercolor", Santa Barbara Preprint (1981). M. Dine, W. Fischler and M. Sredrucki, "Supersymmetric Technicolour" Princeton preprint (1981).

90. P. Fayet, Proceedings of XVIth. Rencontre de Moriond.
    G.R. Farrar and P. Fayet, Phys. Lett. 76B(1978) 575; 79B (1978) 442. G.R. Farrar, ref. 95 .
    P. Fayet, Phys. Lett. 78B (1978) 417; ibid 84B (1979) 421.

91. P. Fayet, Proceedings of the Conference on the Unification of the Fundamental Particle Interactions, Erice (Italy) 1980 (Plenum Pub. Corp., N.Y.) 587.

92. P. Alibran et al., Phys. Lett. 74B 134 (1978); T. Hansl et al., Phys. Lett. 74B, 139 (1978). P.C. Bosetti et al., Phys. Lett. 74B, (1978) 143.
93. G. Barbiellini et al., "Supersymmetric particles at LEP", DESY preprint 79/67 (1979).
94. Campbell, J. Ellis, S. Rudaz, to appear.
95. G.R. Farrar, Rutgers University prprint (1981) RU-81-07.
96. P. Fayet, Phys. Lett. 95B (1980) 285; 96B (1981) 83 and CERN preprint TH.2972 to be published in Nucl. Phys. B; see also ref. 90 .
97. H. Krasemann and S. Ono, DESY preprint 79/09 (1979).
98. J. Ellis, M.K. Gaillard and D.V. Nanopoulos, Nucl. Phys. B106, 292 (1976); J. Ellis, Proc. 1978 SLAC Summer Institute, Ed. M. Zipf (SLAC-215, 1978), p. 69; M.K. Gaillard, Comm. Nucl. Part. Phys. 8, (1978) 31; G. Barbiellini et al., DESY preprint 79/27 (1979).
99. M.A.B. Beg, H.D. Politzer, P. Ramond, Caltech preprint (1980).
100. M. Wise, Harvard University preprint (1981).
101. R.N. Cahn, M.S. Chanowitz and N. Fleishon, Phys. Lett. 82B (1979) 113.
102. J.D. Bjorken, Proceedings of the 1976 SLAC Summer Institute on Particle Physics, (SLAC-198, 1977).

NEXT GENERATION OF EXPERIMENTS

D I S C U S S I O N S

CHAIRMAN: G. G. ROSS

Scientific Secretary: R.D.C. Miller

DISCUSSION 1

- JOVANOVICH:

Earlier you alluded to a difficulty with the K_L-K_S mass difference, could you expand on this point ? Considering the fact that Δm (K_L-K_S) was used to correctly predict the c quark mass it seems to me that the mass difference is well explained without the t quark, so how can Buras use this process to estimate the t quark mass ?

- ROSS:

The difficulty alluded to was that encountered in a theory without the c quark. As in the rare process $K \to 2\mu$ phenomenological predictions were too large. The introduction of the c quark naturally leads to a suppression of these amplitudes due to the unitarity of the mixing matrix (the GIM mechanism) bringing theory into line with experiment. The K_L-K_S mass difference offered a method for determining the c quark mass giving a value in the vicinity of 2 GeV, this was subsequently confirmed experimentally. Buras reconsidered these rare processes. He does not strictly 'predict' the t quark mass but places a stringent bound upon it, $m_t \lesssim 30$ GeV. This clearly results from requiring consistency with the c quark contributions.

I should stress though that Buras employs a Bag model estimate of four fermion operators; if we use a valence quark estimate his bounds disappear.

- HERTEN:

You expressed doubts that the gluon has actually been firmly established experimentally. In your opinion what are the weak points in the experimental analysis ?

- ROSS:

What I actually mean here is that QCD has no real competitor. In such a situation it is difficult for one to be totally "objective" in testing QCD. Personally I would prefer to wait and see whether there is really no alternative to interpreting experimental evidence for the "3 gluon vertex" before being thoroughly convinced.

- GATHERAL:

You said that you expect experiments will eventually come to some agreement upon the numerical value of $\Lambda_{\overline{ms}}$; why?

- ROSS:

Provided that the renormalization scheme is specified Λ is a well-defined renormalization group invariant quantity. Clearly if QCD is the correct theory, experiments will have to converge to a single value for Λ in a given scheme. A recent review by Roberts gives a strong indication towards this trend.

- KARLINER:

(a) Could you tell us how many Centauro events have been seen so far and how similar are they in the total energy, multiplicity and final particle content. (b) Do we know what to do in order to see more such events and if we do is anything going to be done for that purpose ?

- ROSS:

Not being an expert in this area I cannot answer all your questions but I can tell you what I do know about Centauro events. Centauro events are very high energy processes with total energy content in the range of 10^3 TeV. They were, of course, discovered in cosmic ray detectors. In all, I believe, ten or so such events have been detected (but many more events with smaller energies ~10 TeV have been observed).

To discuss multiplicities consider the experimental set up and a typical event.

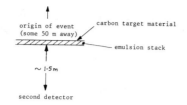

In the first detector multiplicities are roughly of the order of 6-7. In the lower detector this increases to values of the order of 10^3. The great predominance of charged debris (very few neutrals are observed) poses a difficult question. The evidence so far suggests that maybe new physical processes are occurring at these large energies, but our understanding of the events is quite limited.

- BERKELMAN:

I think I can help Karliner regarding the second part of his question. There is an experiment, called UA5, being set up at the CERN SPS $\bar{p}p$ collider, which will, in the next few months, search for 'Centauro-type' events with a streamer chamber in $\bar{p}p$ collisions at 540 GeV c.m. energy.

- MONDAL:

You have suggested that Centauro events may signal new physics in the 100-1000 TeV range. I would like to point out that there are rather successful alternative explanations such as viewing them as fluctuating air shower events. In this case we do not need to invent 'new physics' to explain them.

- GOLLIN:

How would one look for experimental evidence for the three gluon coupling required by QCD ?

- ROSS:

The Q^2 - evolution of the QCD running coupling constant depends strongly on the three gluon coupling. A measurement of this Q^2 dependence is thus one approach to verifying a gluon self-coupling.

More immediately we could measure the Q^2 - evolution of the gluon jet (or of the gluon distribution function) as this is dependent on the triple gluon coupling in first order. One may also hope to measure the triple gluon coupling in, for example, pp $\to$ gluon jet + X where the gluon jet is produced from the triple gluon coupling involving gluon constituents of the proton.

- DUNCAN:

If the Higgs particle is found how would one know whether it is the Higgs of spontaneous symmetry breaking (SSB) or possibly the pseudo-Goldstone boson of a theory with dynamical symmetry breaking, such as technicolour ?

- ROSS:

In the literature a number of distinguishing tests have been proposed. Let us recall that in technicolour the pseudo-Goldstones have an axial coupling while the Higgs of SSB have a scalar coupling; this obviously will lead to different physical effects. I might point out though that we may take the standard model with minimal Higgs scalars and introduce another Higgs doublet. In such a case we have one neutral Higgs with axial couplings in a theory with SSB. Logically then if we find a Higgs scalar with a scalar coupling this is bad news for technicolour and on the other hand if it should have a pseudoscalar coupling we could not immediately conclude this to be a success for technicolour. In the latter event it would be a difficult task to distinguish between the two theories. But, of course, there are other ways to make the distinction. For example, technicolour induces strangeness changing couplings. Again I can construct a model which has SSB and yet mimics this feature.

DISCUSSION 2   (Scientific Secretary: R.D.C. Miller)

- BANKS:

This morning we discussed alternatives to the standard model in which Higgs particles were not necessary. In particular you mentioned the Abbott-Farhi model. Apart from the fact that the charged and neutral weak vector bosons, the W's, of this model have the same mass what other experimental differences would you expect ?

- ROSS:

Because of the larger coupling constant of this model the W mass is no longer 80 GeV but somewhere nearer 120 GeV. Should the W not turn up at 80 GeV we will at that stage be at a sufficiently high

Q^2 to investigate the Q^2 - dependence of the quark structure function required in the composite model of Abbot and Farhi.

- LUCHA:

Models in which fermions are placed in higher than the fundamental representation of SU(2) have been constructed. You did not mention these. Have they already been ruled out experimentally ?

- ROSS

The success of neutral current phenomenology does suggest that the left-handed doublet structure is the right one for the light quarks. However, clearly experimental uncertainties could allow for small admixtures of higher dimensional representations but there is at present no reason to suspect this is so. Pure higher dimensional representations I would say are now ruled out.

- BERKELMAN:

Would you please make some remarks concerning models which connect mixing matrix angles with quark masses ?

- ROSS:

To get predictions for mixing angles we want to restrict the most general form of the mass matrix. For example suppose we restrict ourselves to a four quark theory, the mass term appearing in the Lagrangian will look like,

$$(\bar{u}\ \bar{c})_L\ \underline{\underline{M}}\ (u\ c)_R^t$$

where $\underline{\underline{M}}$ is a non-diagonal mass matrix (we have a similar expression for the s-d charge - $\frac{1}{3}$ pair) Wilczek, Zee and Fritzsch imposed discrete symmetries (embedded in $SU(2)_R \times SU(2)_L \times U(1)$) to reduce $\underline{\underline{M}}$ to a two parameter dependence. These parameters can be determined by the u-c masses giving a prediction for the Cabibbo angle θ_c.

$$\tan \theta_c \doteq \left| \sqrt{\frac{m_d}{m_s}} - \sqrt{\frac{m_u}{m_c}}\ e^{i\eta} \right|$$

which is surprisingly good. Cabibbo et al. have shown such predictions cannot be obtained from $SU_L(2) \times U(1)$ models with discrete symmetries. Recently B. Pendleton and I have shown that these results may follow approximately in an $SU_L(2) \times U(1)$ theory if one takes into account radiative corrections. Now the generalisation of this approach to the Kobayashi-Maskawa 6 quark model shows that off-diagonal elements of the mixing matrix $\underline{V}$ are typically $V_{ij} = \sqrt{m_i/m_j}$ (i < j). This in itself is intuitively what one would expect, i.e., a flavour prefers to couple to a nearby flavour.

- KARLINER:

You mentioned the intriguing possibility that the photon may acquire mass below 2.7°K. How could we test this ? (perhaps with liquid helium black-body radiation). If photons are massive at these low energies what would be the impact on our current theoretical framework ?

- ROSS:

I believe experiments are being designed along the lines you suggest but I am not familiar with the details. On the theoretical side of course we do know how to achieve such a phenomenon. We are all familiar with the idea that a symmetry may be broken in different phases (as in GUT's) and it is really this fact which aroused interest in the question of photon mass generation at low temperatures. As to the impact of such a discovery there is no doubt that it would have a great effect on our way of thinking. But at present this is speculation.

- WITTEN:

I do not quite see how the photon could acquire a mass at low temperatures. In order to generate mass for a gauge boson as a consequence of small changes in the parameters we require one of two scenarios: One of them is that there should be an extremely light charged scalar which under small changes in the parameters goes from having a small positive mass squared to a small negative mass squared. Now it is hard to know experimentally whether the photon mass is exactly zero but we do know with confidence that there is no charged scalar with mass < 1 eV; at least I hope we know that! If not it is something which should urgently be experimentally verified, even more so than experiments on photon mass.

Another way to generate mass for the gauge boson following small variations in parameters would be the presence of a first order phase transition. This too I feel is unpalatable if we consider what it means. One vacuum which is stable at 3°K becomes unstable as we approach 0°K. Although this is possible in principle, it implies the existence of two almost degenerate vacua. I am happy to say that the barrier to such a transition would probably be very large and transitions could not be achieved in the laboratory. I am happy to draw this conclusion since if we could induce such a transition in the laboratory the expansion of the new vacuum might bring us to the end of this world.

- YAMAGISHI:

Prof. Ross you mentioned fermion mass generation as a consequence of the existence of infrared fixed points in $SU(3)_c \times SU(2) \times U(1)$. As far as I know this theory does not have any infrared fixed points. Could you comment on this?

- ROSS:

Yukawa theory by itself is not asymptotically free. Non-Abelian gauge theory on the other hand does exhibit asymptotic freedom. In a theory with both couplings the beta function of the Yukawa coupling has two contributions. One part being infrared free and positive the other being asymptotically free and negative. It is clear that a non-zero fixed point is possible and by explicit calculation in leading order we may actually calculate it (something which Pendleton and I did over a year ago).

- CREUTZ:

Is the Abbott-Farhi model an example of the mechanism discussed in the classic paper of Fradkin and Shenker, where with a fundamental representation of Higgs they had shown confinement, and Higgs phases are smoothly connected. In particular I am confused by the residual SU(2) symmetry which seems to make their model distinguishable from the standard model.

- ROSS:

I am not familiar with the paper you have quoted and cannot therefore give an honest answer to your question.

- BANKS:

I would like to make a few remarks in this connection. I think that the residual SU(2) symmetry is the same as the one in the standard model, that is, the one which enforces the relation $m_W = m_Z \cos\theta_W$, in the presence of the Higgs mechanism. The only difference is that in this case there is no mixing between SU(2) and U(1) so the mass relation becomes $m_{W0} = m_{W^+}$. Other than this I believe this model is indeed related, in the Fradkin-Shenker sense, to the SU(2) part of the standard (S-W) model.

- COHEN:

Pursuing the discussion on the Abbott-Farhi model I would like to know how established calculations such as those of the magnetic moment of the muon or electron, are affected ?

- ROSS:

I think people had a wrong impression this morning. I presented the Abbott-Farhi model as an <u>example</u> of a composite model. Many of the consequences of this model have not been worked out and I do not wish to imply that it is a strong competitor of the standard model. I agree with what you say; such calculations ought to be done to ensure that the model does not contradict what we already know about radiative corrections.

DISCUSSION 3   (Scientific Secretary: R.D.C. Miller)

- HERTEN:

Earlier you displayed results for supersymmetry calculations of m_x, τ_p and $\sin\theta_w$. The corrections due to technicolour are very small. Yet there are problems with technicolour and we really require a new technicolour at high energies to explain the masses of techniquarks. Since no one knows how many techniquark generations there are between 10^2 and 10^{15} GeV, I would like to know how one can possibly make reliable predictions such as those I have mentioned ?

- BANKS:

May I make a comment here; techniquarks have a dynamical mass. We <u>don't</u> need anything else to give them a mass.

- KARLINER:

Can we learn anything new about the generation puzzle from supersymmetry ?

- ROSS:

The presentation of supersymmetry I gave casts no new light on this problem. The solution might be that all the generations themselves form a representation of a larger GUT group in which the (3, 2, 1) supersymmetric gauge group is embedded. In such a case supersymmetry would permit this to be a viable low energy theory.

- ETIM:

I would like to make a remark. A theorem will soon be published regarding supersymmetric models in which only three generations are allowed. I know of this theorem, it definitely exists. Of course, certain conditions must be satisfied but the proof is quite rigorous.

- ROSS:

You are referring to extended supersymmetry (SS). As I have just mentioned the model presented this morning was one with global SS. Imagine the following physics - energy plot.

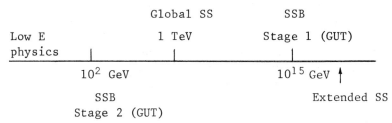

I place extended SS at this scale because their fundamental representations cannot accommodate particles such as the W's or the μ. If one is really interested in such theories I believe the way out is to view them as constituent theories. We then ask whether these theories may have a low energy spectrum, corresponding to what we see. There are models which do this and your comment refers to one of these. I agree that such models can lead to restrictions on the number of allowed generations.

- ZICHICHI:

Who are the authors of this theorem?

- ETIM:

S. Ferrara and C. Savoia.

- MIKHOV:

This morning you considered a model which was a direct product of supersymmetry and gauge symmetry. Your motivation I believe was the difficulty of fitting the known particles into supermultiplets. I feel the interesting case is really that in which one unifies the two. A consequence of this would be larger supermultiplets, but otherwise we would obtain more stringent constraints on the allowed processes. Do you know how one could look for observable evidence of such a unification?

- ROSS:

Well if you could predict reliably the low energy or bound state structure of such a theory that would undoubtedly be observable, and distinct from that of the cross product unification. You must remember that these theories are yet in their infancy and most of their consequences remain unexplored. I am not really aware of much more than Etim has already remarked upon.

- DUNCAN:

In the extended technicolour mechanism you Fierz transform your current-current interaction $\sim \bar{F}\gamma^\mu f \bar{f} \gamma_\mu F$ (f the ordinary fermions and F technifermions) and obtain four terms. The first term $\sim \bar{f}f\bar{F}F$ below the technifermion threshold corresponds to fermion mass generation since we assume that $\bar{F}F$ can develop a non-zero vacuum expectation value $<\bar{F}F>$. Similarly a second term $\sim \bar{f}\gamma^5 f \bar{F} \gamma^5 F$ generates the pseudo-Goldstone boson axial coupling. Why is it that the third and fourth terms, $\sim \bar{f}\gamma^\mu f \bar{F} \gamma_\mu F$ and $\bar{f}\gamma^\mu \gamma^5 f \bar{F} \gamma_\mu \gamma^5 F$ are never discussed.

- ROSS:

These last two terms are phenomenologically irrelevant at the moment, vector and axial vector terms not being able to develop non-zero vacuum expectation values. Furthermore there is no reason (based on the analogy with QCD) for there to be light vector bound states.

- FORGÁCS:

What is the most promising process for detecting the Higgs particle ?

- ROSS:

The Higgs couples mainly to the heaviest objects around. If toponium happens to be heavier than the Higgs then we should expect it to be readily produced in $t\bar{t}$ decay, as this rate is proportional to m_t^2.

Another method of production could be through the Z, again if the Z is heavier than the Higgs. It has also been suggested that they may readily be produced in strong interactions due to the two gluon coupling to the Higgs via virtual loops of heavy fermions.

- FORGÁCS:

In which facility might they be detected ?

- ROSS:

Well if the above processes are achievable in LEP then certainly we would hope to see Higgs production there.

- FORGÁCS:

Can a lower bound on the Higgs mass be established ?

- ROSS:

This is a problem. Since the Higgs coupling is proportional to the fermion mass then for the light fermion processes such as K decay it happens that the rate for light Higgs production is very small. As such we cannot yet rule out the existence of a light Higgs mass. Going to higher energies with heavy quark or Z production is thus a positive benefit in looking for the Higgs even if it is not very heavy.

- HIMEL:

You presented two theories (SU(5) and supersymmetry) which predict $\sin^2\theta_W$ = 0.23 and 0.21 respectively. You said that the forthcoming M_Z measurement will determine $\sin^2\theta_W$ to an accuracy of 0.001 thus allowing one to determine which theory is correct. Are theoretical estimates really that accurate ?

- ROSS:

The SU(5) prediction you mentioned is accurate up to this order as a result of a detailed theoretical calculation including $O(\alpha_{EM})$.

For the other theories I mentioned, only first order corrections have been made. I agree that more work must be done here to achieve the accuracy obtained in the SU(5) calculation. This is not really such a difficult task as a number of methods are available. The difficult task is computing the radiative corrections to experimental measurements. This has recently been done for eD and νN processes as I mentioned in my lecture.

- GOLLIN:

You mentioned that there might be manifestations of technicolour or supersymmetry at mass scales of the order of ∼ 1 TeV. What kind of things should we look for in p$\bar{\text{p}}$ colliders as definite signs of technicolour or supersymmetry ?

- ROSS:

The spectrum of supersymmetric particles hasn't been calculated as yet, so it's hard to give an answer for that case. The lightest technicolour particles, the $P^\pm$ and P^0 should be light enough to be produced. The $P^\pm$ should already have been seen if their mass $\lesssim 13$ GeV. P^0 may be produced in, e.g., pp collisions via a virtual fermion loop. At $\sqrt{s} = 540$ GeV, their production cross section is about 3×10^{-38} cm^2. $\Delta S \neq 0$ F^0 decays might provide some kind of signature but a Higgs sector with two or more doublets could mimic this.

- ZICHICHI:

Why is this cross section so small?

- ROSS:

This is due to the fact that the production rate is proportional to the inverse technicolour scale,
$$F_{\pi_{TC}}^{-1} \sim \Lambda_{TC}^{-1}$$

- KAPTANOĞLU:

This is a comment rather than a question. Certainly there is no shortage of difficulties in technicolour theories, but I would disagree with you that the flavour changing neutral currents is the most problematic. This problem may be solved as any other in technicolour theories, with the risk perhaps of introducing a new problem. The problem as I see it is one of obtaining a model in which we can eliminate all difficulties simultaneously. As an example I can offer a method which overcomes the strangeness changing neutral current problem alluded to. All we need to do is break the extended technicolour group in stages. The first stage gives masses to the gauge bosons mediating neutral current, flavour changing, processes amongst the light quarks. This mass scale is around 100 TeV or higher. The second stage occurring around 20 - 30 TeV gives masses to the gauge bosons mediating quark-techniquark transitions, thus providing the current quark masses. I do not suggest this as a serious alternative as I know it has additional problems but maybe it illustrates my point that problems with technicolour are more deeply rooted and non-specific.

- ROSS:

Your comments remind me of a remark I should have made earlier. One advantage of supersymmetry in the strangeness changing decay sector is that, provided I generate my quark masses using a single Higgs multiplet for up and for down quarks, then it is exactly the same situation as in the standard model, i.e., no strangeness changing neutral currents in first order. Yes, I too would be interested to know of any viable alternative technicolour theory.

- ETIM:

I find it hard to accept fermion mass generation from a single Higgs multiplet in supersymmetry. Is this really possible ?

- ROSS:

The Higgs multiplet structure we discussed does automatically have this property. The first doublet gives masses to the top quarks and the second to the bottom quarks.

- ETIM:

But then you did not include the gravitino ? You did not put in spin $3/2$ objects ?

- ROSS:

You are talking about extended supersymmetry, here I am discussing only global supersymmetry.

- ZICHICHI:

May I make a remark here. The most impressive prediction of supersymmetry is the prediction of baryons with bosonic characteristics and mesons with fermionic characteristics. Surely these supersymmetric objects would have profound cosmological implications for the structure of the early universe. The baryons would introduce a lot of lumpiness in nuclear matter because of allowed condensations.

- WILCZEK:

I do not think they would necessarily have such a significant effect. We could examine it in detail with a particular supersymmetric model if there were a viable one.

SUPERSYMMETRIC UNIFIED MODELS*

S. Dimopoulos[†] and Frank Wilczek

Institute for Theoretical Physics
Santa Barbara, California, 93106

ABSTRACT

Supersymmetric unified models are proposed, in which one obtains a light Higgs doublet as required for $SU(2) \times U(1)$ breaking whose colored partner automatically remain superheavy. We regard this as a long step toward solving the gauge hierarchy problem. A minimal model uses the group $SU(7)$ broken in an unconventional way. Models based on $SO(10)$ and $E(6)$ are constructed.

INTRODUCTION

A notorious feature of existing unified models of strong, electromagnetic, and weak interactions is the gauge hierarchy problem.[1] It shows up as follows, in the standard $SU(5)$ model:[2,3] the renormalization group analysis of coupling constants requires that the full $SU(5)$ symmmetry breaks at a scale $\sim 10^{15}$ GeV, whereas the weak-electromagnetic subgroup $SU(2) \times U(1)$ is known to break at $\sim 10^2$ GeV. To implement the second breaking, and generate quark and lepton masses, one requires an $SU(2)$ Higgs doublet which in the full $SU(5)$ becomes a vector ($\underline{5}$) multiplet. Thus together with the desired Higgs doublet we find an extra triplet of colored Higgs particles. These particles mediate baryon-number violating processes, and on phenomenological grounds must be very heavy, $\gtrsim 10^{15}$ GeV. The necessity of breaking a single $SU(5)$ multiplet

---
*Talk given by one of us (S.D.) in the Erice Summer School, Aug. 81.
†Address after June 1, 1981 - Department of Physics, University of Michigan, Ann Arbor, MI 48109

into two pieces with vastly different mass parameters is very
awkward from a group-theoretic point of view, and requires
delicate cancellations among a priori unrelated coupling
parameters. This is one aspect of the problem; another is that
typically mass terms for scalars contain ultraviolet divergences
and the required cancellations must take place among renormalized
coupling constants, which is contrary to the hope that the laws of
physics take their simplest form at short distances (involving
bare couplings).

It may be possible to address these problems in a stepwise
fashion, as follows:

i) Insure in a natural way that the Higgs doublet is
massless, while its colored partners obtain mass $\sim 10^{15}$
GeV, at the unification scale.

ii) Protect this pattern from large radiative corrections in
moving down to the scale $\sim 10^2$ GeV.

iii) Generate as small ($\sim 10^4$ GeV^2) but nonvanishing effective
(mass)^2 for the Higgs doublet as required in SU(2) × U(1)
phenomenology.

Step (ii) is easy to achieve in supersymmetric theories[4,5] - one
can have chiral symmetries which insure that masslessness of
scalars is preserved by radiative corrections. In this paper we
shall show how step (i) may be achieved in certain models.
Supersymmetry plays a crucial role in our proposals for this step,
also. Step (iii) involves breaking the supersymmetry. This can
be done explicitly (softly), which is formally acceptable but
seems ad hoc and has little predictive power, or perhaps
spontaneously.

UNCONVENTIONAL UNIFICATION IN SU(7):

A simple, though unrealistic, example of our idea is as
follows. Consider a supersymmetry SU(5) gauge theory with scalar
multiplets ϕ^i, ϕ^{ij} transforming as the vector $\underline{5}$ and antisymmetric
$\underline{10}$ representations. The superpotential can contain, among other
terms, the following

$$V_I = \varepsilon_{ijk\ell m} \phi^{ij} \phi^{k\ell} \phi^m \tag{1}$$

which will lead in the ordinary potential to a term

$$V_I = |\varepsilon_{ijk\ell m} \phi^{k\ell} \phi^m|^2 + \text{(irrelevant)} \tag{2}$$

Now if $\phi^{k\ell}$ acquires a vacuum expectation value

$\langle \phi^{k\ell} \rangle \propto \delta^{k1}\delta^{\ell 2} - \delta^{k2}\delta^{\ell 1} = \varepsilon^{k\ell 345}$ (3)

i.e. one component non-vanishing, as is not difficult to arrange - then (2) will generate a mass term of the kind we need in the hierarchy problem. That is, the triplet ϕ^3, ϕ^4, ϕ^5 acquires mass while the doublet ϕ^1, ϕ^2 remains massless.

The SU(5) model is not realistic because the vacuum expectation value (3) breaks SU(5) → SU(3) × SU(2), with no room for the physical hypercharge U(1). It is easily extended, however, to an interesting model based on SU(7).[6] We now require two 3-index antisymmetric tensors ϕ^{ijk}, ϕ'^{ijk} in addition to the vector ϕ^i. The superpotential $v_I = \varepsilon_{ijk\ell mnp}\phi^{ijk}\phi'^{\ell mn}\phi^p$ gives in the ordinary potential a term $|\varepsilon_{ijk\ell mnp}\phi^{ijk}\phi^p|^2$, and if $\langle \phi^{ijk} \rangle \propto \varepsilon^{ijk4567}$ we find a split multiplet. The symmetry breaking is SU(7) → SU(4) × SU(3), the the $\underline{7}$ multiplet breaks up into a heavy (4,1) and a massless (1,3).

This structure may be the skeleton of a phenomenologically acceptable model (with a fully natural gauge hierarchy). In the breakdown SU(7) → SU(4) × SU(3), color SU(3) sits inside the SU(4) while weak SU(2) sits inside the SU(3). The physical hypercharge matrix will be

$$Y = \begin{pmatrix} 1/3 & & & & & & \\ & 1/3 & & & & 0 & \\ & & 1/3 & & & & \\ & & & -1 & & & \\ \hline & & & & 0 & & \\ & 0 & & & & -1 & \\ & & & & & & +1 \end{pmatrix} \quad (4)$$

SU(4) × SU(3) can be broken down to SU(3) × SU(2) × U(1) by a rank-2 (or 5) antisymmetric tensor with $\langle \phi^{ij} \rangle \propto \varepsilon^{ij12356}$ and finally into SU(3) × U(1) with the light triplet part of the vector Higgs field. If matter fermions are assigned to antisymmetric tensor representations $\underline{7}$, $\underline{35}$, $\underline{21}$ with 1, 3 and 5 indices then triangle anomalies cancel. After symmetry breaking the fermions consist of two ordinary families plus two families with unusual electric charges shifted from normal by ±1 unit.

Remarkably, if the breakdown SU(4) × SU(3) → SU(3) × U(1) occurs at low energies, there is a simple prediction for the Weinberg angle.

$$\frac{1}{\tan^2 \theta_W} = 3 + \frac{8}{3} \left(\frac{g_W}{g_S}\right)^2 \tag{5}$$

which is not incompatible with experiment. Of course, to implement this we must keep the multiplet ϕ^{ij} light. As long as ϕ^{ij} does not couple to ordinary quarks in the superpotential this will give no troublesome B-violation. Notice that supersymmetry makes this condition natural, since the superpotential is unrenormalized.

CONVENTIONAL UNIFICATION, ADJOINT REPRESENTATIONS

More conventional unified models make use of Higgs multiplets in the adjoint representation by symmetry breaking. This leads to interesting features as we now analyze.

Let's first determine what answer we want. The adjoint ϕ^i_j will couple to the two vectors ϕ^i, ϕ_i through the superpotential

$$V_I \propto \phi_i \, \phi^i_j \, \phi^j \tag{6}$$

(Two vector Higgs fields, of opposite chirality, are necessary to give masses to quarks in a supersymmetric theory). The ordinary potential then has terms like

$$V_I \propto |\phi_i \, \phi^i_j|^2 + |\phi^i \, \phi^j_i|^2 + \text{irrelevant} \tag{7}$$

Now if we can arrange that the vacuum expectation value of the adjoint representation after diagonalization for SU(5), or reduction to canonical form for SO(10), vanishes in the weak SU(2) directions then the mass terms induced by (5) will split the vector representation and generate the desired hierarchy.

Can the vacuum expectation values be so arranged? For SU(5) it seems very difficult. We want supersymmetry to be broken only at low energies, so the breakdown of SU(5) at the unification scale must lead to a vanishing value for the auxiliary F^i_j-field associated with the adjoint. The F^i_j-field is formed as the derivative of the superpotential with respect to ϕ^i_j, so this constraint will take the form in matrix notation:

$$0 = F = a(A^2 - \frac{1}{5} \operatorname{tr} A^2) + bA + (\text{driving terms}) \tag{8}$$

where a,b parameters of the superpotential and the driving terms come from couplings of A to other fields in the superpotential. We want tr $A^2 \neq 0$, so the SU(5) breaks, but also that (with A diagonal) zero entries for A occur. Clearly this is impossible for SU(5) - we are done in by the trace condition. In extended unitary groups zero entries could only occur by virtue of delicate cancellations between the driving terms and the trace[7] - precisely the kind of feature we have gone to such lengths to avoid. Altogether, the unitary groups with adjoint breaking do not look very promising in this regard.

Remarkably, the orthogonal groups yield equations with a very different structure. Indeed, for them the canonical form of the adjoint representation is

$$\phi_{ij} = \begin{pmatrix} \lambda_1 \eta & & & \\ & \lambda_2 \eta & & \\ & & \ddots & \\ & & & \end{pmatrix} \quad \eta = \begin{pmatrix} 0 & 1 \\ -1 & 0 \end{pmatrix} \quad (9)$$

The fact that $\langle\phi\rangle$ is antisymmetric, while $\langle\phi^2\rangle$ is symmetric, means that there are no $\langle\phi^2\rangle$ terms in the F equation; while the trace automatically vanishes. Thus the F equation reduces to

$$0 = \langle F \rangle = \langle\phi\rangle + \text{(driving term)} \quad (10)$$

We will get the desirable form for $\langle\phi\rangle$, which generates a hierarchy, if the driving term D.T. takes the form (for SO(10))

$$\text{D.T.} = \lambda \begin{pmatrix} \eta & & & & \\ & \eta & & & \\ & & \eta & & \\ & & & 0 & \\ & & & & 0 \end{pmatrix} \quad (11)$$

A first attempt at such a driving term involves adding a complex rank-three antisymmetric tensor ϕ^{ijk} with a term

$$v_I = \phi_{ij} \phi_{imn} \tilde{\phi}_{jmn} \quad (12)$$

(our notation is $\phi_{ijk} = \phi^1_{ijk} + \phi^2_{ijk}$, $\tilde{\phi}_{ijk} = \phi^1_{ijk} - i\phi^2_{ijk}$, where ϕ^1, ϕ^2 are chiral superfields) in the superpotential. This gives a contribution to the F equation of the form

$$D.T. = \frac{\delta\sigma_I}{\delta\phi_{ij}} \propto \phi_{imn}\tilde{\phi}_{jmn} - (i \leftrightarrow j) \tag{13}$$

Now if the vacuum expectation value of ϕ_{ijk} has non-vanishing entries proportional to the complex ε-symbol $\varepsilon_{i+i2,3+i4,5+i6}$ ($\varepsilon_{135} \equiv 1$, $\varepsilon_{235} \equiv i$, $\varepsilon_{146} = -1$, etc.) then the driving term will be of the desired form. This is not quite good enough however - the complex ε-symbol also breaks ordinary hypercharge. We can correct this by taking a 6-index tensor totally antisymmetric in its first three and in its last three entires, with appropriate vacuum expectation values (schematically, $\varepsilon_{i+i2,3+i4,5+i6} \times \varepsilon_{1-i2,3-i4,5-i6}$).

The structure of the F-equation for the adjoint representation of E_6 is similr to that for $SO(10)$. We are presently investigating the possible driving terms.

E_6 has a special feature that may be very significant for the possibility of spontaneous supersymmetry breaking. This is, E_6 contains a $U(1)$ subgroup commuting with $SO(10)$ under which all ordinary fermions have the same charge (and the $SU(2) \times U(1)$ Higgs field has the opposite charge). In E_6 the ordinary fermions and the Higgs field are assigned to a 27 which decomposes $27 \to \underline{16}(+1) + \underline{10}(-2) + \underline{1}(+4)$ under $E_6 \to SO(10)_{U(1)}$ and the extra $U(1)$ charge distinguishes the different representations. Now in spontaneously broken supersymmetry we have mass relations between fermions and bosons in the same supermultiplet of the type

$$m_F^2 - m_S^2 \propto q_1 \langle D_1\rangle + \text{other terms} \tag{14}$$

where q_1 is the $U(1)$ charge of the multiplet and D_1 the expectation value of the auxiliary D_1 field for the $U(1)$. The extra $U(1)$ in E_6 thus has the potential to lift the scalar $SO(10)$ 16 partners of light fermions and the fermionic 10 partners of the Higgs field (with a negative bare mass squared) to a high mass characteristic of the supersymmetry breaking. This enables us to avoid the theorem of Dimopoulos and Georgi[5] who show that in most theories, without a suitable $U(1)$ symmetry, these mass relations are disastrous.

COMMENTS

1. To summarize, we have shown that in several supersymmetric models it is possible to arrange in a natural way for incomplete Higgs multiplets, thus easing the hierarchy problem. Possibly the simplest example is an unconventional $SU(7)$

model, which has striking consequences for low energy ($\lesssim$ few TeV) physics. That the single requirement of a hierarchy might allow us to determine the underlying theory in such detail (like a paleontologist constructing a dinosaur from a single tooth) is striking, and certainly models of the kind deserve much further study. Models based on orthogonal groups seem possible but more contrived, while E_6 is an intriguing possibility we have not fully investigated.

2. The role of supersymmetry in our proposal for the hierarchy problem is absolutely essential. In ordinary theories one will always have terms in the effective potential of the form $|\phi_1|^2 |\phi_2|^2$ (products of quadratic operators), and without control on these any large vacuum expectation value for $|\phi_2|$ will induce a large scale for $|\phi_1|$ unless delicate adjustments are made. In supersymmetric theories, the effective potential is derived from a superpotential of very restricted form, and these dangerous terms do not arise.

3. Another possible advantage of the SU(7) model regards magnetic monopoles. In standard unified models, estimates of superheavy monopole production in the big bang give too large a value by many orders of magnitude[8]. In SU(7), electromagnetic U(1) is embedded in the compact group SU(4) × SU(3) at relatively low energies, the monopoles are light and both cosmological estimates for their production and the observational limits are altered. If the production is thermal, too many monopoles are produced[9]. However, monopoles are extended coherent objects difficult to produce in collisions, so a better estimate might be roughly one monopole per horizon volume at the temperature where SU(4) × SU(3) → SU(3) × U(1). If this is correct, the member of monopoles produced can be very small ($\sim 10^{-36}$ per baryon) and not inconsistnet with observation.

ACKNOWLEDGEMENTS

We thank S. Barr, M. Einhorn and H. Georgi for useful conversations.

REFERENCES

1. E. Gildener, see also reference 3.
2. H. Georgi, S. Glashow, Phys. Rev. Lett. 32, 438 (1974).
3. A. Buras, J. Ellis, M. Gaillard, D. Nanopoulos, Nucl. Phys. B135, 66 (1978).
4. For a review of supersymmetry see Fayet, Ferrara, Physics Reports 32C.
5. For recent attempts to use supersymmetry in unified models:

E. Witten (Princeton preprint, 1981)
S. Dimopoulos, S. Raby (ITP preprint, 1981)
M. Dine, W. Fischler, M. Srednicki (Princeton preprint, 1981)
S. Dimopoulos, H. Georgi (Harvard preprint, 1981)
6. We are grateful for H. Georgi for informing us of this sort of model, which has been first proposed by J. Kim, Phys. Rev. Letters $\underline{45}$, 1916 (1980).
7. Or through a=0.
8. J. Preskill, Phys. Rev. Lett. $\underline{43}$, 1365 (1979).
9. M. Kholpov, Ya. Zeldovich, Phys. Lett. $\underline{B79}$, 362 (1979).

DISCUSSION

CHAIRMAN: S. DIMOPOULOS

Scientific Secretary: M.J. Duncan

DISCUSSION

- **ROSS:**

Could you clarify your statement that proton decay is rather insensitive to the number of incomplete SU(5) multiplets?

- **DIMOPOULOS:**

The value I quoted of $\sim 10^{16}$ GeV for the unification scale assumes only two doublets of Higgs mesons. If you had four doublets the unification scale moves down. The prediction was made for the minimal model.

- **ANTONELLI:**

Is it true that the hierarchy problem arises when one takes into account radiative corrections to scalar particle masses, because then there is no symmetry to prevent dangerous mixings between heavy and light scalars?

- **DIMOPOULOS:**

Consider any unified model with light and heavy Higgs, S and B respectively. At the tree level there are no symmetries preventing a coupling of the type $S^*S\ B^*B$ so that the effective potential has to take this coupling into account. The general term contributing to the mass of S at the tree level is $S^*S\ (\mu^2 - B^*B)$. When the heavy B field acquires an expectation value $<B^*B> \sim 10^{34}$ GeV^2 we find we have to adjust μ^2 to give S a mass of order 300 GeV. Thus μ^2 has to be adjusted to around 30 decimal places to cancel the large contribution!

Note that this has occurred at the tree level so one need not consider radiative corrections to see the problem.

- *SEIBERG:*

Do you think that introducing this zoo of new particles in order to make a small number natural is a simplification?

- *DIMOPOULOS:*

The only theories that solve the gauge hierarchy problem have a zoo of new particles around 1 TeV. These particles ought to be seen in the next decade. If they are not seen then the theories are disproved.

- *MONDAL:*

In your theory the unification mass is increased to 10^{17} GeV. How does this affect the baryon to photon ratio in the universe?

- *DIMOPOULOS:*

Prof. Wilczek has already discussed the "drift and decay" mechanism for baryon number generation in his lectures. If the mass of the X boson is larger, then it is easier for them to get out of thermal equilibrium than in the standard model. Thus the baryon to photon ratio becomes larger than in the standard model.

- *DUERKSEN:*

In your superunified SU(7) model you quoted a bare value of $\tan^2\theta_w$ which increases under renormalization to the correct value. In the standard Weinberg-Salam model the value falls under renormalization. Why is the behaviour different in the case of SU(7)?

- *DIMOPOULOS:*

$\tan \theta_w$ is defined by the ratio of hypercharge to weak coupling constants:

$$\tan^2\theta_w = g_y^2/g_w^2.$$

In the SU(7) model the weak group is contained in the SU(3) subgroup, hence $g_w = g_3$. Furthermore hypercharge can come from both the SU(3) and SU(4) subgroups. Intuitively then assigning weights p_3 and p_4 to the contributions to the hypercharge of the SU(3) and

SUPERSYMMETRIC UNIFIED MODELS

and SU(4) subgroups respectively, that is:

$$g_y^2 = p_3 g_3^2 + p_4 g_4^2$$

one finds
$$\tan^2\theta_w = p_3 + p_4 g_4^2/g_3^2$$

As you come down in energy g_4 increases faster than g_3 so that $\tan^2\theta_w$ increases. In the standard model the ratio is:

$$\tan^2\theta_w = g_1^2/g_2^2$$

and since g_1 decreases and g_2 increases then $\tan^2\theta_w$ decreases with energy.

- KARLINER:

Can one construct supersymmetric theories with composite quarks and leptons ?

- DIMOPOULOS:

To have composite fermions one would have to have some extra-strong group which binds the constituents. You would have then to include this new group in your theory. What such theories would be like with supersymmetry is not clear at present.

- DUNCAN:

SU(5) with three families provides a good standard model of grand unification. However in SU(5) - supersymmetric theories with three families, the SU(2) subgroup has a vanishing beta function. Is this a coincidence or could there be something fundamental in this result ?

- DIMOPOULOS:

The result I quoted was for a pure SU(5) theory with three supersymmetric families plus gauge bosons and fermions. When one introduces supersymmetric Higgs multiplets the result does not hold, so it appears to be a coincidence.

- D'HOKER:

In considering the breakdown of E(6) to SO(10) x U(1) you

decomposed a $\underline{27}$ of $E(6)$ into $(16,1)_{family} + (10,-2)_{Higgs} + (1,4)$.
Could you explain where the $U(1)$ assignments come from as they are
significant in considering mass splittings of bosons and fermions.

- *DIMOPOULOS*:

Suppose the $U(1)$ assignments for the $\underline{16}$, $\underline{10}$ and $\underline{1}$ of $SO(10)$
were x,y,z, and then we look at two $U(1)$ properties, namely,
$Tr\ U(1) = 0$ and the vanishing of the triangle anomaly for three $U(1)$
currents, these give the following two algebraic equations

$$16\ x\ +\ 10\ y\ +\ 1\ z\ =\ 0$$
$$16\ x^3\ +\ 10\ y^3\ +\ 1\ z^3\ =\ 0$$

It is easy to see that multiples of $(x,y,z) = (1,-2,4)$ are solutions
of this system of equations. Then we fix the normalization by the
normalization of $E(6)$.

- *D'HOKER*:

What are the problems with technicolour and extended technicolour ?

- *DIMOPOULOS*:

The short answer is that technicolour theories have too large
a symmetry group. The number of left and right handed fermion
fields are respectively conserved so fermion masses are not allowed.
In extended technicolour we introduce gauge bosons which change
fermions into technifermions. The technifermions condense and so
we can obtain fermion mass terms through the process

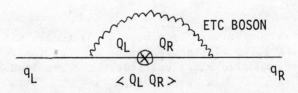

However, interactions which involve these extended technicolour
gauge bosons influence flavour changing neutral currents at a
level which is unacceptable.

- CLARE:

Is there any basic understanding of why there are three families?

- DIMOPOULOS:

No.

- SPIEGELGLAS:

You said there were problems in extended technicolour models with flavour changing neutral currents. Do these problems occur in other theories?

- DIMOPOULOS:

It is a characteristic problem in extended technicolour theories that the $K_L - K_S$ mass difference is wrong. In supersymmetric models we do not have this problem in general because the internal particles are almost degenerate in mass and so the GIM cancellation is operative.

ERICE LECTURES ON COSMOLOGY

F. Wilczek

Institute for Theoretical Physics
University of California
Santa Barbara, CA 93106

LECTURE 1: REVIEW OF BIG-BANG COSMOLOGY

A. Dynamic Equations of Cosmology

The correct equations for development of a homogeneous, isotropic universe follow from simple Newtonian considerations (justified by general relativity):

Conservation of energy, applied to test mass:

$$\frac{\dot{R}^2}{2} - \left(\frac{4\pi R^3}{3}\rho\right)\frac{G}{R} = \pm k^2$$

$$\boxed{\left(\frac{\dot{R}}{R}\right)^2 - \frac{8\pi}{3}G\rho = \pm \frac{k^2}{R^2}} \qquad (1)$$

Conservation of energy, applied to fluid: $dU = -pdV$

or: $\quad \dfrac{d}{dt}\left(\dfrac{4\pi}{3}\rho c^2 R^3\right) = -p\,\dfrac{d}{dt}\left(\dfrac{4\pi}{3} R^3\right)$

or: $\quad \boxed{\dot\rho = -3(\rho + p/c^2)\dot R/R}\;.$ \hfill (2)

Comments: i) These 2 eqns., plus eqn. of state, determine evolution completely -- 3 eqns. for 3 unknowns. ii) $\pm k^2$ is an integration constant. + sign => eventual expansion, - sign => ultimate collapse.

B. Relation to Hubble Parameters

For expansion of homogeneous, isotropic space we have the Hubble law

$\boxed{\vec v = H(t)\vec r}$ (at one instant of <u>cosmic</u> time). \hfill (3)

This is inevitable as no other scales exist or from a simple picture

. . . . .

$\downarrow \Delta t \qquad\qquad \Delta x \propto x \qquad$ (note: $H = \dot R/R$).

. . . . .

Variations of $H(t)$ with t are measureable even without waiting, since far-away objects are observed at smaller t (finite speed of light). They are parametrized by deceleration,

$q_o \equiv (1/H)\dot{} - 1 = -\ddot R\, R/\dot R^2$ \hfill (4)

$q_o > 0$ on general principles.

For non-relativistic matter, i.e. $p/c^2 \ll \rho$, (1) and (2) yield

$$H^2(2q_o - 1) = \mp k^2/R^2 \quad \text{(n.r. matter)} \tag{5}$$

and

$$2q_o = \left(\frac{8\pi G}{3H^2}\right)\rho \equiv \rho/\rho_c \quad \text{(n.r.)} \tag{6}$$

$$\rho_c = \frac{3H^2}{8\pi G} \quad \text{(n.r.)} \tag{7}$$

<u>Comment</u>: $q_o = 1/2$ divides cases of eternal expansion ($q_o < 1/2$) from eventual collapse ($q_o > 1/2$). q_o may in principle be determined directly by determining speed of distant objects (H); or by measuring ρ - counting up mass.

C. Simple Solutions

i) <u>Pressure-free</u> ($p/c^2 \ll \rho$); k = 0 (Einstein-deSitter).

$\boxed{4\pi/3\ \rho R^3 \equiv M = \text{const. from (2)}}$ ←|| for all p=0 models

$R = (9/2\ GM)^{1/3}\ t^{2/3}$ from (1)

$$H = \dot{R}/R = 2/3t \tag{8}$$

$$q_o = 1/2 \tag{9}$$

ii) <u>Pressure-free, + k^2</u>. Eqn. (1) separates and one can solve by quadratures. Result is

$$t = \frac{GM}{k^3}(\sinh k\eta - k\eta)$$

$$R = \frac{GM}{k^2}(\cosh k\eta - 1)$$

As $k \to 0$ this reduces to E - d.S.; also as $t \to 0$ for any k.

$$H = \frac{k^3}{GM} \frac{\sinh k\eta}{(\cosh k\eta - 1)^2} \quad 1/t \quad \text{as } t \to \infty \quad \text{always } H > 2/3t$$

$$q_o = \frac{1}{2(\cosh k\eta/2)^2} < 1/2$$

iii) <u>Pressure-free, $-k^2$</u>.

$$t = \frac{GM}{k^3} (k\eta - \sin k\eta)$$

$$R = \frac{GM}{k^2} (1 - \cos k\eta)$$

reduces to E - d.S. as $k \to 0$ or $t \to 0$, and becomes singular (Big Crunch) at

$$\eta_{end} = \frac{2\pi}{k} \qquad t_{end} = \frac{2\pi GM}{k^3}$$

$$H = \frac{k^3}{GM} \frac{\sin k\eta}{(1 - \cos k\eta)^2} < \frac{2}{3t} \quad ; \quad \text{note change in sign at } \eta = \pi/k \;!$$

$$q_o = \frac{1}{2(\cos k\eta/2)^2} > 1/2$$

Lifetime of universe: $t_{end} = \frac{2\pi}{H} \frac{q_o}{(2q_o - 1)^{3/2}}$.

<u>Comments</u>: i) Atomic clocks measure t, so determination of age put bounds on acceptable values of H and q_o. In particular if objects are found with ages $t > 2/3H$, the universe will always expand.
ii) The parameter η, appearing here as an aid to quadrature, has direct physical significance as we shall soon see.

iv) <u>Radiation-dominated, $k = 0$ (early universe)</u>. For radiation, $p/c^2 = 1/3 \rho$. Eqn. (2) says ρR^4 = const. Then from (1)

ERICE LECTURES ON COSMOLOGY

$$R = \left(\frac{32\pi \, G\lambda}{3} \right)^{1/4} t^{1/2} . \qquad (10)$$

<u>Comment</u>: Since ρR^4 = const. for radiation; ρR^3 = const. for non-relativistic matter; the radiation dominates ρ for small R; n.r. matter dominates for large R.

D. Global Structure of the Universe I: Fixed Cosmic Time

The three types of homogeneous isotropic spaces are:

sphere S^3

flat space

hyperboloid: this can be realised as the surface
$x_4^2 - x_1^2 - x_2^2 - x_3^2 = 1/k^2$ in Minkowski space.

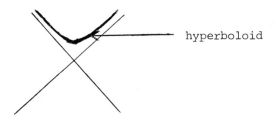
hyperboloid

From the picture it is clearly space-like; to find the metric use hyperspherical coordinates χ, θ, ϕ

($x_4 = 1/k \cosh k\chi$

$x_3 = 1/k \sinh k\chi \cos \theta$

$x_2 = 1/k \sinh k\chi \sin \theta \sin \phi$

$x_1 = 1/k \sinh k\chi \sin \theta \cos \phi$;

$0 \leq k < \infty$

$0 \leq \theta < \pi$

$0 \leq \phi < 2\pi$) .

We find the distance between nearby points:

$$d\ell^2 = dx_1^2 + dx_2^2 + dx_3^2 - dx_4^2$$

$$= d\chi^2 + \left(\frac{\sinh k\chi}{k}\right)^2 (d\chi^2 + \sin^2\theta\, d\phi^2).$$

The hyperboloid is homogeneous and isotropic by rotations and boosts in the underlying Minkowski space. The 3-sphere may be parametrized similarly; all the metrics take the form:

$$d\ell^2 = d\chi^2 + \Psi(\chi)^2(d\theta^2 + \sin^2\theta\, d\phi^2) \tag{11}$$

with

$$\Psi(\chi) = \begin{cases} \sin k\chi/k & \text{sphere} \\ \chi & \text{flat space} \\ \sin k\chi/k & \text{hyperboloid} \end{cases} \tag{12}$$

E. Global Structure II: Dynamics

If homogeneity and isotropy are to be maintained only an overall change of scale is possible in time. Thus the space-time metric will be

$$ds^2 = dt^2 - R(t)^2(d\chi^2 + \Psi(\chi)^2\, d\Omega^2). \tag{13}$$

General relativity relates this to the dynamical equation as follows:
* The field equations for a fluid with density, pressure ρ, p in the space-time metric (13) evolves according to the dynamical equations (1),(2). $+ k^2$, the eternal expansion, corresponds to the hyperboloid; $- k^2$, eventual collapse, corresponds to the sphere; $k = 0$ corresponds to flat space.*

Although the t = const. slices of these model universes look very different, as seen in real time through telescopes they are not so different. To study light rays introduce a new time variable η, with $d\eta = dt/R(t)$. Then the metric is

ERICE LECTURES ON COSMOLOGY

$$ds^2 = R^2(d\eta^2 - (d\chi^2 + \Psi(\chi)^2 \, d\Omega^2))$$

and light rays at a fixed angle θ, ϕ = const. travel along $\eta \pm \chi$ = const. The η variable appearing in the parametrizations above is consistent with this notation; for Einstein-deSitter we have the limiting case $t = GM \, \eta^3/6$, $R = GM \, \eta^2/2$.

Example: mass of visible universe. Volume of space first becoming visible to us (at $(\eta, \chi) = (\eta_o, 0)$) in interval η, $\eta + d\eta$ of cosmic time is bounded by light cones $\eta_o = \eta + \chi$, $\eta_o = \eta + d\eta + \chi$; i.e. $\eta_o - \eta \le \chi \le \eta_o - \eta - d\eta$.

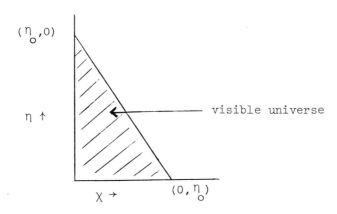

Its volume is

$$dV = R(\eta) \, d\chi \int R(\eta) \Psi(\eta_o - \eta) \, d\theta \, R(\eta) \Psi(\eta_o - \eta) \sin\theta \, d\phi$$

$$= 4\pi \, R^3(\eta) \Psi(\eta_o - \eta)^2 \, d\eta \; .$$

Mass inside is

$$d\mu = 4\pi \, \rho(\eta) R^3(\eta) \Psi(\eta_o - \eta)^2 \, d\eta \qquad\qquad \rho(\eta) R^3 = \text{const.}$$

$$= 3M \, \Psi(\eta_o - \eta)^2 \, d\eta \; .$$

Total mass is

$$\mu = 3M \int_0^{\eta_o} \Psi(\eta_o - \eta)^2 d\eta$$

which is perfectly finite and smoothly varying as a function of k. For Einstein-deSitter

$$\mu = M \eta_o^3 = 6t/G \qquad \text{(in c.g.s. units, } 6tc^3/G\text{)}.$$

Numerically for $t = 10^{10}$ yrs. $\mu = 10^{57}$ gm. which is roughly Avogadro's number times the mass of the sun $M_\odot$.

F. Red-shift

Since distant galaxies are receding spectral lines from them will be shifted toward longer wavelengths by Doppler's effect. A more exact approach is through the wave eqn., with

$$\hbar\omega = E \to i\frac{\partial}{\partial t} \qquad p \to -i\frac{\partial}{\partial \vec{\ell}} \quad \leftarrow \text{normalized!}$$

The wave eqn. is just $(E^2 - p^2)\phi = 0$. So

$$\left(\left(\frac{\partial}{\partial t}\right)^2 - \frac{1}{R^2(t)}\left(\frac{\partial}{\partial \chi}\right)^2\right)\phi = 0 \qquad (14)$$

which admits plane-wave solutions

$$\phi(\chi,t) = \exp\left(i\left(\int^t \frac{dt'}{R(t')} \pm \chi\right)\right) = \exp(i(\eta \pm \chi)).$$

The local frequency is $\partial/\partial t$ of this and thus is $\propto 1/R$:

$$\frac{\omega(t_o)}{\omega(t)} = \frac{R(t)}{R(t_o)} \equiv (1+z)^{-1}. \qquad (15)$$

z is the astronomers' "redshift" parameter which is directly observable. For small separations r in space

$$R(t) = R(t_o) + (t-t_o)\dot{R} = R - r/c\,\dot{R}$$

so

$$H = \frac{cz}{r} \quad . \tag{16}$$

For larger separations we should be able to extract $\ddot{R}$ and hence q_o. But we must face up to the question of what is meant by "distance" observationally. Two canonical measures are decrease in angular size or luminosity of "standard candles" (physical objects with standard properties independent of their distance).

Angular size: Physical size $\Delta\ell$ is fixed; in terms of observed angles $\Delta\theta$ this says

$$R(\chi)\Psi(\eta_o-\eta)\Delta\theta = \text{fixed}$$

$$\Delta\theta \propto \frac{1}{R(\eta)\Psi(\eta_o-\eta)} \quad . \tag{17}$$

Note this has a minimum value, then blows up as $\eta \to 0$!

Some straightforward but tedious algebra gives an expansion for small z:

$$\Delta\theta \propto \frac{1}{z - \frac{1}{2}(3+q_o)z^2} \quad \text{small } z$$

Luminosity: Radiation from point source at $(\eta,0)$ will spread to $(\eta_o, \eta_o-\eta)$ at present η_o. Area of this surface $\propto R^2(\eta_o)\Psi(\eta_o-\eta)^2$. Luminosity is also decreased by red-shift of photon energy and dilation of time between pulses; that is by $R^2(\eta)/R^2(\eta_o)$. Altogether then

$$L \quad \frac{R^2(\eta)}{\Psi(\eta_o-\eta)^2} \tag{18}$$

$$\sqrt{L} \propto \frac{1}{z + \frac{1}{2}(1-q_o)z^2} \quad \text{small } z \quad .$$

G. Observational Handles on Closure Question

o) H: Observations give 50 km/sec Mpc $\leq H \leq$ 100 km/sec Mpc. This translates to 10×10^9 yr $\leq 1/H \leq 20 \times 10^9$ yr.

i) <u>Ages</u>: Recall for EdS the age is $2/3\, H^{-1}$; longer ages indicate an open universe.

Stellar age: Massive stars are the most luminous and exhaust their hydrogen fuel most quickly. When this happens they turn off the "main sequence", developing an anomalous temperature-luminosity relation. By observing populations formed at one time (globular clusters) to find the turn-off luminosity, using the theory of stellar evolution one can date the population. Estimates for oldest clusters are $12 \div 15 \times 10^9$ yr.

Radioactive age: $^{235}U/^{238}U$ are believed to form in supernovae with initial abundance ratio $\simeq 1.5$; observing present ratio gives us estimate of how long ^{235}U has had to decay away. This and other similar measurements give $\tau \geq 8.7 \times 10^9$ yrs, with a concordant age 13.5-15.5×10^9 yrs according to recent reviews.

ii) <u>Mass</u>: $\rho_c = 3H^2/8\pi G \simeq 10^{-28}$ gm/cm^3. The universe is closed if $\rho > \rho_c$. Estimates of "luminous" matter give $\lesssim .06\, \rho_c$, but there is evidence for non-luminous matter as we shall review.

iii) <u>Direct measures</u>: q_o can be obtained from red-shift vs. angular size or luminosity curves as above. However so far "standard candles" have proved elusive. There is high hope for using the space telescope to exploit supernovae.

<u>Summary</u>: The universe cannot be very closed; open seems slightly favored but EdS is not definitely excluded.

LECTURE 2: CONTENTS OF THE UNIVERSE, ESPECIALLY NEUTRINOS

A. Notable Events in Universal History

The previous lecture set up the stage for cosmology; now we introduce the script and the actors. A main task will be to explain the following time-line:

temperature [GeV]	time [sec]	event
10^{19}	10^{-44}	Planck time; strong gravity?
10^{15}	10^{-36}	baryon synthesis?
10^{2}	10^{-10}	$SU_2 \times U_1$ breaking
10^{0}	10^{-6}	quarks → hadrons
10^{-2}	10^{-2}	neutrinos decouple
10^{-4}	10^{2}	$d + {}^4He$ synthesis
10^{-9}	10^{12}	plasma → atoms; microwave bkgd. $\rho_{non-rel.} > \rho_{radiation}$
	$10^{16} - 10^{17}$	galaxy formation
	$3 - 6 \times 10^{17}$	F.W. lectures on cosmology

The dominant theme that runs through the history is that the content of the universe is essentially a thermal equilibrium ensemble at extremely high temperatures, for times close to the Big Bang. The increase in temperature is easy to understand as the inverse of the redshift. Photons of energy $\hbar\omega_o$ now had energy $\hbar\omega_o R_o/R(t)$ at time t so the present thermal distribution $e^{-\hbar\omega_o/T_o} = \rho(\omega_o)$ in terms of the old energy is $\exp[-(\hbar\omega_o R_o/R(t))/(R_o T_o/R(t))]$, which is a thermal distribution with temperature $T(t) = R_o T_o/R(t)$.

At extremely high energies and densities we can expect (and will demonstrate, in B.) that reaction rates are very rapid so that thermal equilibrium is attained except for very weak interactions. This vastly simplifies the description of the contents of the universe, since we can apply statistical mechanics. Moreover, in asymptotically free gauge theories (on which all our description of the very early universe rests) the couplings are weak at high energies so we can treat the cosmological fluid as an ideal gas to first approximation. We then know the equation of state, and can solve the dynamical equations for the temperature as a function of time. The result is

$$\frac{dT}{2T^3} = -\left(\frac{45}{\pi^3 nG}\right)^{-1/2} dt \tag{1}$$

where $n = N_B + \frac{7}{8} N_F$ is the number of boson degrees of freedom plus seven-eighths the number of fermion degrees of freedom which are "effectively massless" ($T \gg m$). For rough purposes we just put $n = 100$ in constructing the time-line, so

$$t = \left(\frac{45}{\pi^3 nG}\right)^{1/2} \bigg/ T^2 \simeq \frac{10^{-6}}{T^2 [\text{GeV}]} \text{ sec}.$$

Here is a little summary of the statistical mechanics of relativistic ideal gases:

$$\text{energy density } \rho = T^4 \frac{\pi^2}{30} (N_B + \frac{7}{8} N_F) \quad T^4 \frac{\pi^2}{30} n \tag{2}$$

$$\text{entropy density } S = T^3 \frac{2\pi^2}{45} (N_B + \frac{7}{8} N_F) \tag{3}$$

$$\text{particle number density } n = T^3 \frac{\zeta(3)}{\pi^2} (N_B + \frac{3}{4} N_F) \tag{4}$$

where N_B, N_F are the number of boson and fermion degrees of freedom respectively -- e.g. $N_B = 2$ for a photon gas (2 helicities), $N_F = 4$ for relativistic electrons and positrons (2 helicities each).

With these ideas in mind, let us discuss the events listed in the timeline. At $T = 10^{19}$ GeV, $GT^2 = 1$ so gravitational cross-sections for typical energies T approach the unitarity limit - gravity has become a strong interaction. Ordinary general relativity is non-renormalizable so we lose control of the theory at these temperatures.

According to standard unified theories (e.g. SU(5)) the full gauge symmetry is restored at $T > 10^{15}$ GeV; in any case the vector bosons associated with exotic SU(5) interactions will be effectively light and hence abundant at such temperatures. Since such bosons mediate baryon number violation and their decays violate baryon number, in equilibrium above this temperature the net baryon number

ERICE LECTURES ON COSMOLOGY

would be zero. The present asymmetry between matter and antimatter could be generated, as we shall discuss in detail in Lecture 3, when the B-violating interactions are too feeble to maintain equilibrium but not so rare that they are negligible, e.g. in SU(5) at temperatures close to 10^{15} GeV.

At much lower temperatures we have SU(2) x U(1) symmetry breaking and the quark-hadron phase transition. These raise fascinating problems of principle but so far as is known leave no observable relics. (A possible effect is that if these transitions occur far from equilibrium they might generate lots of entropy, diluting the baryon number previously produced.)

At $T \lesssim 10$ MeV, the interactions of neutrinos are so feeble that they for the most part have not scattered between then and now. We say they are decoupled; however their gravitational effects can be very significant as we shall see.

Above $T \gtrsim 1$ MeV, the binding energy of deuterium is not sufficient to allow much deuterium to stick together in the face of a constant bombardment with energetic γ's. Protons and neutrons are present as free particles. Below 100 keV they stick together to form deuterium; the deuterium undergoes further strong interactions producing mainly the very stable nucleus ^{4}He. For more on this, see E. and F.

In a similar fashion, protons (and helium nuclei) do not combine with electrons into atoms until $T \simeq 1$ eV. Before they combine, the ever-present photons interact with these changed particles readily, but after they combine the photons are decoupled - most of them have not interacted since then until the present day (and never will, in an open universe!). It is this sea of relic photons that is detected as the microwave background radiation.

At about the same time, apparently as a numerical accident, the mass density in non-relativistic matter (baryons, or possibly neutrinos with mass > 1 eV) becomes greater than that in relativistic matter. The pressure-free models we discussed in detail above are adequate from that time forward.

Some time around z = 5-10 it is believed galaxies formed. The last entry requires no comment.

B. Neutrino Decoupling

As an example of the computation of reaction rates and the concept of decoupling we now consider neutrino reaction rates.

The interesting temperatures will turn out to be 10 MeV << 100 GeV so we can use ordinary 4-fermion theory to describe weak interactions. The rate of interaction for a neutrino will be $n\sigma c = n$ where n is the density of things to interact with. For the period of interest ($\sim$ 10 MeV) these are electrons, other neutrinos, and their antiparticles, and $n \simeq \frac{7\pi^2}{240} N_F T^3 \simeq 3T^3$ from statistical mechanics. $\sigma \simeq \frac{1}{\pi} G_F^2 T^2$ for weak processes so $n\sigma \simeq G_F^2 T^5$. On the other hand in this radiation-dominated era $dt = -\frac{dT}{2T^3}\left(\frac{\pi^3 Gn}{45}\right)^{1/2}$ from (1) so the number of interactions for one neutrino between temperatures T_1, T_2 is

$$I = \int_{T_1}^{T_2} G_F^2 T^2 \left(\frac{45}{4\pi^3 Gn}\right)^{1/2} dT . \qquad (5)$$

We notice immediately that this integral is highly convergent for low temperatures, so that if our neutrino fails to interact before the universe gets too cool then it most likely never will. (Actually we would have to modify the formula once the universe becomes matter dominated but the conclusion remains the same.) The number of reactions becomes substantial when T_2 is so large that

$$\frac{G_F^2 T_2^3}{3}\left(\frac{45}{4\pi^3 Gn}\right)^{1/2} \gtrsim 1$$

which gives $T_2 \gtrsim 20$ MeV, as advertised. Above $T \simeq 20$ MeV, each neutrino will interact many times. Moreover, if there were no neutrinos they would be rapidly produced in weak processes. Evidently, arguments like this justify the assumption of thermal equilibrium also with respect to other types of interactions in the early universe and tell us when it breaks down.

C. Density of the Neutrino Gas

As we have discussed, the Big Bang picture predicts that photons are present in thermal equilibrium above $T \gtrsim 1$ eV; below this the universe is transparent to them. The relic photons, highly redshifted, have of course been detected in the form of the 2.7°K microwave background radiation. Statistical physics tells us this corresponds to

$$\frac{2\zeta(3)}{\pi^2} T^3 \simeq 600 \; \gamma\text{'s/cm}^3 . \tag{6}$$

We also have the relic sea of neutrinos, which as we have just calculated decoupled at $T \simeq 20$ MeV. What is the density of these neutrinos? This density cannot quite be read off from the photon density (multiplying by 3/4 to take into account the difference between the ideal Fermi and Bose gases). Between ν decoupling at $\sim$20 MeV and γ decoupling at $\sim$1 eV there occurred the annihilation of hordes of e^+e^- pairs; which were common for $T \gtrsim m_e$ but rare in equilibrium according to the Boltzmann factor for $T < m_e$. Considerations like those we just presented in B. show that the annihilation process is very rapid and efficient (except in its very last stages) so equilibrium is maintained to a good approximation, and the entropy in a comoving volume should remain constant. At the end of the annihilation we have according to Eqn. (3) a photon gas with an entropy S'

$$S' = \frac{2 + \frac{7}{8}(4)}{2} S = \frac{11}{4} S \tag{7}$$

11/4 times what it would have been without electron annihilation. The factor 7/8 is the relative entropy per degree of freedom for an ideal Fermi versus Bose gas; the electron, positron and their two polarizations represent 4 degrees of freedom.

Of course the neutrinos, being decoupled, do not tap the additional entropy supplied by the e^+e^- annihilation (i.e. few $\nu\bar{\nu}$ pairs are thus produced). So in going from γ to ν densities we must supply not only the statistical factor 3/4 but also the 4/11 as just computed, yielding

$$n_{\nu+\bar{\nu}, \text{ per family}} \simeq \frac{3}{4} \cdot \frac{4}{11} \cdot 600/\text{cm}^3 \simeq 160/\text{cm}^3 \quad (8)$$

for the average density of neutrinos plus antineutrinos, for each family type (e.g. e, μ, ... neutrinos).

D. Mass Limit on Cosmologically Stable ν's

From the age results and gravitational measures of cosmological mass distribution (virial theorem applied to galaxy clusters, etc.) it seems safe to conclude $\rho/\rho_c \lesssim 2$. Putting this together with the computed density $n_{\nu+\bar{\nu}}$, we can bound the mass of cosmologically stable neutrinos, those with lifetimes $\gtrsim 10^{17}$ sec. (This degree of stability is very plausible as we shall discuss in G.) The limit is simply

$$\left(\sum_{\text{families}} m_\nu \right) n_{\text{per family}} \lesssim 2 \rho_c = \frac{3H^2}{4\pi G}$$

or, numerically

$$\sum_f m_\nu \lesssim 130 \, h^2 \text{ eV} \quad (9)$$

where we have borrowed the astrophysicists' fudge factor $H \equiv 100 \, \frac{\text{km/sec}}{\text{Mpc}} \times h$.

This bound covers neutrinos which were abundant at decoupling, so it requires $m_\nu \lesssim 20$ MeV. There is another allowed mass range at much higher masses, $m_\nu \gtrsim 1$ GeV, though whether such objects should be called "neutrinos" or be expected to have conventional weak interactions is unclear to me.

E. Nucleosynthesis

The great quantitative prediction of Big Bang cosmology, supplementing the qualitative prediction of the existence of a background photon gas, concerns the relative abundance of ^{4}He and possibly d nuclei. ^{3}He abundance may also be a significant relic.

The ^{4}He abundance is observed to be 25% by weight in a number of astrophysical environments (solar atmosphere, planetary nebulae, etc.). Such a large abundance cannot have been produced in stars - the integrated luminosity of these stars would be far larger than what is observed, so such a large abundance would be difficult to reconcile with the much smaller amounts of heavier elements, etc. Deuterium, on the other hand, is so fragile that it is dissociated or cooked in stellar environments, so it is generally accepted that the ^{4}He abundance is a cosmological relic, and probably also d abundance.

Numerical integrations following nucleosynthesis through the hot Big Bang give fairly good agreement with the observed values. I will not try to reproduce these results quantitatively, but just discuss the qualitative picture so we can understand how different assumptions will modify the picture.

Above $T \gtrsim 100$ keV d nuclei are rare, as discussed, and nucleosynthesis cannot proceed. Below these temperatures, the d gets formed and most of it rapidly converts into the more stable ^{4}He in subsequent collisions. Thus the crucial parameter for computing the ^{4}He abundance is the ratio P/N of protons to neutrons at $T \sim 100$ keV. The helium abundance by weight will then be

$$f = \frac{4\ ^4\text{He}}{4\ ^4\text{He} + p} = \frac{4(N/2)}{4(N/2) + (P-N)} = \frac{2N}{N+P} = \frac{2}{1 + (P/N)} . \quad (10)$$

In thermal equilibrium at high temperatures the proton-neutron mass difference is negligible and P/N = 1. If the protons and neutrons were in chemical equilibrium at 100 keV we would have P/N = $e^{\Delta/100 \text{ keV}} \simeq e^{13}$ where Δ is the n-p mass difference, and hence very little helium.

Reality is somewhere in between. The neutron-proton balance is maintained by reactions like $\nu + n \leftrightarrow e^- + p$. If these reactions were fast enough, chemical equilibrium would be maintained and we would get very little helium. However, as we have seen, in an expanding universe feeble interactions "decouple" becoming very rare below a certain critical temperature. Neutrinos for the most part last scattered at $T \sim 20$ MeV. Here we are asking a closely analogous question: when did neutrons and protons essentially cease interconverting? The weak interactions are again involved, and we could do a calculation very like that in B. to compute the relevant temperature. The answer is slightly modified, because the cross-section for $\nu + n \leftrightarrow e^- + p$ is larger than pure lepton x-sections like $\nu_\mu + e \leftrightarrow \nu_e + \mu$. Also the lepton abundance is larger than the nucleon abundance by $\sim 10^{10}$ at those temperatures (Lecture 3) so the probability for a given nucleon to interact with any neutrino is much larger than the probability for a given neutrino to interact with any nucleon. As a result, the effective temperature at decoupling is smaller, $T^* \simeq .7$ MeV. This gives P/N = $e^{\Delta/T^*} \simeq 7$, and hence $f \simeq 1/4$ as observed. In calculating the critical temperature we must know the rate of expansion, which in turn depends on the number of species n - compare Eqn. (5). The standard, successful calculation assumes 3 neutrino families.

For the calculation of d abundance the crucial question is how much d escapes conversion into the more tightly bound nuclei that are favored in equilibrium. The answer is critically dependent on the density of nucleons. High densities will make nuclear burning more efficient and wipe out the deuterium. Agreement with observed abundances is possible for $\rho_{baryon}/\rho_c \simeq .06$, but is impossible for much larger baryon densities. This is an argument that the "visible" baryon mass density is all there is - the "missing mass" should be something else.

F. Effect of Neutrino Flavors

We now treat the effect of neutrino flavors on ^{4}He abundance more quantitatively. From the analog of (2) but with $G_F^2 m^2 T$ replacing $G_F^2 T^2$ in the x-section we have an equation of the type

$$\int_0^{T^*} G_F^2 mT \left(\frac{45}{4\pi^3 Gn}\right)^{1/2} dT = 1 \tag{11}$$

for the decoupling temperature at which P/N abundance froze in. Thus $T^{*2} \propto n^{1/2}$. Some arithmetic gives us the relation between δf and δn:

$$\delta f = -\frac{2}{(1+P/N)^2} \delta(P/N) \simeq -\frac{1}{32} \delta(P/N) = -\frac{1}{32} (P/N) \, \delta(\Delta/T^*)$$

$$= \left(-\frac{1}{32} \cdot 7\right)\left(-\frac{\Delta}{T^*} \frac{\delta T^*}{T}\right) = \frac{7 \cdot \ln 7}{32}\left(\frac{1}{4}\frac{\delta n}{n}\right) \simeq \frac{1}{9}\frac{\delta n}{n} \, . \tag{12}$$

Now we are talking about $T^* \simeq .7$ MeV so electrons are just barely relativistic. Counting them fully, together with photons and ℓ families of leptons, we have $n = 2 + \frac{7}{8}(4 + 2\ell)$. In changing $\ell = 3$ to $\ell = 4$ we find $\frac{\delta n}{n} = \frac{7/4}{43/4} \simeq \frac{1}{6}$. Thus each neutrino species adds about 2% to the helium abundance. It is very difficult to reconcile ≥ 5 species with the ^{4}He observations.

G. Neutrino Stability

There are very powerful astrophysical constraints on radiative neutrino decay $\nu' \to \nu + \gamma$. In quasi-orthodox weak interaction models this is the leading branch for light (< 1 MeV neutrinos; such processes as $\nu_1 \to \nu_2 + \nu + \bar{\nu}$ require flavor-changing neutral currents and are expected to be heavily suppressed. (Not that unorthodox models featuring this decay, or decay of neutrinos into other hypothetical light particles, have not been proposed!)

An extremely powerful limit comes from the theory of white dwarfs. Each white dwarf is expected to cool by emitting 10^{58} 100 keV neutrinos. Those would decay giving x-rays. Folding in the formation rate of white dwarfs, and comparing the experimental limits on x-ray background, one finds $\tau_o/m_\nu \gtrsim 10^{17}$ sec/eV.

Radiative decay of cosmological background neutrinos can also wreak havoc. Remembering the number density of ν' is about 10^9 of baryons, $\nu' \to \nu + \gamma$ decays with $E_\gamma \gtrsim$ few eV can be very efficient in ionizing atoms, distorting or destroying the microwave background. Roughly we must have

$$\text{\# of decays} = \frac{10^{12} \text{sec}}{\tau} \leq 10^{-9} \quad ; \quad \tau \gtrsim 10^{21} \text{sec} \tag{13}$$

to be safe. This argument can be tightened; Rephaeli and Szalay claim $\tau \gtrsim 10^{23-25}$ sec.

We would expect $\nu' \to \nu\gamma$ to go by the diagrams

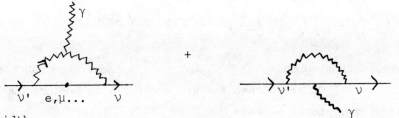

with width

$$\Gamma = \alpha G_F^2 \, m_\nu^5 \left[\frac{\Delta m_\ell^2}{m_\omega^2} \sin \theta \right]^2 \tag{14}$$

or lifetime

$$1/\Gamma = \tau \simeq 10^{37}/m_\nu^5 \, [\text{eV}] \, [\ ]^2 \, \text{sec} \tag{15}$$

with the [ ] factor not much greater, and probably much smaller, than unity.

Altogether then, both theoretically and observationally, lack of cosmological stability would be a big surprise.

ERICE LECTURES ON COSMOLOGY

H. Missing Mass

An important motivation for considering that neutrinos may be massive is that mass in the form of non-luminous matter is required in several astrophysical contexts:

i) Numerical experiments by Ostriker, Peebles and others demonstrate that matter distributed proportional to the light intensity in spiral galaxies is unstable to the formation of bars:

The instability can be relieved if there is additional spherically distributed mass (halo) around the disc, with total mass at least comparable to the mass in the stellar disc.

ii) Observations of the rotation curves - speed of revolution of gas clouds around the galaxy as a function of their distance from the center - through the Doppler shifts in the 21 cm emission of hydrogen indicate that gravitating matter extends well beyond the luminous matter in most galaxies. If the matter were localized inside a radius R then as a function of distance for $r > R$ we would observe Keplerian velocities

$$v \propto r^{-1/2} \quad (r > R) . \tag{16}$$

Instead it is observed that the velocities are remarkably constant with radius, even for distances well beyond where the luminosity is localized. Constant v corresponds to a mass density $\rho(r) \propto 1/r^2$ in the halo. Existing observations require a few times as much non-luminous as luminous mass; since the integrated mass

$$\int_0^D \rho(r) \, 4\pi r^2 dr \propto D \text{ for } \rho(r) \propto 1/r^2$$

continued flatness of the rotation curve would indicate even more non-luminous mass.

iii) There is a long-standing discrepancy in the application of the virial theorem to large clusters of galaxies (in particular, the Coma cluster). Using the masses of constituent galaxies inferred from their luminosity and from the mass/luminosity ratio expected for galactic stellar populations, the virial theorem indicates that this cluster cannot be gravitationally bound. This is very difficult to reconcile with the visual appearance (regularity) of the cluster. For binding, a factor of 10 or 20 in mass is required.

For the missing mass on a galactic scale to be in neutrinos, the neutrino mass must be $\gtrsim$ 30 eV. This is because Fermi statistics limits the number density of neutrinos that can exist with velocities less than the escape velocity from the galaxy, and hence the mass density:

$$\rho \leq m \left(\frac{1}{6\pi^2} m^3 v^3 \right) . \qquad (17)$$

For the observed ρ and v, $m \gtrsim 30$ eV is indicated.

I. Neutrino Mass and Galaxy Formation

More detailed considerations show that $m_\nu \simeq 30$ eV would be desirable for cosmology. The main argument goes as follows:

i) Fluctuations in baryon density cannot grow until the baryons combine (at $T \simeq 10$ eV) into neutral atoms. Before this, the photons gas forms a viscous medium through which charged particles cannot accelerate. Moreover, if the fluctuations in baryon number density are correlated to the fluctuations in photon number - so-called adiabatic fluctuations - then whatever fluctuations occurred at 10 eV would be imprinted on the microwave background radiation. Present observations severely limit the anisotropy of the radiation, so the amplitude of such fluctuations are strongly limited ($\leq 10^{-4}$). We must then ask ourselves if such small fluctuations can grow sufficiently to form galaxies in the required time. Detailed computations show that growth can be rapid enough only if the density of the universe is nearly the critical value ρ_c.

By the way, adiabatic fluctuations are suggested by the theories of baryon number generation we shall discuss in Lecture 3. Since in these models n_B/n_γ has a microphysical origin, it should be a universal constant even if the numerator and denominator separately vary.

ii) Although large densities $\rho \simeq \rho_c$ seem desirable for the theory of galaxy formation, calculations of nucleosynthesis indicate that the baryon density should be considerably smaller. High baryon densities lead to over-production of d and ^{3}He. So the extra density required for galaxy formation must be in a non-baryonic form. For neutrinos to do any good (supply enough density) their mass must be $\simeq 30$ eV.

LECTURE 3: MATTER-ANTIMATTER ASYMMETRY

A. n_B/n_γ as a Relic

We now turn to the question of the origin of the asymmetry between matter and antimatter in the universe. For reasons that will become clear, there is at present no definitive quantitative theory of this asymmetry. Still, there has been great progress in recent years and there is now a conceptual framework in which the asymmetry can be sensibly discussed. Accordingly our discussion will approach the problem from different angles - we hope to surround the answer although we cannot quite capture it.

In this section, I assume that matter predominates over antimatter everywhere; i.e. that distant galaxies are composed of baryons rather than antibaryons. This assumption will be discussed briefly in B.

A quantitative measure of asymmetry is the ratio n_B/n_γ of baryon to photon number densities. The numerator can be estimated from measures of galactic masses and from nucleosynthesis calculations as discussed in Lecture 2; the denominator is known from the temperature of the microwave background radiation. The observed ratio n_B/n_γ is in the range 10^{-9}-10^{-11}. The first important thing to notice about n_B/n_γ is that it is a relic - it has changed little since shortly after the Big Bang. The numerator, the baryon number, is unchanged in a comoving volume insofar as baryon number is conserved. We have of course excellent evidence that B-violating processes are very rare at present. If B-violating interactions occur at all they must be associated with exchange of very heavy or very weakly coupled particles. In either case, they cease to become important as the universe becomes dilute and cool, just as the usual weak interactions did in our calculation of neutrino decoupling. The denominator, the number of photons, is essentially constant (within modest factors) as long as the expansion of the universe is an adiabatic process. This is because the entropy of the photon gas is proportional to the number of photons, and the entropy in a comoving volume is (trivially) constant in adiabatic expansion.

As an exercise, let us calculate the change in n_B/n_γ as we cool from $T = 10^{15}$ GeV to now in a minimal SU(5) model, assuming adiabatic expansion. The principles are the same as in our calculation of n_ν/n_γ in Lecture 2. We have 28 light boson degrees of freedom (the vector bosons of SU(3) x SU(2) x U(1), each with 2 helicites, plus 4 for a complex Higgs doublet) and 90 fermion degrees of freedom (6 quarks x 3 colors x 2 helicities x 2 (for particle-antiparticle), plus 3 charged leptons x 2 x 2 plus 3 neutrinos x 2) which are effectively massless at $T \simeq 10^{15}$ GeV. All these degrees of freedom are in equilibrium due to ordinary weak interactions until $T \lesssim 20$ MeV. At this stage the total entropy density of the original gas, $S \propto (N_B + 7/8\ N_F) = (24 + 7/8\ (90))$ is concentrated in the remaining effectively light particles - photons, electrons, and neutrinos. This gas of light particles has been heated by the annihilation of the heavier ones so that it preserves the entropy; its entropy density is therefore $[24 + 7/8\ (90)]/[2 + 7/8\ (4 + 3 \cdot 2)]$ times what it would otherwise be. At $T \simeq 20$ MeV the neutrinos decouple and their entropy is lost to the photon gas; on the other hand the electrons and positrons annihilate into photons and do contribute to the entropy in the photon gas. The photon entropy is therefore multiplied by a further factor $(2 + 7/8\,(4))/2$. So altogether the photon entropy, and hence the photon number, has been enhanced by a factor

$$\frac{24 + 7/8\,(90)}{2 + 7/8\,(10)} \cdot \frac{2 + 7/8\,(4)}{2} \simeq 31$$

as particles annihilated between $T \simeq 10^{15}$ GeV and the present. Of course the baryon number density is unaffected by these (baryon number conserving) decays so n_B/n_γ has gone down a factor 31.

If the universe went through any highly non-equilibrium process between the time of baryon number synthesis and the present, such as supercooling at a first-order phase transition, then n_B/n_γ may have been diluted by a large factor.

B. Evidence for Asymmetry

It is not easy to assure ourselves that distant galaxies are composed of baryons (not antibaryons). The basic problem is that almost all astronomical information comes from electromagnetic radiation, and the photon is its own antiparticle.

There is some evidence from cosmic rays, namely that no antinuclei have been observed although many ordinary nuclei are seen. However the origin of cosmic rays is not clearly understood and in fact there seems to be an excess of antiprotons compared to standard models (these may of course be secondary products of high-energy collisions involving protons in the atmosphere or near the original source). Also regions where matter and antimatter impinge would be prolific sources of high-energy γ-rays from e^+e^- annihilation and no appropriate sources are seen. If antimatter existed it would have to be in "islands" separated from matter "islands". Since galaxy clusters are known to be filled with gas they must be all matter or all antimatter; the segregation would have to occur on very large scales. It is difficult to arrange this consistent with homogeneity and isotropy of the microwave background.

To get an idea of the difficulty, consider what would happen if there were no net baryon number and everything were homogeneous and isotropic. The protons and antiprotons annihilate down to a small residual density, then decouple. Decoupling occurs when the probability for annihilation after T_d becomes less than one:

$$1 \gtrsim \int_0^{T_d} n\sigma \, dt \simeq \int_0^{T_d} ((mT)^{3/2} e^{-m/T}) \frac{1}{m_\pi^2} \left(\frac{dT}{G^{1/2} T^3} \right) \tag{1}$$

where $(mT)^{3/2} e^{-m/T}$ is the number density for nonrelativistic particles (protons or antiprotons) of mass m at zero chemical potential. The integrand will be very large for $T \simeq m$ because $1/(G^{1/2}) = M_{p\ell} \gg m$ but drops rapidly as T decreases because of the exponential Boltzmann factor. We would thus need roughly $e^{-m/T_d}(M_{p\ell}/m) \simeq e^{-m/T_d} 10^{19} \simeq 1$ for the integral to be close to unity. More precise calculations show that at decoupling the ratio of protons (or antiprotons) is in fact nearly

$$\frac{n_p}{n_\gamma} \simeq \frac{(mT_d)^{3/2} e^{-m/T}}{T_d^3} \simeq 10^{-19}. \tag{2}$$

This misses the observations by many orders of magnitude. Therefore in baryon symmetric cosmology the separation of protons and antiprotons must occur at higher temperatures in large domains. Within each domain we have the same old problem of explaining the asymmetry between matter and antimatter; moreover we must find a way to stabilize the domains, arrange for the microwave background to come out smooth, and explain why the direction of the asymmetry varies in space. If this is possible at all it must be pretty complicated, and in the following I will discuss only the situation within one domain which I pretend (and suspect) is the whole observed universe.

C. Requirements for a Theory of Asymmetry

We will try to make a theory in which a state with B = 0 evolves into one with B ≠ 0 as the universe expands. There are several obvious requirements such a theory must meet:

i) $\Delta B \neq 0$ processes must occur.

ii) C violating processes must occur (since B is C-odd, and the expansion of the universe does not violate C).

iii) CP violating processes must occur (since B is CP-odd,...).

Notice that (fortunately!) CPT-violating processes are not required even though B is CPT-odd, since in an expanding universe both T and CPT are violated by the boundary conditions - the expansion away from the Big Bang fixes an arrow of time.

C and CP violation have been seen in the laboratory; B violation has become very plausible theoretically. We shall try to put these elements together shortly, but first consider two slightly more subtle requirements:

iv) Thermal equilibrium must be violated (if $\Delta B \neq 0$ processes occur then in thermal equilibrium B = 0 - since the thermal equilibrium state is unique it is CPT invariant).

v) Massive particles (masses not $\ll M_{p\ell}$) must occur.

This is closely related to iv). Consider the Boltzmann equation for evolution of a gas of massless particles. The distributions satisfy

$$\frac{dn_i(p)}{dt} = \text{(collision term)} + \text{(expansion term)} . \qquad (3)$$

If the initial distribution is thermal, then the expansion term is taken care of by rescaling the temperature, as in Lecture 2a. For a thermal distribution, however, the collision term vanishes. So, self-consistently, we get a solution by taking the distributions to be always thermal with the appropriately redshifted temperature. The <u>distribution</u> remains thermal even though the reactions may be too slow to maintain thermal equilibrium.

Insofar as particles do have mass we would expect that positive powers of $m/M_{p\ell}$ occur in any asymmetry developed, since the asymmetry must vanish as $m \to 0$ and $(M_{p\ell})^{-1}$ sets the time scale of expansion. Unless m is not much less than $M_{p\ell}$, this will rapidly bring us below the desired $n_B/n_\gamma = 10^{-10}$.

For massive particles it is the momentum which is redshifted (cf. the discussion to Lecture 2) and the thermal distribution, which follows from the Boltzmann factor involving energy, is destroyed by expansion. This also achieves condition iv) - expansion of a gas through temperatures near the mass of some constituent particle is an inherently nonequilibrium process (bulk viscosity).

D. A Simple Scenario - Drift and Decay

A very simple scenario which is not quite realistic but illustrates the principles is the following:

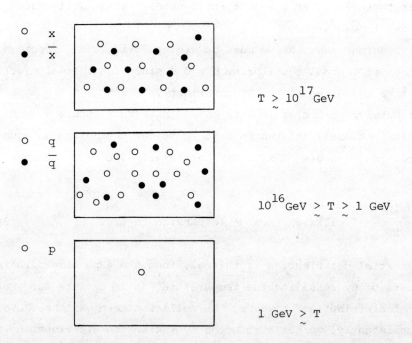

x and $\bar{x}$ are supposed to be massive bosons ($M_x \simeq 10^{17}$ GeV, say) which are so weakly coupled that they cease to interact significantly while they are relativistic (say at $T \simeq 10^{18}$ GeV). They then simply

drift until they finally decay. If the decays violate C, CP, and B
it is possible that the decay products of an equal number of x's and
$\bar{x}$'s produces more quarks than antiquarks. Eventually the quarks and
antiquarks annihilate and combine into ordinary baryons.

The "weak coupling" condition of the previous paragraph amounts
to $M_x \gg M_{P\ell}/\alpha_x$ where α_x is the "fine-structure constant" for x
couplings. If this condition is satisfied, the simple drift and
decay scenario is essentially correct. Standard unified models do
not quite satisfy the condition, so more reactions have to be taken
into account and the kinetics is more complicated.

By the way it is easy to locate the source of disequilibrium in
the drift and decay scenario. As the temperature goes below M_x the
x's and $\bar{x}$'s do not become rare as $e^{-M/T}$, instead they simply drift
until (maintaining a constant density) they finally decay. Before
the decays, they are overabundant compared to their equilibrium
distribution.

E. A Little Microscopics

It is instructive to analyze drift and decay more micro-
scopically.

The decay of x and $\bar{x}$ bosons is assumed to occur after they have
become non-relativistic. They produce a gas of energetic, effectively massless, particles. If the density of x's is n_x, the
temperature of this gas is determined by the conservation of energy
to be $T^4 \propto \rho = M_x n_x$. If the average net baryon number produced is
a per decay of x and $\bar{x}$, then the baryon number density is

$$n_B = a n_x . \qquad (4)$$

The baryon to photon ratio is then approximately

$$n_B/n_\gamma \simeq a n_x/T^3 \simeq a n_x^{1/4} / M_x^{3/4} .$$

To determine n_x, we realize that when $T \gtrsim M_x$ the density of x-bosons is essentially T^3. In expansion from $T = M$ to the time t_d of decay, the density becomes

$$n_x = M_x^3 (R/R_o)^3$$

where R, R_o are the final and initial radii respectively. R/R_o is determined from the equation for expansion in a matter-dominated era:

$$\dot{R}/R \simeq G^{1/2} M^2 (R/R_o)^3$$

so

$$(R_o/R)^{3/2} - 1 \simeq G^{1/2} M^2 t_d .$$

The decay time is approximately $t_d \simeq 1/(\alpha_x M_x)$. If $t_d \lesssim 1/(G^{1/2} M^4)$, i.e. $M_x \lesssim \alpha M_{P\ell}$, then R_o/R will be close to one and n_B/n will be essentially a. If $M_x \gg \alpha_x M_{P\ell}$ there is a suppression and

$$n_B/n_\gamma \simeq a(\alpha_x M_{P\ell}/M_x)^{1/2} . \tag{5}$$

It is remarkable that this says that if n_B/n_γ is to be reasonable M_x cannot be much larger than $M_{P\ell}$; on the other hand we have seen that M_x cannot be much smaller than $M_{P\ell}$ if we are to get out of equilibrium.

It remains to calculate a. Compare x and $\bar{x}$ decaying into conjugate channels:

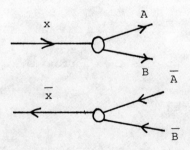

ERICE LECTURES ON COSMOLOGY

The amplitude for $x \to A+B$ will be an expression of the form

$$a_{x \to A+B} = a_1 I_1(i\varepsilon) + a_2 I_2(i\varepsilon) + \ldots$$

where the a_i are couplings associated to different Feynman graphs and the I_i are the corresponding integrals, which may require the $i\varepsilon$ prescription for their definition, with $I_i(i\varepsilon)^* = I_i(-i\varepsilon)$. The corresponding expression for $\bar{x} \to \bar{A}+\bar{B}$ is

$$a_{\bar{x} \to \bar{A}+\bar{B}} = a_1^* I_1(i\varepsilon) + a_2^* I_2(i\varepsilon) + \ldots$$

The difference in rates is therefore of the form $|a_x|^2 - |a_{\bar{x}}|^2 \propto \text{Im } a_1^* a_2 \text{ Im } I_1 I_2^* + \ldots$ The necessity for $\text{Im } a_1^* a_2 \neq 0$ if we want an asymmetry of course reflects the need for CP violation. If the Feynman integrals were well-defined without the $i\varepsilon$ prescription then our two amplitudes would simply be complex conjugates and the corresponding rates would be equal; the requirement $\text{Im } I_1 I_2^* \neq 0$ shows that a non-zero asymmetry requires interference between absorbtive and non-absorbtive processes.

The simplest possible contribution would arise from interference of the tree and (absorbtive) one-loop graphs

In practice for realistic theories more complicated graphs are usually required. The fundamental reason is that gauge theories do not like to break CP. Arbitrary Yang-Mills theories coupled to fermions do not violate CP (except for the θ-angle); the required CP violation must come from the poorly understood Higgs sector.

A very important negative result is that the minimal SU(5)

model, with quarks and leptons getting their masses from a single Higgs field, leads to a very small value for a. The non-vanishing contributions to a require many small Higgs couplings. Since this minimal model can adequately represent the phenomenology of CP violation in the K-meson system, we learn that the presently observed CP violation parameters are not sufficient to predict n_B/n_γ quantitatively.

F. Thermalization

Although lack of knowledge concerning CP violation at ultrahigh energies prevents a quantitative calculation of n_B/n_γ, there is an instructive piece of the calculation that can be done. Suppose that, by whatever means, an asymmetry between particles and antiparticles has been generated, at say $T \simeq 10^{14}$ GeV. Baryon number violating exchange of virtual heavy bosons will tend to restore thermal equilibrium, to wipe out the asymmetry. Given enought time thermal equilibrium, with $B = 0$, would be approached at any fixed temperature. However in an expanding universe there is a competition between reaction rates and the rate of expansion.

The return to equilibrium does not depend on CP violation. It occurs in tree graphs by exchange of superheavy gauge bosons, whose couplings we know (if we know anything about superunified theories). The calculations are straightforward but rather involved. They indicate that with standard values of the parameters for superunified theories the thermalization is an efficient process. Any asymmetry existing at $T \simeq 10^{14}$ GeV will be severely suppressed in general.

An interesting exception to this occurs if there are conserved or approximately conserved quantum numbers. As an example, suppose that although B is violated by heavy boson exchange, baryon minus lepton number B-L is conserved (this occurs in many superunified models). There are reactions, $q + q \leftrightarrow \bar{q} + \bar{\ell}$, $\bar{q} + \bar{q} \leftrightarrow \bar{q} + \ell$; which can change B but leave B-L unchanged. We can track the

development of asymmetries from these reactions; for instance from
$q + q \to \bar{q} + \bar{\ell}$ we have

$$\frac{dn_q}{dt} \propto -2n_q^2$$

$$\frac{dn_{\bar{q}}}{dt} \propto n_q^2$$

$$\frac{dn_{\bar{\ell}}}{dt} \propto n_q^2$$

(Actually these equations are slightly modified by Fermi statistics.)
Some algebra allows us to find, adding all four reactions,

$$\frac{d(3\lambda+\eta)}{dt} \propto -(3\lambda+\eta)$$

$$\frac{d(3\lambda-\eta)}{dt} \propto 0$$

where n_q, $n_{\bar{q}}$, n_ℓ, $n_{\bar{\ell}} = (1+\lambda)$, $(1-\lambda)$, $(1+\eta)$, $(1-\eta)$ times an average density and λ, η are small quantities. These equations tell us that B+L is thermalized, while B-L is conserved. If the reactions are efficient, B+L will rapidly approach zero. If the initial asymmetries are B_i, L_i then for the final asymmetries we will find

$$B_f + L_f = 0$$

$$B_f - L_f = B_i - L_i$$

so $B_f = (B_i - L_i)/2 = -L_f$. This example shows both the importance of conserved quantum numbers and the transformation of asymmetries into different forms by thermalization.

LECTURE 4: MYSTERIES IN THE SKY

A. Horizons

Our generic metric is $ds^2 = d\eta^2 - (d\chi^2 + \Psi(\chi)^2 d\Omega^2)$ so the light cones look just like flat space (this is why η is a useful variable). The plane slice through $\theta = \pi/2$ looks as follows:

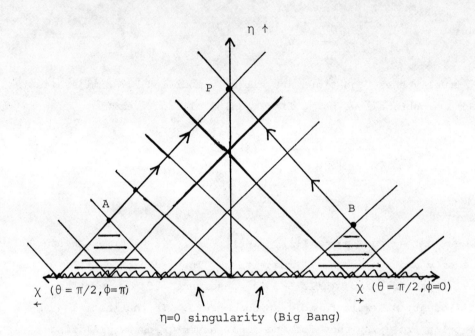

The important new thing is that we cannot penetrate past the singularity at $\eta = 0$. This effects the causal structure of the universe. An example of the implications is shown in the diagram. The observer at P receives signals from events at A and B which have never been in causal contact. That is, the light cones drawn back in time from A and B do not intersect. This means that no single localized event could have influenced both A and B.

It may seem peculiar that even though in the early stages the radius $R \to 0$, the observers being squished together cannot communicate. This is because even though R is getting small, t is getting small even faster. There is no time to exchange light signals. The causal disconnection will occur whenever $R \to t^p$, $p < 1$, near the singularity. In particular, it occurs for Einstein-de Sitter where $p = 2/3$ or for the radiation dominated case $p = 1/2$. Only drastic modifications of the equation of state (or of gravity theory) could undo this feature.

The closed universe presents some additional features with tragic consequences for its inhabitants. In this case the universe ends in a Big Crunch symmetric with the Big Bang. Also χ is a periodic variable, so our diagram is modified to this:

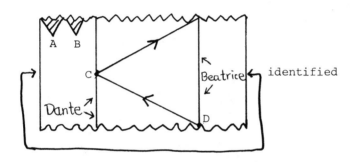

One can imagine an observer, call him Dante, at C first catching sight of the beautiful maiden Beatrice in her early youth at D. She lives half-way across the universe. Immediately upon catching sight of Beatrice, Dante writes a poem declaring his love. He sends the poem as quickly as he can by radio signals toward Beatrice. She is just about to receive it when - Crunch! This is an example of an event horizon - beyond which lie events which we can never know about. Another tragedy is in the upper left corner. A and B can no longer communicate; these former friends are now causally disconnected. And in the end each observer must face the Crunch alone.

B. Background Radiation

A profoundly disturbing example of causal disconnection concerns the microwave background radiation.

Consider events at "time" η. If we observe these at time η_o, what portion of the sky is causally connected? Well, a typical event at $(\eta,0)$ generates a backward light cone including $(0,\chi)$ with $\chi \leq \eta$; propagating this forward we find it causally connected with all (η,χ) for $\chi \leq 2\eta$. On the other hand all events (η,χ) with $\chi \leq \eta_o - \eta$ are visible from $(\eta_o, 0)$.

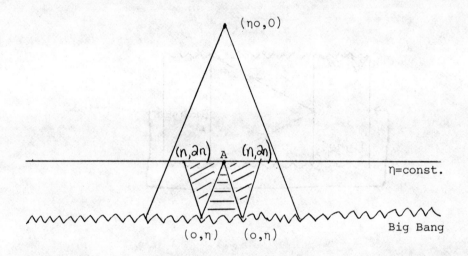

The fraction f of the sky causally connected is just the ratio of areas:

$$f = \frac{\Psi^2(2\eta)}{\Psi^2(\eta_o - \eta)} . \qquad (1)$$

For definiteness let us evaluate this for E-dS; the answer is

$$f = \frac{4\eta^2}{(\eta_o - \eta)^2} = \frac{4}{(\eta_o/\eta - 1)^2} \qquad (2)$$

or remembering $\eta \propto (1/1+z)^{1/3}$; $f = \dfrac{4}{((z+1)^{1/3} - 1)^2}$.

Another interesting region is the region affected causally by a local disturbance in the very early universe; you can easily convince yourself that to find this we just substitute $2\eta \to \eta$ in the numerator of (1).

For the microwave background we look back to decoupling at $z \simeq 1000$, so $f \simeq 1/25$. Thus the radiation from a given direction could not have been correlated by any local physical event with the raidation 1/5 of the way across the sky. Yet it is observed that all directions have the same intensity, to 10^{-4}! It is a big mystery. It perhaps suggests that <u>non-local</u> events in the early universe created the correlations. An example of this could be nucleation of true vacuum out of a metastable background, which would take place coherently over an extended region ("bubble"). Or perhaps the universe was indeed, as suggested by statistics, once infested with black holes which then decayed away by the Hawking process. These speculations will be made more attractive in d)-e).

C. A Large Entropy

It is always suspicious when large dimensionless numbers occur in the physical world. We will now consider some whoppers.

Let us consider the entropy within our horizon.

Our horizon includes $\chi \leq 2\eta_o$ which is a volume $4\pi R^3 \int_0^{2\eta_o} \psi^2(\chi) d\chi$. For E-dS, which is an adequate approximation; this is $32\pi/3\, \eta_o^3\, R^3(\eta_o)$. The entropy in a comoving volume is constant, so we can simply multiply by $s = (4\pi^2)/45\, T_o^3$ to get the total entropy

$$S = \frac{128\pi^3}{135} \eta_o^3 R_o^3 T_o^3 = \frac{256\pi^3}{45} R_o^3 T_o^3 \frac{t}{GM} = \frac{64\pi^2}{15} t \frac{T_o^3}{G}$$

$$= \frac{512\pi^3}{45} T_o^3 \frac{t}{H^2} = \frac{1024}{135} \pi^3 \left(\frac{T_o}{H}\right)^3 \quad (3)$$

where various E-dS identities have been used to express everything in terms of H. The real point, which could have been forseen on dimensional grounds, is the dependence on the ratio T_o/H which is numerically enormous: $kT_o/\hbar H \simeq 10^{29}$, so $S \simeq 10^{89}$.

The existence of this large number is sometimes called the "horizon problem"; perhaps this phrase is better reserved for the problem discussed in B. In any case, it is very peculiar.

D. Another Large Entropy

A similar large number arises if we ask for the entropy within one "curvature volume" of the universe; i.e. within $\chi \lesssim 1/k$. Assuming generous bounds for q_o, we find from our expressions $q_o = \frac{1}{2(\cosh k\eta/2)^2}$ or $q_o = \frac{1}{2(\cos k\eta/2)^2}$ for respectively open and closed universes that $k\eta \lesssim 2$. Hence the curvature volume is of order $(R/k)^3$ is $> R^3\eta^3$. But this reduces us to the previous case, $S \gtrsim 10^{89}$. Another way of seeing this is simply to note that in the visible universe we see little sign of curvature, so the curvature volume cannot be much less than the horizon volume.

The large amount of entropy within one curvature volume is calles the "flatness problem". It says the universe is much less curved than one might expect a priori (i.e., the Planck scale 10^{-33} cm!) The entropy inside a curvature volume, like the entropy in any comoving volume, is approximately constant. So this very large number is a relic of the early universe.

E. A Very Small Entropy

From another point of view the universe has a very small entropy. The possible entropy in black holes is enormous. In fact, a black hole of mass M has entropy

$$S = 4\pi GM^2 = 4\pi(M/M_{P\ell})^2.$$

For the sun S would be $\sim 10^{79}$ if it became a black hole. Beside this its thermal entropy $\sim R_\odot^3 T_\odot^3 \simeq 10^{48}$ is paltry indeed. This means the sun is in a very statistically unlikely state. It is worse for larger systems since the gravitational entropy grows as M^2, faster than additive.

The universe therefore started in a very ordered, statistically unlikely state. Although we commonly assume thermal equilibrium for "matter", the equilibrium is not complete; it does not include gravity. Otherwise we would have lots of big black holes. I think it is just this lack of gravitational entropy in the beginning that ultimately allows matter to become organised at all (reducing its entropy at the cost of driving the universe very slightly toward gravitational thermal equilibrium). But why this state is realised in the beginning is completely mysterious.

The problems C., D., E. may all be phrased in a unified way: there is much more entropy in matter (as opposed to gravity) than we might expect a priori. Perhaps some highly non-equilibrium process in the early universe pumped lots of entropy into matter. One candidate for this is supercooling as proposed by Guth. His observation is that in many quasi-realistic gauge theories the symmetric vacuum is stable at sufficiently high temperatures but only metastable at lower temperatures. In cooling the universe may get "hung up" in the metastable vacuum; we are then driven far from equilibrium and generate lots of entropy, perhaps mostly in matter. This idea seems to have difficulties however; especially it may be difficult to make the metastable → stable transition complete. By the way, there are still other problems of the same sort which are rarely discussed:

The Q problem: Why is the electric charge per unit entropy (a relic of the early universe) so small?

The L problem: Why is the angular momentum (in units of $\hbar$) per unit entropy (another relic) so small? This is a rephrasing of the isotropy of the universe.

F. Monopoles

As another problem let us consider the cosmological production of magnetic monopoles. These are extended objects connected with the non-trivial topology of the manifold of degenerate ground states in a spontaneously broken gauge theory. Their mass is set by the scale at which the U(1) of electromagnetism is first embedded in a compact group G. Since the monopoles are extended coherent objects they may be very difficult to produce in ordinary collisions. Let us suppose they are produced approximately one per horizon volume at the temperature corresponding to their mass.

If this temperature T occurs at time coordinate η the corresponding horizon volume is $\propto R(\eta)^3 \eta^3$ and the entropy $S \propto R^3 \eta^3 T^3$. For a radiation-dominated universe

$$d\eta = \frac{dt}{R} \propto \frac{dt}{G^{3/4} t^{1/2}} \text{ so } \eta \propto \frac{t^{1/2}}{G^{1/4}} \propto \frac{R}{G^{1/4}} = R M_P \propto \frac{M_P}{T}$$

so $S \propto (M_{P\ell}/T)^3$. Thus very roughly if it is true that one monopole is produced per horizon volume we have, remembering entropy $\propto$ # of photons from statistical mechanics,

$$r = \frac{\text{\# of monopoles}}{\text{\# of photons}} = (T/M_{P\ell})^3 \quad .$$

For the standard SU(5) model $T \simeq 10^{15}$ GeV so this gives $r = 10^{-12}$ or 10^{-3} monopoles per baryon which is absurd (monopole mass density $\gtrsim 10^{11} \rho_c$). On the other hand if U(1) is embedded in a compact group at 10^4 GeV (say), $r = 10^{-45}$ which is quite safe.

G. Cosmological Constant

A last problem concerns the energy of the vacuum. In a spontaneously broken gauge theory the normal vacuum is an energy minimum separated from the symmetric vacuum by energy densities $b \sim (10^{15} \text{GeV})^4$ or $b \sim (10^3 \text{GeV})^4$ for superunified respectively electroweak symmetry breaking. The vacuum would then be a fluid with necessarily

$\langle T_{\mu\nu}\rangle = -bg_{\mu\nu}$ for relativistic invariance. ($T_{\mu\nu}$ = energy-momentum tensor.) In more traditional units $b_{super} \sim 10^{78}$ gm/cm^3, $b_{electrowk} \sim 10^{42}$ gm/cm^3, both much larger than $\rho_c \sim 10^{-29}$ gm/cm^3.

What is the principle that normalizes the zero of energy at the present asymmetric vacuum? Or has the effective b always been zero, even when the universe was in the symmetric phase? Could there be a mechanism here analagous to the axion scheme which can potentially explain the smallness of an analogous parameter in the strong interactions?

ACKNOWLEDGMENT

This work was supported in part by the National Science Foundation, Grant No. PHY77-27084.

REFERENCES

Here is a brief list of sources with which you can follow up these notes:

Lecture 1:

General references: S. Weinberg, Gravitation and Cosmology (Wiley, 1972); J. Silk, The Big Bang (Freeman, 1980).
Nuclear chronology: E. Symbalisty and D. Schramm, Rep. Prog. Phys. 44, 293 (1981).
Supernovae as standard candles: R. Wagoner, Comm. Astr. and Astr. 8, 121 (1979).

Lecture 2:

Neutrinos in cosmology: A. Doroshkevich et al., in XI Texas Symposium on Relativistic Astrophysics (to be published).

Lecture 3:

Popular account: F. Wilczek, Sci. Am. 243 #6, p 81 (1980).
Microscopic calculations: S. Barr, G. Segre, A. Weldon, Phys. Rev. D20, 2494 (1980); D. Nanopoulos and S. Weinberg, Phys. Rev. D20, 2484 (1980).
Thermalization: S. Treiman and F. Wilczek, Phys. Lett. 95B, 222 (1980).

Lecture 4:

R. Penrose, in General Relativity, ed. Hawking, Isreal (Cambridge, 1979); A. Guth, Phys. Rev. D23, 347 (1981).

A much fuller treatment of the subjects in these notes is being prepared by the author and will appear as a book published by Princeton University Press.

DISCUSSIONS

CHAIRMAN: F. Wilczek
Scientific Secretary: E. D'Hoker

DISCUSSION SECTION 1

D'HOKER: Is the big crunch in contradiction with the second law of thermodynamics?

WILCZEK: The answer is fundamentally no. There is an apparent paradox in that the universe is in its maximum entropy state near $t=0$; the entropy density in a comoving volume can only increase with time, yet it again approaches the same (maximum entropy) states near $t=t_{end}$. The resolution of this is that our assumption of thermal equilibrium for matter does not really correspond to the maximum entropy state; we forgot to include gravity. Absolute homogeneity and isotropy corresponds to a very low gravitational entropy; it is very improbable (but nevertheless the universe is this way!)

In reality we would expect formation of many big black holes near the Crunch, so the initial state is not really reproduced.

Related matters will be discussed further in Lecture 4.

ROBERTS: Is there some deep significance to the equaling of cosmology derived classically, and derived through General Relativity?

WILCZEK: The Newtonian approximation can be justified, at least for non-relativistic matter. For general relativity, the equivalent of Newton's iron-sphere theorem is true. If you have a spherically symmetric distribution of matter, you can ignore all that goes on outside and only look inside a small region. This is called Birkhoff's theorem. Once we know that we can study the equations on very small scales, then the curvature of space is negligible and the Newtonian approximation should work. Using this theorem, you can justify the Newtonian approximation, at least for non-relativistic matter. I do not know how you would justify this for radiation.

TELLER: You justified the situation that from a state of apparent maximum entropy you got into a state that is not a maximum by the action of gravitation. In some very early stage you have the situation of uniform temperature, apparently locally maximum entropy, and yet, half an hour later you are left with a lot of deuterium, which today I am trying to use as a fuel. Explanation. That is not due to gravitation, because the gravitation in that process was assumed to be uniform. It appears, looking at the universe from inside, that having started from a state of maximum entropy possible at that time, you develop into a state which is not maximal entropy at that time.

WILCZEK: The fact that you go from a state of maximal entropy at one time to a state that is not maximal possible entropy at a later time does not mean that the entropy in a comoving volume has not increased. It has increased less than it could have but it is still increasing.

COHEN: If I consider a universe filled with only E.M. radiation $R(t) \propto t^{1/2}$, it means the universe is open (for any value of the initial radiation) density. Then is the presence of matter what gives the possibility of a closed universe?

WILCZEK: No, for radiation there are still these three possibilities of the k^2 coefficient being greater than, equal to, or less than zero. Only the equation of state is different. The radiation dominated universe is most interesting as a practical matter for the very early stages where R is small and the curvature is negligible; this is the only case I worked out in the lecture.

HERTEN: The basic assumption in your lecture was that the mass distribution in the universe is homogeneous and isotropic. Are there measurements which show that this assumption is good?

WILCZEK: Yes, there is one very good indication of the isotropy and homogeneity of the universe which is the microwave background radiation at roughly $2.7^{\circ}K$. The fact that it looks the same in all directions of space indicates that when those photons originated, when they last interacted, the universe was homogeneous and isotropic. So we have very good evidence that down to redshifts of a thousand the universe was homogeneous and isotropic. For earlier times, if the universe were very non-uniform, that would propagate to the background radiation observed today. So isotropy and homogeneity were probably also characteristic of the universe at redshifts larger than a thousand.

WIGNER: 1) What is the meaning of the maximum entropy statement? The existence of stars, galaxies, of our life, seem to contradict this. 2) Can you also tell us why the entropy would not have assumed its maximum value if the world is "infinitely old"? Or is it not?

WILCZEK: 1) What we assume in the standard cosmology is that in the early universe the matter became so dense and was so hot that any kind of interactions which are now far from equilibrium, were in equilibrium, like e.g. the weak and electromagnetic interactions. So the assumption is that in the early universe we had thermal equilibrium for all interactions except gravity and then we follow the evolution of that. As the universe expands and becomes cooler, neutrinos will have a more difficult time interacting, ceasing roughly when $T < 20$ MeV (Lecture 2). So the expansion of the universe has driven us out of equilibrium and we are no longer in a state of maximum entropy. On the other hand the entrpy per unit of comoving volume has increased all along. 2) There would be no explanation if the world were infinitely old.

CASTELLANI: Another tacit assumption is the constancy of G (Newton's constant). What if, as Prof. Dirac suggested a long time ago, G varies with time, e.g. decreases in time? Couldn't that provide an amusing explanation for the missing mass? Indeed in this case we are observing galaxies governed by a bigger G than ours, so that a lesser mass is necessary to close them!

WILCZEK: The missing mass problem exists on many scales, including galactic ones. That is, even nearby galaxies seem to contain lots of non-luminous matter (Lecture 2). These galaxies are observed at recent historical epochs, so G would have to change ridiculously fast to account for this missing mass.

COLEMAN: In the case of the galactic clusters, 1 million lightyears away from us, then one would have a gravitational constant changing by a factor of 2 within 1 million years. By your kind of reasonings that is clearly impossible, since the earth would have been at only half the distance from the sun, and we would all have been burnt 1 million years ago.

If you assume that the universe is homogeneous and isotropic and that all observers in the universe are going to the same history and you want to explain the missing mass problem on the basis of a time-changing gravitational constant, uniformly for all the scales of missing mass you are in great trouble, because there are nearby galaxies where mass is missing.

TELLER: I think the situation is now quite clear if you assume that the gravitational constant is changing fast enough to cure the missing mass, you would be in all kinds of troubles within the solar system. That is true.

WILCZEK: Let me add that the missing mass problem is not just a problem of magnitude of the mass, but also the form of galactic distribution of the mass which cannot be solved even in principle assuming a change in the gravitational constant. There is non-luminous matter there from rotation curves.

ETIM: There is a new paper by Canuto, dedicated to Prof. Dirac on the occasion of his 80th birthday. This paper takes into account the change in the gravitational constant as indicated by Dirac. But he also finds that even taking this change into account in the Friedman universe, there is a problem of missing mass.

WIGNER: Our present physics assumes a fundamental and sharp difference between initial conditions and laws of nature. Do you think this sharp difference has and will have unrestricted validity?

WILCZEK: It is always progress when we absorb what used to be boundary conditions into the laws of physics. This has been the case for nuclear abundances and for baryon number density as we shall discuss in Lectures 2 and 3. Other things of this nature which we would like to understand will be discussed in Lecture 4. So one main theme is to obliterate the distinction, yes.

DISCUSSION SECTION 2

D'HOKER: You mentioned that the angular size for an object of fixed physical size has a minimum. Is this minimum physical, and if so where does it occur?

WILCZEK: The formula for angular size is $\Delta\theta \propto \frac{1}{R(\eta)\psi(\eta_o - \eta)}$. For Einstein-de Sitter $R(\eta) \propto \eta^2 (\propto 1/(1+z))$ and $\psi(\eta) = \eta_o - \eta$ so the minimum angle occurs for $\eta^2(\eta_o - \eta)$ = maximum, or $\eta = 2/3\, \eta_o$. This means $1/(1+z) = 4/9$ or $z = 5/4$. This is a very interesting sort of red-shift, in the realm of observed quasars. The exact position of the minimum could in principle determine q_o.

TELLER: Is it true that the maximum angle when you look at the complete sky is 4π, once you have passed the minimum?

WILCZEK: Yes, it would fill the whole sky.

D'HOKER: You discussed the wqve equation for massless particles. What happens with massive particles?

WILCZEK: Extending our consideration in the lecture, the wave equation is

$$\left(-\left(\frac{\partial}{\partial t}\right)^2 + \frac{1}{R^2(t)}\frac{\partial^2}{\partial \chi^2}\right)\phi = m^2\phi \;.$$

It can be solved by $\phi = e^{i\lambda\chi}f(t)$ where

$$f'' + \left(\frac{\lambda^2}{R^2(t)} - m^2\right)f = 0 \;.$$

The momentum $p \propto 1/R\, \partial/\partial\chi\, \phi$ goes like λ/R; it is redshifted. The energy is not simply redshifted - the equation for f is not so trivial. In the quantum mechanical theory, this corresponds to scattering off a time-dependent potential which can lead to particle creation, etc.

HERTEN: In order to calculate the abundance of ^{4}He the abundance of p and n is used at the time this abundance froze in. But it is also possible that ^{4}He is produced later in stars or during supernova explosions. How big is this amount of ^{4}He?

WILCZEK: It is believed to be small. Stars of course produce lots of ^{4}He by fusion but this is locked in the core. In supernovae ^{4}He is produced along with heavier elements. So we cannot easily explain the factor > 100 disparity in abundances between ^{4}He and heavier elements if we rely on supernovae for the helium.

DUERKSON: As an indication of the possibility (or impossibility) of detecting the cosmic background neutrinos, could you compare the present energy and flux of these neutrinos to that of the solar neutrinos captured by chlorine in the solar neutrino flux experiments?

WILCZEK: The neutrinos may be concentrated around galaxies rather than uniformly distributed. This gives a density of about $10^8/cm^3$ which is down by a factor of 10^{-2} compared to the solar neutrinos. Furthermore, the cosmic neutrinos are virtually stationary, as opposed to the solar neutrinos' energy which is typically about 1 MeV. Since these neutrinos are so unenergetic, I cannot think of an experimental means of detecting them.

KARLINGER: You mentioned the old and the new missing mass problems. Suppose we try to solve the new one by assuming a suitable neutrino mass distribution in the galaxies. To what extent does that solve the old missing mass problem?

WILCZEK: The magnitude of the mass deficit is roughly a factor 10 for galaxy clusters (and binary galaxies); possibly as large for single galaxies as discussed in the lecture. So a unified explanation seems possible.

ROBERTS: The dynamical equations have a parameter k with units of velocity; does anything wonderful happen when it appraoches or exceeds the velocity of light?

WILCZEK: No, it is unphysical.

DUNCAN: The table you showed at the beginning of the lecture today had a lot of great events in the history of the universe. There was nothing between 10^2-10^{15} GeV. What happens if you have oases in this desert?

WILCZEK: Of course it depends on exactly what they are sowing at the oases, but since monopoles and baryons are the only known relics from earlier times the main effect anticipated would be a dilution of the density of these objects (relative to entropy density).

BANKS: With regard to the Ostriker-Peebles missing mass problem: does the numerical experiment give any time scale for the instability? In other words, is it possible to say that spiral galaxies exist without a stabilizing halo, simply because the universe is in some sense young?

WILCZEK: The time scale is very rapid on the cosmological scale. I do not remember the exact number, but it must be a modest multiple of the rotation period.

TELLER: One very simple question and then one a little more complicated. You said this morning that more than 5 kinds of neutrinos cannot be tolerated. If there were 6 what would happen?

WILCZEK: We would produce more ^{4}He than observed.

TELLER: What if there were 3?

WILCZEK: Then enough ^{4}He could still be produced.

TELLER: If you look far enough back, you see things coming in with an angle of 4π and that is what you do see, as you have said with an accuracy of 10^{-4} in the $2.7°$K background radiation.

WILCZEK: Yes, but the 2.7 degree radiation is not associated with objects of a fixed size.

TELLER: What happened later was the formation of galaxies. If someone would look for the formation of galaxies, these should be lumpy in the beginning and should make in these 10^{-4} fluctuations some deviations. Is there any hope to see this formation of galaxies?

WILCZEK: Yes. The full answer to this question depends on your theory of galaxy formation, but some models are tightly constrained by the isotropy. This is an area where slightly more accurate observation might be very fruitful.

DISCUSSION SECTION 3

GOTTLIEB: You mentioned the numerical experiment of Ostriker and Peebles which indicates that spiral galaxies are unstable to the collapse of the spiral arms to a bar. What is the observational data on bar galaxies; do they exhibit anomalous rotational spectra?

WILCZEK: I'm sorry, I do not know. That is a good question.

D'HOKER: Could you explain which are the relevant parameters that enter the problem of the damping of fluctuations away from B=0 and why it is so large in most standard models?

WILCZEK: The relevant parameter is the rate of reactions which violate B compared to the expansion rate. If the reactions were infinitely fast no asymmetry would remain. The parameter of smallness is $M_{Planck}/M_{unified}$ since the reaction rate goes inversely with M, the expansion rate as $M_{p\ell}$. Because this number is large, the situation is dangerous. It is somewhat relieved by the smallness of the coupling constant, kinematic factors, etc.

TELLER: You have given the time at which the neutrinos go out of equilibrium at 20 MeV.

WILCZEK: Something like 1 second.

TELLER: Time of galaxy formation is about 10^{16} sec. In between, the neutrinos can move freely, except if they are restrained by gravitation. Do you therefore postulate that between 1 sec and 10^{16} sec the neutrinos have been confined by gravitation?

WILCZEK: Not confined, but their distribution has been effected by gravitation. In particular, if you start out with a small fluctuation of neutrinos so you have a higher gravitational potential in some region then that fluctuation tends to grow because more neutrinos will be attracted to the place where they are already dense. In other words, the expansion rate is slower there.

TELLER: I would like to claim the opposite. If the neutrinos do not collide, and if in a region there are an excess of neutrinos, then they will diffuse out of this region. Additional neutrinos that go through that region will be speeded up and come even close to light velocity and cross that region with extra speed. One does not get an instability. One does not get an increasing lumpiness unless one has collisions and without collisions, neutrino fluctuations created at an early time will long ago have disappeared before galaxies can be formed.

WILCZEK: The effects you talk about do exist, but it is a quantitative question which is more important. This is a classic problem in galaxy formation and it is found that fluctuations can grow when the neutrinos become nonrelativistic.

SEIBERG: There are models where baryon number violating processes take place by the exchange of several heavy particles rather than just one. In such models the very high mass scale may be much smaller than 10^{15} GeV without causing contradictions with the measured limit of the lifetime of the proton. What are the implications of such models on the baryon number of the universe?

WILCZEK: I think they may have difficulties, because the baryon number violating processes will be in equilibrium down to low temperatures and B will be thermalized. Although proton decay may require exchange of several particles, some other more elementary B-violating processes will be much faster.

DUNCAN: You talked about net baryon number of the universe as indication of matter-antimatter asymmetry. Can you ever get information about any net lepton number from astrophysical data (good to find B-L)? What about isotropy of B nonconservation from cosmic rays?

WILCZEK: It is difficult to observe since it requires subtracting a large number of ν's from a large number of $\bar{\nu}$'s. Cosmic rays are deflected by magnetic fields in our galaxy so anisotropy is expected (except for the highest energies perhaps).

COHEN: Could you tell how sensitive the theoretical prediction on the actual asymmetry of the baryon-antibaryon depends on the CP violation parameters associated with the new families of quarks?

WILCZEK: The crucial element to know for baryon number violation is CP violation at temperatures of the order of 10^{15} GeV and then there are probably many more interactions that contribute.

FORGACS: Is it conceivable that primordial black holes played an important role in generating baryon asymmetry?

WILCZEK: Yes, although there is no reliable way to calculate it.

DISCUSSION SECTION 4

D'HOKER: You mentioned the problem of the cosmological constant. It is often said that supersymmetry could offer a solution to this question because by construction the vacuum energy is zero. However supersymmetry has to be broken at energies of at least 1 GeV. Are there any realistic models where this could be made consistent with the small cosmological constant?

WILCZEK: Supersymmetry sharpens the problem since the vacuum energy is a calculable finite quantity in such theories. In ordinary theories it is simply an infinite subtraction term. Unfortunately the finite result you get in supersymmetric theories is much too large; remember the scale we need for the cosmological constant is at most 10^{-12} GeV.

KARLINER: With regard to causally disconnected regions of the universe: you said we do not understand why they should be in the same temperature. What about the direction of symmetry breaking, which seems to be the same everywhere?

WILCZEK: This is a gauge-dependent remark; the direction of symmetry breaking is not of direct physical significance.

BANKS: Could you discuss the cosmological model of Guth, which attempts to deal with some of the problems you discussed this morning?

WILCZEK: The main idea is that of a non-local, highl- nonequilibrium event, which I mentioned. Guth's original idea was that, sometime between monopole production and baryon production, the universe was in an unstable vacuum of some sort, and gets far out of equilibrium. Getting out of the false vacuum is a problem, but if it happens it would involve a large increase in entropy which could account for what is now observed.

TELLER: Doesn't the assumption of maximum local entropy everywhere require that all regions look the same and therefore remove the problem in understanding why the microwave background is isotropic?

WILCZEK: No. Concretely, what is not excluded is that there could be modulations of the local temperature over scales of a horizon size. Each region would be in equilibrium, but at different temperatures.

COHEN: Why is the density of the universe so close to the critical density? Could you comment on the possibility that the explanation is biological?

WILCZEK: This is exactly the flatness problem. I do not find the biological explanation acceptable. You could then ask why we exist and you would just have moved the problem from one place to another.

WITTEN: One can quantitatively refute the biological explanation as follows. The biological explanation only makes sense if you consider an ensemble of universes and you can show that life only developed in those where $\rho \simeq \rho_{cr}$. Those were very rare, but those are the ones in which intelligent beings finally studied the universe. If you look at it that way for every universe like the one we have, there should be thousands with a ρ different from ρ_{cr} by a little bit more, and an infinitesimal number where ρ is a little bit less. Now it is clear that ρ has to be close to ρ_{cr} for us to exist, but not as close as it actually is and if you believe that could have been as close to ρ_{cr} as it actually is, then even the biological explanation says that we are really observing an incredible fluctuation.

WILCZEK: Conditions for life are not very restrictive; we could have an oasis of solar system size in a very inhomogeneous, unflat universe.

CASTELLANI: Take two regions of the sky with disconnected light cones that extend to the singularity (as in your figure). Why do you say that these two regions have to be uncorrelated? After all, both their (inverted) light cones hit the same singularity. Very close to the singularity, the temperature exceeds 10^{19} GeV, and thus quantum gravity has to enter the stage, maybe providing some correlation mechanism.

WILCZEK: The statement is that with reasonable equations of state and gravity theory extrapolated right back to the beginning you get a spacelike singularity. Of course the extreme conditions at this time might very well require different physics, that cannot be excluded. It is misleading though to say that they hit the same singularity and could thereby be correlated; the conventional singularity is not a local physical event.

INTRODUCTION TO SUPERSYMMETRY

Edward Witten[*]

Joseph Henry Laboratories
Princeton University
Princeton, New Jersey 08544

I. INTRODUCTION

Supersymmetry is a remarkable subject that has fascinated particle physicists since it was originally introduced.[1] Although supersymmetry is no longer a new idea, we still do not know in what form, if any, it plays a role in the proper description of nature.

The present status of supersymmetry might be compared very roughly to the status of non-Abelian gauge theories twenty years ago. It is a fascinating mathematical structure, and a reasonable extension of current ideas, but plagued with phenomenological difficulties.

In these lectures, I will present an introduction to supersymmetry, or at least to some aspects of this extensive subject. I also will describe some recent results. Supersymmetry has the reputation of a subject that is difficult to learn. I will try to at least partially dispel this unjustified impression.

In these lectures we will discuss only global supersymmetry, not supergravity.

---

[*]Supported in part by NSF Grant PHY80-19754.

A number of useful reviews of supersymmetry are available, which approach the subject from several points of view.[2] Many aspects of supersymmetry that I will not be able to discuss are treated in these review articles.

II. BOSE-FERMI SYMMETRY

If one considers a theory of two decoupled Bose fields, ϕ_1, and ϕ_2, so

$$\mathcal{L} = \int d^4x \frac{1}{2}((\partial_\mu \phi_1)^2 + (\partial_\mu \phi_2)^2) \tag{1}$$

it is possible to combine ϕ_1 and ϕ_2 into a conserved current $J_\mu = \phi_1 \partial_\mu \phi_2 - \phi_2 \partial_\mu \phi_1$ even though they are not interacting. Actually, since, for example, $\partial_\alpha \phi_1$ or $\partial_\alpha \partial_\beta \phi_1$ satisfies the Klein-Gordon equation just as ϕ_1 does, one can form in the free field theory additional conserved currents such as

$$J_{\mu\alpha} = \partial_\alpha \phi_1 \partial_\mu \phi_2 - \phi_2 \partial_\mu \partial_\alpha \phi_1$$

$$J_{\mu\alpha\beta} = \partial_\alpha \partial_\beta \phi_1 \partial_\mu \phi_2 - \phi_2 \partial_\mu \partial_\alpha \partial_\beta \phi_1 \tag{2}$$

It is easy to check that these are conserved because $\nabla^2 \phi_1 = \nabla^2 \phi_2 = 0$.

The reason that J_μ is far more important than $J_{\mu\alpha}$ or $J_{\mu\alpha\beta}$ is that J_μ can still be conserved in interacting theories. One can add to $\mathcal{L}$ a term like $V(\phi_1^2 + \phi_2^2)$ that is invariant under $\delta\phi_1 = \phi_2$, $\delta\phi_2 = -\phi_1$. J_μ is still conserved.

However, when interactions are included, $J_{\mu\alpha}$ and $J_{\mu\alpha\beta}$ are no longer conserved. Moreover, it is not possible to redefine them (by adding extra terms to allow for the interactions) so that they will still be conserved. One may readily verify this in special cases. In general, it is a consequence of the Coleman-Mandula theorem.[3] Coleman and Mandula showed (basically by S matrix theory alone) that in a theory with non-zero scattering amplitudes in more than 1+1

INTRODUCTION TO SUPERSYMMETRY

dimensions the only possible conserved quantities that transform as tensors under the Lorentz group are the following. The usual space-time symmetries are certainly allowed: the energy-momentum operator P_μ and the Lorentz transformations $M_{\alpha\beta}$ commute with the S matrix in all of our usual theories. We may also have arbitrary Lorentz-invariant conserved quantum numbers Q_i (electric charge, baryon number, etcs.). Finally, if all particles are massless, the Coleman-Mandula theorem allows conformal invariance, which is not usually realized, however, in field theories with interactions. The Coleman-Mandula theorem forbids "exotic" conservation laws — conservation laws other than the usual space-time symmetries which do not commute with Lorentz transformations.

For a proof, I refer you to reference (3); I will just give a rough idea. The basic idea is that conservation of P_μ and $M_{\alpha\beta}$ leaves only the scattering angle unknown in (say) a two body collision. Additional, exotic conservation laws would determine the scattering angle, leaving only a discrete set of possible angles. Since the scattering amplitude is always an analytic function of angle, it would actually then have to vanish at all angles.

Note that the argument obviously does not apply in 1+1 dimensions. In 1+1 dimensions, the only possible angles are 0 and π; there is no such thing as analyticity as a function of scattering angle. This is why, in 1+1 dimensions, it is possible to have interacting systems, such as the sine-gordon equation, with exotic conservation laws.

To illustrate the argument by a concrete example, suppose we have a conserved traceless symmetric tensor $Q_{\beta\gamma}$. This is an exotic conservation law because, transforming as a tensor, it does not commute with Lorentz transformations. Its matrix element in a one particle state of momentum p and (for simplicity) spin zero would have to be $\langle p | Q_{\beta\gamma} | p \rangle = p_\beta p_\gamma - \frac{1}{4} g_{\beta\gamma} p^2$, by Lorentz invariance.

Applied to a two-body collision, with incident particles of momentum p_1, p_2, and outgoing particles of momentum q_1, q_2 (figure (1)), the conservation of $Q_{\beta\gamma}$ would tell us $p_{1\beta} p_{1\gamma} + p_{2\beta} p_{2\gamma} = q_{1\beta} q_{1\gamma} + q_{2\beta} q_{2\gamma}$.* This is possible only if the scattering angle is zero. With more effort, the same type of argument works even if the particles have non-zero spin.

Now going back to the exotic conserved currents that we defined in free field theory (equation (2)), the corresponding conserved charges

$$Q_\alpha = \int d^3x\, J_{o\alpha}$$
$$Q_{\alpha\beta} = \int d^3x\, J_{o\alpha\beta} \qquad (3)$$

are forbidden by the Coleman-Mandula theorem for theories with a non-trivial S matrix in more than 1+1 dimension. This is why it is impossible to add interactions to the free field theory in a way that preserves the conservation of $J_{\mu\alpha}$ and $J_{\mu\alpha\beta}$.

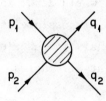

Figure 1

---

*It is here assumed that the matrix element of $Q_{\beta\gamma}$ in the two particle state $|p_1 p_2\rangle$ is the sum of the matrix elements in the states $|p_1\rangle$ and $|p_2\rangle$. This is true if $Q_{\beta\gamma}$ is the integral of a local current density, or, more generally, if $Q_{\beta\gamma}$ is defined in a way that is "not too non-local".

INTRODUCTION TO SUPERSYMMETRY

Instead of two non-interacting scalars, consider next a free massless charged scalar ϕ and a free two component (left handed) fermion ψ_α. The Lagrangian is

$$\mathcal{L} = \int d^4x \, (\partial_\mu \phi^* \partial^\mu \phi + \bar{\psi} \, i \, \slashed{\partial} \, \psi) \tag{4}$$

Again — although the fields are non-interacting — one can define conserved currents connecting them. One of the simplest is*

$$S_{\mu\alpha} = (\partial_\sigma \phi^* \, \gamma^\sigma \, \gamma_\mu \, \psi)_\alpha \tag{5}$$

It is easy to show $\partial_\mu S^\mu_{\ \alpha} = 0$:

$$\partial_\mu S^\mu_{\ \alpha} = (\partial_\mu \partial_\sigma \phi^* \, \gamma^\sigma \, \gamma^\mu \, \psi)_\alpha + (\partial_\sigma \phi^* \, \gamma^\sigma \, \gamma^\mu \, \partial_\mu \psi)_\alpha \tag{6}$$

The second term vanishes since $\gamma^\mu \partial_\mu \psi = 0$. For the first term, note $\partial_\mu \partial_\sigma \psi$ is symmetric in μ and σ, so we may replace $\gamma^\sigma \gamma^\mu$ by $1/2(\gamma^\mu \gamma^\sigma + \gamma^\sigma \gamma^\mu) = g^{\mu\sigma}$. We then have $g^{\mu\sigma} \partial_\mu \partial_\sigma \phi^* = 0$.

Again, additional conserved currents can be defined. The only property of ψ that we used was the Dirac equation, which is also satisfied by $\partial_\gamma \psi$, so

$$S_{\mu\gamma\alpha} = (\partial_\sigma \phi^* \, \gamma^\sigma \, \gamma_\mu \, \partial_\gamma \psi)_\alpha \tag{7}$$

is also conserved, that is, $\partial_\mu S^\mu_{\ \gamma\alpha} = 0$.

The exciting fact that makes supersymmetry interesting is that, although conservation of $S_{\mu\gamma\alpha}$ is always ruined when interactions are included, it is possible to add interactions in such a way that $S_{\mu\alpha}$ is conserved. A simple example is

$$\mathcal{L} = \int d^4x \, (\partial_\mu \phi^* \partial^\mu \phi + \bar{\psi} \, i \, \slashed{\partial} \, \psi - g^2 |\phi|^4 - g(\phi \, \psi_\alpha \, \psi^\alpha + h.c.)) \tag{8}$$

which is known as the "massless Wess-Zumino model". Although $S_{\mu\alpha}$

---

*$S_{\mu\alpha}$ is here a left-handed Weyl spinor, like ψ itself. We will soon go over to a Majorana basis.

as previously defined is not conserved in this model, by adding an extra term

$$S_{\mu\alpha} \to S_{\mu\alpha} + ig\,\gamma_\mu\,\phi^{*2}\,\psi^* \qquad (9)$$

One preserves conservation of the current in the presence of interaction. (Here ψ^* is a <u>right-handed</u> two component spinor, the hermitian conjugate of ψ, which is <u>left-handed</u>. See, for example, the notes of Wess and Bagger[2] for more details.)

Why can $S_{\mu\alpha}$, but not $S_{\mu\gamma\alpha}$, be conserved in the presence of interactions? We can find out by studying the conserved charges

$$\hat{Q}_\alpha = \int d^3x\, S_{o\alpha}$$

$$\hat{Q}_{\gamma\alpha} = \int d^3x\, S_{o\gamma\alpha} \qquad (10)$$

We cannot apply the Coleman-Mandula theorem directly to the $\hat{Q}_\alpha$ and $\hat{Q}_{\gamma\alpha}$, because the Coleman-Mandula theorem deals with conserved charges that transform as Lorentz tensors, while the $\hat{Q}$ transform as spinors (or vector-spinors). However, the Coleman-Mandula theorem can be applied to the bosonic conserved charges that can be formed from the anticommutators of the $\hat{Q}$. (It is the <u>anti</u>-commutators of the $\hat{Q}$ that we should consider, because $S_{\mu\alpha}$ and $S_{\mu\gamma\alpha}$, being linear in fermi fields, <u>anti</u>-commute at spacelike separation.)

Now $\hat{Q}_\alpha$, being a left-handed spinor, transforms as $(1/2, 0)$ under Lorentz transformations. Its hermitian adjoint, $\hat{Q}^*_\sigma$, transforms as $(0, 1/2)$. The anticommutator of $\hat{Q}$ with its adjoint $\hat{Q}^*$, which cannot vanish (since the anticommutator of any operator with its hermitian adjoint is non-zero), transforms as $(1/2, 1/2)$ under Lorentz transformations. The Coleman-Mandula theorem permits the conservation of precisely one operator that transforms as $(1/2, 1/2)$, namely the energy-momentum operator P_μ. But $\hat{Q}_{\gamma\alpha}$, a vector-spinor, has components of spin up to $3/2$. The anti-commutator $\{\hat{Q}_{\gamma\alpha}, \hat{Q}^*_{\sigma\tau}\}$, which cannot vanish, for the reason noted above, has components of spin up to three. Since $\{\hat{Q}_{\gamma\alpha}, \hat{Q}^*_{\sigma\tau}\}$ is conserved if $\hat{Q}_{\gamma\alpha}$ is, and since the Coleman-Mandula theorem does not permit conservation of an operator of spin three in an interacting theory, $\hat{Q}_{\gamma\alpha}$ cannot be conserved in an interacting theory.

The Coleman-Mandula theorem does permit conservation of the $\hat{Q}_\alpha$. Changing basis from the $\hat{Q}_\alpha$ (not hermitian as I have defined them) to a basis of four hermitian operators Q_α that transform as a (real, Majorana) Lorentz four-spinor, the algebra of the Q_α turns out to be

$$\{Q_\alpha, \bar{Q}_\beta\} = \gamma^\mu_{\alpha\beta} P_\mu \tag{11}$$

One may readily check that this is so in, for example, the free field theory (4).

The right-hand side of (11) certainly contains only operators that are permitted by the Coleman-Mandula theorem, but one might ask whether there are more general possibilities. This question was answered by Haag, Sohnius, and Lopuszanski.[4] Since Q_α transforms as (1/2, 0) + (0, 1/2), its anti-commutator with itself might contain pieces transforming as (1, 0), (0, 1), or (0, 0), apart from the (1/2, 1/2) term we have already encountered. By the Coleman-Mandula theorem, the only conserved operators transforming as (1, 0) or (0, 1) are the Lorentz generators $M_{\mu\nu}$; however, Haag, Sohinius, and Lopuszanski showed that it is impossible to introduce $M_{\mu\nu}$ in (11) without violating the Jacobi identity. As for the possibility of adding to (11) operators that transform as (0, 0) (in other words, operators that commute with Lorentz transformations), this is possible, but only in the "extended supersymmetry" theories in which there are several conserved spinors $Q_{\alpha i}$. Beautiful though those theories are, I believe that they are too restrictive to describe the physics we know at energies much less than the Planck mass.[*] At such high

---

[*] The basic problem is that they require the charged fermions to form a "real representation" of the gauge group.

energies, they may, of course, be relevant. In these lectures we will only consider the simplest theories with a single conserved spinor Q_α and the basic algebra (11).

The algebra (11) has some dramatic consequences. Since Q_α is hermitian, $\bar{Q}_\beta = Q_\sigma \gamma^o_{\sigma\beta}$; so (11) can be written

$$\{Q_\alpha, Q_\sigma\} \gamma^o_{\sigma\beta} = \gamma^\mu_{\alpha\beta} P_\mu \tag{12}$$

If we multiply by $\gamma^o_{\beta\alpha}$, sum over β and α, and use the facts $(\gamma^o)^2 = 1$, $\mathrm{Tr}\, \gamma^o \gamma^\mu = 4\delta^{o\mu}$, we get

$$4 P_o = \sum_\alpha \{Q_\alpha, Q_\alpha\} \tag{13}$$

Here P_o is, of course, the Hamiltonian H. For any operator A, $\{A, A\} = 2A^2$. Equation (13) is equivalent to

$$H = \frac{1}{2} \sum_\alpha Q_\alpha^2 \tag{14}$$

which is one of the keys to understanding supersymmetry.

It follows immediately from equation (14) that if supersymmetry is not spontaneously broken — if the Q_α annihilate the vacuum $|\Omega\rangle$ — then the energy of the vacuum is zero. If $Q_\alpha |\Omega\rangle = 0$ then obviously

$$H |\Omega\rangle = \frac{1}{2} \sum_\alpha Q_\alpha^2 |\Omega\rangle = 0 \tag{15}$$

If conversely, supersymmetry is spontaneously broken, that is, if $Q_\alpha |\Omega\rangle \neq 0$, then

$$\langle\Omega| H |\Omega\rangle = \frac{1}{2} \sum_\alpha \langle\Omega| Q_\alpha^2 |\Omega\rangle = \frac{1}{2} \sum_\alpha |Q_\alpha |\Omega\rangle|^2 > 0 \tag{16}$$

so in this case the vacuum energy is positive.[*]

---

[*] Usually, one is free to add a constant to the Hamiltonian, but here the zero of energy is fixed in a natural way by equation (14). When asking what is the energy of the vacuum, we always have in mind the definition (14) of H.

Combining these remarks, we see that supersymmetry is spontaneously broken if and only if the energy of the vacuum is greater than zero. This is often illustrated by the diagram of figure (2). In figure (2(a)) a scalar field has a vacuum expectation value, possibly breaking some internal symmetry. However, supersymmetry is unbroken because the ground state energy — the minimum of the potential — is zero. In figure (2(b)), the expectation value of the scalar field is zero, but supersymmetry is spontaneously broken because the energy of the ground state is greater than zero.

If supersymmetry plays any role in nature it must be spontaneously broken. This follows immediately from the fact that the Q_α have spin 1/2, so acting on any particle they change its spin by $\pm 1/2$:

$$Q_\alpha |\text{spin s}> \sim |\text{spin s} \pm 1/2> \tag{17}$$

Since Q_α commutes with the Hamiltonian it does not change the mass of the particle on which it acts. Since we do not observe in nature the degeneracies among particles of different spin that would be predicted by (17), supersymmetry must be spontaneously broken if it is relevant to nature.

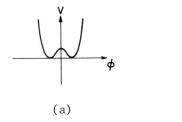
(a)

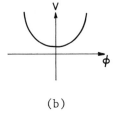
(b)

Figure 2

A puzzle immediately presents itself. Since broken supersymmetry means a positive vacuum energy, why does not a cosmological constant arise as soon as a supersymmetric theory is coupled to gravity? No good answer is known to this question, and it certainly is one of the most important questions that must be answered. Actually the positive vacuum energy of global supersymmetry does not necessarily become a positive cosmological constant after coupling to gravity, because the coupling to gravity introduces extra terms in the scalar potential; these extra terms are not positive definite. While a cancellation can occur,[5] leaving zero cosmological constant, such a cancellation apparently depends on a special choice of parameters, for which there is no known rationale.

Now, the subject of spontaneously broken supersymmetry has a special flavor, which arises from the fact that (leaving gravity aside), supersymmetry is spontaneously broken if and only if the ground state energy is positive. Suppose that in some theory an approximate calculation gives an effective potential such as the one shown in fig. (3). In this approximation, the minimum of the potential is zero, and supersymmetry is unbroken. Even if the approximation in question is very accurate, one cannot be certain, just from this result, that supersymmetry really is unbroken, because the errors in one's calculation may shift the potential slightly away from $V = 0$ (figure (4)) (the approximate calculation and the exact result are indicated in figure (4) by the solid and dotted line, respectively).

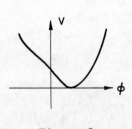

Figure 3

If, instead, an approximate calculation shows that supersymmetry <u>is</u> spontaneously broken, one can be confident in the result if

INTRODUCTION TO SUPERSYMMETRY

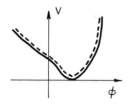

Figure 4

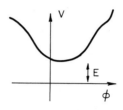

Figure 5

(figure (5)) the calculated vacuum energy E is much bigger than the uncertainty ΔE in the calculation.

On the other hand, for ordinary symmetries, a reliable approximation (like perturbation theory, in a weakly coupled theory) can reliably indicate whether the symmetry is spontaneously broken.

I have discussed these matters in much more detail in a recent paper.[6] In that paper I also described a simple quantum mechanics model in which tiny quantum effects shift the energy slightly away from zero, as in figure (4).

III. SUPERSYMMETRY AND THE LARGE NUMBERS

Supersymmetry is a very beautiful idea, but I think it is fair to say that no one knows what mysteries of nature (if any) it should explain. To fix ideas, I will make some definite assumptions in the next few lectures about what problems supersymmetry might solve.

Certainly, the ultimate applications may be in a very different area.

Perhaps supersymmetry is spontaneously broken at very high energies — energies that may be as high as the Planck mass 10^{19} GeV. We will assume instead in most of our discussion that supersymmetry is, more or less, within reach of the new generation of accelerators.

What would we like to explain with supersymmetry? Of the many possible answers, I will focus on Dirac's "problem of the large numbers,"[7] which nowadays is often called the "gauge hierarchy problem." As posed by Dirac, the question is why the Planck mass M_{Pl} is so much larger than the nucleon mass M_N: $M_{Pl}/M_N \sim 10^{+19}$. In a contemporary version,[8] one might ask why the mass scale M_X of presumed grand unification is so much larger than the mass scale M_W of weak interactions: $M_X/M_W \gtrsim 10^{+13}$. Starting with Dirac, most physicists have believed that such large numbers should not be postulated arbitrarily but must have a definite explanation.

We now know that the "normal" masses — the quark, lepton, and W and Z masses — are determined by SU(2) x U(1) breaking. Specifically, they are all proportional to the expectation value $<\phi>$ of the Higgs boson. So the question is really why $<\phi>$ is so tiny compared to the "large" masses — the mass scale of grand unification, or the Planck mass.

Looking at the standard Higgs potential

$$V(\phi) = \frac{\lambda}{4} \phi^4 - \frac{m_\phi^2}{2} \phi^2 \tag{18}$$

we have $<\phi> = m/\sqrt{\lambda}$, so the real problem is to explain why $m_\phi << M_{Pl}$.

Why would supersymmetry be relevant to this?

As we now understand it, bare mass terms for the quarks and leptons are forbidden by SU(2) x U(1) gauge invariance. Left-handed quarks and leptons transform as doublets of SU(2) x U(1), but right-

INTRODUCTION TO SUPERSYMMETRY

handed quarks and leptons are singlets. So bare masses are impossible, and the quarks and leptons get masses only from SU(2) x U(1) breaking. We do not know why $<\phi>$ is so small (relative to the large masses of physics) but if this could be explained, the lightness of the quarks and leptons would follow.

The problem with the Higgs boson is that its bare mass does not violate SU(2) x U(1), or any other gauge symmetry. Given any Higgs multiplet ϕ^i, the mass term $\sum_i \phi_i^* \phi^i$ is always gauge invariant. We would like a <u>symmetry</u> violated by m_ϕ, to explain why ϕ is so light. The symmetry will hopefully be spontaneously broken, but only on a very small mass scale, to explain why $<\phi>$ is tiny but non-zero.

Supersymmetry can do this if it relates the Higgs doublet $\binom{\phi^o}{\phi^-}$ to fermions like $\binom{\nu_e}{e^-}_L$ whose masses violate SU(2) x U(1). Then $m_\phi = 0$ as long as supersymmetry and SU(2) x U(1) are unbroken. And m_ϕ is small, solving the old problem of the "large numbers," if we can understand that supersymmetry and SU(2) x U(1) are <u>weakly</u> broken.

So that we can discuss these matters in a more tangible way, we must have available the explicit form of some supersymmetric Lagrangians. Let us consider first supersymmetric theories with fields of spin zero and spin one half only. In such theories, one may have an arbitrary number of complex scalar fields A^i, their supersymmetric partners being left-handed spinor fields ψ_L^i. We introduce a function W that depends only on the A^i, not on their complex conjugates A_j^* — in other words, W is an analytic function of the A^i. W is usually called the "superspace potential".

The Lagrangian is $\mathcal{L} = \mathcal{L}_{kinetic} + \mathcal{L}_{scalar} + \mathcal{L}_{Yukawa}$, where

$$\mathcal{L}_{kinetic} = \partial_\mu A_i^* \partial_\mu A^i + \bar{\psi}_i i \not{\partial} \psi^i$$

$$\mathcal{L}_{scalar} = -\sum_i \left|\frac{\partial W}{\partial A^i}\right|^2$$

$$\mathcal{L}_{Yukawa} = -\left(\frac{\partial^2 W}{\partial A^i \partial A^j} \psi_\alpha^i \psi^{\alpha j} + h.c.\right) \quad (19)$$

For how this construction was discovered, I refer you to the literature.[2] The easiest way to show that this describes a supersymmetric theory is to demonstrate that the supersymmetry current S^μ is conserved. This current is

$$S^\mu = \partial_\alpha A_i^* \gamma^\alpha \gamma^\mu \psi^i + i \frac{\partial W^*(A^*)}{\partial A_i^*} \gamma^\mu \psi_i^* \qquad (20)$$

The first term in (20) we have already seen in free field theory; the second term is added due to the interactions.

Note that, for a renormalizable theory, W should be at most a cubic function of the A^i. If W is cubic, then (19) contains only terms of dimension four or less, and this corresponds to a renormalizable theory.

For our purposes, the most important part of (19) is the formula for the scalar potential,

$$V(A^i, A_j^*) = \sum_i \left| \frac{\partial W}{\partial A^i} \right|^2 \qquad (21)$$

Let us ask, under what conditions is supersymmetry spontaneously broken at the tree level? Evidently, if for some value of the A^i the equations

$$\frac{\partial W}{\partial A^i} = 0 \qquad (22)$$

are simultaneously satisfied, then for this value of the fields the potential energy vanishes, classically. Supersymmetry is unbroken. On the other hand, if the equations (22) are inconsistent — if they are not satisfied for any choice of the A^j — then the minimum of the potential is strictly positive, and supersymmetry is spontaneously broken.

The simplest example is the O'Raifeartaigh model.[9] There are three fields, A, X, and Y, and the superspace potential is

$$W(A, X, Y) = g A Y + \lambda X (A^2 - M^2) \qquad (23)$$

INTRODUCTION TO SUPERSYMMETRY

Here g, λ, and M are constants. This theory is technically natural, because of global symmetries.* The scalar potential is

$$V(A, X, Y) = \left|\frac{\partial W}{\partial A}\right|^2 + \left|\frac{\partial W}{\partial X}\right|^2 + \left|\frac{\partial W}{\partial Y}\right|^2$$

$$= g^2 |A|^2 + \lambda^2 |A^2 - M^2|^2 + |gY + 2\lambda AX|^2 \qquad (24)$$

This potential is strictly positive, and supersymmetry is spontaneously broken. In fact, $\partial W/\partial Y = 0$ only if $A = 0$, and $\partial W/\partial X = 0$ only if $A = \pm M$; these requirements are clearly inconsistent. If $g/\lambda M$ is large enough, the minimum of the potential is at $A = 0$ and the vacuum energy is $\lambda^2 M^4$, plus quantum corrections. Expanding around the minimum of the potential (and making use of our previous formulas for the Yukawa couplings as well as the scalar interactions), it is easy to see that the bosons and fermions have unequal masses, which is expected, since the positive ground state energy indicates spontaneous breaking of supersymmetry.

We will have more to say about this model later, but for the moment let us consider a model of another kind. Let us consider a theory with SU(5) symmetry, the only scalar field being a complex field $A^i{}_j$ in the adjoint representation of SU(5) (so Tr A = 0). The most general choice of W would be

$$W = \frac{g}{3} \text{Tr } A^3 + \frac{M}{2} \text{Tr } A^2 \qquad (25)$$

where g and M are constants. The equations $\partial W/\partial A^i{}_j = 0$ give

$$g((A^2)^i{}_j - \frac{1}{5} \delta^i{}_j \text{Tr } A^2) + M A^i{}_j = 0 \qquad (26)$$

---

*The theory is invariant under $A \to -A$, $Y \to -Y$ and under $Y \to e^{i\alpha}Y$, $X \to e^{-i\alpha}X$. It should be noted that any transformation under which W changes only by an overall phase is a symmetry operation (the phase of W cancels out of the scalar potential and can be removed from the Yukawa couplings by a chiral transformation).

If we assume that A can be diagonalized by a unitary transformation,*
then it is easy to see that there are three solutions:

$$A^i{}_j = 0$$

$$A^i{}_j = \frac{M}{3g} \begin{pmatrix} 1 & & & & \\ & 1 & & & \\ & & 1 & & \\ & & & 1 & \\ & & & & -4 \end{pmatrix}$$

$$A^i{}_j = \frac{M}{g} \begin{pmatrix} 2 & & & & \\ & 2 & & & \\ & & 2 & & \\ & & & -3 & \\ & & & & -3 \end{pmatrix} \qquad (27)$$

These three solutions correspond to the unbroken gauge groups SU(5), SU(4) x U(1), and SU(3) x SU(2) x U(1), respectively. They each correspond to unbroken supersymmetry, and they are exactly degenerate at zero energy at least in this approximation (figure (6)), because they were all found by requiring $\partial W/\partial A^i{}_j = 0$.

Now, what really is the physics of this theory? It depends entirely on the nature of the quantum corrections to the effective potential. If really E = 0 for each of the three vacuum states, as appears to be the case in the classical approximation, then this one theory describes three different, inequivalent worlds. In one world, the strong gauge group is SU(3) and a baryon is made from three quarks; in one world, the strong gauge group is SU(4) and baryons are bosons, made from four quarks; in one world the strong gauge group is SU(5) and a baryon is made from five quarks.

If the quantum corrections break supersymmetry in, say, two of

---

*It is really when SU(5) is made into a gauge symmetry that this assumption is justified, because of extra terms that are then present in the scalar potential. Since A is complex and not hermitian, it cannot necessarily be diagonalized by a unitary transformation.

INTRODUCTION TO SUPERSYMMETRY

the three worlds (figure (7)), then the true vacuum is the one in which E = 0 and the supersymmetry is not spontaneously broken.

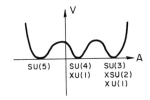

Figure 6

If supersymmetry is spontaneously broken, and E ≠ 0, in each of the three worlds, then (figure (8)) the true vacuum is the one in which supersymmetry is broken most weakly and the vacuum energy is least.

One might usually guess that such a degeneracy would be resolved, in perturbation theory, by loop diagrams. Perhaps the vacuum energy is of order α, coming from a one loop diagram, or of order α^2, coming from a two loop diagram. The loop diagrams would then be inducing supersymmetry breaking. Some particles whose masses violate super-

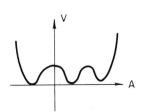

Figure 7

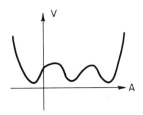

Figure 8

symmetry would get masses. Some scalars would get positive mass squared; some would get negative mass squared. If the W boson gets a mass from a Higgs doublet that gets an expectation value in this way, we would get a hierarchy of some sort, with $(M_W/M_X)^2$ perhaps of order α or α^2. This clearly would fall far short of the experimentally observed hierarchy.

However, a most remarkable theorem[10] states that this does not occur. It states that if at some point in field space the classical potential vanishes, then the effective potential vanishes at that point to all finite orders of perturbation theory. Our degeneracy is not lifted in perturbation theory.

This fact is an example of the "non-renormalization theorems" of supersymmetry. Here are some other examples:

(1) There is no renormalization of W in perturbation theory — neither finite nor infinite renormalization. (There is wave function renormalization. There are quantum corrections to the effective potential. There is no renormalization of W, if properly defined in terms of the coefficients of certain operators in the effective potential.)

(2) As long as supersymmetry is unbroken, any particle that is massless at the tree level is massless to all finite orders of perturbation theory — even if it was massless at the tree level only because of an arbitrary adjustment of parameters.

These theorems have been proved — on the basis of details of perturbation theory — but they are not well understood.

Returning to our model with the three degenerate vacuum states, if the relation $\langle \Omega | \ H \ | \Omega \rangle = 0$ were to break down in perturbation theory, and if this breakdown were the origin of the "light" masses, the resulting mass ratios would, as we have said, not be small enough to solve Dirac's problem of the large numbers. However, if the mysterious theorem concerning non-renormalization of $\langle\Omega| H |\Omega\rangle$ were to break down non-perturbatively — as it does in a quantum mechanics model discussed in ref. (6) — then we might obtain a solution to the large number problem. We might get a formula of the form

$$\frac{M_W}{M_X} \sim \exp -\frac{1}{\alpha} \tag{28}$$

and this could solve the "large number" or "gauge hierarchy" problem.

We are thus imagining a theory with two mass scales. At the large mass scale, the scale of grand unification, a unified group breaks down to SU(3) x SU(2) x U(1), or some other phenomenologically acceptable group, but supersymmetry remains unbroken. At a vastly smaller mass scale, the non-renormalization theorems break down, supersymmetry is spontaneously broken, and the Higgs boson gets an expectation value, breaking SU(2) x U(1) down to U(1).

What mechanism might be responsible for the tiny, non-perturbative effects that make $\langle\Omega| H |\Omega\rangle$ non-zero and break supersymmetry? With the present state of our knowledge, there are two obvious candidates. The obvious candidates are strong gauge forces, and grand unified instantons.

Let us discuss first the effects of strong gauge forces. Consider, as an example, the SU(5) theory that we discussed earlier. Assume that the SU(5) coupling is gauged; this can be done in a way compatible with supersymmetry and with our previous remarks. In each of the three vacuum states that exist in perturbation theory, there is an unbroken non-Abelian interaction that will become strong at low enough energies. For instance, in the vacuum $A^i{}_j = 0$, this is the full SU(5) group. As we have learned in the last few years, strong gauge forces cause a variety of non-perturbative effects, including confinement, mass generation, and chiral symmetry breaking. Perhaps in supersymmetric theories strong gauge forces also cause supersymmetry breaking, presumably with the binding of a color singlet Goldstone fermion.[6,11]

Of course, in this SU(5) theory, the vacuum of most phenomenological interest is the one in which the unbroken gauge group is SU(3) x SU(2) x U(1). In this vacuum there are strong SU(3) gauge forces. But if <u>they</u> were responsible for supersymmetry breaking, we would get bose-fermi mass differences of order Λ_{QCD} (since that is the energy at which SU(3) becomes strong), and this is clearly unsatisfactory.

In the same theory a more promising mechanism would be "grand unified instantons," that is, instantons of the SU(5) group which do not lie in the unbroken SU(3) x SU(2) x U(1) subgroup. Such instantons are not afflicted with infrared divergences. They have a natural mass scale, the mass M_X of grand unification, and a natural, small coupling, the coupling α_G of grand unification. If such instantons were responsible for supersymmetry breaking, we would get bose-fermi mass splittings roughly of the order

$$m_B^2 - m_F^2 \sim M_X^2 \exp - \frac{2\pi}{\alpha_G} \qquad (29)$$

which might be reasonable, for α_G of about 1/12.

Actually, the instanton mechanism seems to work in 2+1 dimensions, but, apparently, not in 3+1 dimensions.[6] This point is in need of further clarification.

Clearly, we must learn to analyze supersymmetry breaking in a non-perturbative way. Tomorrow, I will describe some steps in that direction. We will be able to derive some constraints on the possibility of supersymmetry breaking by strong gauge forces.

IV. $\text{Tr}(-1)^F$

In our previous discussions we have seen that, in general, it is difficult to determine from an approximate calculation whether supersymmetry is spontaneously broken. Even if the supersymmetry seems to be unbroken in some approximation, it is always possible that a small, non-zero vacuum energy is induced by the corrections to the approximation in question.

To show that supersymmetry is unbroken in a given theory one must prove that the ground state energy is <u>exactly</u> zero. Since an approximate calculation will not accomplish this, in general, we must search for other methods. I will now describe an indirect method which in some theories can be used to prove that the ground state energy is <u>exactly</u> zero. Of course, a theory in which supersymmetry is <u>not</u> broken is not a candidate for describing nature. The purpose of proving that some theories do not break supersymmetry is to cut down on the range of options that must be considered.

As a technical convenience, we will formulate our theories in a finite volume, with periodic boundary conditions (such boundary conditions respect supersymmetry). Our goal is to find criteria under which we can prove that the ground state energy of a theory $E_o(V)$ is zero in any finite volume V. Since the large V limit of zero is zero, this implies that the ground state energy is also zero in the infinite volume limit, and therefore that supersymmetry is <u>not</u> broken in the infinite volume theory which is of real interest.

Supersymmetry implies that every state of $E = 0$ also has $\vec{P} = 0$. (Supersymmetry implies $E \geq 0$ in each frame; by Lorentz invariance this requires $E \geq |P|$; so a state of zero energy necessarily has $\vec{P} = 0$.) So, in trying to determine whether the ground state energy vanishes, we lose nothing by restricting ourselves to the sector of Hilbert space consisting of states of $\vec{P} = 0$.

Let Q be any one of the (hermitian) supersymmetry generators. There are several Q_α, of course, but we will only need one. In the $\vec{P} = 0$ sector, the supersymmetry algebra is particularly simple. It is simply

$$Q^2 = \frac{1}{2} H \qquad (30)$$

(For nonzero $\vec{P}$, $\vec{P}$ would appear on the right-hand side of (30).)

Since $Q^2 = \frac{1}{2} H$, Q annihilates any states of zero energy. States of non-zero energy are not annihilated by Q. Rather, they are paired by the action of Q. Given any boson state $|b\rangle$ of non-zero energy E, Q acting on $|b\rangle$ gives a fermion state of $|f\rangle$. Q acting on $|f\rangle$ gives back $|b\rangle$. To be precise

$$Q|b\rangle = \sqrt{\frac{E}{2}} |f\rangle$$

$$Q|f\rangle = \sqrt{\frac{E}{2}} |b\rangle \qquad (31)$$

if phases are chosen properly. (Because $Q^2 = \frac{1}{2} H$, the factors of $\sqrt{\frac{E}{2}}$ are consistent with $\langle b|b\rangle = \langle f|f\rangle = 1$, and the fact that Q is hermitian.)

I should explain that when I refer to a boson state $|b\rangle$, I do not mean a one particle state (a concept that in a finite volume is not really well defined) but any state of integral angular momentum. Likewise, $|f\rangle$ is any state of half integral angular momentum. Here, "angular momentum" refers to the 90° rotations such as $\exp -\frac{i\pi}{2} J_z$ which are well defined in the finite volume theory. If we define the operator

$$(-1)^F = \exp - 2\pi i J_z \qquad (32)$$

which distinguishes bosons from germions, then a boson state $|b\rangle$ is any state that obeys $(-1)^F |b\rangle = |b\rangle$, and a fermion state $|f\rangle$ is any state that obeys $(-1)^F |f\rangle = -|f\rangle$. A boson state could be, for

INTRODUCTION TO SUPERSYMMETRY

example, any state that in the infinite volume limit goes over to a configuration of 92 neutrons.

Although the states of non-zero energy form bose-fermi pairs as indicated in equation (31), this is certainly not true for the states of zero energy. Any zero energy state, boson or fermion, is just annihilated by Q:

$$Q|b, E = 0\rangle = Q|f, E = 0\rangle = 0 \tag{33}$$

Again, this is so because $Q^2 = \frac{1}{2} H$. The zero energy states are singlets — one dimensional representations of supersymmetry.

The general form of the spectrum of a supersymmetric theory is indicated in figure (9). The states of non-zero energy are in bose-fermi pairs. The zero energy states are not paired in general since each one is separately annihilated by Q. They need not be equal in number. In the figure there are two boson states of zero energy, and one fermion.

What happens when we change the parameters of this theory? (By "parameters", I mean the volume, the bare mass, and the coupling constant.)

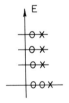

Figure 9

Circles indicate bosons; an "x" indicates a fermion.

Under the change in parameters, the states of zero energy will, of course, move around in energy. However, they will move around in bose-fermi pairs. It may happen that as we change the parameters a bose state will move down to zero energy. If so, it will always be accompanied (figure (10)) by a fermi state moving down to zero

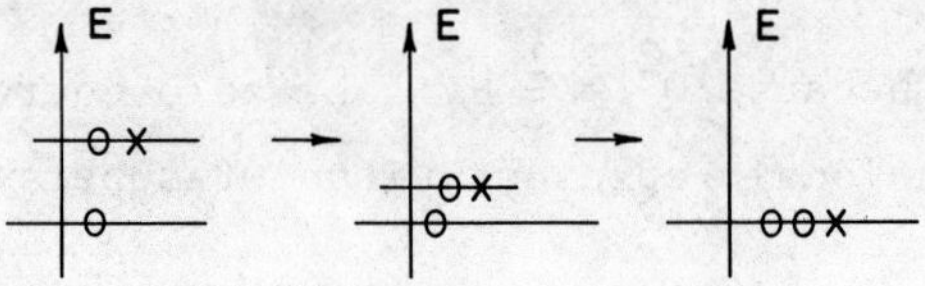

Figure 10

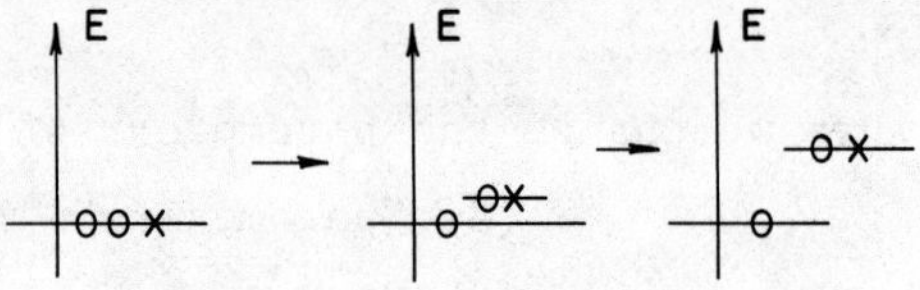

Figure 11

Circles indicate bosons; an "X" indicates a fermion.

energy. Conversely, as we change the parameters, a zero energy state may get a non-zero energy. If so, as soon as it gets non-zero energy, it must have a partner (figure (11)), because states of non-zero energy are always in bose-fermi pairs. It is not possible however, for a zero-energy state to simply appear or disappear. In quantum mechanics states always move around continuously in energy.*

In the process of figures (10) and (11), the number of zero energy states changes. However, the <u>difference</u> between the number $n_B^{E=0}$ of zero energy states that are bosons and the number $n_F^{E=0}$ that are fermions does not change. This will be our basic tool.

Formally, the difference $n_B^{E=0} - n_F^{E=0}$ may be interpreted as the trace of the operator, $(-1)^F$, that distinguishes bosons from fermions. Formally, the states of non-zero energy cancel out of the trace of $(-1)^F$ because they come in bose-fermi pairs. The trace of $(-1)^F$ can therefore be evaluated among the zero energy states only, and equals $n_B^{E=0} - n_F^{E=0}$. We will henceforth refer to $n_B^{E=0} - n_F^{E=0}$ as $\text{Tr}(-1)^F$. However, this is only a definition, since $\text{Tr}(-1)^F$ is not absolutely convergent.

The fact that $\text{Tr}(-1)^F$ does not change when the parameters of a supersymmetric theory are changed is an important fact, for the following reasons:

(1) If $\text{Tr}(-1)^F \neq 0$, supersymmetry is definitely not spontaneously broken.

(2) $\text{Tr}(-1)^F$ can be calculated reliably even in quite complicated theories.

---

*As discussed in detail in reference (12), there is really an important caveat to be imposed here. One may not consider changes in parameters that overwhelm the terms already present in the Hamiltonian.

Let me explain these points:

If $\text{Tr}(-1)^F \neq 0$ then $n_B^{E=0} \neq 0$ or $n_F^{E=0} \neq 0$ or both. In any case, there are <u>some</u> zero energy states. Hence the ground state energy is zero and supersymmetry is unbroken. (Since $\text{Tr}(-1)^F$ is independent of the volume, a non-zero value of $\text{Tr}(-1)^F$ means that the ground state energy is zero for any V and hence also as $V \to \infty$.)

$\text{Tr}(-1)^F$ can be calculated reliably because it is independent of the parameters. We can calculate $\text{Tr}(-1)^F$ in some convenient limit, such as small volume, large bare mass, and weak coupling. Almost any theory simplifies enough in this limit (or some analogous limit) that $\text{Tr}(-1)^F$ can be calculated reliably.

The results can then be applied to the situation of interest — large volume, physical mass, and physical coupling — because $\text{Tr}(-1)^F$ is independent of all parameters.

It is important to realize that $\text{Tr}(-1)^F$ can be calculated reliably even though, in general, we may be unable to tell which states have exactly zero energy. Suppose that in some approximation we find in some theory the spectrum of figure (12(a)). In this approximation we have $n_B^{E=0} = 2$, $n_F^{E=0} = 1$, and $\text{Tr}(-1)^F = 1$. Even if the approximation is excellent we cannot be sure we have calculated $n_B^{E=0}$ or $n_F^{E=0}$ correctly. As a result of small errors in any approximation, the true answer might be that of figure (12(b)). This corresponds to $n_B^{E=0} = 1$, $n_F^{E=0} = 0$, and again $\text{Tr}(-1)^F = 1$. The hypothetical corrections to our approximation gave a small, non-zero energy to one boson and therefore (by supersymmetry) also to one fermion. The original approximation gave $n_B^{E=0}$ and $n_F^{E=0}$ incorrectly, but it gave $\text{Tr}(-1)^F$ correctly, essentially because the extra boson at $E = 0$ has no potential fermion partner and so no way to get $E \neq 0$.

Let me now explain the calculation of $\text{Tr}(-1)^F$ in a simple model. I will consider the Wess-Zumino model, which we have discussed

INTRODUCTION TO SUPERSYMMETRY

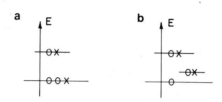

Figure 12

earlier. This model contains a single complex field ϕ (and fermi partner ψ) with

$$V(\phi) = g^2 \left| \phi^2 - \frac{a^2}{g^2} \right|^2 \tag{34}$$

(There also is a Yukawa coupling, $L_{Yuk} = g \phi \psi_\alpha \psi^\alpha + h.c.$). At the tree level ϕ and ψ are massive, their masses being equal to a, in lowest order. Supersymmetry is unbroken at the tree level since $V = 0$ at $\phi = \pm a/g$. Because of the non-renormalization theorems that we discussed yesterday, it is known that supersymmetry is unbroken to all finite orders of perturbation theory. Let us now prove that this is true independently of perturbation theory by showing that $Tr(-1)^F \neq 0$.

Actually, nothing could be easier. The potential has two minima, at $\phi = \pm a/g$. In each minimum of V, there is one zero energy state in perturbation theory — the "vacuum". It is a bosonic state, with zero angular momentum. Because ϕ and ψ have non-zero mass, all other states have (at least for weak coupling) a non-zero energy, at least equal to the mass of the ϕ and ψ particles. They do not contribute to $\text{Tr}(-1)^F$. $\text{Tr}(-1)^F$ receives one contribution from the vacuum at $\phi = +a/g$ and one contribution from the vacuum at $\phi = -a/g$. Altogether $\text{Tr}(-1)^F = 2$, so supersymmetry is unbroken.

Suppose instead that we wish to calculate $\text{Tr}(-1)^F$ in the massless Wess-Zumino model — that is, with $a = 0$ in (34). The potential is now simply $V(\phi) = g^2|\phi|^4$. Now $m_\phi = m_\psi = 0$, in perturbation theory. In addition to the "vacuum" (which now is a trickier concept), one can have in perturbation theory states of approximately zero energy by adding ϕ or ψ quanta to the "vacuum" in momentum eigenstates of $\vec{p} = 0$. (Such states are normalizable in finite volume.) It is difficult to count these states because any number of ϕ particles may have $\vec{p} = 0$ and it is difficult — because of the non-linearity — to know which of these states have $E = 0$ exactly, or to count them. This makes it difficult to calculate $\text{Tr}(-1)^F$.

The easiest way to calculate $\text{Tr}(-1)^F$ in the $a = 0$ model is to remember that $\text{Tr}(-1)^F$ is independent of a. So even if your interest is $a = 0$, you can calculate $\text{Tr}(-1)^F$ by considering a to be non-zero and large. This makes the calculation easy. So $\text{Tr}(-1)^F = 2$. This illustrates the utility of knowing $\text{Tr}(-1)^F$ to be independent of all parameters.

Going back to the massive model, with $a \neq 0$, there actually is a simpler argument to prove that supersymmetry is unbroken for small enough g. For supersymmetry to be spontaneously broken,[2] there must be a massless Goldstone fermion. In this model, for $a \neq 0$ and small enough g, the elementary fermion is certainly not massless. Also, for small enough g the elementary ϕ and ψ will certainly not form massless bound states.

INTRODUCTION TO SUPERSYMMETRY

So for small enough g, supersymmetry must be unbroken, there being no candidate Goldstone fermion. However, one might have believed that for large enough g the fermion mass goes to zero and supersymmetry is broken. Such behavior is very common in the case of global symmetries, but the fact that $Tr(-1)^F = 2$ shows it cannot happen here — supersymmetry is unbroken even for strong coupling.

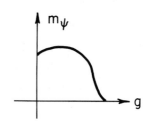

Figure 13

What cannot happen in supersymmetric theories: the fermion becomes a Goldstone fermion for large enough g.

Also, the a = 0 model has a massless fermion in perturbation theory. Could not this particle become (at the non-perturbative level, since it has been proved not to happen in perturbation theory) a Goldstone fermion? The fact that $Tr(-1)^F = 2$ shows that this does not occur.

Now a few simple comments:

(1) If supersymmetry is broken spontaneously at the tree level, then at the tree level $n_B^{E=0} = n_F^{E=0} = 0$, there being no zero energy states of any kind. So $Tr(-1)^F = 0$.

(2) If $Tr(-1)^F \neq 0$ supersymmetry is definitely unbroken. But if $Tr(-1)^F = 0$ we do not know. We may have $n_B^{E=0} = n_F^{E=0} = 0$, and supersymmetry broken, but we may equally well have $n_B^{E=0} = n_F^{E=0} \neq 0$, and supersymmetry unbroken.

(3) As in the $a \neq 0$ Wess-Zumino model, it is easy to calculate $Tr(-1)^F$ in any model where supersymmetry is unbroken at the tree level and all particles have mass. It is easy because, there being

no massless particles, only the "vacuum states" have E = 0 in perturbation theory. The vacuum states are all (spin zero) bosons so $Tr(-1)^F$ is equal to the number of zeros of the classical potential (there are some since supersymmetry is assumed unbroken at the tree level). So in all theories with no massless particles at the tree level, $Tr(-1)^F$ is positive, and supersymmetry is not spontaneously broken.

Finally, let us discuss a case in which these methods really yield interesting results — supersymmetric non-Abelian gauge theories. In the simplest such theory, the only fields are the gauge field A_μ^a and its partner, the fermi field ψ_α^a, also in the adjoint representation of the gauge group. The Lagrangian is

$$\mathcal{L} = \int d^4x \, (-\tfrac{1}{4}(F_{\mu\nu}^a)^2 + \tfrac{1}{2}\bar{\psi}^a i \not{\partial} \psi^a) \tag{35}$$

The easiest way to show that this is a supersymmetric theory is to show that the supersymmetry current

$$S_\mu = \sigma_{\alpha\beta} F^{\alpha\beta a} \gamma_\mu \psi^a \tag{36}$$

is conserved. This is readily demonstrated, with the aid of some Dirac algebra and the use of fermi statistics.

This theory is tricky to deal with because of the zero momentum modes of the massless particles.

The main problem is the gauge field. Only a finite number of $\vec{p} = 0$ fermions can fit in the box, and we could count those states. But the $\vec{p} = 0$ mode of the gauge field is a problem.

In infinite volume the $\vec{p} = 0$ mode of the gauge field can gauged away. If $A_\mu = C_\mu$ (the C_μ being constants), then $A_\mu = \partial_\mu \varepsilon$, with $\varepsilon = C_\mu x^\mu$. Here the constant can't be gauge away because the gauge parameter ε isn't periodic.

INTRODUCTION TO SUPERSYMMETRY

More generally, the guage invariant

$$\text{Tr } P \exp i \oint A_\mu \, dx^\mu \qquad (37)$$

for a contour (figure (14)) that runs "around the box" is different at A = constant from its value at A = 0, proving that the zero momentum mode can't be gauge away.

It is possible to come to grips with this problem,[12] but in the case of the gauge group SU(N), there is a much simpler approach. One can choose boundary conditions that the zero momentum mode does not satisfy. Such boundary conditions are the "twisted boundary conditions" of 't Hooft.[13] In a special case that is general enough for our purposes, the twisted boundary conditions, in a box of length L, mean that

$$\begin{aligned}\phi(x,y,z) &= P\phi(x+L,y,z) P^{-1} \\ &= Q\phi(x,y+L,z) Q^{-1} \\ &= \phi(x,y,z+L) \qquad (38)\end{aligned}$$

where ϕ may be A_μ or ψ. For P = Q = 1 this would describe conventional periodic boundary conditions. 'T Hooft instead requires P and Q to be constant matrices that obey

$$PQ = QP \exp 2\pi i/N \qquad (39)$$

Explicit matrices P and Q that obey (39) are easily found. For instance, we may take

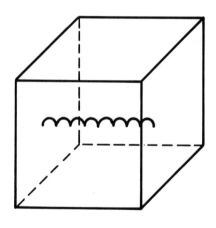

Figure 14

$$P = \alpha \begin{pmatrix} 0 & 1 & & & \\ & 0 & 1 & & \\ & & 0 & 1 & \\ & & & \ddots & \ddots \\ 1 & & & & 1 \\ & & & & & 0 \end{pmatrix} \quad Q = \beta \begin{pmatrix} 1 & & & & \\ & e^{i\delta} & & & \\ & & e^{2i\delta} & & \\ & & & \ddots & \\ & & & & e^{(N-1)i\delta} \end{pmatrix} \tag{40}$$

where $\delta = 2\pi/N$, and α and β are constants chosen to ensure det P = det Q = 1.

We are entitled to adopt the twisted boundary conditions because, in the large volume limit, the physics is expected to be independent of the boundary conditions. If we can show that, with twisted boundary conditions, the ground state energy vanishes for any value of the volume, then, in the large volume limit, the energy vanishes, and supersymmetry is unbroken, for any choice of the boundary conditions.

Accoarding to 't Hooft, the theory formulated as in (38) describes a world with "one unit of magnetic flux in the z direction." The motivation for that terminology is not essential for our purposes. For our purposes the key point is that the twisted boundary conditions eliminate the $\vec{p} = 0$ mode. It just doesn't satisfy the boundary condition.

The zero momentum mode ϕ_o of a field ϕ that satisfies (38) would have to obey

$$\phi_o = P \phi_o P^{-1} = Q \phi_o Q^{-1} \tag{41}$$

It is easy to see that for ϕ_o in the Lie algebra of SU(N), (41) requires $\phi_o = 0$. (To commute with Q, ϕ_o must be diagonal. But a traceless, diagonal matrix that commutes with P must vanish.) Therefore, the twisted boundary conditions remove the zero momentum mode.

With this accomplished, evaluating $Tr(-1)^F$ is as easy as in the Wess-Zumino model. Only the "vacuum states" have $E = 0$. Other states have energy at least equal to the lowest allowed momentum for modes of non-zero momentum in the box of length L.

One can expand around $A_\mu^a = 0$; this "vacuum" contributes one to $Tr(-1)^F$. There are other vacuum states that can be obtained from $A_\mu^a = 0$ by topologically non-trivial gauge transformations. In a world defined by a given value of the vacuum angle θ there are still N sectors of configurations of $F_{\mu\nu}^a = 0$ (they are created from $A_\mu^a = 0$ by the gauge transformations which, according to 't Hooft, measure the electric flux in the z direction). So altogether, for SU(N), $Tr(-1)^F = N$, and supersymmetry is not spontaneously broken.

Twisted boundary conditions may be introduced for any group which has a non-trivial center. However, for groups other than SU(N), the twisted boundary conditions are not very useful in calculating $Tr(-1)^F$. The problem is that, for other groups, the twisted boundary conditions do not eliminate the zero momentum mode because the condition analogous to (41) does not imply $\phi_o = 0$.

However, it is possible to calculate $Tr(-1)^F$ with untwisted boundary conditions, by quantizing the zero momentum modes in a Born-Oppenheimer approximation. One finds[12] that for a simple non-Abelian Lie group of rank r, $Tr(-1)^F = r+1$. Thus, spontaneous supersymmetry breaking does not occur in these theories.

What happens if additional matter fields are added? If the additional fields are in a real representation of the gauge group, bare masses m_i are possible for all of the matter fields. In this case all states containing the new quanta have non-zero energy, equal to or greater than the smallest of the m_i. The new fields do not contribute to $Tr(-1)^F$. So supersymmetry is unbroken, just as if the new fields were not present.

This argument shows that the ground state energy is zero for any non-zero values of the m_i. Taking now the limit as $m_i \to 0$, the ground state energy must remain zero. The only assumption needed here is that the zero mass limit should exist. We conclude that supersymmetry is unbroken for a theory with massless charged matter fields, as long as they lie in a representation of the gauge group such that they <u>could</u> have had bare masses.

In the very interesting case of theories with massless fields in a complex representation of the gauge group — so that gauge invariant bare masses are impossible — I do not know how to calculate $\text{Tr}(-1)^F$. For reasons explained elsewhere,[12] there are difficulties even in formulating the problem.

I should also mention that for a theory with gauge group U(1), $\text{Tr}(-1)^F = 0$, as long as all charged fields have or could have had bare masses. It is nonetheless possible to prove by a variant of the concept of $\text{Tr}(-1)^F$ that dynamical supersymmetry breaking does not occur in any U(1) gauge theory in which the Hamiltonian commutes with charge conjugation invariance (or in which charge conjugation invariance is broken only by Yukawa couplings, and not by the gauge couplings or the Fayet-Iliopoulos D term).

Likewise, in a theory with a simple gauge group that is spontaneously broken at the tree level to a subgroup such as SU(3) x SU(2) x U(1) that contains a U(1) factor, $\text{Tr}(-1)^F = 0$.

Clearly, these results impose significant restrictions on the possibility of supersymmetry breaking by strong gauge forces. However, loopholes remain, such as the question of matter fields in a complex representation of the gauge group. It also remains for the future to determine whether these methods can shed light on the apparent difficulties in 3+1 dimensions in supersymmetry breaking by instantons.

V. ANOTHER APPROACH TO MASS HIERARCHIES

In this last lecture I would like to describe a different approach to the gauge hierarchy problem.

It is usually assumed, in thinking about the hierarchy problem, that the "large" masses (the mass scale of grand unification and the Planck mass) are the fundamental ones. The problem then is to explain why the "small" masses (the masses of ordinary particle physics) are so small. One may seek to explain this by finding a non-perturbative mechanism that generates tiny masses. It was this possibility that motivated the discussion of dynamical supersymmetry breaking in the last few lectures.

Another possibility is to assume that the small masses are the fundamental ones. One must then explain why the large masses are so large.

This approach was actually followed by Dirac[7] in his original approach to the problem of the large numbers. Dirac assumed that the electron and proton masses were the basic ones. To account for the fact that the Planck mass is enormously larger, Dirac postulated that the Planck mass is not constant in time but increases in time as the universe grows older. The enormous present value of the Planck mass was thus attributed by Dirac to the fact that (in elementary particle units) the present universe is extremely old. Unfortunately, this beautiful idea has some serious empirical difficulties, which I will mention later.

Today I will be describing a class of supersymmetric theories[14] which spontaneously generate a mass scale vastly larger than the mass scale postulated in the Lagrangian. In these theories it is conceivable that the fundamental mass scale of nature could be comparable to the mass scale of the Weinberg-Salam model. All larger masses would be dynamically generated. As we will see, these theories have something in common with Dirac's approach, and we can, but need not, retrieve Dirac's idea that the "constants" of nature are slowly varying functions of time.

In our previous discussions we emphasized dynamical (non-perturbative) symmetry breaking. Now, however, we will consider theories in which supersymmetry is spontaneously broken at the tree level (classically). There will be no mystery about supersymmetry breaking; we will simply choose a scalar potential whose minimum is not invariant under supersymmetry. The problem will be to extract the consequences of supersymmetry breaking.

To understand the idea, let us consider one of the simplest models with supersymmetry spontaneously broken at the tree level — the O'Raifeartaigh model. As we discussed before, in this model there are three complex fields A, X, and Y of spin zero. The superspace potential is $W(A,X,Y) = \lambda X(A^2 - M^2) + g Y A$, and the ordinary potential energy is $V(A,X,Y) = |\partial W/\partial A|^2 + |\partial W/\partial X|^2 + |\partial W/\partial Y|^2$ or

$$V(A,X,Y) = \lambda^2 |A^2 - M^2| + g^2 |A|^2 + |2\lambda AX + gY|^2 \qquad (42)$$

This potential is strictly positive, so supersymmetry is spontaneously broken.

We determine A by minimizing the first two terms. For $(g/\lambda M)^2 > 2$ one finds $A = 0$; for $(g/\lambda M)^2 < 2$ one finds $A = \sqrt{M^2 - g^2/4\lambda^2}$. Supersymmetry is spontaneously broken in either case.

Although minimization of the potential determines A, it does not determine X and Y uniquely. Since the only term in the potential that depends on X and Y is $|2\lambda A X + g Y|^2$, we can choose any X, as long as

$$Y = -2 \lambda A X / g \qquad (43)$$

Classically, X may be arbitrarily large; the energy is minimized as long as (43) is satisfied.

This degeneracy is not a property of this one model alone. Such degeneracies occur in many models in which supersymmetry is spontaneously broken at the tree level.

Although the energy is independent of X, the particle masses are not. A glance at (42) shows that for X >> M the mass of the A particle is approximately $2\lambda X$. If X >> M, the theory has at least two different energy scales. The small scale is the mass M in the Lagrangian. The large scale is the vacuum expectation value of X, which is undetermined by the classical Lagrangian. In an extended version of this model, we may try to interpret these as the scales of weak interactions and of grand unification, respectively.

However, we should not <u>arbitrarily</u> assume X >> M. We should <u>determine</u> X by calculating quantum corrections that remove the classical degeneracy.

Let us recall how this goes.[15] For bosons, the zero point energy per unit volume is positive and has the form

$$\frac{1}{2}\hbar \sum_i \omega_i = \frac{1}{2}\hbar \sum_i \int \frac{d^3k}{(2\pi)^3} \sqrt{k^2 + M_i^2(X)} \tag{44}$$

The sum on the right-hand side of (44) is a sum over the spin states of the various bosons; the M_i are the masses of the bosons, which depend on X. For fermions we have instead the <u>negative</u> energy from filling the Dirac sea. It is

$$-\frac{1}{2}\hbar \sum_i \int \frac{d^3k}{(2\pi)^3} \sqrt{k^2 + M_i^2(X)} \tag{45}$$

where now the sum runs over the fermion spin states.

If supersymmetry is unbroken at the tree level, the bosons and fermions have the same masses. Then (44) and (45) cancel. This is an illustration of the "non-renormalization" theorems: the ground state energy is zero to all finite orders if it is zero classically. We are, however, interested in cases in which supersymmetry is spontaneously broken at the tree level; the integrals do not cancel.

Both integrals (44) and (45) are quartically divergent. The quartic divergence cancels because the bosons and fermions have the same number of spin states. The quadratic divergence cancels because of a sum rule of supersymmetric theories. Finally, the logarithmic divergence is removed by the renormalization counterterms that make the Green's functions finite.

After renormalization, and after doing the integrals, the final, finite expression for the $O(\hbar)$ corrections to the energy is

$$\Delta V(X) = \hbar \sum_i \frac{(-1)^F}{64\pi^2} M_i(x)^4 \ln \frac{M_i^2(x)}{\mu^2} \qquad (46)$$

where $(-1)^F$ is plus one for bosons, and minus one for fermions, and where μ is a renormalization mass.

Equation (46) shows that for large X, the main contribution to ΔV comes from the particles that become heavy as $X \to \infty$, namely A and its supersymmetric partner ψ_A. It is not difficlut to evaluate their contribution. Assuming for simplicity that $g/\lambda M \gg 1$, the potential to this order, including the lowest order result and the one loop correction, is

$$V(X) = \lambda^2 M^4 \left(1 + \frac{\lambda^2}{8\pi^2} \ln |X|^2/\mu^2\right) + O(1/|X|^2) \qquad (47)$$

This result was first derived by Huq.[16]

The logarithmic correction in (47) is characteristic of renormalizable theories. Its positive coefficient means that X cannot become large — the energy increases with $|X|$. This model develops no big mass hierarchy from a spontaneous large vacuum expectation value.

The logarithmic equation (47) always arises, and always has a positive coefficient, in theories of particles of spins 0 and $\frac{1}{2}$ only.

INTRODUCTION TO SUPERSYMMETRY

This fact has a simple renormalization group interpretation. The logarithm in equation (47) can be understood as a replacement of the bare coupling λ^2 with an effective coupling $\bar{\lambda}^2(X) = \lambda^2 (1 + \frac{1}{8\pi^2} \ln |X|^2/M^2)$. The coefficient of the logarithm is positive because theories of spin 0 and spin $\frac{1}{2}$ fields only are not asymptotically free. The effective coupling $\bar{\lambda}(X)$ increases with X.

As this reasoning suggests, a different result can be obtained by enlarging the O'Raifeartaigh model to include non-Abelian gauge fields. I first wish to discuss this in a qualitative way. In a gauge invariant generalization of the O'Raifeartaigh model, the one loop corrections have the form

$$V(X) = a\lambda^2 M^4 (1 + (b\lambda^2 - ce^2) \ln |X|^2/M^2) \qquad (48)$$

where λ is a scalar coupling, e is the gauge coupling, and a, b, and c are positive constants, of order one.

We see that if $b\lambda^2 - ce^2 < 0$, a runaway behavior occurs. It is favorable for X to become large since the potential is a decreasing function of X (figure (15)). In fact, (48) seems to show that the potential V becomes negative as $X \to \infty$. This is impossible; in supersymmetric theories $V(X) \geq 0$ for all X. The fact that (48) becomes negative for very large X just means that, as one might expect, perturbation theory breaks down when $e^2 \ln |X|^2/M^2 \sim 1$.

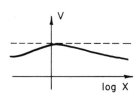

Figure 15
The classical potential (dotted line) and effective potential (solid line). The horizontal scale is logarithmic.

What happens for very large $|X|$ when perturbation theory breaks down? One possibility is that for very large $|X|$, the potential eventually ceases to decrease and begins increasing (figure (16a)). The potential would thus have a stable minimum at $e^2 \ln |X|/M \sim 1$, or equivalently at $X \sim M \exp +\frac{1}{\alpha}$. The theory would have at least two vastly different mass scales, M and X. As mentioned earlier, one might try to interpret these as the mass scales of weak interactions and of grand unification, respectively.

Another possibility is that the potential might decrease indefinitely as X becomes larger. Since V is bounded below by zero, it would have to approach a limit for large X (figure (16b)); the limit could be positive or zero. As we will discuss later, this corresponds to a theory in which, as envisaged by Dirac, the constants of nature are slowly changing functions of time. In most of this talk, we will assume that the potential has a stable minimum at large X.

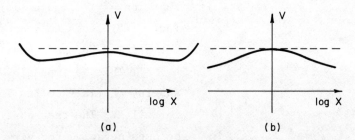

Figure 16

The potential may develop a stable minimum for very large X or may decrease indefinitely ((a) and (b) respectively). The horizontal scale is again logarithmic.

INTRODUCTION TO SUPERSYMMETRY

More complicated possibilities can also be imagined. Conceivably, in some theories the potential has an oscillatory dependence on X for large X.

It is possible to use the renormalization group to improve upon the one loop approximation of equation (48). In this way one can, to a large extent, decide between the various possibilities just described. One finds that the behavior of figure (16a) and also that of figure (16b) occur in a significant class of examples.[17]

I will now describe a simple model[14] which exhibits such runaway behavior. Let us consider an SU(5) theory with two complex fields $A^i{}_j$ and $Y^i{}_j$ in the adjoint representation of SU(5) and one singlet X. The superspace potential we choose to be

$$W(A,X,Y) = \lambda \, \text{Tr} \, A^2 Y + gX \, (\text{Tr} \, A^2 - M^2) \tag{49}$$

This is the most general choice of W compatible with certain global symmetries, analogous to those of O'Raifeartaigh model.

Supersymmetry is spontaneously broken because the equations $\partial W/\partial X = 0$ and $\partial W/\partial Y^i{}_j = 0$ are inconsistent. As in the O'Raifeartaigh model, minimization of the potential uniquely determines A. One finds

$$A = \frac{gM}{\sqrt{\lambda^2 + 30g^2}} \begin{pmatrix} 2 & & & & \\ & 2 & & & \\ & & 2 & & \\ & & & -3 & \\ & & & & -3 \end{pmatrix} \tag{50}$$

However, X is once again undetermined at the tree level. One may choose any X as long as

$$Y = \frac{g}{\lambda} X \begin{pmatrix} 2 & & & & \\ & 2 & & & \\ & & 2 & & \\ & & & -3 & \\ & & & & -3 \end{pmatrix} \quad (51)$$

If X becomes large, the large expectation value of Y will strongly break SU(5) down to SU(3) x SU(2) x U(1).

To determine X, we perform a one loop calculation. This calculation is again dominated by the particles whose masses are, for large X, proportional to X. With the one loop correction included, we find

$$V(X) = V_o \left(1 + \frac{g^2}{g^2 + \lambda^2/30} \frac{29\lambda^2 - 50e^2}{80\pi^2} \ln |X|^2/M^2\right) + O(1/|X|^2) \quad (52)$$

where V_o is the lowest order potential. We see that X will become large if $29\lambda^2 - 50e^2 < 0$.

This theory has two mass scales. The mass scale of SU(5) breaking is of order X because of the large vacuum expectation value of Y. However, supersymmetry is broken only at a mass of order M. In fact, only <A> violates supersymmetry and the ground state energy is of order M^4. Perhaps the large ratio X/M is related to the "big numbers" of physics.

Although we have obtained a hierarchy of symmetry breaking at vastly different mass scales, we have not yet addressed the original hierarchy problem. We wish SU(2) x U(1) symmetry breaking to occur as part of the low energy symmetry breaking.

To allow for this, we must include additional fields. A simple approach is to add an extra singlet Z and fields C^i and D_j in the 5 and $\bar{5}$ representation of SU(5). To the superspace potential we add a new term

$$\Delta W = \tilde{g} \, Z \, (C^i D_i - \tilde{M}^2) + \tilde{\lambda} A^i{}_j \, C^j \, D_i \qquad (53)$$

We assume that $\tilde{M}$ is of order M. Minimization of the potential forces C and D to obtain vacuum expectation values, spontaneously breaking SU(2) x U(1) down to U(1).

In this model, SU(5) is broken to SU(3) x SU(2) x U(1) at the energy scale X, but supersymmetry is broken, and SU(2) x U(1) is broken down to U(1), at the far lower scale M. We thus obtain the desired pattern of symmetry breaking. However, there are some severe difficulties.

Perhaps the most severe difficulty is related to the fact that the color triplet components of C^i and D_j are relatively "light", with masses of order M. After introducing quarks and leptons in the model, we must couple the SU(2) x U(1) doublet components of C and D to the quarks and leptons, in order to account for their bare masses. However, SU(5) then requires that we also couple the color triplet components of C and D to the quarks and leptons. Unfortunately, these color triplets then mediate a very rapid decay of the proton, the lifetime being a small fraction of a second.

This particular problem can be avoided if we replace the coupling $A^i{}_j \, C^j \, D_i$ in equation (53) with a term $Y^i{}_j \, C^j \, D_i$. In this case, for reasons explained in reference (14), the color triplet components of C and D automatically become superheavy without any special adjustment of parameters. We can now couple C and D to quarks and leptons, so as to account for the quark and lepton masses, without inducing a rapid proton decay.

Unfortunately,[14] the new coupling also modifies the pattern of symmetry breaking. The vacuum expectation value of Y is changed. We now find that SU(5) is strongly broken down to SU(3) x U(1) x U(1) at the large mass scale X; SU(3) x U(1) x U(1) is then broken to SU(3) x U(1) at energies of order M. In this model the neutral currents have roughly the strength that they actually have in nature, but the charged currents are greatly suppressed.

It is difficult to simultaneously obtain the right pattern of symmetry breaking and a suitably long proton lifetime. Recently, Georgi has described[18] an attempt to do this, based on a model with fields Y^{ij}_{kl} in the 75 dimensional representation of SU(5).

There is another crucial fact which must be borne in mind in thinking about the phenomenology of these models. It turns out that in the models as I have written them, the Goldstone fermion decouples in the limit as X becomes very large.* This means that although supersymmetry is broken, the bosons and fermions are degenerate to within terms of much less than M. Clearly, this is a very unacceptable state of affairs if M is interpreted as the mass scale of weak interactions.

It is possible to avoid this decoupling of the Goldstone fermion, and obtain models with bose-fermi splittings that really are of order M, by adding additional fields. However, the resulting models seem somewhat contrived.

It is equally interesting to explore the physical content of our models such as (49) more carefully, bearing in mind the decoupling of the Goldstone fermion. A more careful study of the model based on equation (49) shows that, at the tree level, many particles have masses much less than M. The SU(3) x SU(2) x U(1) gauge mesons and their fermionic partners are massless at the tree level. All components of Y except the component with a vacuum expectation value have masses of order M^2/X. (As X becomes large, A becomes heavy and decouples; as A decouples, Y becomes massless.) One would naively expect that these light particles would receive masses of order αM

---

*I wish to thank T. Banks for convincing me of this.

or perhaps $\alpha^2 M$ from one or two loop diagrams. Because of the decoupling of the Goldstone fermion and the non-renormalization theorems of unbroken supersymmetry, this is not true. The particles whose masses are of order M^2/X or less at the tree level receive mass corrections at most of order $\alpha M^2/X$ from the loop diagrams.

These theories thus have <u>three</u> mass scales, X, M, and M^2/X, if we do not attempt to tamper with the decoupling of the Goldstone fermion that occurs for large X in the simplest models. The fascinating possibility exists that it is the <u>smallest</u> of these scales, the scale M^2/X, which is the mass scale of particle physics as we know it. One reason that this possibility is interesting is that at energies of order M^2/X, supersymmetry appears to be <u>explicitly</u> (but softly) broken. To write an effective Lagrangian describing this system in which the supersymmetry is <u>spontaneously</u> broken, one must re-introduce some of the degrees of freedom with masses of order M.

Of course, soft, explicit breaking of supersymmetry greatly weakens the constraints associated with supersymmetry and thereby alleviates the problems in finding a realistic model. The question of finding a realistic model in which the mass scale of weak interactions is the <u>smallest</u> scale, M^2/X, is therefore very much worthy of attention. I will return to this matter elsewhere.[19] One must, of course, decide on an appropriate identification of X and M. Plausibly, in this approach, X might be of order the Planck mass; M would then have to be about 10^{11} GeV.

Let us now conclude by discussing a few general questions.

In a theory in which the fundamental mass scale of nature is relatively "small" and the mass of grand unification is a derived quantity, we do not wish to assume that the Planck mass M_{Pl} is a fundamental constant of nature. If it is, we are left with the large and unexplained ratio M_{Pl}/M. Rather, we should assume that the large value of the Planck mass is spontaneously generated along with the large value of the mass scale of grand unification.

This possibility is not as bizarre as it might sound. The usual kinetic energy of the gravitational field is $M_o^2 R$, M_o being the bare Planck mass and R the Ricci scalar. Another possible coupling is the dimension four Brans-Dicke coupling $|X|^2 R$, X being the scalar field that will eventually obtain a large vacuum expectation value. (Such couplings actually arise almost inevitably when theories of global supersymmetry are coupled to gravity.) We thus imagine the gravitational action to be

$$\mathcal{L}_{Grav} = M_o^2 R + \lambda^2 |X|^2 R \tag{54}$$

where λ is a constant of order one.

Clearly, the observed Planck mass would be $M_{Pl}^2 = M_o^2 + \lambda^2 |X|^2$. For large X, the Planck mass is simply λX, unrelated to the original constant M_o.

Let us now return to the question of how the effective potential behaves when X becomes very large. As we discussed earlier, while it is possible that a stable minimum develops for large X, it is equally possible that the potential decreases indefinitely with increasing X, as in figure (16b).

This possibility corresponds to a world without a stable vacuum. However, it is possible to expand about a "cosmological solution" in which X is regarded as a function of time. For large times, X would change very slowly, because the potential (being bounded below) is nearly independent of X for large X.

For instance, if $V(X) \sim |X|^{-p}$ for some p, then, by solving the cosmological equations (with the gravitational action taking the form (54)), it is easy to show that for large t, $X \sim t^{2/(2+p)}$.

Since the masses and couplings of the elementary particles depend on X, this corresponds to Dirac's idea that the "constants" of nature are slowly evolving functions of time. For instance, Newton's constant $G = 1/M_{pl}^2$ scales with X as $1/X^2$, because $M_{pl} \sim X$. So $G \sim t^{-4/(2+p)}$, t being the age of the universe. This may be compared to the behavior $G \sim 1/t$ which Dirac advocated in order to account for the value of G in today's universe.

Present day experiments aimed at measuring $\dot{G}/G$ are not yet sensitive enough to confirm or refute Dirac's suggestion. However, there are two good reasons to doubt that G is changing at this rate. First, if G scales like a power of t, then G was vastly larger in the early universe. This would ruin the successful calculations of nucleosynthesis in the big bang. Second, the measurements of $\dot{\alpha}/\alpha$ (α is the ordinary fine structure contant, $\alpha = e^2/\hbar c$), are far more accurate than the measurements of $\dot{G}/G$. (A recent review has been given by Dyson.[20]) A change in X would produce a change in α through renormalization effects,[14] at a rate which is ruled out by experiment if X is changing as rapidly as Dirac's approach to the hierarchy problem would require.

It is therefore most prudent to assume that the potential energy V(X) has a stable minimum at some large but finite X.

All of this, however, requires a note of caution. In particle physics discussions we usually ignore gravity. We argue that the Planck mass is larger than other masses of interest; but the real reason for ignoring gravity is, of course, that we simply do not know how to take gravitational effects into account. In the approach I have been discussing, it is not at all clear that it is valid to neglect gravity. As I have explained, in this approach the "bare" Planck mass should probably be of order M, the fundamental mass scale of the theory. The observed Planck mass is then larger in today's universe because of a sort of renormalization effect — the large, spontaneously generated expectation value of X. But the mass scale of grand unification is also determined by X, in this approach. So when we discuss grand unification we are really working at energies above the underlying energy scale of gravitation. It is not at all clear that it is sensible, under these conditions, to assume that gravity is a small effect. However, for the time being, it is the best that we can do.

Finally, I would like to draw attention to a possible consequence of this framework.

As I have mentioned, in the simplest models of this type, there are many particles with masses of order M^2/X — so many that one may wish to interpret M^2/X as the mass scale of particle physics as we know it. I believe that this is the most promising approach.

However, there is another approach, which in fact was the first possibility raised at the beginning of this lecture (and suggested in reference (14)). In variants of these models, arranged so that the Goldstone fermion does not decouple as X becomes large, almost all of the particles have masses of order M, perhaps multiplied by a small power of the fine structure constant α. In this case, it is plausible to assume that the fundamental mass scale M of the theory is not too much larger than the mass scale of the Weinberg-Salam theory.

But there is one particle whose mass is always in the range M^2/X. This is the X particle itself. Indeed, the effective potential depends on X only very weakly. It has the general form

$$V(X) = M^4 \, F \, (\alpha \ln |X|^2/M^2) \tag{55}$$

where F is a dimensionless function of the dimensionless variable $\alpha \ln |X|^2/M^2$ which arises in loop diagrams. To one loop order, F is given (in a particular model) in equation (52).

INTRODUCTION TO SUPERSYMMETRY

Taking the second derivative of (55) with respect to X, we see that the mass of the X particle is of order $\alpha M^2/X$. If M^2/X is the mass scale of weak interactions, this is not remarkably small, and the very weak couplings of the X particle (discussed below) would render it unobservable. However, if the fundamental mass scale M is identified as the scale of weak interactions, then the X particle is <u>extremely</u> light, so light that its Compton wave-length λ_X is macroscopic. For 10^{15} GeV $<$ X $< 10^{19}$ GeV we would expect 10^2 cm $> \lambda_X > 10^{-3}$ cm.

The X particle is a scalar and can have coherent couplings to matter. Actually, at the tree level, X does not couple to ordinary, light particles. X has such a large expectation value that anything that couples to X at the tree level is not light! However, coherent couplings of X to matter are generated by loop diagrams. The most important effect is a coupling of X to gluons, induced by a diagram containing heavy, color-bearing particles (figure (17)). This diagram induces a coupling of X to $\frac{\alpha_s}{<X>}$ Tr $F_{\mu\nu} F^{\mu\nu}$, where $F^a_{\mu\nu}$ is the gluon field strength. As α_s Tr $F_{\mu\nu} F^{\mu\nu}$ is essentially the trace of the energy momentum tensor in QCD, whose matrix element in a nucleon state is the nucleon mass m_N, we see that the coupling of X to a nucleon is of order $m_N/<X>$.

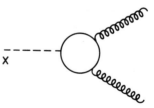

Figure 17

This may be compared to the gravitational coupling, which involves m_N/M_{Pl}. We see that if $<X> \sim M_{Pl}$, the new force is comparable to gravitation, at distances less than λ_X. If $<X> \sim 10^{15}$

GeV the force due to X exchange may be 10^8 times stronger than gravity at distances less than $\lambda_X \sim 10^{-3}$ cm!

New, coherent, short-range forces have been suggested in the recent past, on several different grounds.[21] Laboratory measurements of gravity at short distances place interesting bounds[22] but certainly do not exclude the possibility of such a new force. For instance, neither these experiments nor measurements of the Casimir effect[23] rule out the possible existence of a new force 10^8 times stronger than gravity at a distance of 10^{-3} cm.

These comments really refer to the <u>magnitude</u> of the X field. It may be shown that the <u>phase</u> of the X field behaves rather like a Peccei-Quinn-Weinberg-Wilczek axion, even if there is no underlying U(1) axial symmetry.

A previous version of these lectures was given at the International Center for Theoretical Physics July, 1981. Sections II and IV of these notes are adapted from notes circulated there. I would like to acknowledge the hospitality of the ICTP as well as that of the school on sub-nuclear physics at Erice.

REFERENCES

(1) Y.A. Gol'fand and E.P. Likhtam, JETP Lett. <u>13</u> (1971) 323;
   D.V. Volkov and V.P. Akulov, Phys. Lett. <u>46B</u> (1973) 109;
   J. Wess and B. Zumino, Nucl. Phys. <u>B70</u> (1974) 39.
(2) P. Fayet and S. Ferrara, Phys. Rep. <u>32C</u> (1977) 249; A. Salam and J. Strathdee, Phys. Rev. <u>D11</u> (1975) 1521, Fortschritte der Physik <u>26</u> (1978) 57; J. Wess and J. Bagger, preprint, Institute for Advanced Study (1981); P. van Nieuwenhuysen, Phys. Rept.
(3) S. Coleman and J. Mandula, Phys. Rev. <u>159</u> (1967) 1251.
(4) R. Haag, M. Sohnius, and J. Lopuszanski, Nucl. Phys. <u>B88</u> (1975) 257.

(5) B. Zumino, Coral Gables Lectures (January, 1977), Lectures at the 1976 Scottish Universities Summer School (Edinburgh University Press, Edinburgh, Scotland, 1976); E. Cremmer, B. Julia, J. Scherk, P. van Nieuwenhuysen, S. Ferrara, and L. Girardello, Phys. Lett. 79B (1978) 231.
(6) E. Witten, Nucl. Phys. B185 (1978) 513.
(7) P.A.M. Dirac, Proc. Roy. Soc. (London) A165 (1938) 199.
(8) S. Weinberg, in Proc. of the XVII Int. Conf. on High Energy Physics, 1974, ed. J.R. Smith (Rutherford Laboratory, Chilton, Didcot, Berkshire, England, 1974).
(9) L. O'Raifeartaigh, Nucl. Phys. B96 (1975) 331.
(10) M.T. Grisaru, W. Siegel, and M. Rocek, Nucl. Phys. B159 (1979) 429.
(11) M. Dine, W. Fischler, and M. Srednicki, Nucl. Phys. B189 (1981) 575; S. Dimopoulos and S. Raby, Nucl. Phys. B192 (1981) 353.
(12) E. Witten, Princeton preprint, 1982.
(13) G. 't Hooft, Nucl. Phys. B153 (1979) 141.
(14) E. Witten, Phys. Lett. 105B (1981) 267.
(15) S. Coleman and E. Weinberg, Phys. Rev. D7 (1973) 788.
(16) M. Huq, Phys. Rev. D14 (1976) 3548.
(17) H. Yamagishi, Princeton preprint, 1982.
(18) H. Georgi, Harvard preprint, 1982.
(19) E. Witten, to appear.
(20) F. Dyson, in Current Trends in the Theory of Fields, ed. J.E. Lannutti and P.K. Williams (AIP, 1978).
(21) J.E. Kim, Phys. Rev. Lett. 43 (1979) 103; G. Feinberg and J. Sucher, Phys. Rev. D20 (1979) 1717; J. Scherk, Phys. Lett. 88B (1979) 265.
(22) R. Spero, J.K. Hoskins, R. Newman, J. Pellman, and J. Schultz, Phys. Rev. Lett. 44 (1980) 1645; H.-T. Yu, W.-T. Ni, C.-C. Hu, F.-H. Liu, C.-H. Yang, and W.-N. Liu, Phys. Rev. D20 (1979) 1813.
(23) J.N. Israelachveli and K. Tabor, in Progress and Surface and Membrane Science, Vol. 7, ed. J.F. Danielli, M.D. Rosenberg, and D.A. Cadenhead (Academic Press, 1973).

DISCUSSIONS

CHAIRMAN: E. WITTEN
Scientific Secretary: H. Yamagishi

DISCUSSION 1

— *KARLINER*:

We have some kind of idea as to how non-zero temperature affects conventional symmetries. Can you tell us what happens in supersymmetric theories when $T \neq 0$?

— *WITTEN*:

For ordinary symmetries, the vacuum state is not the only state which is invariant under the symmetry. But for supersymmetry, all excited states are non-symmetric since $H = \frac{1}{2}\sum_\alpha Q_\alpha^2$. Of course, at non-zero temperature, we are always dealing with excited states and so supersymmetry is not a symmetry.

— *COLEMAN*:

Hey, hold it. Thermodynamic states are density matrices, not single states. The fact that the Hamiltonian does not annihilate the excited states doesn't mean that we lose time translation invariance at finite temperatures.

— *WITTEN*:

It doesn't matter. Let me give a formal argument to satisfy Sidney. First of all, what would it mean to say that supersymmetry is broken at zero temperature? One way is to say that $\langle\Omega|\{Q_\alpha,\chi\}|\Omega\rangle\neq 0$ for some χ. So, presumably what you mean by supersymmetry breaking at finite temperature is that $\text{Tr}\, e^{-\beta H}\{Q_\alpha,\chi\}\neq 0$. But the left-hand side is equal to $2\,\text{Tr}\, e^{-\beta H} Q_\alpha\chi$ which is generally non-zero.

- CASTELLANI:

In the model you have mentioned, the cancellations between bosonic and fermionic contributions hold to all orders in perturbation theory. But non-perturbatively, supersymmetry may become broken. Now in supergravity, such cancellations are considered essential for the finiteness of the theory. Are they also just artifacts of perturbation theory?

- WITTEN:

I am not in any case very impressed with the cancellations demonstrated up to now in perturbation theory for supergravity. My personal guess is that none of these theories are finite. To answer your question, however, one might believe that perturbation theory could at least expose infinities. If it does turn out that some of these theories are finite to all orders, the non-perturbative effects should also be finite.

- SEIBERG:

Why cannot supersymmetry be explicitly broken? It is not a gauge symmetry.

- WITTEN:

There are two answers to that question. First, if supersymmetry is explicity broken, it is not clear, at least to me, that it is interesting. Second, although it is true that in these lectures I have been talking of supersymmetry as a global symmetry, if it is really relevant to nature, it must be a local symmetry because of gravity. In the presence of gravity, the definition $Q_\alpha = \int d^3x\, S_{0\alpha}$

doesn't make sense because if you parallel transport a spinor around, the spinor index α changes. In equations, $D_\mu \varepsilon_\alpha = 0$ gives $0 = [D_\mu, D_\nu] \varepsilon_\alpha = R_{\mu\nu\sigma\tau} \sigma^{\sigma\tau} \varepsilon_\alpha$ so $\varepsilon_\alpha = 0$ and there is no constant spinor (charge).

- SEIBERG:

What are the other terms you can add to the right-hand side of $\{Q_\alpha, \bar{Q}_\beta\} = \gamma^\mu_{\alpha\beta} P_\mu$?

- WITTEN:

Since $\{Q_\alpha, \bar{Q}_\beta\}$ has an integer angular momentum, the Coleman-Mandula theorem permits us to have the Poincaré generators P_μ, $M_{\mu\nu}$ and Lorentz invariant charges B_i. (I am leaving aside conformal invariance.) If Q_α gives an irreducible representation of the Lorentz group, you learn that only P_μ and $M_{\mu\nu}$ are allowed due to the statistics. But then Haag, Sohnius and Lopuzanski have shown that the $M_{\mu\nu}$ term does not satisfy the (generalized) Jacobi identity. However, the same authors have shown that if Q_α is reducible, you can also have the B_i's, which are called central charges. I am not going to discuss theories with central charges because while they are beautiful, they are all too restrictive to describe nature, at least below the Planck mass.

- MIKHOV:

Could you elaborate on why the theories with central charges are too restrictive?

- WITTEN:

Apparently, one of the basic things in particle physics is that left-handed fermions have different quantum numbers from the right-handed fermions. But this is impossible to have in N>1 (extended) supersymmetry. Basically, these theories have two helicity raising operators $Q^{(1)}_+$ and $Q^{(2)}_+$. The quarks for instance would either be

INTRODUCTION TO SUPERSYMMETRY

in a representation $|-\frac{1}{2}\rangle\,|0\rangle\,|\frac{1}{2}\rangle$ or $|0\rangle\,|\frac{1}{2}\rangle\,|1\rangle$. (The labels are helicities.) For the former, the left-handed quarks and the right-handed quarks would be in the same supersymmetry multiplet and consequently would have the same SU(3)xSU(2)xU(1) quantum numbers. For the latter, the quarks would be related to the gauge bosons, which must be in the adjoint representation. But that is always real, which again means that the left-handed quarks and the right-handed ones have the same quantum numbers.

- *FORGACS:*

What is the role of space-time dimension in the Coleman-Mandula theorem?

- *WITTEN:*

In one sentence the theorem says that exotic quantum numbers, if they existed, would fix the scattering angles to discrete values. But in 1+1 dimensions they are fixed anyway, the only possible angles being zero and π.

- *FORGACS:*

And is there any connection with the existence of an infinite number of conservation laws?

- *COLEMAN:*

The same answer. They exist in only 1+1 dimensions.

- *FORGACS:*

There still might be conservation laws which are non-local in higher dimensions.

- *COLEMAN:*

Our theorem is strictly an S-matrix theorem, so the only kind of locality it uses is that under the associated symmetry transformation. A multi-particle state transforms as the product of single particle states. So you would need something which was hideously non-local. Of course, if you solve the equations of motion and write the fields at t=0 (or t=-(χ)) as functions of fields at time t, that always gives you a non-local conservation law.

- TELLER:

Are the phases of corresponding scattering processes the same in supersymmetry theories?

- WITTEN:

Not necessarily. There are only fixed relations between them, just as in πN scattering.

- GIPSON:

You've said that supersymmetry was a symmetry in search of an application. But Franco Iachello has used supersymmetry to connect energy levels of nuclei differing by half-integer spin.

- WITTEN:

Yes, your comment is quite true. On the other hand, the algebra is quite different and does not have a relativistic extension.

- SPIEGELGLAS:

You have said that supersymmetry is broken or unbroken according to the behavior of W at infinity. Now such behavior is invariant under W → W+const. Is there any meaning to this?

- WITTEN:

The connection between supersymmetry breaking and the asymptotic behavior of W is a kind of statement about the topology in field space, and that has many generalizations. Some of which I hope to explain in the following lectures. I don't think there is anything special to adding a constant to W.

- SPIEGELGLAS:

Given a quantum mechanical Hamiltonian, is there a simple method of seeing whether it can be written as a sum of squares corresponding to an underlying supersymmetry?

- WITTEN:

No. I think it would be difficult.

- COLEMAN:

May I return to the first question? The question is whether there are at least two phases which turn into each other under supersymmetry transformations, which I think is what we usually mean by spontaneous supersymmetry breaking at finite temperatures.

- WITTEN:

That isn't the case in general. What I take it to mean is that boson masses are not equal to fermion masses at finite temperature. For instance, in the massless Wess-Zumino model, the one-loop contribution to the boson mass is greater than the correction to the fermion mass.

- COLEMAN:

That's better. Thank you.

DISCUSSION 2   (Scientific Secretary: H. Yamagishi)

- CASTELLANI:

How general is the theorem that the vacuum expectation value of H does not get renormalized to all orders in perturbation theory? In particular, does it hold for any supersymmetric Yang-Mills theory, regardless of the gauge group? And does it hold for local supersymmetry?

- WITTEN:

For global supersymmetry, there is only one exception. For groups which are not semi-simple and contain a fundamental U(1), there may be a term in the potential like $\sum_i (e_i \phi_i^* \phi_i + \mu^2)^2$. The μ dependent part is called the Fayet-Illiopoulos D term. Apart from wave function renormalization, this is the only renormalization we have in supersymmetry. Recently, it was proved by six authors from SLAC (Susskind et al.) that the renormalization of D occurs only at the one-loop level. Actually, it is not apparent, but this term is parity-violating, so you don't have to worry about it in parity-conserving theories. But for a parity-violating theory with a U(1) component of the gauge group, the D term will be generated at the one loop level with a quadratically divergent coefficient. So in this case, the loop correction can break supersymmetry. (The D term often does this.) But you should realize that this is just the case where you have to worry about naturalness in the first place. As to your second question, the nature of the proofs suggest to me that they should also go through for supergravity (with perhaps some exceptions), although I am not aware of any definite theorem.

- KARLINER:

We usually prove renormalizability order by order in perturbation theory. Since some strange things happen in supersymmetric theories to all orders, could it indicate some problems with renormalizability?

- WITTEN:

Even for ordinary theories which are not asymptotically free, you have to worry about what happens to renormalization theory non-perturbatively. As far as I can see, supersymmetric theories are more likely better, than worse, in this respect, since some of the renormalization constants are finite.

INTRODUCTION TO SUPERSYMMETRY

- KARLINER:

Would it be a realistic project to do Monte-Carlo calculations for supersymmetric theories, since non-perturbative effects are important?

- WITTEN:

The problem is to put fermions on a lattice and preserve supersymmetry. I may have given the wrong impression that important effects are all non-perturbative. There are also interesting questions in perturbation theory, as I would like to show in the last lecture.

- SEIBERG:

Is the Lagrangian based on W the most general form for renormalizable supersymmetric theories?

- WITTEN:

For renormalizable theories with only spin 0 and 1/2, yes. Of course, if you want to discuss all supersymmetric operators of any dimension which can appear in the effective Lagrangian, W is not sufficient. (It is best to use the superfield formalism.) With spin one (gauge fields) it is possible to include supersymmetric gauge couplings. Renormalizable theories with spin greater than one are not known (and I doubt that they exist) with or without supersymmetry.

- GIPSON:

The conserved charge is a spinor. Does this mean that we can't see it?

- WITTEN:

In supergravity, there is a field coupled to it, so you can measure its matrix elements in principle. But since the field is fermionic, there are no macroscopic effects, unlike mass or electric charge.

- BANKS:

In the supersymmetric SU(5) theory, you found three (nearly) degenerate vacua. If we want to make a grand unified theory of the real world out of this, with SU(3)xSU(2)xU(1) at low energies, which vacuum should we start from at high energies?

- WILCZEK:

Presumably at finite temperature, one vacuum has a lower free energy than the others, so you should choose that one for cosmology.

- TELLER:

Why can't $\partial W/\partial A_i$ be perturbed, in which case the zero of $\partial W/\partial A_i$ may just shift to a new position?

- WITTEN:

In general, quantum corrections can't be written as a change in W.

- MOTTOLA:

I am a bit confused. You said that "grand-unified instantons" may lift the degeneracy in the SU(5) model. On the other hand, you said that there is no tunneling between the vacua. How can that be?

- WITTEN:

Oh, the tunneling occurs along the winding-number axis which I haven't shown.

- SPIEGELGLAS:

How do you find the supercurrent in general?

- WITTEN:

The best way is to use superfields.

- *SPIEGELGLAS*:

Does supersymmetry impose any constraints on the symmetry group and its representations?

- *WITTEN*:

You can just about choose any group and any fermion representation which then gives you the scalars and the gauge bosons. A more restrictive framework would be extended supersymmetry, but that we talked about already.

DISCUSSION 3   (Scientific Secretary: N. Seiberg)

- *BANKS*:

Could you clarify the distinction between massless theories in which the mass doesn't have to be zero and those in which it does?

- *WITTEN*:

Consider a theory with massless particles. In such theories the vacuum plus a zero momentum particle may have zero energy. The computation of $\mathrm{Tr}(-1)^F$ becomes tricky provided that the massless particles are inevitably massless. On the other hand if the theory with mass terms for these massless particles makes sense, the calculation of $\mathrm{Tr}(-1)^F$ is easier. You add the mass terms to the Lagrangian and calculate $\mathrm{Tr}(-1)^F$ in the presence of the mass. Now the vacuum is the only zero state and $\mathrm{Tr}(-1)^F > 0$. Since $\mathrm{Tr}(-1)^F$ is independent of the mass, it remains positive even when m=0. If the masses could have been non-zero, $\mathrm{Tr}(-1)^F$ is always positive and supersymmetry is unbroken. If on the other hand as in non-Abelian gauge theories the massless theory is not the limit of a massive theory the method described above for calculating $\mathrm{Tr}(-1)^F$ is invalid.

- KARLINER:

If we start from $n_F^{E=0}=1$, $n_B^{E=0}=2$ and then change some parameter so that the fermion acquires a mass, this fermion must take a bosonic partner along. How does it decide which one to take?

- WITTEN:

It depends on the dynamics. It is not at all strange that a particle that was a singlet when it was massless becomes a part of a doublet when it is massive. Consider

$$Q = \begin{pmatrix} 0 & \lambda \\ \lambda & 0 \end{pmatrix} \qquad H = \tfrac{1}{2} \begin{pmatrix} \lambda^2 & 0 \\ 0 & \lambda^2 \end{pmatrix}$$

For $\lambda = 0$ there are two singlets and for $\lambda \neq 0$ there is one doublet.

- COLEMAN:

Consider, in one spatial dimension, the group generated by space translations and parity. An eigen-state of the momentum operator with non-zero eigen-value belongs to a two-dimensional representation while the zero momentum state is a singlet. One situation can be the continuous limit of the other and nothing is strange here.

- WITTEN:

That is a good example.

- DUERKSEN:

You described an SU(5) GUT which exhibited supersymmetry breaking at the scale of 10^3 GeV and the breaking of SU(5) symmetry to $SU(3)_c \times SU(2) \times U(1)$ at 10^{15} GeV. How does one understand the ratio 10^{12} between these two mass scales?

- WITTEN:

We do not know the exact supersymmetry breaking mechanism - no dynamical supersymmetry breaking has been demonstrated yet. However, a good candidate for this might be SU(5) instantons which live

INTRODUCTION TO SUPERSYMMETRY

partly in the SU(3) and partly in the SU(2) group. If one assumes the SU(5) symmetry breaking occurs with mass scale μ_x, then the scale for the supersymmetry breaking μ_w in this scenario will be

$$\mu_w \cong \mu_x \, e^{-\frac{\pi}{\alpha_{GUT}}}$$

which gives the type of ratio we have encountered. More details can be found in my paper "Dynamical Breaking of Supersymmetry" to be published in Nucl. Phys. B.

- WIGNER:

What is the physical meaning of a particle with zero energy?

- WITTEN:

In the infinite volume limit the existence of such a particle has no measureable consequences. Because of technical reasons I decided to work in a finite volume where the momentum and, therefore, the energy is discrete. In the infinite volume limit the momentum as well as the energy become continuous but the masses remain discrete.

- ETIM:

You can define a modified trace by

$$\text{Tr}(-1)^F \exp -\beta H \, .$$

Such an operation can always be defined and analyzed and we can study the thermodynamical properties of a gas of supersymmetric objects.

- WITTEN:

The density matrix defined this way is not positive definite. In fact, I can calculate the partition function

$$\text{Tr}(-1)^F \exp -\beta H = \text{Tr}(-1)^F.$$

This is true simply because for every non-zero energy value there are two states with opposite values of $(-1)^F$, so the states of $E \neq 0$ do not contribute.

- GOTTLIEB:

Please explain the difference between U(1) and SU(N) gauge theories, which allows the former to evade your theorem showing that SU(N) may not break supersymmetry.

- WITTEN:

For U(1) we have $\text{Tr}(-1)^F = 0$. That is because we may eliminate the zero mode of the gauge field. For the Abelian theory, $A_i \to A_i + \partial_i \varepsilon$ and we may eliminate $A_i = c_i$ by setting $\varepsilon = -\vec{c}\cdot\vec{x}$. This doesn't work for the non-Abelian theory because of the additional $[A_i, \varepsilon]$ term in the gauge transformation. This term destroys the periodic boundary conditions. Hence, for the Abelian case only, we may eliminate the zero energy excitation. There are four zero energy states: the normal vacuum, vacuum plus zero momentum spin up fermion, vacuum plus spin down fermion and a state with two fermions coupled to spin zero added to the vacuum. The first and last are bosonic states, the other two are fermionic. Thus, $\text{Tr}(-1)^F = 0$.

- CASTELLANI:

You calculated $\text{Tr}(-1)^F$ in your SU(5) example. The result was $\text{Tr}(-1)^F = 5$ - supersymmetric SU(5) cannot be spontaneously broken.

- WITTEN:

I can give you an example of an SU(5) supersymmetry theory with $\text{Tr}(-1)^F = 0$. I introduce a scalar field Y and I take

$$W = \tfrac{1}{3} \bar{g} \, \text{Tr} \, A^3 + \tfrac{1}{2} g \, Y \, (\text{Tr} \, A^2 - \mu^2).$$

The equations are $\quad \frac{\partial W}{\partial A} = 0 \; ; \; \frac{\partial W}{\partial Y} = 0 \quad$ i.e.

$$\text{Tr} \, A^2 - \mu^2 = 0 \; ; \; \bar{g} \, [(A^2)_i^{\;j} - \tfrac{1}{5} \delta_i^{\;j} \, \text{Tr} \, A^2] + g \, Y A_i^{\;j} = 0.$$

Obviously A = 0 is not a solution. The minima of V correspond to

INTRODUCTION TO SUPERSYMMETRY

the vacua with the symmetry SU(4)×U(1) or SU(3)×SU(2)×U(1) and, therefore, Tr $(-1)^F$ = 0.

DISCUSSION 4   (Scientific Secretary: N. Seiberg)

- SPIEGELGLAS:

Where was supersymmetry needed in today's models?

- WITTEN:

I have used the non-renormalization theorems very heavily. Also the type of degeneracy we used at the tree level depends on supersymmetry.

- D'HOKER:

In the last model that you described today, there were two mass scales; 1 TeV and 10^{-14} GeV. The lower scale is remarkably close to the scale of the bound on the cosmological constant. Is this a coincidence?

- WITTEN:

It is the numerical coincidence

$$\mu_{Planck} \cdot \Lambda^{\frac{1}{4}} \sim \mu_W$$

where Λ is the upper bound on the possible value of the cosmological constant. No one knows if it is a coincidence.

- SEIBERG:

Could you explain why σ is so small in your model?

- WITTEN:

The model I have described possesses a Peccei-Quinn symmetry, if W is purely cubic. If the coefficients of the linear and quadratic terms in W (a_i and a_{ij} respectively) are not zero, this symmetry is not exact. Although a_i is of order μ^2 and a_{ij} is of order μ, X is anomalously big –

$$X \sim a_i \exp \frac{1}{\alpha}$$

and the Peccei-Quinn symmetry is almost exact on the scale X.

- GLASHOW:

What are the mass and coupling of the axion?

- WITTEN:

It is light and its coupling is extremely small. This is because the value of its F_n analog is very big.

- KARLINER:

You have shown us today a supersymmetric model, which has some extremely interesting consequences. However, this model is only semi-realistic. To what extent would the interesting effects persist in more realistic models?

- WITTEN:

Changing of G_n with time is not a firm prediction. It depends on the non-existence of a stabel vacuum. The prediction of corrections to gravity at short distances is much more model independent since the X field has generally a very small mass – $\mu_x \sim \alpha \frac{\mu^2}{x}$. Even so, it is difficult to give an exact prediction of its mass because it depends on the value of μ.

- HERTEN:

One prediction of your model is the time dependence of the fine structure constant. Could you explain how the limit on this dependence can be obtained?

- WITTEN:

One measures the light coming from distant galaxies. The light was emitted billions of years ago. If the fine structure constant α

INTRODUCTION TO SUPERSYMMETRY

was different at that time, the atomic spectral lines should be different too. The best bound comes from the measured 21 cm line of hydrogen, and one obtains

$$\frac{\dot{\alpha}}{\alpha} < 10^{-16}/\text{year}.$$

Other (stronger) bounds come from observations of the abundances of the nuclei.

- *CASTELLANI*:

In your model asymptotic freedom seems to be important for the stability of the vacuum. Is this a general feature of these models?

- *WITTEN*:

The stability of the vacuum depends on the sign of $b\bar{\lambda}^2 - c\,\bar{e}^2$. This term may be positive or negative at large X with or without asymptotic freedom.

NEUTRINO PHYSICS AT FERMILAB

Drasko Jovanovic

Fermi National Accelerator Laboratory
Batavia, Illinois USA

The neutrino program at Fermilab can be best described by giving preliminary results from the three recently finished experiments. These three are: (1) measurement of charmed particle lifetimes, (2) measurement of total neutrino and antineutrino cross sections, and (3) some new limits on neutrino oscillations. All three experiments represent different experimental techniques: emulsions, counter-detectors and bubble chambers. Although there are many more experiments which were conducted as a general neutrino program, these three have a certain unity which addresses the recent new interest in neutrino masses and cosmology.

NEUTRINO PHYSICS AT FERMILAB

"Secondo la puoposta di Pauli si puo per esempio ammettere l'esistenza di una nuova particella, il cosi detto <<NEUTRINO>>, avente carica electrica nulla e massa dell'ordine di grandezza di quella dell'elettrone o minore.

E. Fermi
Nuovo Comento II 1-19(1934)

It is quite fitting that a report on Neutrino physics experiments from Fermilab be given here in Italy where seminal ideas about the neutrino were born and where the "Godfathering" or christening happened in October 1931 during the "Convegno di Fisica Nucleare" in Rome.

Fifty years from its birth, so to speak, neutrino has come of substantial "experimental maturity" but as you shall see still bringing to the field substantial new vigor quite unprecedent for its ripe 50 year old age. In this report we shall discuss the preliminary results of the three resently completed experiments and give a progress report on the Fermilab experiment still in data taking stage.

1. E-531: Charm particle production and lifetimes; produced by a neutrino beam in nuclear emulsion. Collaboration: Fermilab, Ohio State Univ., Aichi Univ., Kobe, Korea Univ., McGill, Nagoya Okayama, Osaka, Ottawa, Inst. CR Res Tokyo, Toronto, Yokahama.

2. E-616: Measurement of neutrino and antineutrino total cross sections. Narrow band beam experiment by the collaboration: Columbia Univ., Fermilab, Rockefeller, Rochester.

3. E-53A: A limit on the neutrino oscillation "15 foot B.C. experiment" Columbia BNL collaboration.

These three experiments illustrate the diversity of methods and the topics to which one can employ neutrino beams. Yet at the same time these experiments demonstrate the unity of the field since the results of these experiments put direct or indirect bounds on each other.

E-531

Measurement of charm particle lifetimes
AFKK McGOOOTTY-Collaboration

The objective of experiment E-531 is to measure the lifetime of charmed hadrons. (Charmed hadrons are hadrons which contain one charmed quark as a constituent.) Rather early it was recognized that the charmed particle production represented 5 to 15% of the charged current neutrino interactions:

$$\frac{\nu_\mu + N \to \mu^- + \text{charmed hadrons}}{\nu_\mu + N \to \mu^- + \text{hadrons}} \approx 0.10.$$

Since the branching ratio of leptonic charm decays is ~10% and the measured di-muon yield in neutrino experiments is ~ 0.5×10^{-2} at neutrino energies of 20 GeV or higher. Thus neutrino interactions yield charm particles with a signal to noise ratio of 1:10 as contrasted with photoproduction or hadronic production where it is much lower:

	neutrino	photon	hadron
$\sigma_{ch}/\sigma_{total}$ =	~0.1	~0.01	~.001

The enhancement of charm production in neutrino interactions thus suggests neutrino experiments to be a good way to investigate charm providing sufficient ν flux is available. To meet the objective of measuring lifetimes as short as 5×10^{-13} sec one must measure distance in the range,

$$L = \gamma\beta c\tau = 100\mu - 1000\ \mu$$

Obviously one needs a detector with vertex resolution about one order of magnitude better, i.e 10-100 microns. Fine grain photographic emulsions are the only detector known to fulfill this criteria. This reasoning was recognized and M.K. Guillard pleaded in 1978 (CERN - Neutrino workshop) for more emulsion experiments. Both CERN and at Fermilab responded by approving a number of such experiments; E-531 seems to be the most productive thus far. The essence of the experiment is best understood by referring to Figure 1. Neutrino interactions which occur anywhere in the 23 liters of the nuclear emulsion are tracked back by using frequently changeable emulsion sheets immediately downstream of the main stack. The charged particle are tracked and momentum analyzed by the spectrometer, a combination of drift chambers and an open magnet. Particle identification is accomplished by digitizing each charged particles time of flight with hodoscopes (I and II) before and after the drift chambers. Finally, muons are identified by a range hodoscope and penetration through ~ 2m of steel. Lead glass and iron calorimeter complement the reconstruction of the events where applicable. A one to one correlation between particles identified in the spectrometer and the emulsion tracks is established by comparing angles with respect to the neutrino direction as measured in the front drift chambers with those measured in the emulsion. In such a manner it can be easily established whether a muon origatates at the primary vertex or several hundred microns downstream in a secondary vertex.

A particle's identity is achieved by a comparison of momentum and time of flight. The resolution of the time of flight (TOF) hodoscopes together with the spectrometer is given in Fig. 2 for a sample of tracks with momentum <2 GeV/c.

In all emulsion experiments conducted thus far one of the principal difficulties was vertex finding based on projecting spectrometer tracks back into the emulsion stack. In addition to a careful survey and precise location of the emulsion stack with respect

NEUTRINO PHYSICS AT FERMILAB

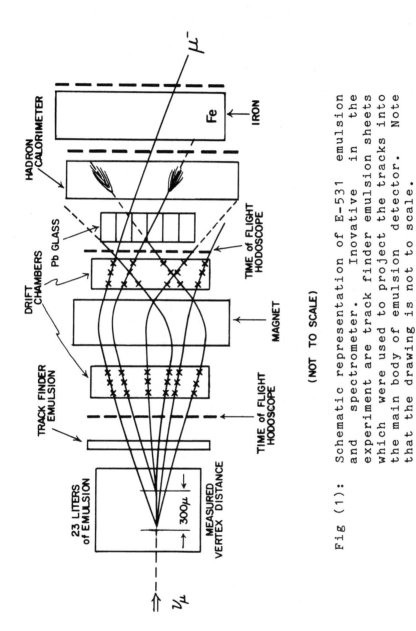

Fig (1): Schematic representation of E-531 emulsion and spectrometer. Inovative in the experiment are track finder emulsion sheets which were used to project the tracks into the main body of emulsion detector. Note that the drawing is not to scale.

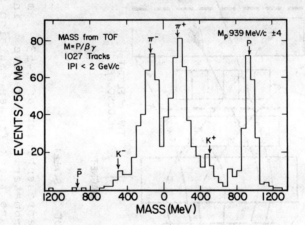

Fig (2): Mass distributions of some 1000 tracks as measured by the spectrometer and the time of flight counters.

to the spectrometer, a clever new inovation known as track finder sheets was used in this experiment. These sheets could be 50_μ thick emulsion plates which could be changed periodically to avoid integrated confusion of all cosmic ray and background tracks.

The first run of the experiment E-531 took place in the early 1979 using a wide band (Horn focused) neutrino beam.

Run Parameters were

1) 350 GeV incident proton energy

2) 7×10^{18} total protons in the horn target.

3) 1 ms spill (21% electronic dead time)

This run resulted in 2200 neutrino events with the vertex originating in the emulsion stack. Of these 1800 were in the proper fiducial volume. The emulsion scan resulted in finding 1170. Thus the "scanning" efficiency in the emulsion was 75%. Once the neutrino vertex was found the detailed scan for charm candidates began. Three methods were used:

1) Track following: Follow all tracks with $\theta < 200$ mr for neutral decay

2) Neutral decay: Scan cylinder $100 \mu m \times 300 \mu m$ diam. for neutral decay;

3) "Scanback": Follow back spectrometer tracks with $p > 0.7$ GeV/c to within 1mm of vertex back into emulsion.

A total of 84 candidates were classified as found of which 40 were classified as "kinks" and 44 as multiprongs. Careful matching of spectrometer information, particle identification and momentum fits resulted in the following sample of charm events:

18 D^0 events

6 D^+ events

3 F^+ events

6 Λ_c^+ events

3 Ambiguous events

As an example of the event sample the relevent parameters of the 18 D^o events are given in Table I. The whole set of events found D^o, D^+ thend F^+ yields the following table of lifetimes of charmed mesons; while Fig. 3 gives the plot of D^o lifetime.

Table II

$\tau_{D^o} = 3.1 ^{+1.1}_{-0.7} \times 10^{-13}$ sec $M_{D^o} = 1856 \pm 15$ MeV

$\tau_{D^+} = 9.5 ^{+6.5}_{-3.3} \times 10^{-13}$ sec $M_{D^+} = 1851 \pm 20$

$\tau_{F^+} = 2.0 ^{+1.8}_{-0.8} \times 10^{-13}$ sec $M_{F^+} = 2026 \pm 20$

$\tau_{\Lambda_c^+} = 1.7 ^{+0.9}_{-0.5} \times 10^{-13}$ $M_{\Lambda_c^+} = 2265 \pm 30$

This experiment together with the results from SLAC[1] e^+e^- colliding measurements provides first evidence about the difference of D^+, and D^o lifetimes. The SLAC experiment measures the branching ratio of semileptonic to total decay modes while this experiment measures the lifetimes directly. Clearly this result, if corroborated by further measurements now in progress, provides evidence that the non leptonic enhancements play a rather prominent role in the decay of charmed mesons. (See C. Quigg: Z Physik 4, 55-62 (1980). Following the rather successful interpretation of non leptonic hyperon decays (Gaillard, Gaillard - Weak Interactions) the four quark model may throw some new

TABLE 1

EVENT	DECAY MODE Observed	DECAY LENGTH (Microns)	P(GeV/c) 3-Cfit	MASS MeV(2C)
1)	$D^0 \rightarrow \pi^-\pi^-\pi^+\pi^+\pi^-\pi^+(\pi^0)$	125 ± 6	8.85±.05	—
2)	$D^0 \rightarrow \pi^-\pi^+K^-\pi^+$	256 ±10	12.30±0.39	1816±40
3)	$D^0 \rightarrow \pi^+\pi^-\bar{K}^0$	326 ±10	11.23±.16	1911±76
4)	$D^0 \rightarrow \pi^+K^+\pi^-\pi^-\pi^0$	27 ± 3	9.18±1.3	1766±48
5)	$D^0 \rightarrow \pi^+K^-\pi^0\pi^0$	116 ±7	30.08±1.04	1939±117
6)	$D^0 \rightarrow \pi^-\pi^+\mu^+K^-(\nu_\mu)$	5472±30	36-59	—
7)	$D^0 \rightarrow \pi^+\pi^-\pi^+K^-\pi^0$	4053±50	23.6 ± .4	1859±38
8)	$D^0 \rightarrow \pi^-K^-\pi^+\pi^+$	749 ±10	13.53±.49	1876±122
9)	$D^0 \rightarrow \pi^-K^-\pi^+\pi^+\pi^0$	42 ± 3	15.43±.32	1855±43
10)	$D^0 \rightarrow \pi^+\pi^-$	67 ±5	11.3±.4	—
11)	$D^0 \rightarrow e^+K^-(\nu_e)$	4374±50	30/70	—
12)	$\bar{D}^0 \rightarrow \pi^-\pi^+\pi^0\pi^0$	184±5	20.89±.6	1857± 70
13)	$D^0 \rightarrow \pi^+\pi^+K^-\pi^-\pi^-\pi^+$	6.5±10	19.2±.31	1925±46
14)	$D^0 \rightarrow \pi^-\pi^+K^0\pi^0$	703±50	12.35±.21	1859±34
15)	$D^0 \rightarrow K^-\mu^+(\nu_\mu)$	2646±30	23/39	—
16)	$\bar{D}^0 \rightarrow K^+\pi^-(\pi^0)$	187±5	6.8/9.5	—
17)	$D^0 \rightarrow \pi^+K^-\pi^0$	590±30	13.23±.24	2000±130
18)	$\bar{D}^0 \rightarrow \pi^-K^+$	3170±50	47.8±2.2	1893±151

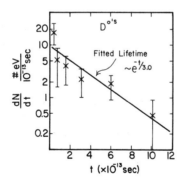

Fig (3): Differential lifetime plot of the eighteen D^0 events.

and possibly unexpected light on the structure of the weak non leptonic hamiltonian. For an experimentalist a simplified approach would be to appeal to a set of quark diagrams: (S.P. Rosen, Phys. Rev Lett. 44, 4, 1980).

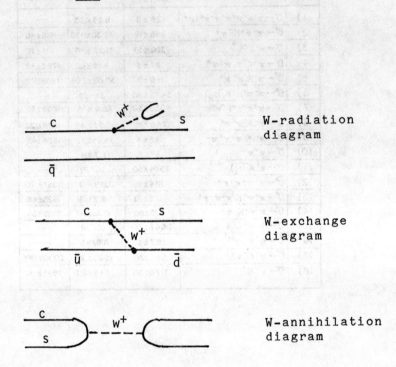

One argues that the processes in diagram (1) contribute equally for all charmed mesons decay whereas (2) is not allowed in D^+ decays (Cabbibo allowed decays) and (3) is present only in F^+ decays.

Currently the E-531 experimental group has completed an additional exposure of 6×10^{18} protons with the emulsions in the same apparatus. The number of events one expects from this exposure is to increase threefold due to the improvements in the apparatus, the beam (400 rather than 350 GeV), and better shielding.

E-616
Neutrino and Antineutrino Total Cross Sections
CFRR - collaboration

The second experiment reported here is the measurement of the total neutrino/antineutrino cross sections. It is conducted in the narrow-band neutrino beam with a massive, ~ 1100 ton neutrino detector. The detector and the beam were first commissioned in the summer of 1978 while the major data run took place from July 1979 to January 1980. The experiment is an outgrowth of an earlier CALTECH - Fermilab detector which pioneered the narrow band beam ideas in the early days of Fermilab.

The experiment collected data for seven months accumulating a total of 10^{19} 400 GeV protons on the target. Its primary aim is directed toward investigation of nucleon structure functions and in particular a measurement of the R-parameter, the ratio of transverse to longitudinally aligned boson propagator. The analysis is now in progress with some 160,000 events collected in both neutrino and antineutrino runs. What is reported here is the first step in the data analysis, measurements of σ_ν and $\sigma_{\bar\nu}$ on the partial sample of the data. Bear in mind that the total cross section magnitudes enter linearly into the values of structure functions F_2 and xF_3.

The neutrino or antineutrino-nucleon total cross sections have a simple linear energy dependence:

$$\sigma_T(E_\nu) = S_\nu E_\nu$$

$$\sigma_T(E_{\bar\nu}) = S_{\bar\nu} E_{\bar\nu}$$

where $E_{\nu/\bar\nu}$ is the neutrino energy and $S_{\nu/\bar\nu}$ is a slope parameter. It is customary therefore to express neutrino (antineutrino) cross sections as a measurement of $S_{\nu/\bar\nu}$ slope parameter assumed to be a constant independent of neutrino energy. Deviation from the linear rise is expected to occur at substantially higher energies as caused by propagator effects for currently adopted masses of W and Z^o. A departure from the QCD predicted scaling violation may also show up as a deviation from a straight line slope parameter. Finally B. Pontecorvo has conjectured, that neutrinos may have a finite mass and be mutually coupled in a manner similar to the quarks which are coupled through Cabbibo angle. Such coupling would then lead to a phenomenon called neutrino oscillations which may in turn modulate the neutrino flux and hence result in the apparent modulation of the slope parameter.

Measurement of the neutrino cross section is conceptually straight forward:

$$\sigma_T = \frac{n(events)}{n(target) \, N(neutrinos)}$$

The numerator, the number of detected events, is free from substantial systematic errors to the order of few percent for any well designed detector of reasonable acceptance. The number of target nuclei is also a well known number provided one keeps a track of the weight of the detector material and the geometry of the fiducial volume. It is the third quantity, the number of neutrinos or neutrino flux, which is the most difficult to measure. There are two ways to deduce the neutrino flux: (1) one is to measure the flux, momentum and divergence of the neutrino parent particles, pions and kaons. The other (2), is to measure the flux of muons accompanying the decay of pions and kaons. Either method has intrinsic advantages and disadvantages. The Fermilab experiment E-616 has adopted the first method (1), namely the hadron flux monitoring while CERN experiments depend mostly on the other method.

Figure (4) is the schematic representation of the Fermilab Dichromatic (narrow band) beam line. The primary proton beam, incident from the left on the diagram, strikes a BeO target of the 50 cm

NEUTRINO PHYSICS AT FERMILAB

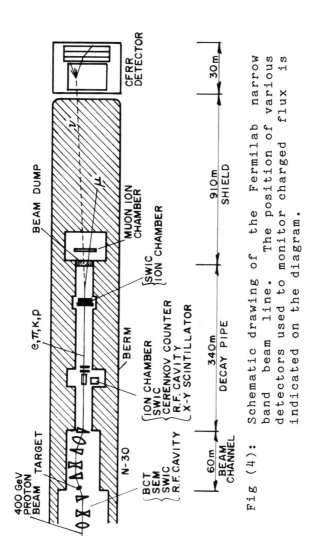

Fig (4): Schematic drawing of the Fermilab narrow band beam line. The position of various detectors used to monitor charged flux is indicated on the diagram.

length. The secondaries are momentum and sign selected and directed as a parallel beam into a decay pipe. Protons which did not interact are deflected some 12 milliradians and stopped by a beam element designed for that purpose. The beam of secondaries had a momentum spread of ± 10% $\Delta p/p$ and the rms angular divergence of ± 0.15 milliradians. Both the momentum spread and angular divergence are measured quantities which are in good agreement with the Monte Carlo calculation of the beam optics. The beam optics were checked out with the 200 GeV extracted primary proton beam exploring both momentum and beam stop apertures. The <u>absolute</u> intensity of the secondary beam is measured by two He-filled ionization chambers. During all data taking the output of both ionization chambers was digitized and recorded on a pulse by pulse basis. Stability of both chambers was very good, never exceeding ±2% deviation in response during 7 months of data taking. The two chambers were located at two different distances in the decay pipe, one at 50 meters and the other at 300 m from the entrance of the 340 m long 1.9 m diameter evacuated decay pipe. The absolute response of the ionization chambers was calibrated by four independent methods:

(1) Foil activation; counting Na^{24} produced in Cu.

(2) Direct counting   275 GeV/c π^- beam

(3) RF-cavity   RF-structure of the beam

(4) BCT (beam current transformer)

Figure (5) illustrates the result of the whole absolute calibration procedure. The overall agreement is ±3%. There was an attempt to use a special 50 cm opening diameter BCT detector in the decay pipe but its response was deemed unreliable mostly due to the decay muon flux passing through the steel armature of the transformer.

The linearity of both chambers was checked over several orders of magnitude different beam

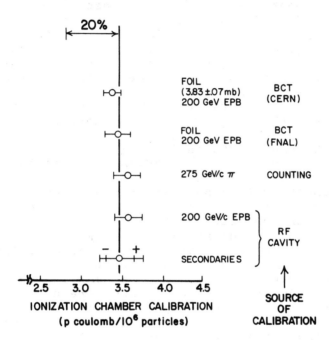

Fig (5): The figure illustrates diagramatically the agreement between several independent calibration measurements of the He-ionization chamber.

intensities. Calibration runs were repeated several times for both 1m sec fast spill and the slow 0.5 sec durations slow spill. All of these runs agree to within 0.7%. The foil activation cross section was measured against BCT and gave the result of 3.90 ± 0.13 mb, a result in good agreement with the values measured independently at CERN or elsewhere. The set of calibrations outlined above provide the basis of the absolute measurement of the charged particle flux in the decay pipe consisting of: e, π, k, p and μ's. Only one detector provides however, the information about the fraction of these particles in the beam. That detector is a differential Cerenkov counter. It is a unique, stand alone device which integrates rather than counts individual particles which traverse the length of it. Typical flux of charged particles passing through the He-gas was 10^9-10^{10} particles/msec during positive (neutrino) runs and $\sim 10^8$ for negative (antineutrino) settings. The length of the counter was 2 meters and the cerenkov light was detected between 0.7 and 1 mrad angles, with the differential width of the iris corresponding to 0.3 milliradiams. To measure the complete cerenkov curve the pressure was varied in small increments with a shutter providing data on the light detected in the glass envelope of the photomultiplier or outside the active volume. Some typical cerenkov curves are shown in Fig. 6 and 7. The absolute response of the counter was established by sending 200 GeV primary beam through it and also testing the counter response to a δ-like beam in momentum. Cerenkov curves were taken frequently and during data taking ensuring that the device was stable and reproducible. The particle fractions as experimentally determined as shown in Fig. 8. These fractions are in good agreement with the published value at 200 GeV measured at CERN and also with the particle production measurements done elsewhere.

Finally, split electrode ionization chambers were also installed in the decay pipe providing pulse by pulse information about the beam position and angle. Care was taken that the beam always pointed at the center of the detector to within 5 cm, throughout the seven month run. The beam steering information was written on the data tapes so that

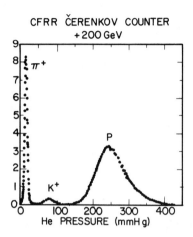

Fig (6): Typical Cerenkov counter pressure curve at +200 GeV beam. Clean separation of the beam components is apparent.

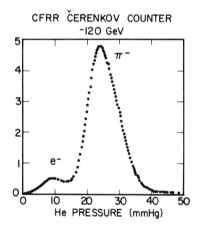

Fig (7): Partial view of the Cerenkov curve at -120 GeV. The separations of electron and pion peaks can be clearly discerned.

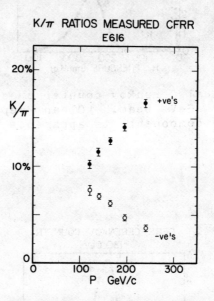

Fig (8): Measured K/π ratios for the narrow band beam settings used in the run.

mis-steered pulses could be subsequently eliminated from the data sample.

In conclusion, neutrino flux is derived from the measurement of the charged flux, a pencil beam ~10 cm diameter traveling in the decay pipe, and the particle composition data as measured by the differential cerenkov counter. All the values of neutrino flux for various settings of the beam were not statistics limited. Rather, the errors attributed to those measurements reflect the experimental estimate of the systematic and other measurement errors. Therefore it is believed that the overall flux normalization is known to ±5% and the point to point setting error to a value better than 5%.

Detection of neutrino interactions was accomplished by a 690 ton target calorimeter followed by a ~400 ton muon spectrometer being an instrumented, magnetized iron toroid. The detector is shown in Fig. (9) and Fig. (10). The ionization sampling in the front calorimeter was every 10cm while spark chambers (magnetostrictive readout) provided the digitization of the charge particle tracks every 20cm of iron. The relevant resolution parameters of the detector were:

$$\Delta E_h/E_h \; 0.89/\sqrt{E_h} \; (GeV)$$

$$\Delta p_\mu/p_\mu = \left[(0.114)^2 + (8.33 \times 10^{-5} p)^2\right]^{1/2} ; p(GeV/c)$$

$$\Delta \theta_\mu = 0.14 + 57/p \; (GeV/c) \; mrad.$$

The first two parameters were experimentally determined using π, p and μ beam of the precisely known momentum incident on the detector.

The separation of events due to neutrinos originating by pion or kaon decay, pion and kaon

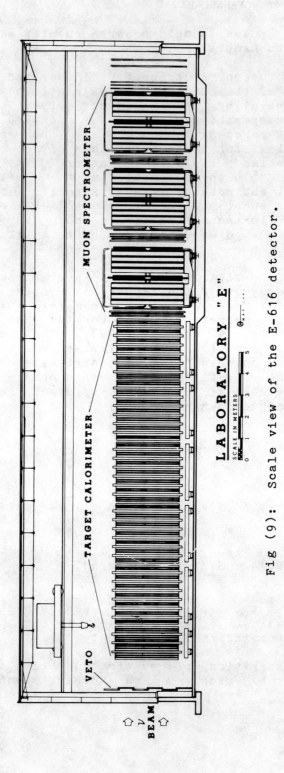

Fig (9): Scale view of the E-616 detector.

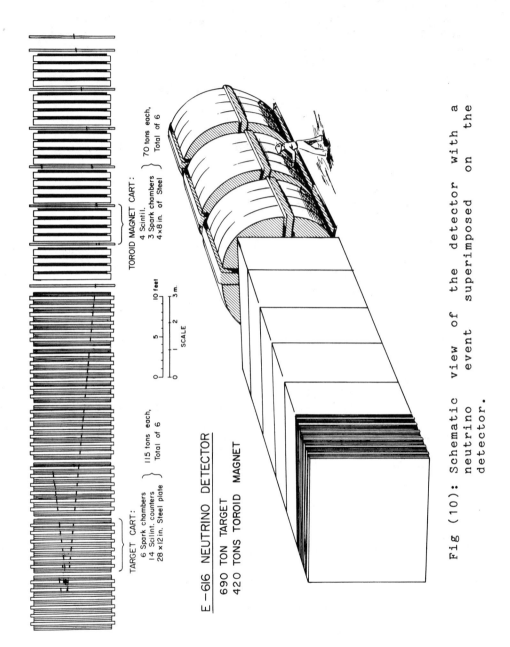

Fig (10): Schematic view of the detector with a neutrino event superimposed on the detector.

events for short, was trivial, as can be seen from Fig. (11). Due to the different Q values of $\pi\to\mu\nu$ and $K\to\mu\nu$ decay the events are easily separable in energy-radius plots. Data were taken at the beam settings of 120, 140, 168, 200 and 250 GeV/c. The cross sections reported here constitute one third of neutrino events and all of the antineutrino sample. Rather than using Monte Carlo for generating the events in order to find the acceptance of the aparatus, each event was weighted by the geometrical efficiency obtained by rotating each event around the beam direction axis. Beside acceptance other important corrections to data sample were:

(1) wide band background (measured quantity) 1%-15%

(2) cosmic ray contamination (measured quantity) 1%

(3) three body kaon decay (calculated) 1-2%

(4) acceptance correction for high x, and high y 2%

The consistency checks done were: (1) Consistency between measured and predicted energy at a given radius of the event, (2) trigger stability throughout the run; at the same flux dose - for a given setting same number of neutrino events should be observed. (3) Consistency of the total cross sections in the regions of overlap and (4) the equality of $d\sigma/dy$ at y=0 for neutrino and antineutrino events.

Figures (12) and (13) show the final slope parameters for neutrino and antineutrino data respectively. The figures show also the previously published CDHS data. The experimental points do not correspond to the individual energy settings but are rather the combined average of the data where substantial overlap occurs. By fitting a straight horizontal line through the data points one obtains the average value of the slope parameters in the energy region 40-230 GeV:

$S_\nu = 0.719 \pm 0.006 \pm 0.036 \times 10^{-38} cm^2/GeV/nucleon$

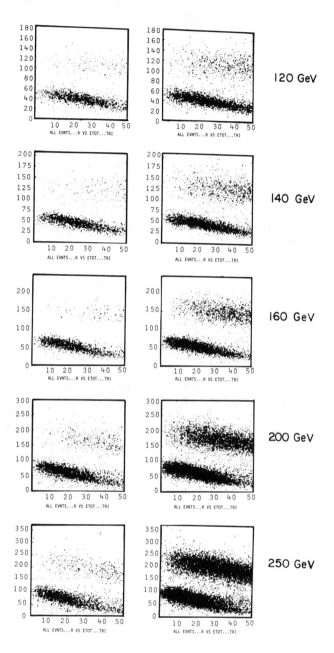

Fig (11): Total measured neutrino event energy versus radial position with respect to the charged beam direction for all beam settings. The two bands are due to the kinematic separation of events due to neutrinos originating from pion or kaon decays.

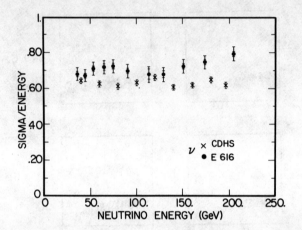

Fig (12): Measured slope parameter values for the total neutrino cross sections. The two sets of errors in the E-616 data are statistical (small error bars) and systematic (larger error bars). Also indicated are published CERN - WA-1 experimental points.

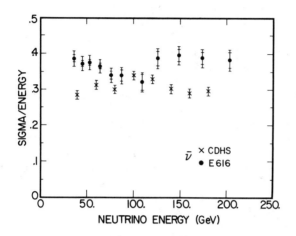

Fig (13): Measured antineutrino slope parameters. Same comments of Fig 12 apply.

$$S_{\bar{\nu}} = 0.371 \pm 0.004 \pm 0.019 \times 10^{-38} \text{cm}^2/\text{GeV}/\text{nucleon}$$

The first error is statistical while the second quoted error is the estimated systematic error. The values listed above differ by 15-20% from the previous published data by CDHS and others.

As already emphasized, this measurement comprises a substantial redundancy of the flux monitoring. Systematic uncertainties in the results of this experiment appear unlikely to be able to explain this discrepancy. Considering the fact that both neutrino and antineutrino slope parameters are higher by approximately the same amount one cannot ascribe this 20% shift to the large proton contamination of the charged flux. Only in the neutrino case the charged flux contained the large fraction of protons whereas in the antineutrino case the charged flux was composed of mainly pions and kaons. Similar set of arguments can be used to point out that the cerenkov counter particle separation had to perform in very different ambience for either positive or negative beam. What may be of more consequence for this discrepancy is the overall normalization information stemming from Beam Current Transformers (BCT), foil activation RF cavities and such devices. All of these independent monitors are at present under careful scrutiny at Fermilab. Whether the similar discussion applies to the CERN beam monitors: SOLID STATE ionization chambers emulsion calibration and BCT-devices, it is hard to tell. Suffice it to say that much further work on the neutrino flux monitoring will be necessary on both sides of the Atlantic to remove this discrepancy.

Next point to examine is the interesting, but not very statistically significant energy dependence of the slope parameters. With the systematic errors folded in the χ^2, for the hypothesis of slope being constant with energy, is quite acceptable. A better fit is obtained however using simple two neutrino oscillation parameters: $\Delta m^2 - 350 eV^2$ and $\sin^2 2\alpha = 0.18$. The same parameters seem to fit also the antineutrino data. Rather intriguing is the "out of phase" appearance of the CDHS data taken at the same energies. This can be taken into account bearing in mind that the CDHS data were taken at a distance of about 750 meters (distance from the center of the decay pipe to the detector) whereas the CFRR data were taken at a mean distance of ~1100 meters (see Fig. 14). Same neutrino oscillation parameters would support this appearance of slope parameter modulation for all four sets of data but not the discrepancy in the overall normalization

*) We are aware that the more recent data from CDHS - group presented at Moriond Conference last April do not exhibit any energy modulation.

The CFRR group has proposed and approved for a measurement to follow up this possible effect using two detectors of almost identical construction. The two detectors will be located at 720 meters (new detector) and 1100 meters (present detector). With such a setup the flux monitoring problems are removed to the first order since the two detectors see simultaneously the same neutrino flux. It is expected that the new detector will be assembled by Fall 1981 and data taken in late 1981 and early 1982.

Finally, if true, the suggestive energy dependence of the slope parameter may be caused by effects other than neutrino oscillation. Modulation may be due to the onset of other neutrino channels charm production, QCD - non scaling phenomena etc. In this case the modulation may be real and not apparent as being caused by the modulation of the neutrino flux due to the neutrino oscillations.

E-53A
New Limits on the neutrino oscillations
Columbia-BNL collaboration

The large wide band run with the 15-foot B.C 64% Ne fill of 100,000 pictures was recently augmented with yet another run for the similar number of

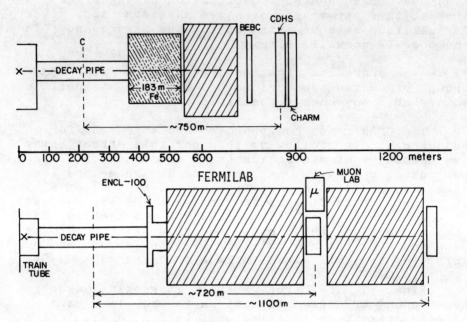

Fig (14): The mean distance: middle of decay pipe to detector is illustrated for two experiments. The upper part indicates CERN-CDHS experiment mean distance of 750 m whereas the E-616 mean distance was 1100 meters.

pictures. The neutrino event rate in these two exposures was roughly 0.6 events per picture. Careful scan for the electron events is now complete for the both sets but we are reporting the results from the first scan. There were

ν_μ + Ne ± μ^- + hadrons   67,000

ν_e + Ne ± e^- + hadrons   942±53

In the wide band beam one predicts the presence of ν_e-neutrinos from the K_{e3} decays to be of the other 1.5%. On that basis one would predict that the number of electron events be 1005 ± 206. Having such a sample of e^- charge current events and the large sample of μ^- charge current events, one can approach the problem of neutrino oscillation in a general way. In what follows we will use simple "2-channel" neutrino mixing and only in the later part re-examine the overall consequence of three channel neutrino mixing.

The probability $P(\nu_i \to \nu_j)$, that a neutrino of the energy E, and of the flavor ν_i, oscillate into a neutrino of flavor ν_j after traveling a distance L from the source is given by:

$$P(\nu_i \to \nu_j) = \sin^2 2\alpha \cdot \sin^2(1.27 \Delta m^2 L/E)$$

where $\sin^2 2\alpha$ is the magnitude of mixing and the second sin - term the kinematic - oscillatory part as a function of L/E and the difference of neutrino mass eigenstates $\Delta m^2 = m_2^2 - m_1^2$. The distance L is in meters, E neutrino energy in MeV units and the Δm^2 in eV^2.

The somewhat simplified arguments now proceeds by letting either one of the two sine terms be unity and letting the other go to zero. Since the wide band beam had the estimated flux mixture.

$\nu_\mu / \bar{\nu}_m / \nu_e / \bar{\nu}_e$ / - 100/3/1/0.2

one notes that the expected number of e events is the one found:

$$\nu_\mu \to \nu_e : (1005 \pm 206) - (942 \pm 53) = 63 \pm 213$$
$$ \text{expected} \quad\quad \text{observed}$$
$$ \text{number} \quad\quad\; \text{number}$$

This results in the overall limit:

$$\frac{\nu_\mu \to \nu_e \to e^-}{\nu_\mu \to \mu^-} \leq 3 \times 10^{-3} \quad 90\% \text{ conf. limit}$$

The $\nu_e \to \nu_\tau$ mixing is analyzed by noting that the number of ν_e events is not depleted. For the upper bound of the number of e events which would have deviated from the "expected", (calculated) number of $\nu_e \to e$ events the authors use: 340 events. One may simply state that this number represents the ~30% uncertainty in accounting how many ν_e are in the beam. Some of the $\nu_e \to \nu_\tau$ oscillated events return to the sample:

$$\nu_\tau + N_e \to \tau^- + \ldots$$
$$ \downarrow$$
$$ e^- + \nu_\tau \nu_e \quad 17\% \text{ branching ratio}$$

Therefore the "340" event deviation should be corrected

$$N_{\nu_e \to \nu_\tau} \leq \frac{340}{1-(0.7)(0.17)} = 386$$

where 0.7 factor accounts for a lower mean ν_τ-charged current cross section in this energy range. This puts not so restrictive a limit on $\nu_e \to \nu_\tau$ oscillation

$$P(\nu_e \tau \nu_\tau) \leq \frac{386}{942 \pm 340} = 0.3$$

A summary of all oscillation data from this experiment can be summarized by following diagrams:

A) $\sin^2 2\alpha = 1$ (maximum mixing)

   Mass Limits

$$\Delta m^2 \leq 0.4 \, eV^2$$

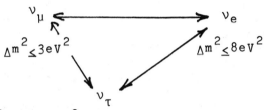

B) $\sin^2(1,27 \, \Delta m^2 L/E) = 1/2$

   (Oscillations average to a mean value)

$$\sin^2 2\alpha \leq 6 \times 10^{-3}$$

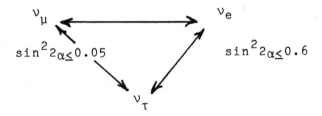

The mixing limits should be compared to the Cabibbo mixing angle: $\sin^2(2\theta)$ Cabibbo) ~ 0.2

In this review of Fermilab a recent neutrino program we presented three different experiments exploiting very different experimental techniques: Emulsions, spark chamber-counter, and bubble chamber filled with neon. Each of these experiments pursued,

what it appears at first glance, a very different
scientific objective. Nonetheless, a very
interesting and crucial overlap exists between these
three experiments. The overlap exists in the domain
of neutrino masses and neutrino mixing. To better
understand this statement let us pursue the set of
assumptions, (the parameters of neutrino mixing), and
see how each of the above experiments imposes a
boundary on such a hypothesis. Firstly, the three
flavor oscillation is described by a 3x3 mixing
matrix where the $P(i \rightarrow j)$ probability containes more
than just two sine terms. Let us further assume that
the Δ_{mij} mixing terms are of the order of 10-100 eV^2.
Whereas the first assumptions is logical because of
the three neutrinos now known ν_e, ν_μ and ν_τ, the
second assumption is only an adhoc guess to
illustrate the point. Using such large mass
differences has the point to make L/E factor of such
magnitude to be applicable in the 20-200 GeV neutrino
energy domain. Under such circumstances the
oscillations between three species is given by:

$$P(e \rightarrow e) = 1 - \sin^2 2\alpha \, \sin^2 \Delta$$

$$P(e \rightarrow \mu) = \sin^2 2\alpha \, \cos^2 \beta \, \sin^2 \Delta$$

$$P(\mu \rightarrow \tau) = \sin^2 2\beta \, \sin^4 \alpha \, \sin^2 \Delta$$

where Δ is the dominant mass difference term. One
can, by judicius choice of these parameters then
explain the CERN dump experiment and yet leave the
global number of neutrino-electron events in Columbia
BNL experiment the same. The qualitation argument
would be as follows:

1) Allow small contribution of $\nu_\mu \rightarrow \nu_\tau$ say of the
order of 5%. Such a mixture in a wide band run
outlined above would result in 5000 ν_τ events.
Actually the number should be diminished by the
ν_τ charge current cross section averaged over
the broad band neutrino spectrum, thus resulting
in $\simeq$ 2500 ν_τ events.

2) Of these 2500 events only 17% would give electrons i.e 425 would increase the observed ~ 900 events in the 15 foot B.C. run.

3) On the other hand one would allow substantial $\nu_e \to \nu_\tau$ transition of the order of 30% which would deplete the observed number of 900 "ν_e" events by ~300. This is consistent with the paper of DeRujula et al.

These three arguments would maintain electron contribution the same in the B.C. runs both at Fermilab and CERN and even in the cases of narrow band experiments.

However the consequence of that argument would be a substantial presence of ν_τ neutrinos in all neutrino beams; of the order of 5%! Such a large fraction would undoubtedly be detectable by emulsion experiments provided the lifetime of the τ-meson is of the same order of magnitude: 5×10^{-13} sec, same as that of charmed events. None have been observed. The experiment E-531 described here places that limit to a number of the order of ~ 2% with 90% confidence limit. Whether the τ-events have yet to be found or the lifetime is much shorter (longer lifetimes would be visible in the bubble chambers) is the open question. The simplest way is to dismiss the mixing. The illustration above is used to show that two channel oscillation analysis has its short comings and that the three experiments we have described have a great deal in common.

Clearly in this repot a certain bias of the reporter prevails. That the neutrino should have mass is not compelling on general grounds. Suffice it to say that the massive neutrinos would most likely remove a sterile (not dynamic) conservation low of lepton conservation. On the other hand, cosmological problems of missing visible mass, stability of galactic arms and the closing of the expanding universe would be much helped with massive neutrinos which have masses just in this interesting domain. If that be so, better arguments of "unity" could not be found.

DISCUSSION

CHAIRMAN: D. JOVANOVICH

Scientific Secretary: Z. Hloušek

DISCUSSION

- WITTEN:

What is the argument that experiment rules out ν_μ and ν_τ mixing ?

- JOVANOVICH:

The emulsion experiment is sensitive to detection of a gap of $L = \gamma\beta c t_0$ ($t_0 \equiv$ life-time) of the order of 10 microns, hence of any ν_τ produced τ - lepton via charge current interaction. Since E - 531 has 1600 neutrino events detected in emulsion, a 5% admixture should give the order of 80 τ - events. They see none.

- WITTEN:

Provided that one believes that the τ life-time is 10^{-13} sec, and presuming that charmed events can go down by a factor of 10 or 20, for example, then it might happen that detector efficiency for τ is too small.

- JOVANOVICH:

As far as I know detection efficiency is big enough and it is about 50%, the same as for charm events.

- WITTEN:

You have said that charm would imply neutrino cross-sections corresponding to about 200 events, and yet you have only a couple of dozen events.

- JOVANOVICH:

To miss a couple of dozen events would mean a too high inefficiency for finding them.

Let me elaborate on this argument. The scan of the events is such that rejections occur based on the fact that one sees just a break or a kink of the tracks somewhere down stream from the primary vertex. Whatever potential candidates are classified as tau events would have a muon coming from this vertex.

One can eliminate muon events by matching angles with that of a spectrometer.

About other scanning inefficiencies I cannot say much.

- HERTEN:

What is the range of L/E ratio which is covered by E - 53-A and which was used for plotting the neutrino oscillations as a function of L/E ?

- JOVANOVICH:

The range of L/E covered by experiments at Fermilab is between 10 $(eV)^2$ to 500 $(eV)^2$. It corresponds to a neutrino energy of about a few hundred GeV and a distance of about a kilometer.

- GOLLIN:

Let us suppose that you wished to explain the sudden change in the behaviour of the ratio of cross-section to energy σ/E in the E - 616 experiment at about 120 GeV as an effect due to the set-up.

What would you say went wrong ? (See your figs. 12 and 13).

- *JOVANOVICH*:

Naturally one would repeat the experiment using two detectors simultaneously. Several experiments of this kind have been proposed; one at CERN running at the PS energy, another at Fermilab, and yet another at Los Alamos at low energy. This is clearly the most logical way to set about it.

If after these checks there is no variation with distance and energy then one would eliminate the effect of improper functioning of the set-up and revert back to other questions, e.g., is the neutrino spectrum flat or is it modulated by other effects ?

HEAVY FLAVOUR PRODUCTION AT THE

HIGHEST ENERGY IN (pp) INTERACTIONS

        M. Basile, C. Bonvicini, G. Cara Romeo, L. Cifarelli,
        A. Contin, G. D'Ali, B. Esposito, P. Giusti, T. Massam,
        R. Nania, F. Palmonari, A. Petrosino, G. Sartorelli,
        G. Valenti and A. Zichichi

        CERN, Geneva, Switzerland.
        Istituto di Fisica dell'Università di Bologna, Italy
        Istituto Nazionale di Fisica Nucleare, Bologna, Italy
        Istituto Nazionale di Fisica Nucleare, LNF, Frascati
        Italy

Presented by A. Zichichi

1. INTRODUCTION AND BASIC POINTS

Let me first remind you that the detection of heavy-flavoured states is much more difficult in proton-proton machines than in e^+e^- machines, where the heavy-flavour production competes well with the light-flavour production. In fact, since in (e^+e^-) annihilation the quark production proceeds according to the quark-charge coupling, once above the threshold energy, the only difference in production rate is between the up-like (u, c, t) and the down-like (d, s, b) quarks, irrespective of their mass.

In hadronic collisions, and therefore in the (pp) case, the phenomenological trend, also expected from QCD, is in terms of the inverse mass squared of the quark to be produced. The production of heavy flavours is therefore depressed, and it is not an easy task to identify its presence in the much higher background of standard hadronic processes.

The basic technique used to disentangle the heavy-flavour signal from the standard hadronic background mentioned above has been to trigger on the lepton coming from the semileptonic decay of one member of the (particle-antiparticle) pair of heavy-flavoured states produced, and to look at the hadronic decay mode of the other member of the pair. This is shown schematically in Fig. 1 for the charm case.

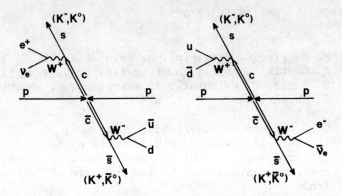

Fig. 1 Schematic diagram of $c\bar{c}$ associated production, with either an e^+ (from $c \to s$ semileptonic decay) or an e^- (from $\bar{c} \to \bar{s}$) in the final state, showing the principle of the experiment: trigger on the semileptonic decay of a heavy-flavoured state (antistate) and look for the hadronic decay of the associated antistate (state).

For example, when looking for (baryon-antimeson) pair production, the hadronic charm detection was based on the proton identification plus a strange particle, whilst the antimeson was detected via its semileptonic decay.

Now let me say a few words on quark nomenclature. Dalitz has sent me a letter suggesting that we stop calling quarks in terms of ladies attributes: charm, strange, beauty, etc. Let me try to make a proposal, even if difficult to accept it in place of the above terms. Quarks can be identified as odd and even:

$$\begin{pmatrix} u \\ d \end{pmatrix} \begin{pmatrix} c \\ s \end{pmatrix} \begin{pmatrix} t \\ b \end{pmatrix} \Bigg\} \text{ former nomenclature}$$

$$\downarrow \quad \downarrow \quad \downarrow$$

$$\begin{pmatrix} 1 \\ 2 \end{pmatrix} \begin{pmatrix} 3 \\ 4 \end{pmatrix} \begin{pmatrix} 5 \\ 6 \end{pmatrix} \begin{matrix} \Leftarrow \text{ Odd } (\equiv \text{ uplike}) \text{ quarks} \\ \Leftarrow \text{ Even } (\equiv \text{ downlike}) \text{ quarks} \end{matrix}$$

The charge formula can simply be written as

$$Q = \frac{1/3 + f_i}{2},$$

where $1/3$ means the baryonic number and f_i is the "flavour" value:

$$f_i = +1 \text{ for odd quarks}, \quad (i = 1, 3, 5),$$

$$f_i = -1 \text{ for even quarks}, \quad (i = 2, 4, 6).$$

From this there follows immediately the sign of the electron in the semileptonic decay of the flavours we are looking for. An (uplike to down-like) transition, for example (c → s), is an (odd-even) transition. It implies f_i going from +1 to -1. The leptonic part of signature must be an e^+; for a (down-up) or (even-odd) transition, such as (b → c), the signature will be an e^-.

Heavy-flavour searches are associated with positive leptons for charm states and negative leptons for beauty states. Anticharm and antibeauty transitions, ($\bar{c} \to \bar{s}$) and ($\bar{b} \to \bar{c}$), will be marked by negative and positive leptons, respectively. So, if you detect an e^+, this might come from a (c → s) or a ($\bar{b} \to \bar{c}$) transition. If you detect an e^-, this might come from a (b → c) or a ($\bar{c} \to \bar{s}$) transition. The quark and antiquark transitions, with the associated lepton charges, are summarized in the following table:

Quarks		
	f_i	Transition and associated lepton charges
Odd (up-like) (u c t)	+1	e^+ ↓ e^- ↑
Even (down-like) (d s b)	−1	

Antiquarks		
	f_i	Transition and associated lepton charges
Odd (up-like) ($\bar{u}\ \bar{c}\ \bar{t}$)	−1	e^- ↓ e^+ ↑
Even (down-like) ($\bar{d}\ \bar{s}\ \bar{b}$)	+1	

2. THE EXPERIMENTAL SET-UP

The experiment I am going to report on has been performed at the CERN Intersecting Storage Rings (ISR) using the Split-Field Magnet (SFM) spectrometer.

Let me just mention the main features of the experimental apparatus, shown in Fig. 2. A detailed description has already been given elsewehere[1-6].

The positive and negative electrons were detected in the 90° region, by the coincidence ($\check{C}_0\check{C}_3$ or $\check{C}_0\check{C}_4$) of two gas Čerenkov counters (with a momentum threshold for π's of $\simeq$ 5.5 GeV/c) and by a minimum energy release (E^e min $\geq$ 500 MeV) in the electromagnetic shower detectors (EMSDs): LG3 and LG4 (lead-glass arrays) and SW3 and SW4 (lead/scintillator hodoscopes). A multiwire proportional chamber (209) with analog read-out ("dE/dx" chamber) near the interaction region provided the information needed to reject off-line the unresolved and resolved (e^+e^-) pairs that are due to external γ conversions and to π^0 and η Dalitz decays.

The multiwire proprotional chamber (MWPC) system of the SFM provided the momentum measurement of the charged secondaries over about 75% of the total solid angle with a momentum accuracy $\Delta p/p \leq 30\%$.

HEAVY FLAVOUR PRODUCTION

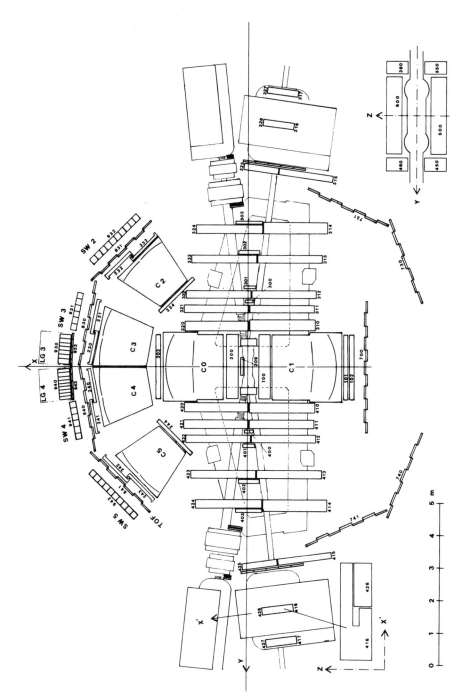

Fig. 2 Top view of the SFM apparatus at the CERN ISR.

Finally, a large time-of-flight (TOF) system allowed, π, K, p separation for momenta ≤ 2 GeV/c over about 10% of the total solid angle.

3. CHARM PRODUCTION

The following reactions were studied[7-10]:

$$pp \rightarrow \Lambda_c^+ + \text{(anticharmed state)} + \text{anything} \quad (1)$$
$$\quad \hookrightarrow pK^+\pi^- \qquad \hookrightarrow e^-$$

$$pp \rightarrow D^+ + \text{(anticharmed state)} + \text{anything} \quad (2)$$
$$\quad \hookrightarrow K^-_{TOF}\pi^+\pi^+ \qquad \hookrightarrow e^-$$

$$pp \rightarrow D^0 + \text{(anticharmed state)} + \text{anything} \quad (3)$$
$$\quad \hookrightarrow K^-\pi^+ \qquad \hookrightarrow e^- + K^+_{TOF}$$

$$pp \rightarrow D^+ + \text{(anticharmed state)} + \text{anything} \quad (4)$$
$$\quad \hookrightarrow K^-\pi^+\pi^+ \qquad \hookrightarrow e^- + K^+_{TOF}$$

$$pp \rightarrow D^0 + \text{(anticharmed state)} + \text{anything} \quad (5)$$
$$\quad \hookrightarrow K^-_{TOF}\pi^+\pi^+\pi^- \qquad \hookrightarrow e^-$$

$$pp \rightarrow D^- + \text{(charmed state)} + \text{anything} \quad (6)$$
$$\quad \hookrightarrow K^+_{TOF}\pi^-\pi^- \qquad \hookrightarrow e^+$$

$$pp \rightarrow \overline{D^0} + \text{(charmed state)} + \text{anything} \quad (7)$$
$$\quad \hookrightarrow K^+_{TOF}\pi^-\pi^-\pi^+ \qquad \hookrightarrow e^+$$

$$pp \rightarrow \Lambda_c^+ + \text{(charmed state)} + \text{anything} \quad (8)$$
$$\quad \hookrightarrow p_{TOF}K^-\pi^+ \qquad \hookrightarrow e^-$$

$$pp \rightarrow \Lambda_c^- + \text{(charmed state)} + \text{anything} \quad (9)$$
$$\quad \hookrightarrow \bar{p}_{TOF}K^+\pi^- \qquad \hookrightarrow e^+$$

HEAVY FLAVOUR PRODUCTION

The time-of-flight (TOF) technique has been used in reactions (2), (5), (6), (7), (8), and (9) in order to identify a particle which enters the invariant mass spectrum. This particle could be a K^- [reactions (2) and (5)] or a K^+ [reactions (6) and (7)], or a proton [reaction (8)] or an antiproton [reaction (9)].

Moreover, the TOF technique has also been used to "trigger" on the decay of the particle produced in association with the one searched for via the analysis of the mass spectra. This has been done with K^+ in reactions (3) and (4).

Finally, a set of "control" processes has been studied. These processes were all characterized by the fact that the choice of the "trigger" was wrong. The purpose of this study was to see if the invariant mass spectrum associated with the "wrong trigger" produced enhancements.

The reactions with the "wrong" trigger were:

$$pp \to (pK^-\pi^+) + e^+ + \text{anything} \tag{1a}$$

$$pp \to (K^-_{TOF}\pi^+\pi^+) + e^+ + \text{anything} \tag{2a}$$

$$pp \to (K^-\pi^+) + e^+ + K^-_{TOF} + \text{anything} \tag{3a}$$

$$pp \to (K^-\pi^+) + e^- + K^-_{TOF} + \text{anything} \tag{3b}$$

$$pp \to (K^-\pi^+) + e^+ + K^+_{TOF} + \text{anything} \tag{3c}$$

$$pp \to (K^-\pi^+\pi^+) + e^+ + K^+_{TOF} + \text{anything} \tag{4a}$$

$$pp \to (K^-_{TOF}\pi^+\pi^+\pi^-) + e^+ + \text{anything} \tag{5a}$$

$$pp \to (K^+_{TOF}\pi^-\pi^-) + e^- + \text{anything} \tag{6a}$$

$$pp \to (K^+_{TOF}\pi^-\pi^-\pi^+) + e^- + \text{anything} \tag{7a}$$

$$pp \to (p_{TOF}K^-\pi^+) + e^+ + \text{anything} \tag{8a}$$

$$pp \to (\bar{p}_{TOF}K^+\pi^-) + e^- + \text{anything} \tag{9a}$$

Here again the TOF technique was used both to identify particles which entered the invariant mass plots [reactions (2a), (5a), (6a), (7a), (8a), and (9a)] and to identify particles used as triggers [reactions (3a), (3b), (3c), and (4a)].

The particle identified for the invariant mass plot was a K^- in reactions (2a) and (5a), a K^+ in reactions (6a) and (7a), a proton in reaction (8a), and an antiproton in reaction (9a).

The particle identified for the "wrong trigger" was a K^- in reactions (3a) and (3b), a K^+ in reactions (3c) and (4a).

The "trigger" (associated with the invariant mass spectrum) could be *wrong*, either because the electron charge was opposite to the one allowed by charm physics [reactions (1a) to (9a), except (3b)] or because, in addition, also the sign of the K meson was wrong [reactions (3a) and (3b)].

In no case does the invariant mass spectrum associated with a "wrong" trigger show an enhancement that is statistically significant.

The study has also been extended to the production distributions. For reaction (1) the study of Λ_c^+ has been extended to the transverse momentum $d\sigma/dp_T$ distribution and to the longitudinal $d\sigma/dx_L$ distribution[11,12].

For reactions (2) and (3) the transverse momentum distributions $d\sigma/dp_T$ of D^+ and D^0 and the longitudinal $d\sigma/dx_L$ distributions have been measured[13,14].

Finally for reactions (1), (2), and (3)[7-9], cross-section estimates have been performed using different models: "central" (I), "flat y" (II), "flat x_L" (III), as specified in Figs. 9, 15, and 20.

Table 1 shows the summary of the results for all reactions investigated, including the control processes with the "wrong" trigger. The *first* column indicates the particle searchef for; the *second* column, the trigger used; the third column, whether a signal has been observed (with the absence of a signal when the trigger is wrong); the last column indicates the set of figures where the data on that state are reported. These data are the mass spectra for all the reactions (1)-(9); in addition, the production distributions $d\sigma/dp_T$ and $d\sigma/dx_L$ with cross-section estimates for reactions (1), (2), and (3) are given.

The results are presented in the set of figures, from 3 to 26, with the relevant information describing the data shown.

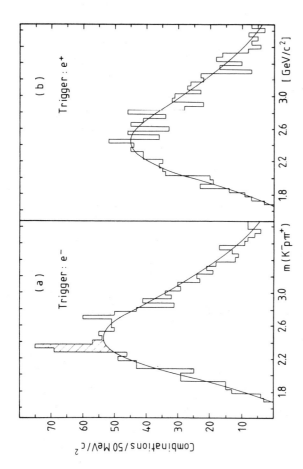

Fig. 3 a) Observation in the $(pK^-\pi^+)$ spectrum of the charmed baryon Λ_c^+, associated with an e^- from the semileptonic decay of an anticharm particle ($\bar{D}^0$ or D^-). Only the proton is identified as a "leading" particle with $x_L = (2p_L/\sqrt{s}) \geqslant 0.3$. The K^- and the π^+ are required to be in the same rapidity hemisphere as the proton and to have $y > 0.1$. A diffractive condition, $\Sigma_i(x_L)_i > 0.5$ or $\Sigma_i(x_L)_i < 0.1$, is also required in the opposite hemisphere.
b) Same as (a) but for e^+ triggered events. No Λ_c^+ is expected here and the solid-line fit represents the background shape.

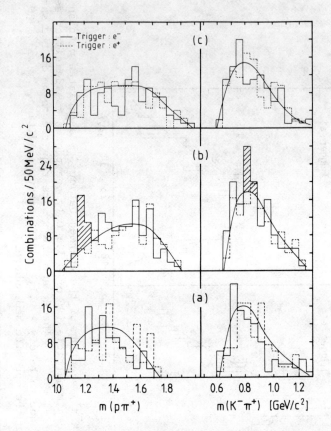

Fig. 4 ($K^-\pi^+$) and ($p\pi^+$) invariant mass spectra corresponding to three different cuts in the ($pK^-\pi^+$) mass of Fig. 3a:

a) $2.18 \leq m(pK^-\pi^+) < 2.28$ GeV/c^2 (OUT$_{below}$);
b) $2.28 \leq m(pK^-\pi^+) \leq 2.38$ GeV/c^2 (IN Λ+ peak);
c) $2.38 < m(pK^-\pi^+) \leq 2.48$ GeV/c^2 (OUT$_{above}^c$).

The dashed-line histograms refer to the e$^+$ triggers and the solid curves are fits to these histograms. Both the Δ^{++} and the $\bar{K}^{*0}$ are observed. The branching ratios are measured to be 0.40 ± 0.17 and 0.28 ± 0.16, respectively, in good agreement with existing measurements.

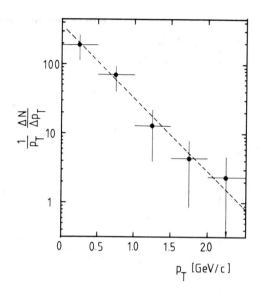

Fig. 5 Experimental p_T distribution of the Λ_c^+ obtained for the events of Fig. 3a, via the "IN-OUT" technique. The dashed line shows the best fit: $(1/p_T)(\Delta N/\Delta p_T) \propto e^{-b p_T}$, where $b = 2.5 \pm 0.4$, in excellent agreement with QCD predictions.

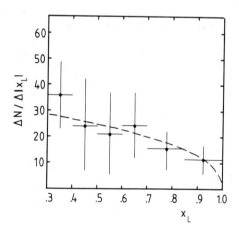

Fig. 6 Longitudinal distribution of the Λ_c^+ in terms of $x_L = 2p_L/\sqrt{s}$ (obtained as in Fig. 5) and observation of the "leading" baryon effect. The dashed line shows the best fit: $\Delta N/\Delta x_L \propto (1-x_L)^{(0.4 \pm 0.25)}$.

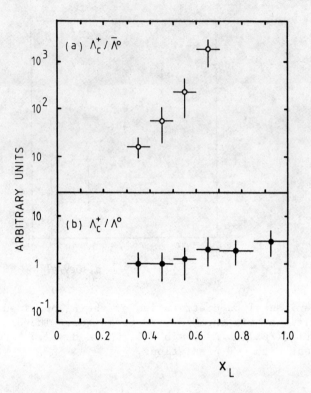

Fig. 7 Comparison of Λ^0, $\bar{\Lambda}^0$, and Λ_c^+ longitudinal distributions.

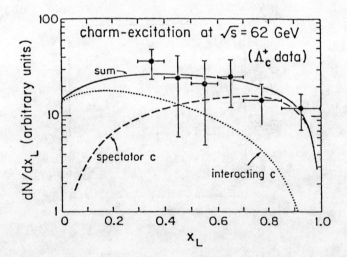

Fig. 8 Fit to the Λ_c^+ results (reaction (1)) according to a very recent QCD prediction by Barger, Halzen and Keung[25].

$\bar{D}$ model			Λ_c^+ model			σ_{tot} (µb)
$E \frac{d\sigma}{dx_L} \propto (1-x_L)^3$		(I)	$E \frac{d\sigma}{dx} \propto (1-x_L)^3$		(I)	4200
$\frac{d\sigma}{dy} = $ const.		(II)	$\frac{d\sigma}{dy} = $ const.		(II)	750
$\frac{d\sigma}{dx_L} = $ const.		(III)	$\frac{d\sigma}{dx_L} = $ const.		(III)	1125
$E \frac{d\sigma}{dx_L} \propto (1-x_L)^3$		(I)	$\frac{d\sigma}{dx_L} = $ const.		(III)	184

Fig. 9a $\bar{D}\Lambda_c^+$ cross-section estimates obtained with different combinations of the following production mechanisms, for the $\bar{D}$ and the Λ_c^+ independently:

model I = "central"; $E(d\sigma/dx_L) \propto (1-x_L)^3$
model II = "flat-y"; $d\sigma/dy = $ const.
model III = "flat-x_L"; $d\sigma/dx_L = $ const.

The inclusive charm cross-section is expressed by $E(d^3\sigma/dp^3)=f(y/y_{max}) \cdot e^{-bp_T}$, where $b^{-1}=0.5$ GeV/c and $f(y/y_{max})$ is parametrized as listed above. The branching ratios used are ~ 0.09 for $\bar{D} \to e^- + $ anything decay and ~ 0.02 for $\Lambda_c^+ \to pK^-\pi^+$ decay. The over-all uncertainty on the cross-section values is $\sim 40\%$.

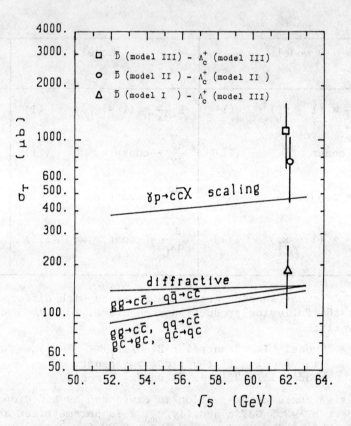

Fig. 9b Comparison with some theoretical predictions of the total cross-section estimates for $\bar{D}\Lambda_c^+$ production. The $\bar{D}$ "central" and Λ_c^+ "forward" models appear to be the most likely.

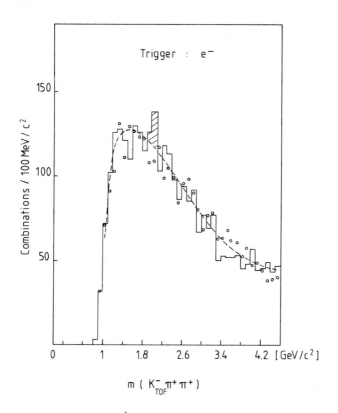

Fig. 10 Observation of a D^+ meson enhancement in the $(K^-\pi^+\pi^+)$ mass spectrum, in association with an e^- from the semileptonic decay of an anticharm meson ($\bar{D}^0$ or D^-). Only the K^- is identified via the TOF analysis (p < 1.5 GeV/c, 90% C.L). The π^+'s are required to have x_L < 0.3. The circles and the dashed-line curve superimposed show the distribution and fit, respectively, of the "event mixing" background.

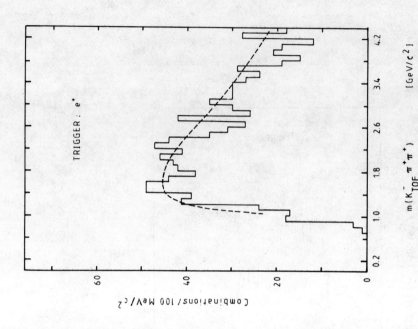
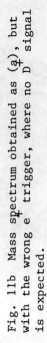

Fig. 11b Mass spectrum obtained as (a), but with the wrong $e^{\pm}$ trigger, where no D signal is expected.

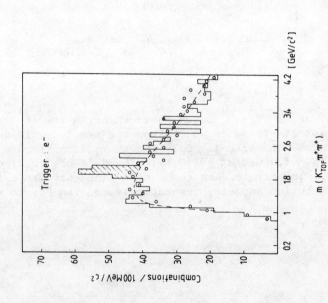

Fig. 11a Same as Fig. 10 when the requirement $p_T(K^-\pi^+\pi^-) \geq 0.7$ GeV/c is also applied. An improvement in the signal/background ratio is clearly observed, following the high-p_T expectation.

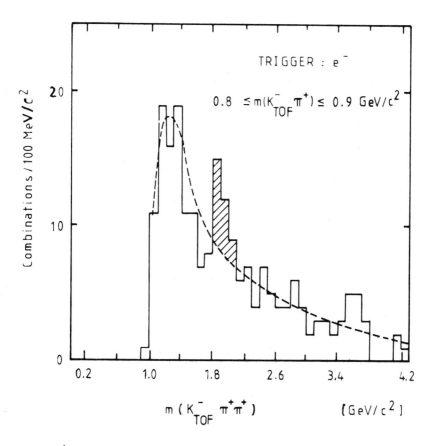

Fig. 12a D^+ signal, observed with e^-, via the decay mode
$$D^+ \to \bar{K}^{*0} \pi^+$$
$$\hookrightarrow K^- \pi^+ .$$

A mass cut $800 \leq m(K^-\pi^+) < 900$ MeV/c^2 and the same conditions as in Fig. 11a were applied. The branching ratio estimate is 0.31 ± 0.16.

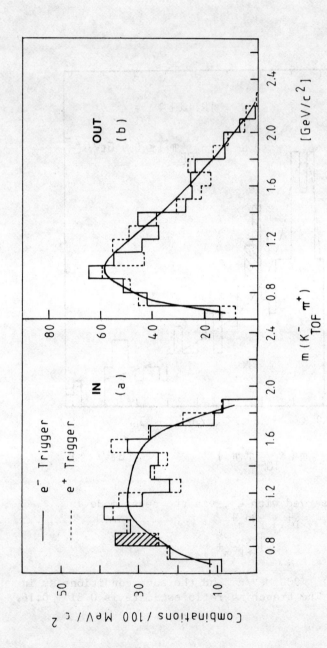

Fig. 12b (K⁻π⁺) mass spectra obtained with two different conditions on the (K⁻π⁺π⁺) mass of Fig. 11a:

$$1.8 \leqslant m(K^-\pi^+_\pi^+) < 2.1 \text{ GeV}/c^2 \quad \text{(IN } D^+ \text{ peak)}$$
$$\left.\begin{array}{l} 1.5 \leqslant m(K^-\pi^+_\pi^+) < 1.8 \text{ GeV}/c^2 \\ 2.1 \leqslant m(K^-\pi^+_\pi^+) < 2.4 \text{ GeV}/c^2 \end{array}\right\} \text{(OUT)}$$

The dashed-line distributions refer to the e^+ triggers and the solid curves are fits to the latter. A $\bar{K}*^0$ enhancement is observed, in good agreement with the 0.31 ± 0.16 branching ratio, quoted in the caption of Fig. 12a.

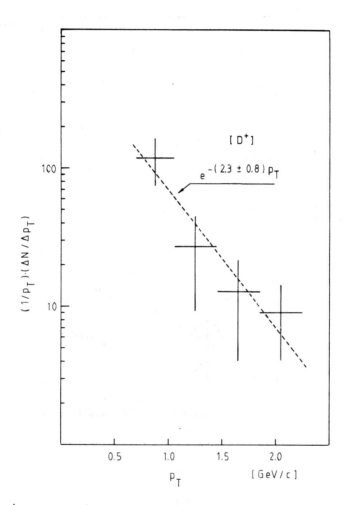

Fig. 13 D^+ p_T distribution obtained via the "IN-OUT" procedure for the events of Fig. 11a. The best fit gives $e^{-(2.3 \pm 0.8)p_T}$, in perfect agreement with the Λ_c^+ result (see Fig. 5).

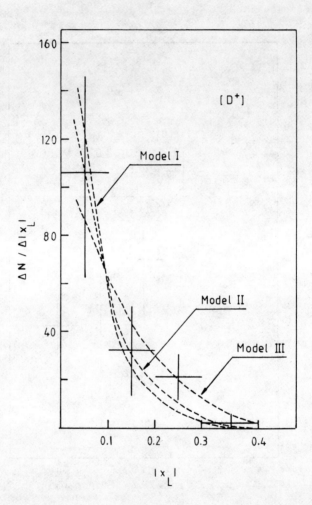

Fig. 14 D^+ x_L distribution (obtained as in Fig. 13). This distribution is strongly biased by the K^- selection (TOF solid angle and momentum cuts). However, the best agreement is found with the Monte Carlo (dashed-line curves) when model I is assumed, i.e. a "central" production with $Ed\sigma/dx_L \propto (1 - x_L)^3$, as expected for a meson.

$\bar{D}$ model			D^+ model			σ_{tot} (μb)
$E \frac{d\sigma}{dx_L} \propto (1 - x_L)^3$	(I)		$E \frac{d\sigma}{dx_L} \propto (1 - x_L)^3$	(I)		305
$\frac{d\sigma}{dy}$ = const.	(II)		$\frac{d\sigma}{dy}$ = const.	(II)		730
$\frac{d\sigma}{dx_L}$ = const.	(III)		$\frac{d\sigma}{dx_L}$ = const.	(III)		> 5000
$E \frac{d\sigma}{dx_L} \propto (1 - x_L)^3$	(I)		$\frac{d\sigma}{dx_L}$ = const.	(III)		1080

Fig. 15 $\bar{D}D^+$ cross-section estimates obtained with the same assumptions as for the Λ_c^+ (see caption of Fig. 9a). The branching ratios used are ~ 0.06 for $D^+ \to K^-\pi^+\pi^+$ and ~ 0.09 for $\bar{D} \to e^- +$ anything. The overall error $\sim 50\%$. The smallest cross-section values are in favour of a "central"-"central" $\bar{D}D^+$ associated production.

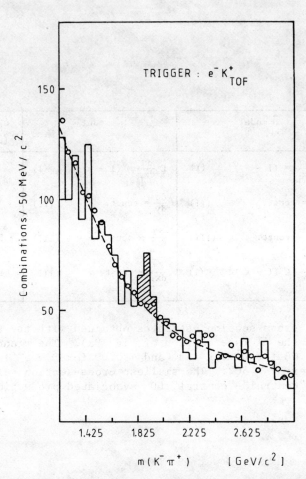

Fig. 16 Observation of a D^o enhancement in the $(K^-\pi^+)$ mass spectrum in association with an e^- and a K^+, both coming from D^o or D^- semileptonic decay. Only the trigger-K^+ is identified by TOF (p < 1.5 GeV/c, 90% C.L.), while the K^- in the invariant mass is not. The π^+ has $x_L < 0.3$. The background superimposed (circles and dashed-line curve) is derived from "event mixing".

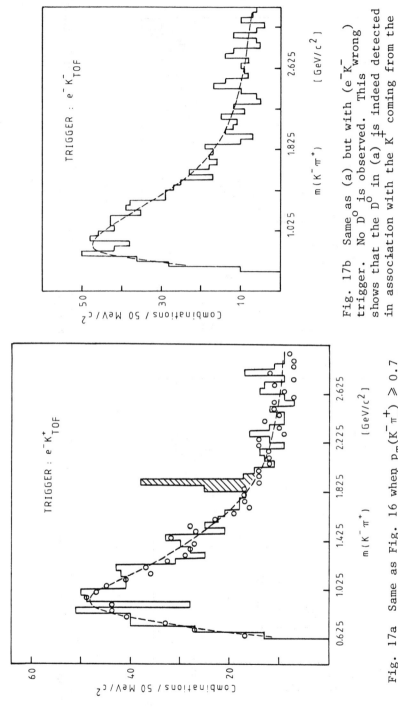

Fig. 17a Same as Fig. 16 when $p_T(K^-\pi^+) \geq 0.7$ GeV/c. As observed for the $D^{\pm}$, the D^0 signal improves remarkably with a high-p_T requirement.

Fig. 17b Same as (a) but with ($e^- K^-$ wrong) trigger. No D^0 is observed. This shows that the D^0 in (a) is indeed detected in association with the $K^{\pm}$ coming from the antimeson.

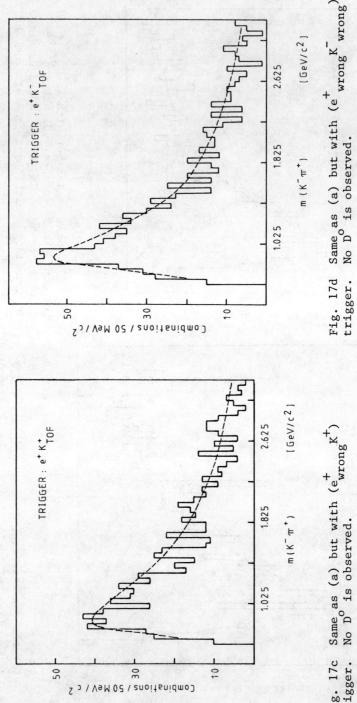

Fig. 17c Same as (a) but with ($e^+_{wrong} K^+$) trigger. No D^0 is observed.

Fig. 17d Same as (a) but with ($e^+_{wrong} K^-_{wrong}$) trigger. No D^0 is observed.

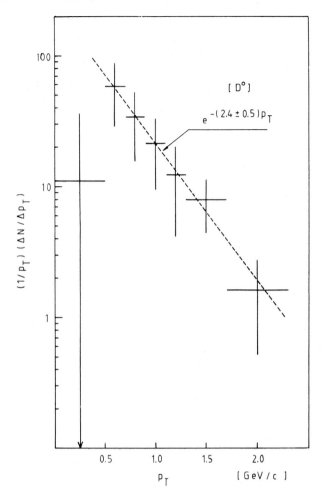

Fig. 18 Experimental D^o p_T distribution obtained for the events of Fig. 16 via the "IN-OUT" difference. The IN region is

$$1.825 \leq m(K^-\pi^+) < 1.975 \text{ GeV/c}^2$$

and the OUT regions are

$$1.675 < m(K^-\pi^+) < 1.825 \text{ GeV/c}^2$$
$$1.975 < m(K^-\pi^+) < 2.125 \text{ GeV/c}^2 \ .$$

The dashed-line best fit gives: $e^{-(2.4 \pm 0.5)p_T}$, in perfect agreement with Λ_c^+ and D^+ p_T slopes (see Figs. 5 and 13).

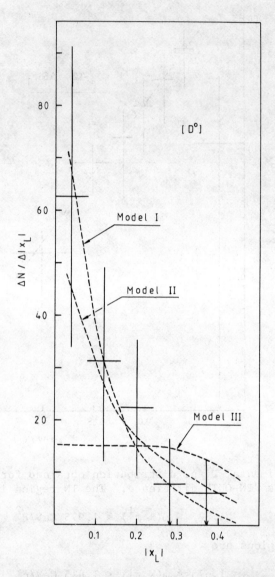

Fig. 19 D^0 x_L distribution (obtained as in Fig. 18). A significant comparison with the Monte Carlo (dashed-line curves) is possible. Model III, i.e. flat-x_L production, is clearly ruled out, while models I and II, i.e. central and flat-y productions, are both compatible within the errors.

D^0 model	$\bar{D}$ model	σ_{tot} (µb)
$E \dfrac{d\sigma}{d\|x_L\|} \propto (1 - \|x_L\|)^3$ (I)	$E \dfrac{d\sigma}{d\|x_L\|} \propto (1 - \|x_L\|)^3$ (I)	575
$\dfrac{d\sigma}{d|y|} = $ const. (II)	$\dfrac{d\sigma}{d|y|} = $ const. (II)	1290
$\dfrac{d\sigma}{d|x_L|} = $ const. (III)	$\dfrac{d\sigma}{d|x_L|} = $ const. (III)	> 5000
$\dfrac{d\sigma}{d|x_L|} = $ const. (III)	$E \dfrac{d\sigma}{d\|x_L\|} \propto (1 - \|x_L\|)^3$ (I)	1000
$E \dfrac{d\sigma}{d\|x_L\|} \propto (1 - \|x_L\|)^3$ (I)	$\dfrac{d\sigma}{d|x_L|} = $ const. (III)	3610

Fig. 20 $\bar{D}D^0$ cross-section estimates obtained with different assumptions. The branching ratios are taken to be ~ 0.045 for $\bar{D} \to e^- K^+ +$ anything and ~ 0.03 for $D^0 \to K^- \pi^+$. Once again the lowest cross-section values are in favour of a "central"-"central" $\bar{D}D^0$ production. The global error amounts to $\sim 50\%$.

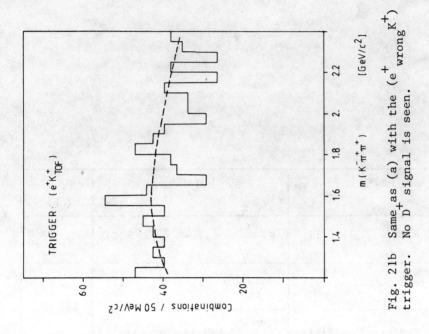

Fig. 21b Same as (a) with the $(e^+_{wrong} K^+)$ trigger. No D^+ signal is seen.

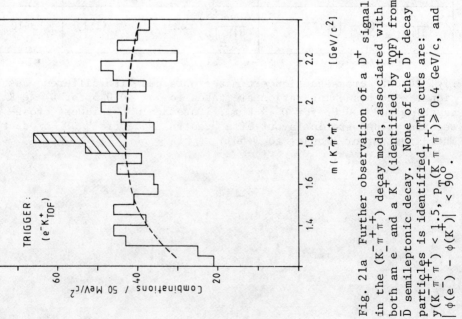

Fig. 21a Further observation of a D^+ signal in the $(K^-\pi^+\pi^+)$ decay mode, associated with both an e^- and a K^+ (identified by TOF) from $\bar{D}$ semileptonic decay. None of the D decay particles is identified. The cuts are: $y(K^-\pi^+\pi^+) < 1.5$, $p_T(K^-\pi^+\pi^+) \geq 0.4$ GeV/c, and $|\phi(e^-) - \phi(K^+)| < 90°$.

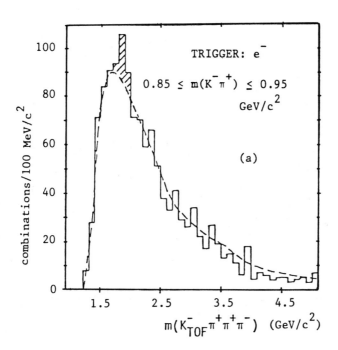

Fig. 22a Further evidence for a D^o via its decay mode:

$$D^o \to \bar{K}^{*o}\pi^+\pi^-$$
$$\hookrightarrow K^-\pi^+$$

in association with e^-. Only the K^- is identified by TOF. The requirements are : $850 \leqslant m(K^-\pi^+) \leqslant 950$ GeV/c^2 and $y(\bar{K}^{*o}\pi^+\pi^-) > 0.8$.

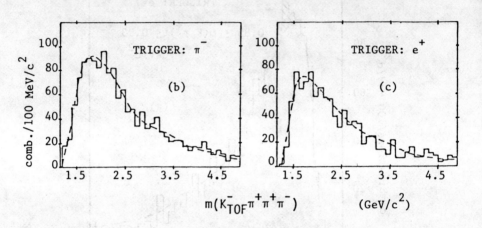

Fig. 22b and c Same as (a) with π^- and e^+ triggers, respectively, showing the background shape.

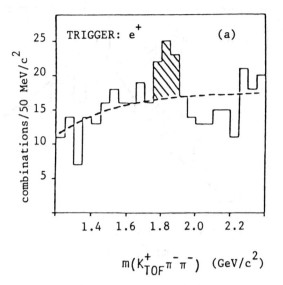

Fig. 23a Preliminary observation of an anticharm meson D^-, decaying into $(K^+\pi^-\pi^-)$ in association with an e^+ from D^0 or D^+ semileptonic decay. The K^+ is identified by TOF and the condition $|\phi(e^+)-\phi(K^+)| > 90°$ is applied.

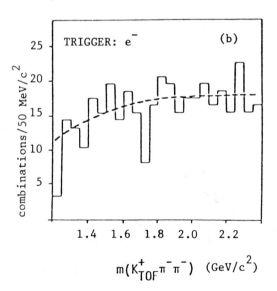

Fig. 23b Same as (a) with the wrong e^- trigger, where no antimeson signal should be seen.

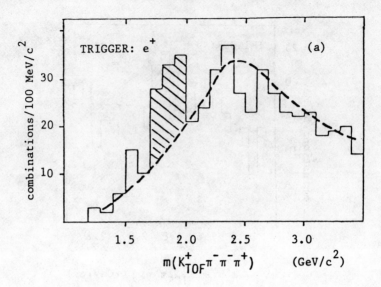

Fig. 24a Preliminary observation of a neutral antimeson decay:

$$\bar{D}^0 \to K^+ \pi^- \pi^- \pi^+$$

associated with an e^+ trigger. The K^+ is identified by TOF and the cuts are: $p_T(e^+) > 800$ MeV/c, $|\phi(e^+) - \phi(K^+)| > 90°$.

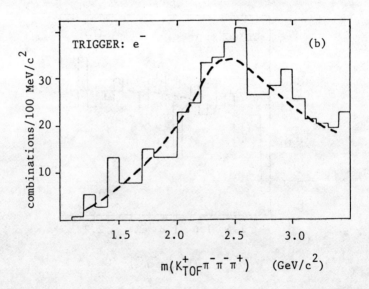

Fig. 24b For comparison, same spectrum as (a), obtained with the "forbidden" e^- trigger.

HEAVY FLAVOUR PRODUCTION

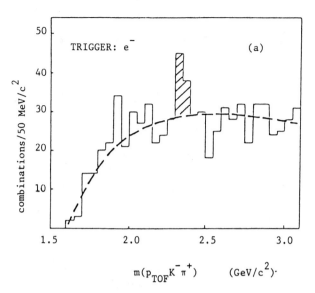

Fig. 25a Further observation of the $\Lambda_c^+ \to pK^-\pi^+$ decay with an e^- trigger. The proton is identified by TOF ($p < 2.5$ GeV/c, 90% C.L.). The cuts are: $p_T(pK^-\pi^+) \geq 0.6$ GeV/c, $y(pK^-\pi^+) < 1$, and $|\phi(e^-)-\phi(p)| > 90°$. The events are also required to have a rather high charge multiplicity (≥ 12) and not to contain any "leading" particle ($x_L > 0.4$).

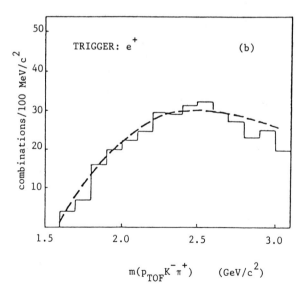

Fig. 25b Same as Fig. 25a with an e^+ trigger: no Λ_c^+ signal appears in this case.

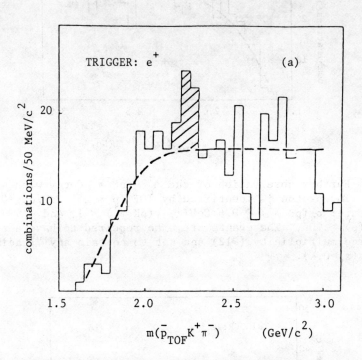

Fig. 26a Preliminary observation of the anticharm baryon $\bar{\Lambda}_c^-$, decaying into $(\bar{p}K^+\pi^-)$ and associated with an e^+ trigger. The $\bar{p}$ was identified by TOF ($p < 2.5$ GeV, 90% C.L.) and the following cuts were applied: $p_T(\bar{p}K^+\pi^-) > 0.7$ GeV/c, $|\phi(e^+)-\phi(\bar{p})| > 90°$, and $p_T(\bar{p}) > 0.4$ GeV/c. The charged multiplicity was required to be $n_{ch} \geq 12$. The absence of any "leading" particle ($x_L > 0.4$) was imposed.

HEAVY FLAVOUR PRODUCTION

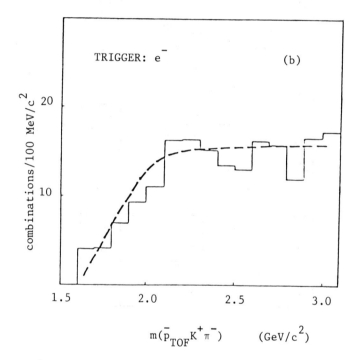

Fig. 26b $(\bar{p}K^+\pi^-)$ background spectrum, obtained as in Fig. 26a but for the "wrong" e^- events, where no Λ_c^- signal shows up.

Table 1

	Particle	Trigger	Signal	Figures
(1)	$\Lambda_c^+ \to pK^-\pi^+$	e^-, $x_L(p) \geq 0.3$	YES	3-9
		e^+, $x_L(p) \geq 0.3$	NO	
(2)	$D^+ \to K^-_{TOF}\pi^+\pi^+$	e^-	YES	10-15
		e^+	NO	
(3)	$D^0 \to K^-\pi^+$	e^-, K^+_{TOF}	YES	16-20
		e^+, K^+_{TOF}	NO	
(4)	$D^+ \to K^-\pi^+\pi^+$	e^-, K^+_{TOF}	YES	21
		e^+, K^+_{TOF}	NO	
(5)	$D^0 \to K^-_{TOF}\pi^+\pi^+\pi^-$	e^-	YES	22
		e^+	NO	
(6)	$D^- \to K^+_{TOF}\pi^-\pi^-$	e^+	YES	23
		e^-	NO	
(7)	$\overline{D^0} \to K^+_{TOF}\pi^-\pi^-\pi^+$	e^+	YES	24
		e^-	NO	
(8)	$\Lambda_c^+ \to p_{TOF}K^-\pi^+$	e^-	YES	25
		e^+	NO	
(9)	$\Lambda_c^- \to \bar{p}_{TOF}K^+\pi^-$	e^+	YES	26
		e^-	NO	

4. BEAUTY PRODUCTION

The following reactions have been studied:

$$pp \to \Lambda_b^0 + \text{(antibeauty state)} + \text{anything} \qquad (10)$$
$$\hookrightarrow pD^0\pi^- \qquad\qquad \hookrightarrow e^+$$
$$\hookrightarrow K^-\pi^+$$

$$pp \to \Lambda_b^0 + \text{(antibeauty state)} + \text{anything} \qquad (11)$$
$$\hookrightarrow \Lambda_c^+\pi^-\pi^-\pi^+ \qquad \hookrightarrow e^+$$
$$\hookrightarrow pK^-\pi^+$$

HEAVY FLAVOUR PRODUCTION

$$pp \to \Sigma_b^{\pm} + \text{(antibeauty state)} + \text{anything} \quad (12)$$
$$\hookrightarrow \Lambda_b^0 \pi^{\pm} \quad\quad\quad \hookrightarrow e^+$$
$$\hookrightarrow pD^0\pi^-$$
$$\hookrightarrow K^-\pi^+$$

The basic cuts applied to the data were the following:

i) the transverse momentum of the positron, $p_T(e^+) \geq 800$ MeV;
ii) at least one positive particle with $|x_L| = 2|p_L|/\sqrt{s} > 0.32$, identified as a proton;
iii) $1.7 \leq m(K^-\pi^+) \leq 2.0$ GeV/c² for reactions (10) and (12); $2.2 < m(pK^-\pi^+) \leq 2.4$ GeV/c² for reaction (11).

4.1 Evidence for $\Lambda_b^0 \to pD^0\pi^-$

Figure 27 shows the invariant mass spectrum for the $p(K^-\pi^+)\pi^-$ system[15], where $(K^-\pi^+)$ satisfies the mass cut 1.7-2.0 GeV/c², and all tracks fit to the interaction vertex and have $\Delta p/p \leq 30\%$. All mass assignments have been made for all tracks, except for those identified by the time-of-flight.

A clear enhancement (29.4 ± 7.4 combinations)[*] is observed in the mass range $5.35 \leq m[p(K^-\pi^+)\pi^-] \leq 5.5$ GeV/c² [16]. On the contrary, if the D⁰ mass cut condition is released, the invariant mass spectrum of the system $(pK^-\pi^+\pi^-)$ appears as shown in Fig. 28 and no enhancement is observed.

A key point now arises. If the enhancement observed in Fig. 27 is really due to the Λ_b^0 decaying into $pD^0\pi^-$, then a corresponding signal for the D⁰ decaying into $K^-\pi^+$ should be observed when a cut in the mass range $5.35 \leq m(pK^-\pi^+\pi^-) \leq 5.5$ GeV/c² is applied to the data from which the mass spectrum, shown in Fig. 28, is obtained, and the $(K^-\pi^+)$ mass spectrum is worked out.

Figure 29a shows the $(K^-\pi^+)$ invariant mass spectrum when the $(pK^-\pi^+\pi^-)$ mass falls inside the above specified mass cut (solid line histogram, IN region) or in two control regions, both 150 MeV/c² wide, immediately above and below this mass range (dashed line histogram, OUT regions). Figure 29b shows the bin-to-bin difference

---

[*] Note that whilst 29.4 is the number of combinations above the background, ±7.4 is the statistical fluctuation of the background level under the observed enhancement. The same applies when discussing the other signals presented in this paper.

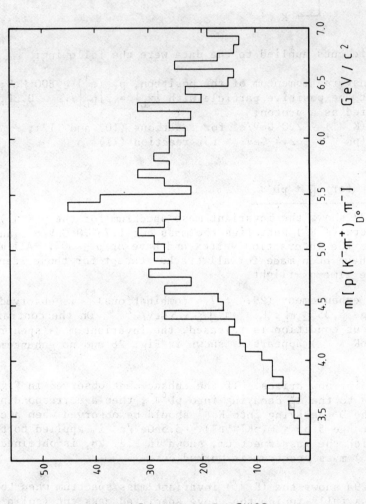

Fig. 27 The $[p(K^-\pi^+)\pi^-]$ invariant mass spectrum obtained from the events satisfying the conditions described in the text and the "wide" D^0 cut $[1.7 \leq m(K^-\pi^+) \leq 2 \text{ GeV}/c^2]$.

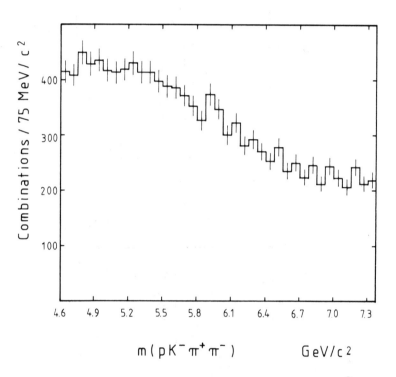

Fig. 28 Same as Fig. 27 but without the "wide" D^o cut.

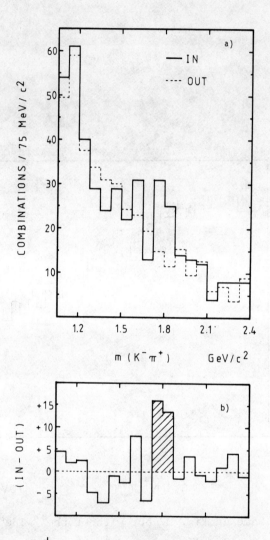

Fig. 29 a) The $(K^-\pi^+)$ invariant mass spectrum when the mass of the system $[p(K^-\pi^+)\pi^-]$ falls in the $(5.35-5.5 \text{ GeV}/c^2)$ mass range (solid line histogram = IN) or in two control regions above and below this range (dashed line histogram = OUT).

b) Bin-to-bin difference of the $(K^-\pi^+)$ mass spectra relative to the IN and OUT regions. A clear D^o enhancement is observed in the mass range $(1.725-1.875 \text{ GeV}/c^2)$.

of the IN and OUT histograms. An enhancement of (28.5 ± 5.5) combinations is indeed observed for the IN sample at a mass about the D^0 mass, $1.725 \leq m(K^-\pi^+) \leq 1.875$ GeV/c^2.

A second very important cross-check can now be done. The Λ_b^0 signal shown in Fig. 27 is obtained applying a wide cut (1.7-2.0 GeV/c^2) around the nominal D^0 mass, while the D^0 signal observed in Figs. 29a and 29b is narrower: 1.725-1.875 GeV/c^2. If the two observed effects are due to physics (i.e. to the production of Λ_b^0, its decay into $pD^0\pi^-$, with the D^0 decaying into $K^-\pi^+$), then, by applying the narrow, experimentally observed mass cut in the $K^-\pi^+$ spectrum, the signal-to-background ratio for the Λ_b^0 signal should improve without loss of events in the observed enhancement.

The $p(K^-\pi^+)\pi^-$ invariant mass spectrum obtained with the cut $1.725 \leq m(K^-\pi^+) \leq 1.875$ GeV/c^2 is shown in Fig. 30. Notice that in this case the signal is given by about 30 ± 5 combinations with a one-to-one signal-to-background ratio, whilst for the case of the wide D^0 mass cut this ratio was only 1 to 2.

In order to be sure that the observed Λ_b^0 signal is not due to instrumental effects such as, for example, detector acceptance, wire chamber misalignment, etc., we have repeated the same analysis:

i) mixing particles from different events;
ii) using as the trigger particle not a positron but a charged hadron;
iii) using as the trigger particle a negative electron.

The results are shown in Fig. 31 for the event-mixing background, in Fig. 32 for the charged hadron background, and in Fig. 33 for the negative electron trigger. No enhancement similar to that obtained with the positron trigger is observed in any of these invariant mass spectra.

Notice that, according to the following process,

$$pp \to \Lambda_b^0 + (\bar{b}\text{-state}) + \text{anything}$$
$$\hookrightarrow pD^0\pi^- \qquad \hookrightarrow (\bar{c}\text{-state})$$
$$\hookrightarrow K^-\pi^+ \qquad\qquad \hookrightarrow e^-$$

a Λ_b^0 signal should be observed also with the negative electron trigger. However, Monte Carlo simulation shows that the momentum spectrum of the electrons from the decay of the anticharm particle is such that the acceptance for these electrons is about four times lower than the acceptance for the positrons because of the antibeauty state decay. Consequently, the number of Λ_b^0 events expected for the e^- trigger is 8 ± 6, which is compatible with what we observed.

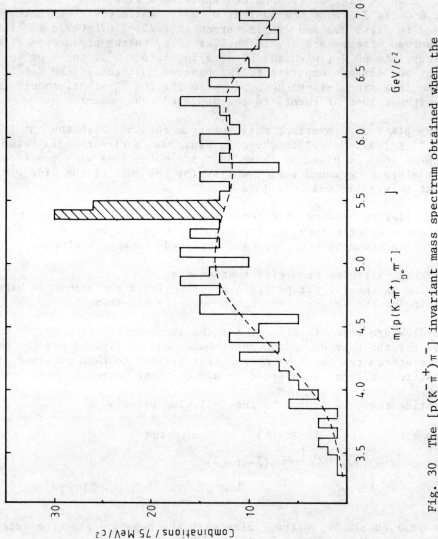

Fig. 30 The $[p(K^-\pi^+)\pi^-]$ invariant mass spectrum obtained when the mass $(K^-\pi^+)_2$ satisfies the "narrow" D^0 trigger $[1.725 \leq m(K^-\pi^+) \leq 1.875 \text{ GeV}/c^2]$.

HEAVY FLAVOUR PRODUCTION

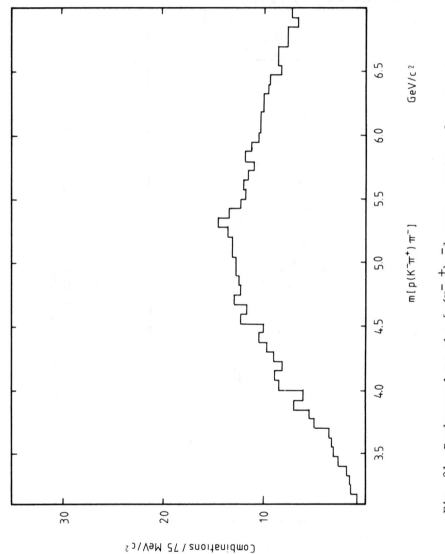

Fig. 31 Background to the $[p(K^-\pi^+)\pi^-]$ mass spectrum from event mixing.

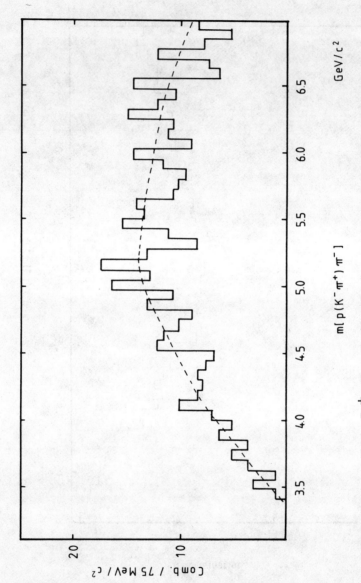

Fig. 32 The $[p(K^-\pi^+)\pi^-]$ invariant mass spectrum obtained when the trigger particle is a charged hadron.

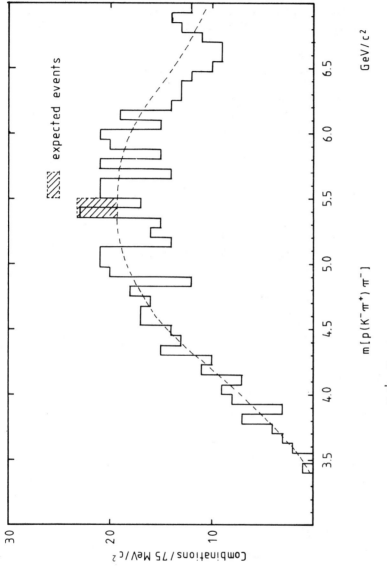

Fig. 33 The [p(K⁻π⁺)π⁻] invariant mass spectrum obtained when the trigger particle is a negative electron. The expected signal is also shown.

4.2 Further studies on the Λ_b^0 signal

The following studies are important not only because they allow an insight into the production mechanism of beauty-flavoured baryons in (pp) collisions, but also because the observation of any difference between the Λ_b^0 events and events without Λ_b^0 increases the statistical significance and the confidence in the observed signal.

In order to perform these studies, we have compared the results obtained from the IN events, i.e. those events where a $p(K^-\pi^+)\pi^-$ combination falls in the Λ_b^0 mass range $\{5.35 < m[p(K^-\pi^+)\pi^-] < 5.5 \text{ GeV}/c^2\}$, with the results obtained from the OUT events, i.e. those events where the above-mentioned particle combination falls into two control regions, 150 MeV/c^2 below and 150 MeV/c^2 above the Λ_b^0 peak.

4.2.1 The longitudinal momentum distribution of Λ_b^0

Figure 34 shows the x_L distribution of the Λ_b^0 [17] obtained, as specified above, by making the (IN-OUT) difference. The full line is the best fit to the data,

$$\Delta N/\Delta|x_L| \propto (1-|x_L|)^\alpha, \text{ with } \alpha = 0.87 \pm 1.26 \ .$$

Notwithstanding the limited statistics, a flat-x_L behaviour is clearly shown by the data. The quantities reported in Fig. 35 are the following ratios:

$$\left(\frac{\Delta N}{\Delta|x_L|}\right)_{\Lambda_b^0} \bigg/ \left(\frac{\Delta N}{\Delta|x_L|}\right)_{\Lambda_c^+} = R_{b/c} \ ,$$

$$\left(\frac{\Delta N}{\Delta|x_L|}\right)_{\Lambda_b^0} \bigg/ \left(\frac{\Delta N}{\Delta|x_L|}\right)_{\Lambda_s^0} = R_{b/s} \ ,$$

$$\left(\frac{\Delta N}{\Delta|x_L|}\right)_{\Lambda_b^0} \bigg/ \left(\frac{\Delta N}{\Delta|x_L|}\right)_{\overline{\Lambda_s^0}} = R_{b/\bar{s}} \ ,$$

as a function of x_L. The sharp increase in the ratio $R_{b/\bar{s}}$ together with the flatness of the ratios $R_{b/c}$ and $R_{b/s}$ clearly indicates that despite the large mass difference between the strange (s), the charm (c), and the beauty (b) quarks, the production of these differently flavoured baryonic states shows the same "leading" effect.

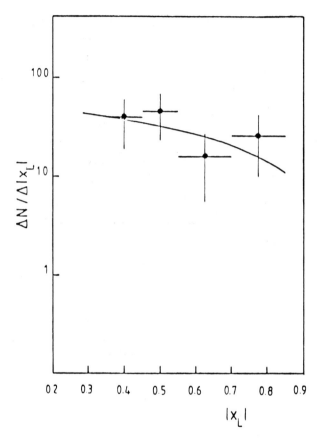

Fig. 34 The x_L distribution of the Λ_b^o.

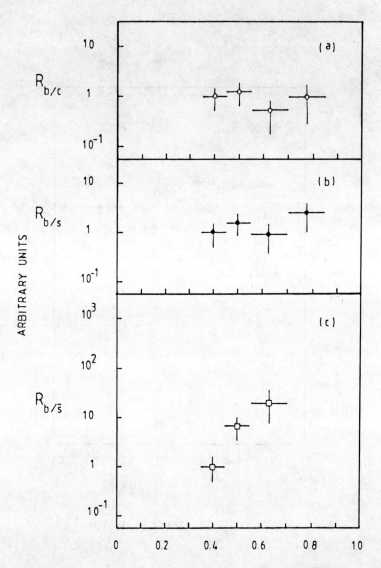

Fig. 35 Showing the ratios $R_{b/c}$, $R_{b/s}$, $R_{b/\bar{s}}$ (defined in the text) as a function of x_L.

HEAVY FLAVOUR PRODUCTION

This can be understood in terms of the fact that all these particles carry two of the original quarks of the incident proton:

$$[(ud)u] = p ,$$

$$[(ud)s] = \Lambda_s^0 ,$$

$$[(ud)c] = \Lambda_c^+ ,$$

$$[(ud)b] = \Lambda_b^0 .$$

4.2.2 The transverse momentum distribution of the positrons from the antibeauty state decay

Figure 36 shows the transverse momentum spectrum of the positrons associated with the IN Λ_b^0 events with respect to the total positron sample and to the positrons associated with the OUT events. Here again a clear difference is observed, indicating the different origin of the positrons associated with the Λ_b^0 signal.

4.3 Possible evidence for $\Sigma_b^\pm \to \Lambda_b^0 \pi^\pm$

We have also investigated the reaction

$$pp \to \Sigma_b^\pm + (\bar{b}\text{-state}) + \text{anything}$$
$$\hookrightarrow \Lambda_b^0 \pi^\pm \qquad \hookrightarrow e^+$$
$$\hookrightarrow pD^0\pi^-$$
$$\hookrightarrow K^-\pi^+$$

Figure 37 shows the mass difference $m[(pD^0\pi^-)\pi^\pm] - m(pD^0\pi^-)$ when the $(pD^0\pi^-)$ mass falls into the Λ_b^0 range (IN region). An enhancement of 14 ± 4.4 events is observed which is not present in Fig. 38, where one sees the same mass difference obtained when the $(pD^0\pi^-)$ mass falls into the OUT region.

This result seems to indicate that indeed some of the observed Λ_b^0 come from the strong decay $\Sigma_b^\pm \to \Lambda_b^0 \pi^\pm$.

4.4 Study of the channel with $\Lambda_b^0 \to \Lambda_c^+ \pi^- \pi^- \pi^+$

Finally we have searched for evidence of Λ_b^0 production in (pp) collisions studying the reaction

$$pp \to \Lambda_b^0 + (\bar{b}\text{-state}) + \text{anything}$$
$$\hookrightarrow \Lambda_c^+ \pi^- \pi^- \pi^+ \qquad \hookrightarrow e^+$$
$$\hookrightarrow pK^-\pi^+$$

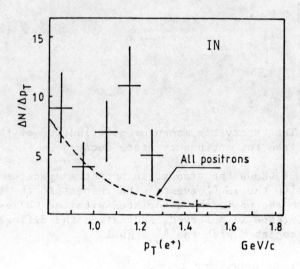

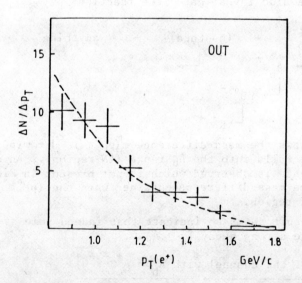

Fig. 36 The transverse momentum spectrum of the positrons associated with the Λ_b^0 compared with the spectrum relative to events that are not Λ_b^0.

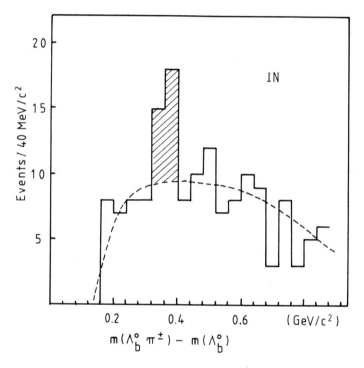

Fig. 37 Possible evidence for the decay $\Sigma_b^\pm \to \Lambda_b^0 \pi^\pm$. The mass difference $\Delta m = m\{[p(K^-\pi^+)\pi^-]\pi^\pm\} - m[p(K^-\pi^+)\pi^-]$ is shown. The histogram corresponds to the case when the $[p(K^-\pi^+)\pi^-]$ mass falls inside the Λ_b^0.

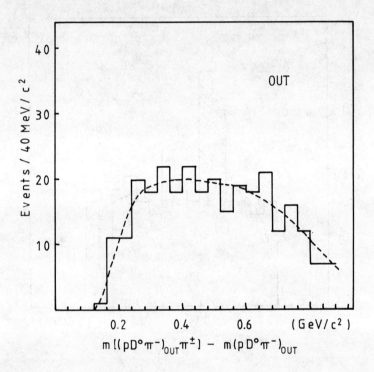

Fig. 38 Showing the same difference as in Fig. 37 for the case when the $[p(K^-\pi^+)\pi^-]$ mass falls in two control regions 150 MeV/c^2 below and above the Λ_b^o peak.

The $(pK^-\pi^+)\pi^-\pi^-\pi^+$ invariant mass spectrum is shown in Fig. 39. An enhancement of 20 ± 5.6 combinations is observed in the mass range $5.35 \leq m[(pK^-\pi^+)\pi^-\pi^-\pi^+] \leq 5.5$ GeV/c².

We have then studied the $(pK^-\pi^+)$ mass spectrum as a function of the invariant mass of the system $pK^-\pi^+\pi^-\pi^-\pi^+$ when the condition $m(pK^-\pi^+) = m(\Lambda_c^+) \pm 100$ MeV/c² is released. Figure 40 shows the $(pK^-\pi^+)$ mass spectrum when the mass of the system $pK^-\pi^+\pi^-\pi^-\pi^+$ satisfies the condition

$$5.35 \leq m(pK^-\pi^+\pi^-\pi^-\pi^+) \leq 5.5 \text{ GeV/c}^2 ,$$

i.e. corresponds to the enhancement observed in Fig. 39 (IN region). An enhancement at a mass corresponding to the Λ_c^+ mass is observed. On the contrary, as shown in Fig. 41, no Λ_c^+ enhancement is observed in the $(pK^-\pi^+)$ mass spectrum when the mass of the $(pK^-\pi^+\pi^-\pi^-\pi^+)$ system corresponds to the two control regions below and above the (5.35-5.5 GeV/c²) enhancement.

Notwithstanding the limited statistics, it is remarkable that here again, as in the case of $\Lambda_b^0 \to pD^0\pi^-$, a charm signal (the Λ_c^+) is observed in association with a positron (which is forbidden by charm physics) only when coupled to an enhancement which can be interpreted as a beauty signal.

4.5 $\underline{\Lambda_b^0 \text{ cross-section estimates and comparison with the } \Lambda_c^+ \text{ cross-section}}$

In order to derive the Λ_b^0 production cross-section σ_b [18] according to the reaction

$$pp \to \Lambda_b^0 + M_{\bar{b}} + \text{anything}$$
$$\quad\quad \hookrightarrow pD^0\pi^- \quad \hookrightarrow e^+$$
$$\quad\quad\quad\quad \hookrightarrow K^-\pi^+$$

where $M_{\bar{b}}$ is an antibeauty-flavoured meson, we have made the following assumptions:

$$\left(\frac{d\sigma}{dp_T}\right)_{\Lambda_b^0, M_{\bar{b}}} \propto p_T \exp(-2.5\, p_T); \quad \left(E\frac{d\sigma}{d|x_L|}\right)_{M_{\bar{b}}} \propto (1-|x_L|)^3; \quad \left(\frac{d\sigma}{d|x_L|}\right)_{\Lambda_b^0} = \text{const.}$$

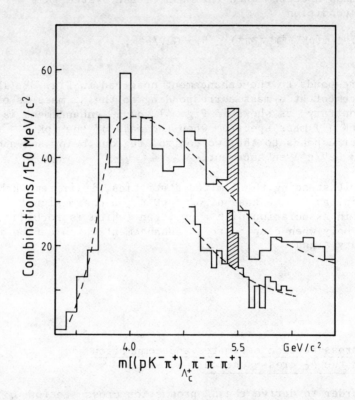

Fig. 39 The $[(pK^-\pi^+)\pi^-\pi^-\pi^+]$ invariant mass spectrum for the events satisfying the conditions specified in the text when the $(pK^-\pi^+)$ mass corresponds to the Λ_c^+ mass $\pm$ 100 MeV/c^2. The enhancement in the mass range (5.35-5.5 GeV/c^2) could be an indication for the decay $\Lambda_b^0 \to \Lambda_c^+\pi^-\pi^-\pi^+$. In the (5-6) GeV/c^2 region, the invariant mass spectrum has also been plotted with 75 MeV/c^2 binning.

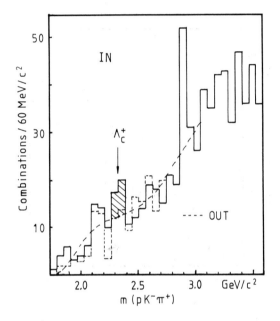

Fig. 40 The $(pK^-\pi^+)$ invariant mass spectrum when the mass of the system $(pK^-\pi^+\pi^-\pi^+)$ corresponds to the possible Λ_b^0 enhancement observed in Fig. 39.

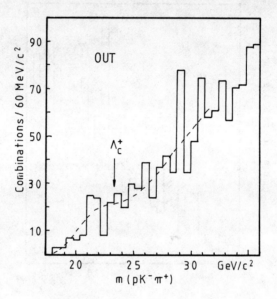

Fig. 41 The $(pK^-\pi^+)$ invariant mass spectrum when the mass of the system $(pK^-\pi^+\pi^-\pi^-\pi^+)$ corresponds to two control regions 150 MeV/c² above and below the enhancement shown in Fig. 39.

For the decay process of $M_b^- \to \bar{D}e^+\nu$, a K_{ℓ_3} matrix element was used, whilst for the decay process $\Lambda_b^0 \to pD^0\pi^-$ three models have been considered, using Lorentz-invariant phase space plus the conditions specified below:

MODEL I: $|x_L|(pD^0\pi^-) \geq 0.32$ and $|y|(pD^0\pi^-) \geq 1.4$;

MODEL II: $p(\text{proton}) > p(D^0) > p(\pi^-)$;

MODEL III: the Λ_b^0 decay is purely isotropic.

Finally, we have taken $B_1(D^0 \to K^-\pi^+) = (3.0 \pm 0.6)\%$[19]; $B_2(M_b^- \to e^+ + \text{anything}) = (13 \pm 6)\%$[20]; and $B_3(\Lambda_b^0 \to pD^0\pi^-) = $ unknown.

The same models were then used to compute the Λ_c^+ production cross-section σ_c corresponding to the reaction[7]

$$pp \to \Lambda_c^+ + \bar{D} + \text{anything}$$
$$\quad\;\;\hookrightarrow pK^-\pi^+ \quad \hookrightarrow e^-$$

studied at the same (pp) c.m. energy and using the same apparatus. For the required branching ratios, we have taken $B_4(D \to e^- + \text{anything}) = (8 \pm 1)\%$[21], and $B_5(\Lambda_c^+ \to pK^-\pi^+) = (2.2 \pm 1)\%$[22].

The results are

Model	$\sigma_b B_3$	σ_c	
I	$2.7 ^{+1.8}_{-1.1}$	$77.5 ^{+57}_{-28}$	
II	$8.2 ^{+6.0}_{-2.8}$	$100 ^{+68}_{-36}$	$\times 10^{-30}$ cm^2
III	$27 ^{+17}_{-11}$	$190 ^{+140}_{-68}$	

From the comparison between σ_c and $\sigma_b B_3$ we can now study how the unknown branching ratio $B_3(\Lambda_b^0 \to pD^0\pi^-)$ depends on the ratio of the cross-sections for charm and beauty (Fig. 42). The most reliable comparison is based on Model I, which requires less extreme extrapolations and corresponds to what has been experimentally observed. Choosing for σ_b/σ_c the value 1/8, which is to be expected if the heavy-flavour production cross-section goes like the inverse masses squared between the beauty and charm quarks[23-24], $B_3(\Lambda_b^0 \to pD^0\pi^-)$ is compatible with a 2% level, within 1.5 standard deviations (Fig. 43).

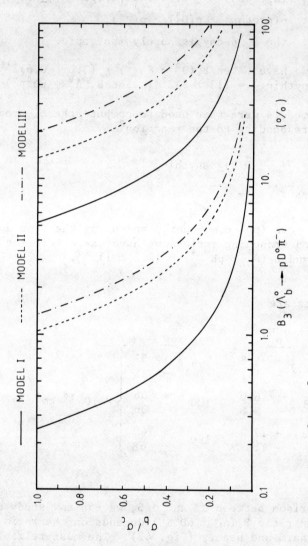

Fig. 42 σ_b/σ_c versus $B_3(\Lambda_b^o \to pD^o\pi^-)$. The 1.5 standard deviation limits are indicated for the three models described in the text.

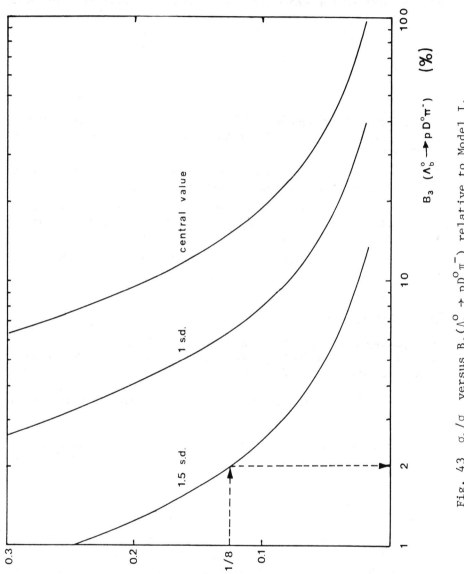

Fig. 43 σ_b/σ_c versus $B_3(\Lambda_b^0 \to pD^0\pi^-)$ relative to Model I.

5. CONCLUSIONS

Charm and beauty production has been studied in (pp) interactions at $(\sqrt{s}) = 62$ GeV at the CERN ISR. From these studies the following features emerge:

Charm production

i) The total cross-section is of the order of one millibarn.

ii) The associated production of "baryon-antimeson" favours "flat x_L", i.e. "forward" production for the baryon and "central" for the antimeson.

iii) The associated production of "meson-antimeson" favours the hypothesis that both are produced "centrally",

iv) The transverse momentum distribution of baryons, mesons, and antimesons follows the expected "low"-exponent behaviour e^{-bp_T} with $b \simeq 2.5$ $(GeV/c)^{-1}$,

v) The production distribution of Λ_c^+ is measured to be "leading", i.e. x_L-flat.

Beauty production

i) The cross-section for Λ_b^0 production is consistent with the ratio being $(\sigma_b/\sigma_c) \simeq (m_c/m_b)^2 \simeq 1/8$, where m_c and m_b are the quark masses for charm and beauty.

ii) The production distribution of the Λ_b^0 is measured to be "leading" i.e. x_L-flat.

iii) The associated "baryon-antimeson" production is consistent with the baryon being produced "forward" and the antimeson "centrally"

Final remark

Let me close with a final remark. Using the same experimental set-up, we have measured, in an extensive range of angles and of transverse momenta, the (e/π) ratio[6]. The value found ($\simeq 1.6 \times 10^{-4}$) is in excellent agreement with previous measurements. However, the production cross-sections for the heavy flavours reported above -- despite the fact that they do represent the lowest values measured at ISR -- should be an order of magnitude lower, in order to be consistent with the measured value of the (e/π) ratio. This puzzle can only be resolved by further, more accurate, studies.

REFERENCES

1. R. Bouclier, R.C.A. Brown, E. Chesi, L. Dumps, H.G. Fischer, P.G. Innocenti, G. Maurin, A. Minten, L. Nauman, F. Piuz and O. Ullaland, Nucl. Instrum. Methods 125, 19 (1975).
2. M. Basile, G. Cara Romeo, L. Cifarelli, A. Contin, G. D'Ali, P. Giusti, T. Massam, F. Palmonari, G. Sartorelli, G. Valenti and A. Zichichi, Nucl. Instrum. Methods 163, 93 (1979).
3. M. Basile, G. Cara Romeo, L. Cifarelli, A. Contin, G. D'Ali, P. Di Cesare, B. Esposito, L. Favale, P. Giusti, T. Massam, F. Palmonari, G. Sartorelli, G. Valenti and A. Zichichi, Nucl. Instrum. Methods 179, 477 (1981).
4. H. Frehse, F. Lapique, M. Panter and F. Piuz, Nucl. Instrum. Methods 156, 87 (1978).
5. H. Frehse, M. Heiden, M. Panter and F. Piuz, Nucl. Instrum. Methods 156, 97 (1978).
6. M. Basile, G. Cara Romeo, L. Cifarelli, A. Contin, G. D'Ali, P. Di Cesare, B. Esposito, P. Giusti, T. Massam, F. Palmonari, G. Sartorelli, G. Valenti and A. Zichichi, Nuovo Cimento 65A, 421 (1981).
7. M. Basile, G. Cara Romeo, L. Cifarelli, A. Contin, G. D'Ali, P. Di Cesare, B. Esposito, P. Giusti, T. Massam, F. Palmonari, G. Sartorelli, G. Valenti and A. Zichichi, Nuovo Cimento 63A, 230 (1981) and 62A, 14 (1981).
8. M. Basile, G. Cara Romeo, L. Cifarelli, A. Contin, G. D'Ali, P. Di Cesare, B. Esposito, P. Giusti, T. Massam, R. Nania, F. Palmonari, G. Sartorelli, G. Valenti and A. Zichichi, preprint CERN-EP/81-125 (1981), submitted to Nuovo Cimento.
9. M. Basile, G. Cara Romeo, L. Cifarelli, A. Contin, G. D'Ali, P. Di Cesare, B. Esposito, P. Giusti, T. Massam, R. Nania, F. Palmonari, G. Sartorelli, G. Valenti and A. Zichichi, Nuovo Cimento 65A, 457 (1981).
10. M. Basile, G. Cara Romeo, L. Cifarelli, A. Contin, G. D'Ali, P. Di Cesare, B. Esposito, P. Giusti, T. Massam, R. Nania, F. Palmonari, G. Sartorelli, G. Valenti and A. Zichichi, internal report CERN-EP/81-02 (1981).
11. M. Basile, G. Cara Romeo, L. Cifarelli, A. Contin, G. D'Ali, P. Di Cesare, B. Esposito, P. Giusti, T. Massam, F. Palmonari, G. Sartorelli, G. Valenti and A. Zichichi, Nuovo Cimento Lett. 30, 481 (1981).
12. M. Basile, G. Cara Romeo, L. Cifarelli, A. Contin, G. D'Ali, P. Di Cesare, B. Esposito, P. Giusti, T. Massam, F. Palmonari, G. Sartorelli, G. Valenti and A. Zichichi, Nuovo Cimento Lett. 30, 487 (1981).
13. M. Basile, G. Cara Romeo, L. Cifarelli, A. Contin, G. D'Ali, P. Di Cesare, B. Esposito, P. Giusti, T. Massam, R. Nania, F. Palmonari, G. Sartorelli, G. Valenti and A. Zichichi, Nuovo Cimento Lett. 33, 17 (1982).

14. M. Basile, G. Cara Romeo, L. Cifarelli, A. Contin, G. D'Ali,
    P. Di Cesare, B. Esposito, P. Giusti, T. Massam, R. Nania,
    F. Palmonari, G. Sartorelli, G. Valenti and A. Zichichi,
    Nuovo Cimento Lett. 33, 33 (1982).
15. M. Basile, G. Bonvicini, G. Cara Romeo, L. Cifarelli, A. Contin,
    G. D'Ali, P. Di Cesare, B. Esposito, P. Giusti, T. Massam,
    R. Nania, F. Palmonari, G. Sartorelli, G. Valenti and
    A. Zichichi, Nuovo Cimento Lett. 31, 97 (1981).
16. A. Martin, Phys. Lett. 103B, 51 (1981).
17. M. Basile, G. Bonvicini, G. Cara Romeo, L. Cifarelli, A. Contin,
    G. D'Ali, P. Di Cesare, B. Esposito, P. Giusti, T. Massam,
    R. Nania, F. Palmonari, G. Sartorelli, G. Valenti and
    A. Zichichi, Nuovo Cimento 65A, 408 (1981).
18. M. Basile, G. Bonvicini, G. Cara Romeo, L. Cifarelli, A. Contin,
    G. D'Ali, P. Di Cesare, B. Esposito, P. Giusti, T. Massam,
    R. Nania, F. Palmonari, G. Sartorelli, G. Valenti and
    A. Zichichi, Nuovo Cimento 65A, 391 (1981).
19. MARK II Collaboration (R.H. Schindler et al.), Phys. Rev. D 24,
    78 (1981).
20. CLEO Collaboration (C. Bebek et al.), preprint CLNS-80/4/5
    (1980).
21. G. Goldhaber and J.E. Weiss, preprint LBL 10652 (1980).
22. MARK II Collaboration (G.S. Abrams et al.), Phys. Rev. Lett. 44,
    10 (1980).
23. G. Gustafson and C. Peterson, Phys. Lett. 67B, 81 (1977).
24. A. Martin, Phys. Lett. 100B, 511 (1981) and private communication.
25. V. Barger, F. Halzen and W.Y. Keung, preprint DO-ER/00881-211
    (1981).

b-QUARK PHYSICS

Karl Berkelman

Laboratory of Nuclear Studies
Cornell University
Ithaca, NY 14853

I. INTRODUCTION

Many physicists had anticipated that the three u, d, s quarks would have to be supplemented by a fourth, before the evidence for the charmed quark actually appeared with the discovery of the ψ.[1] The c quark was needed to account for the absence of strangeness changing neutral week currents,[2] as well as the high value of $R = \sigma(e^+e^- \to \text{hadrons})/\sigma(e^+e^- \to \mu^+\mu^-)$.[3] To most of us the fifth "bottom" quark came as a surprise when the Υ was seen in lepton pair production at Fermilab.[4] But now that we have the b, it is proving to be useful. First, because the b is so massive, the $b\bar{b}$ bound system serves as an convenient laboratory for the study of QCD. And second, the weak decays of b-flavored mesons can help us to solve longstanding problems in the weak interactions. Six quarks may provide the natural explanation for CP violation.[5]

Since the Fermilab dilepton experiments, most of our detailed information on b-quark physics has come from e^+e^- storage rings, the DORIS ring at DESY and the newer CESR ring at Cornell. Table 1 lists a few of the important characteristics of the two machines.

Five experimental facilities have contributed (see Fig. 1). At DORIS the PLUTO detector ran briefly on the $\Upsilon(1S)$ in 1978,

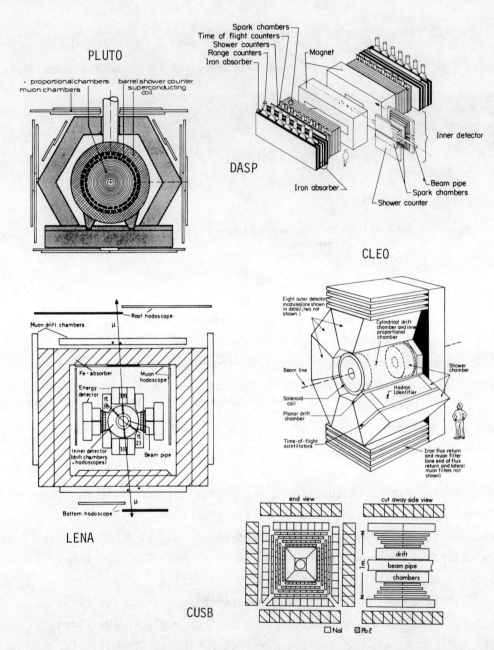

Fig. 1: The five detectors which have contributed to b-quark physics at DORIS and CESR. The LENA detector shows an event of the type $\Upsilon(2S) \to \Upsilon(1S)\pi^+\pi^-$, $\Upsilon(1S) \to \mu^+\mu^-$.

Table 1. Approximate performance figures so far achieved at the e^+e^- rings which have been active in b-quark physics. Improvements in most of these figures are expected within the next year.

	DORIS	CESR
Maximum c.m. energy	10.0 GeV	11.5 GeV
Average luminosity	50 nb^{-1}/day	100 nb^{-1}/day
C.M. energy res. (r.m.s.)	11 MeVx(W/10GeV)^2	4.1 MeVx(W/10GeV)^2
C.M. energy calibration acurracy (r.m.s.)	10 MeV	30 MeV
Measured c.m. energy at Υ(1S)	9462 MeV	9433 MeV

before moving on to PETRA. PLUTO is a solenoidal magnetic detector with shower detectors and muon detectors. That is, it measures charged particle momenta and photon energies over most of the solid angle, and it identifies muons. The DASP detector at DORIS has been operated in the upsilon region by a new group, DASP2. DASP combines a large solid angle shower detector with a double-arm magnetic spectrometer of small solid angle (9% of 4π steradians) but high quality. In the spectrometer one can separate e, μ, π, K, and p. The DESY-Heidelberg nonmagnetic detector, also at DORIS, has been used recently by the LENA group. Built of sodium iodide and lead glass counters, the detector is optimized for photon detection. Charged particles are recorded, but except for showering electrons, their momenta are not measured.

The nonmagnetic detector, CUSB, at CESR is also based on sodium iodide and lead glass. It has excellent photon and electron energy resolution, but does not measure momenta of other charged particles. Finally, the CLEO detector at CESR is a solenoidal magnetic detector, surrounded by hadron identification modules (time-of-flight, gas Cerenkov, and dE/dx) as well as shower detectors and muon detectors. It measures charged and neutral energies over most of the solid angle and identifies e, μ, π, K, and p in favorable momentum and angle ranges.

II. UPSILON SPECTROSCOPY AND QCD
A. Masses

The bound states of $b\bar{b}$ are energetically forbidden to decay into b-flavored mesons. They exist then as very narrow mass states. The ^3S states have $J^{PC} = 1^{--}$, and can therefore couple to e^+e^- annihilating through a virtual photon. The first three in the radial excitation sequence, 1^3S, 2^3S, and 3^3S, are bound and appear as narrow resonances in the e^+e^- total cross section for annihilation into hadrons (Figs. 2, 3).[6-10] Their apparent widths are determined by the beam energy spread (Table 1) induced by synchrotron radiation. The mass differences (Table 2) are measured to a few MeV, but there is an uncertainty of tens of MeV in the overall energy scale calibration of the rings. Eventually this will be accurately established at CESR by a resonant beam depolarization measurement, but for now I would take the DORIS value for the mass of the $\Upsilon(1S)$ and CESR values for the mass differences.

Even before the discovery of the ψ it was suggested that one could treat heavy quark binding using the nonrelativistic Schrödinger equation. The QCD-inspired potentials combine a Coulombic 1/r dependence at small distances, based on single-gluon exchange, and a linear r dependence at large distances, suggested by the string or flux tube picture of confinement. The Υ (and ψ) masses however are most sensitive to intermediate distances, from 0.1 to 3 fm, where one has to make some ad hoc connection between the perturbative small r behavior and the linear large r form. The most sophisticated of these models is the Richardson[15] potential, modified[16] to take second order QCD into account. This potential has only one adjustable parameter which, along with the c and b quark masses, can be chosen to fit very well the mass differences

b-QUARK PHYSICS

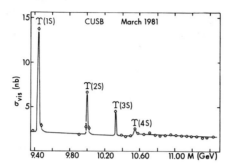

Fig. 2: CUSB data for the uncorrected observed cross section for $e^+e^- \to$ hadrons as a function of center of mass energy.

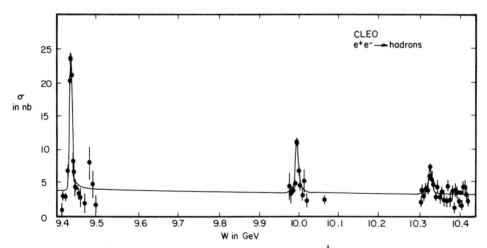

Fig. 3: CLEO data for the cross section for $e^+e^- \to$ hadrons as a function of center of mass energy from 9.3 to 10.5 GeV, showing the $\Upsilon(1S)$, $\Upsilon(2S)$, and $\Upsilon(3S)$.

Table 2: Measured masses and mass differences in MeV, compared with predictions of two potential models.

Quantity	Measured	Expt.	QCD[16]	Power Law[17]
M(1S)	9462±10	DORIS[6-8]	-	-
	9433±28	CESR[9,10]		
M(2S)-M(1S)	553±10	DORIS[7,8]	560*	560*
	560±3	CESR[9,10]		
M(3S)-M(1S)	889±4	CESR[9,10]	890	890*
M(4S)-M(1S)	1114±5	CESR[11,12]	1160	1120
M(χ)**-M(ψ)	426±4	SPEAR[13]	425	425
M(ψ')-M(ψ)	589.07±.10	VEPP4[14]	589	589

*Assumed in fit
**Center of gravity of $^3P_{1,2,3}$ states.

among the ψ states and the Υ states (Table 2). The fitted value of the one parameter implies a QCD scale $\Lambda_{\overline{ms}}$ = 500 MeV with a firm lower bound of 100 MeV. Since the masses are sensitive to the nonperturbative part of the potential, one doesn't know how much to believe the best fit $\Lambda_{\overline{ms}}$. Indeed we should note the fact that one can fit the mass data equally well (Table 2) with a single two-parameter power law potential,[17] which has absolutely no basis in theory. Two conclusions are clear nevertheless. The heavy quark potential is essentially nonrelativistic, and it is flavor independent.

B. Leptonic Pair Widths

Leptonic pair widths are predicted by the van-Royen-Weisskopf[18] formula, with a higher order QCD correction factor[19]:

$$\Gamma_{ee} = 16\pi(\alpha^2 q^2/M^2) |\psi(0)|^2 \cdot (1-16\alpha_s/3\pi).$$

Experimentally, we measure Γ_{ee} by integrating the total hadronic cross section over the resonance, after unfolding the QED radiative effects:

$$\int \sigma_{res} dW = (6\pi^2/M^2)\Gamma_{ee}\Gamma_{had}/\Gamma = (6\pi^2/M^2)\Gamma_{ee}(1-3B_{\ell\ell}).$$

The correction factor $(1-3B_{\ell\ell})$ for lepton pair modes $(B_{\ell\ell} = B_{ee} = B_{\mu\mu} = B_{\tau\tau})$ can be used for the $\Upsilon(1S)$, where $B_{\ell\ell}$ is known, but not yet for the higher upsilons. The factor is probably within 5% of unity for the higher upsilons, so I will neglect it.

The agreement between theory and experiment for the $\Upsilon(1S)$ (Table 3) may be fortuitous in view of the uncertainty in the rather large QCD correction. This uncertainty should not affect the Γ_{ee} predictions for the higher ^3S states if they are normalized to the Γ_{ee} for the $\Upsilon(1S)$. The QCD-inspired potential[16] does

Table 3: Measured and predicted leptonic pair widths, and ratios.

Quantity	Measured	Expt.	QCD[16]	Power Law[17]
$\Gamma_{ee}(1S)$, KeV	1.29±0.09±0.13	DORIS[6-8]	1.1	-
	1.02±0.07±0.15	CESR[10,20]		
$\Gamma_{ee}(2S)/\Gamma_{ee}(1S)$	0.45±0.06±0.02	DORIS[7,8]	0.45	0.43
	0.45±0.03±0.04	CESR[9,10]		
$\Gamma_{ee}(3S)/\Gamma_{ee}(1S)$	0.32±0.03±0.03	CESR[9,10]	0.32	0.28
$\Gamma_{ee}(4S)/\Gamma_{ee}(1S)$	0.24±0.02±0.03	CESR[11,12]	0.26	0.20

extremely well on these ratios, while the power law potential[17] is clearly inferior. I would guess that this is because $|\psi(0)|^2$ is more sensitive than the mass levels to the small distance part of the potential, which is poorly given by the power law.

C. Hadronic Decay of the Υ(1S)

In QCD the only allowed purely hadronic decay of the Υ(1S) proceeds in lowest order through the annihilation of the two heavy quarks into three gluons, which then fragment into hadrons. It is difficult to see three jets in the upsilon data however, because

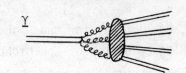

(a) there is a contamination (about 30% at CESR) of two-jet $q\bar{q}$ events from nonresonant background as well as from the virtual photon mediated resonance decays,
(b) the symmetric gluon configuration is expected to be rare, and (c) the transverse momenta in the hadronization process cause the jets to spread and overlap.

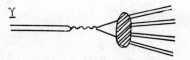

One plots distributions of events in various shape variables—sphericity, thrust, triplicity, jet momenta and angles—comparing the data with simulations based on three gluons, $q\bar{q}$ jets, and phase space.[20] The only theoretically motivated model consistent with all the data is the three gluon mechanism, but if we are not willing to allow theoretical prejudice in the interpretation of the data, we can accommodate an ad hoc model in which the Υ(1S) decays into two jets or via phase space, with about equal probabilities. There is no way yet of convincing a determined skeptic that gluons really do exist in upsilon decays.

If we are willing however to accept that the only strong decay of the Υ(1S) is through three gluons, we can use the measured width to give us what may be the best available value for the strong coupling α_s, or $\Lambda_{\overline{ms}}$. We extract the strong decay width Γ_{dir} of the $\Upsilon(1S)$ from the total width by subtracting the virtual photon contributions:

$$\Gamma_{dir} = \Gamma - (3+R)\Gamma_{ee}.$$

Since the width Γ is much narrower than the experimental energy resoltuion, we have to get it indirectly from Γ_{ee} and measurement of the lepton pair branching ratio

$$\Gamma = \Gamma_{ee}/B_{\ell\ell}$$

(I define $B_{\ell\ell}$ to be B_{ee} or $B_{\mu\mu}$, which I assume to be equal). This is a difficult experiment, because QED electron pairs and muon pairs are much more copious than the resonance decay pairs. Measurements of $B_{\ell\ell}$ are summarized in Table 4. The world average is $B_{\ell\ell} = 3.3\pm0.6\%$, implying $\Gamma = 34\pm8$ keV and $\Gamma_{dir} = 26\pm8$ keV.

For the decay $\Upsilon(1S) \to ggg$ lowest order QCD predicts[28]

$$\Gamma_{ggg} = (160/81)(\pi^2-9)(\alpha_s^3/M^2)|\psi(0)|^2.$$

Reexpressing the wave function at zero $b\bar{b}$ separation in terms of Γ_{ee} gives

$$\Gamma_{ggg} = (10/9\pi)(\pi^2-9)(\alpha_s^3/\alpha^2)\,\Gamma_{ee}$$

The one-loop QCD corrections to Γ_{ggg} have now been calculated[29] with the result that

$$\Gamma_{ggg}/\Gamma_{ee} = (10/9\pi)(\pi^2-9)(\alpha_s^3/\alpha^2)(1 + 9.1\alpha_s/\pi).$$

Assuming $\Gamma_{ggg} = \Gamma_{dir}$, the data imply

$$\alpha_s = 0.137\pm0.009\pm0.007 \text{ and } \Lambda_{\overline{ms}} = 105^{+35}_{-25}\pm20 \text{ MeV}.$$

Table 4: Measurements of $B_{\ell\ell}$, the branching ratio for $\Upsilon(1S)\to\mu^+\mu^-$ (assumed to be equal to the branching ratio for $\Upsilon(1S)\to e^+e^-$), and of $B_{\pi\pi}$, The branching ratio for $\Upsilon(2S)\to\Upsilon(1S)\pi^+\pi^-$

	Measured branching ratio, %	Expt.	Number of events $\pi^+\pi^-\mu^+\mu^-$	$\pi^+\pi^- e^+e^-$
a) $B_{\ell\ell}$	2.2±2.0	PLUTO[21]		
	5.1±3.0	PLUTO[22]		
	2.9±1.3±0.5	DASP2[23]		
	3.5±1.4±0.4	LENA[24]		
	3.0±0.8	Av.from (a)		
	3.4±0.8	Av.from (b),(c)		
	3.2±0.6	World Av.		
b) $B_{\pi\pi}$	26±13	LENA[25]		
	19.1±3.1	CLEO[26]		
	19.6±3.0	Av.from (b)		
	21±6	Av.from (a),(c)		
	20±3	World Av.		
c) $B_{\pi\pi}B_{\ell\ell}$	0.61±0.26	LENA[25]	2	5
	0.63±0.13	CUSB[27]	0	23
	0.68±0.17	CLEO[26]	8	17
	0.64±0.10	Av.from (c)		

The first error is experimental; the second is just a guess at the possible effect of higher order QCD corrections. This may be the best determination of $\Lambda_{\overline{ms}}$ we have. The great advantage of $\Upsilon(1S)$ decay is that the width is proportional to cube of α_s, thereby diminishing the effect of experimental and theoretical uncertainties.

It is worth pointing out that although the value of Λ implied by different kinds of experiments varies from less than 100 MeV to around 500 MeV, it is probably because many determinations are marred by poorly understood higher order and nonperturbative effects and α_s depends only logarithmically on Λ. If QCD or something very much like it were not true, then 20% agreement in the α_s derived from deep inelastic electron scattering, three jet events at PETRA, quarkonium potential models, and the decay rate of the $\Upsilon(1S)$ would have to be considered an extremely unlikely accident.

D. Inclusive Particles in $\Upsilon(1S)$ Decays

If upsilons decay predominantly into three gluons, and continuum annihilations occur primarily through quark-antiquark pairs, we can look for differences between gluon jets and quark jets by comparing upsilon and continuum final states. Charged kaons, selected by time-of-flight, and neutral kaons, identified by $\pi^+\pi^-$ decays, are more frequent (Table 5) in $\Upsilon(1S)$ decays than off resonance. This may be at least partially explained by the higher multiplicities.

A more pronounced effect, first reported by DASP2,[30] is the occurrence of baryons in upsilon decays. In the Lund[31] fragmentation model quark-diquark pairs and quark-antiquark pairs are created from the expanding gluon field in the same way with the same mass dependent suppression factor. The model predicts that the baryon fraction in gluon jets should be somewhat larger than in quark jets, but not enough to explain the DASP2 and CLEO data on protons and lambdas (Table 5).

The same model also predicts that baryons and antibaryons occur closely spaced in rapidity in the same jet. Preliminary CLEO data, however, do not confirm this. Of the 18 $\Upsilon(1S)$ and $\Upsilon(2S)$ decays in which both the baryon and antibaryon are identified only 5 have them on the same side. I would say that we are a long way from understanding baryon production in quark and gluon fragmentation.

Table 5: Preliminary CLEO data on inclusive particle rates in momentum ranges within which the particles can be identified, for $\Upsilon(1S)$ "direct" decays (excluding virtual photon modes) and continuum hadron production.

Particle	Momentum GeV/c	Average number per event $\Upsilon(1S)$ direct	continuum
K^+, K^-	0.5-1.0	0.57±0.05	0.34±0.04
$K^\circ, \bar{K}^\circ$	> 0.3	1.10±0.09	0.78+0.09
$p, \bar{p}$	0.7-1.8	0.46±0.07	0.26±0.04
$\Lambda, \bar{\Lambda}$	>0.4	0.32±0.06	0.12±0.03

E. Hadronic Cascades

If the mass difference is enough, higher ^3S states should be able to decay to lower ones by two-pion emission, just as the ψ' can decay to $\psi\pi\pi$. This mode has been seen by the LENA[25] (Fig. 1), CUSB,[27] and CLEO[26] groups in the readily identifiable topology, $\Upsilon(2S) \to \Upsilon(1S)\pi^+\pi^-$ followed by $\Upsilon(1S) \to e^+e^-$ or $\mu^+\mu^-$. Because $B_{\ell\ell}$ for the $\Upsilon(1S)$ is only three percent, the number of such events is small (Table 4). They provide a measurement of $B_{\pi\pi} \cdot B_{\ell\ell}$, the product of branching ratios for $\Upsilon(2S) \to \Upsilon(1S)\pi^+\pi^-$ and $\Upsilon(1S) \to e^+e^-$. This can be converted into a branching ratio for the two-pion cascade decay using the direct measurements of $B_{\ell\ell}$ (Table 4).

We also have measurements of $B_{\pi\pi}$ without restrictions on the $\Upsilon(1S)$ decay. The LENA group[25] obtains 26±13% by comparing multiplicity distributions in $\Upsilon(2S)$ and $\Upsilon(1S)$ decays. CLEO[26] gets 19±3% from the rate in the $\Upsilon(1S)$ peak in the distribution of masses recoiling against $\pi^+\pi^-$ pairs (Fig. 4). From this measurement and

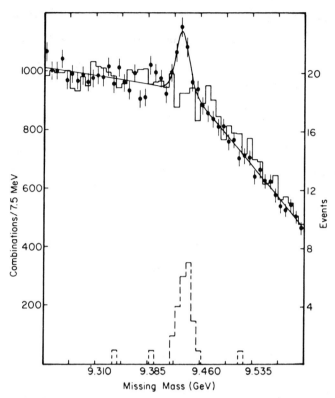

Fig. 4: CLEO data (ref. 26) for the sprectrum of missing masses from $T(2S) \to \pi\pi X$. The points are for opposite sign dipions, the histogram for like sign. The curve is a fit to the background plus $T(2S) \to T(1S)\pi^+\pi^-$. The dashed histogram is for events of the type $T(2S)\pi^+\pi^-\ell^+\ell^-$.

the world average of the product of branching ratios $B_{\pi\pi} \cdot B_{\ell\ell}$ we can derive a value of $B_{\ell\ell}$ for the $T(1S)$ which is as accurate as the direct measurements (Table 4).

If the upsilons are isospin singlets, we expect one half as many $\pi^°\pi^°$ transitions as $\pi^+\pi^-$. The net $T(2S) \to T(1)\pi\pi$ branching ratio is then $3B_{\pi\pi}/2 = 30\pm5\%$. Since we do not yet have a measurement of the total width of the $T(2S)$, the $T\pi\pi$ branching ratio is not immediately convertible to a width. We can estimate the $T(2S)$ total width Γ, however, using some theoretical prejudice. We assume

$$\Gamma = \Gamma_{ggg} + \Gamma_{em} + \Gamma_{\chi\gamma} + \Gamma_{T\pi\pi}.$$

Γ_{ggg} and Γ_{em} (decay via a virtual photon) should both be proportional to Γ_{ee}. Assuming that the proportionality constants for the $\Upsilon(2S)$ are the same as for the $\Upsilon(1S)$, which has no other modes, we can write for the $\Upsilon(2S)$

$$\Gamma_{ggg} + \Gamma_{em} = \Gamma_{ee}/B_{\ell\ell}(1S) = 16\pm 4 \text{ keV}.$$

The width $\Gamma_{\chi\gamma}$ (radiative decays to P states) is not known, but can be estimated from the corresponding charmonium width:

$$\Gamma_{\chi\gamma} = (q_b^2 m_c/q_c^2 m_b)\Gamma(\psi' \to \chi\gamma) = 4 \text{ keV}.$$

Finally, $\Gamma_{\Upsilon\pi\pi}$ can be replaced by $B_{\Upsilon\pi\pi}\Gamma$. Making the substitutions and solving for Γ we obtain for the $\Upsilon(2S)$

$$\Gamma = 27^{+10}_{-7} \text{ keV}$$

and

$$\Gamma_{\Upsilon\pi\pi} = 8^{+3}_{-2} \text{ keV}.$$

In QCD a transition between two heavy quark states proceeds in two steps: the emission of gluons (two in this case), followed by hadronization of the gluons. Since the energy available to the hadrons is small and the gluons are soft, perturbative QCD does not apply. But using a multipole approximation of the gluon field and soft pion techniques one can derive a number of important predictions.[32] The $\Upsilon(2S) \to \Upsilon(1S)\pi\pi$ width for the $b\bar{b}$ system, relative to that for $c\bar{c}$, depends on the relative wave function spreading:

$$\Gamma_{\Upsilon\pi\pi} = (<r^2>_{\Upsilon'}/<r^2>_{\psi'})^2 \cdot \Gamma_{\psi\pi\pi} \approx (1/10) \cdot (108 \text{ keV}) = 11 \text{ keV},$$

a result which is consistent with the data.

This is actually a significant test of QCD; scalar gluons, for example, would have implied a $\Upsilon\pi\pi$ width ten times larger. The theory[32] also predicts that the $\pi\pi$ mass spectrum peaks at high values near the kinematic limit, and that the $\pi\pi$ system is produced isotropically in an s-wave. The CLEO data (Figs. 5 and 6)

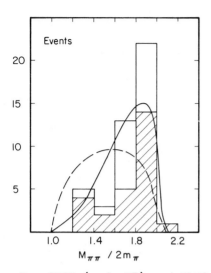 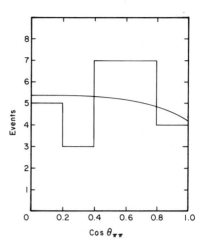

Fig. 5: CUSB (ref. 27) and CLEO (ref. 26, shaded) data for the $M_{\pi\pi}$ spectrum from $\Upsilon(2S) \to \Upsilon(1S)\pi^+\pi^-$, $\Upsilon(1S) \to \ell^+\ell^-$. The solid curve is from ref. 32 and the dashed curve is phase space.

Fig. 6: CLEO data (ref. 26) for the angular distribution of the dipion system with respect to the beam, for events of the type $\Upsilon(2S) \to \Upsilon(1S)\pi^+\pi^-$, $\Upsilon(1S) \to \ell^+\ell^-$. The curve is an isotropic distribution.

confirm both predictions.

Recently CLEO and CUSB have also seen the decay $\Upsilon(3S) \to \Upsilon(1S)\pi^+\pi^-$. CLEO has four (one) events in which the $\Upsilon(1S)$ decays to e^+e^- ($\mu^+\mu^-$). CUSB has three e^+e^-. The preliminary branching ratio for $\Upsilon(3S) \to \Upsilon(1S)\pi^+\pi^-$ is consistent with 4 to 10%.

III. B MESONS

A. Production

The fourth upsilon resonance, seen at CESR[11,12] at a center of mass energy of 10.55 GeV (Figs. 2, 7), has an excitation energy (Table 2) and lepton pair width (Table 3) which make it an obvious candidate for the 4^3S state of $b\bar{b}$, and so we call it the $\Upsilon(4S)$. The observed peak is significantly wider than the three lower energy resonances. If we unfold the effects of machine energy spread and radiative corrections, the CLEO and CUSB data are

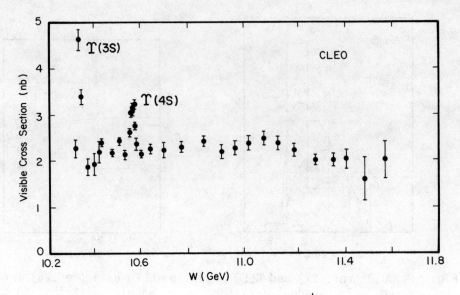

Fig. 7: CLEO data for the cross section for e^+e^- hadrons as a function of center of mass energy from 10.3 to 11.4 GeV, showing the $\Upsilon(3)$ and $\Upsilon(4S)$.

consistent with an intrinsic width $\Gamma_{4S} = 20\pm2$ MeV. This is about three orders of magnitude greater than the intrinsic widths of the three lower upsilon states, indicating that there is between the $\Upsilon(3S)$ and $\Upsilon(4S)$ the threshold for a new strong decay channel.

The lowest lying meson states containing a single b quark should be the pseudoscalars $B^+ = \bar{b}u$ and $B^\circ = \bar{b}d$, and their charge conjugates. The theoretical estimate[33] for the mass (ignoring charge dependence) is

$$M_B = M_D + m_b - m_c + (3/4)(1-m_c/m_b)(M_{D*}-M_D) = 5.2 \text{ GeV},$$

implying a threshold for $B\bar{B}$ production at about 10.4 GeV. The predicted mass difference[33] between the B and the next higher meson, the vector B*, is

$$M_{B*} - M_B = (m_c/m_b)(M_{D*}-M_D) \approx 50 \text{ MeV}.$$

The width Γ_{4S} obviously should depend on the energy of the

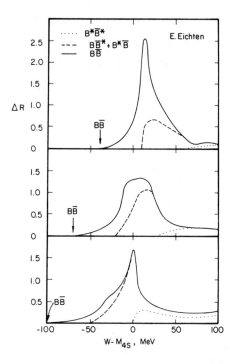

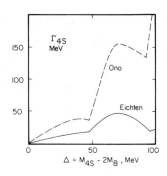

Fig. 9: Total width of the $\Upsilon(4S)$ (refs. 34,35) as a function of the assumed mass above $B\bar{B}$ threshold.

Fig. 8: Predicted cross section (ref. 34) for production of b-flavored mesons using several B mass values.

$\Upsilon(4S)$ relative to the thresholds for $B\bar{B}$, $B\bar{B}^*$ or $B^*\bar{B}$, and $B^*\bar{B}^*$. Fig. 8 shows Eichten's predictions[34] for the $\Upsilon(4S)$ peak shape with three assumed values for $\Delta = M_{4S} - 2M_B$, based on the same coupled channel theory used at charm threshold. Fig. 9 shows Eichten's Γ_{4S} vs. Δ, compared with a calculation by Ono[35] based on a quark pair creation model. The discrepancy between the models prevents us from using Γ_{4S} to measure M_B; the best we can say is that the threshold energy $2M_B$ lies between M_{3S} and M_{4S}, most likely much closer to M_{4S}.

The total cross section for $e^+e^- \to$ hadrons above the $\Upsilon(4S)$ has been measured at CESR up to W=11.5 GeV (Figs. 2, 7). It shows no surprises, no prominent peaks. The average level of $R = \sigma_{had}/\sigma_{\mu\mu}$ is 0.3±0.1 higher at energies above the $\Upsilon(4S)$ than below. Asymtotically one expects $\Delta R = 3q_b^2 = 1/3$ for the $b\bar{b}$ increment.

At the peak of the $\Upsilon(4S)$, however, R above the continuum is about 1.4, making the $\Upsilon(4S)$ the best source of B mesons. The continuum amounts to about R=4. At CESR we now have over 5,000 nb^{-1} integrated luminosity at the $\Upsilon(4S)$ energy, which means about 20,000 events in each of the two experiments, about one third of which should be $B\bar{B}$ events.

B. Inclusive Decays

To investigate many features of the weak decay of B mesons it is not necessary to reconstruct complete decays or even to know which events are $B\bar{B}$ events. If we assume that the excess cross section at the $\Upsilon(4S)$ peak is entirely $B\bar{B}$, then any rate $R_{B\bar{B}}$, as in Table 6, can be extracted as follows:

$$R_{B\bar{B}} = (R_{pk}\sigma_{pk} - R_{co}\sigma_{co})/(\sigma_{pk} - \sigma_{co}),$$

where σ_{pk} and σ_{co} are the annihilation cross sections measured at the peak and at lower energies, and R_{pk} and R_{co} are the measured rates of interest.

Multiplicity. In order to correct the observed charged

Table 6: Preliminary CLEO data on inclusive rates in $\Upsilon(4S)$ decay and in nonresonant hadron production.

Measurement	$\Upsilon(4S)$ decay	Continuum
Obs. charged multiplicity	8.5±0.7	6.6±0.1
Obs. charged sphericity	0.38±0.05	0.23±0.01
K^+,K^- per event, 0.5<p<1.0 GeV/c	0.77±0.27	0.34±0.04
$K^\circ,\bar{K}^\circ$ per event, p>0.3 GeV/c	1.3±0.9	0.78±0.09
$p,\bar{p}$ per event, 0.7<p<1.8 GeV/c	0.1±0.2	0.26±0.04
$\Lambda,\bar{\Lambda}$ per event, p>0.4 GeV/c	-0.06±0.14	0.12±0.03

multiplicity (Table 7) for detector acceptance we make a Monte Carlo, simulating B decays (standard model, Field-Feynman, etc.) and the properties of the CLEO detector. Adjusting the generated multiplicities until the simulated detectable multiplicity distribution matches the real data, we get an average charged multiplicity per B decay, $\bar{n}_B$ = 5.9±0.5.

Photons. If the $\Upsilon(4S)$ is more than 50 MeV above the $B\bar{B}$ threshold, one should expect to produce the B*, which will decay to B plus a 50 MeV photon. The CUSB sodium iodide detector should have good efficiency and 20% FWHM resoltuion at 50 MeV. Fig. 10 shows the CUSB inclusive photon energy spectrum for $\Upsilon(4S)$ decays. There is no evidence for a peak near 50 MeV standing above the π° background. Either the $\Upsilon(4S)$ is less than 50 MeV above $B\bar{B}$ threshold, or the B* yield is a small fraction of the B yield.

Electrons. Both of the CESR groups have measured the yield of inclusive electrons as a function of beam energy W. In the CUSB detector[36] electrons are recognized as showering charged particles. CLEO[37] imposes the additional requirement of a signal in a gas Cerenkov counter and an approximate match between the measured momentum and shower counter pulse height. Fig. 11 shows

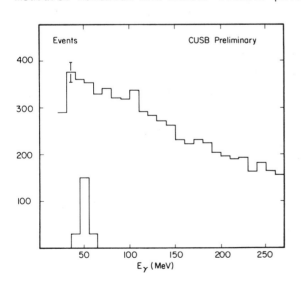

Fig. 10: Preliminary inclusive photon spectrum observed at the $\Upsilon(4S)$ in the CUSB detector.

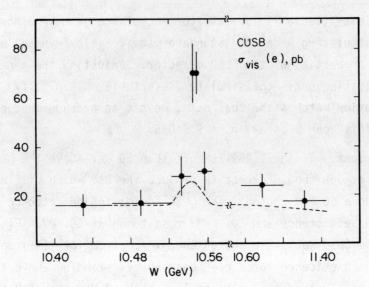

Fig. 11: Uncorrected inclusive cross section for electrons of momentum above 1 GeV/c, from the CUSB experiment (ref. 36). The dashed curve shows the scaled down total cross section in the vicinity of the $\Upsilon(4S)$ resonance. CLEO data for electrons (ref. 37) and muons (ref. 38) look very similar.

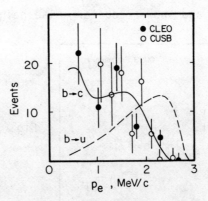

Fig. 12: Momentum spectrum of electrons from $\Upsilon(4S)$ decays from CLEO (ref. 37) and CUSB (ref. 36). The curves show predictions obtained by assuming pure $b \to c\, e^- \bar{\nu}$ (solid) and pure $b \to u e^- \bar{\nu}$ (dashed).

the observed rate of electrons with momenta above 1 GeV/c, as a function of W. Fig. 12 shows the measured momentum spectrum. In both experiments the electron rate increases at the $\Upsilon(4S)$ by a much larger factor than does the total cross section, especially for electron momenta above 1 GeV/c. The low continuum electron yield is in fact consistent with the expected rate from charm and tau decays, QED processes, and misidentified hadrons. The high rate of high momentum electrons at the $\Upsilon(4S)$ implies that the B has an appreciable semileptonic branching ratio, and is the best evidence that the B meson is indeed being produced. Taking the acceptance and electron identification efficiency into account the two groups report a 13% branching ratio for B→Xeν (Table 8). The lower momentum electrons (below 1 GeV/c) are more likely to come from later stages in the decay chain.

Table 7: Measurements of semileptonic branching ratios of B mesons. Only leptons produced in the first generation of the decay are counted. If the charged B and neutral B have different branching ratios, the quoted values apply to the average over charge states produced at the $\Upsilon(4S)$.

Mode	Expt.	Branching Ratio, %
B→Xeν	CUSB[36]	13.6±2.5±3.0
B→Xeν	CLEO[37]	13±3±3
B→Xμν	CLEO[38]	9.4±3.6
B→Xℓν ($\ell=e=\mu$)	Average	12±2

Muons. The CLEO group has made a similar search for muons[38] identifying them with drift chambers outside of a 60 cm thick iron shield surrounding the detector. The minimum momentum which can penetrate the shield is 1.0 to 1.6 GeV/c depending on the direction. The beam energy dependence of the observed rate and the muon momentum spectrum both behave similarly to the case for electrons. The implied branching ratio for $B \to X\mu\nu$ is about 9%, compatible within errors with the electron branching ratio (Table 7).

Kaons. Table 7 shows the observed charged kaon rates in the momentum interval within which kaons reach the CLEO time-of-flight scintillators and are distinguishable from pions and protons. Neutral kaons are observed in $\pi^+\pi^-$ decays for momenta above 300 MeV/c. For both neutral and charged kaons the average number per event is larger in $\Upsilon(4S)$ decays than in the continuum. The momentum spectrum from $\Upsilon(4S)$ decays (Fig. 13) is softer, however.

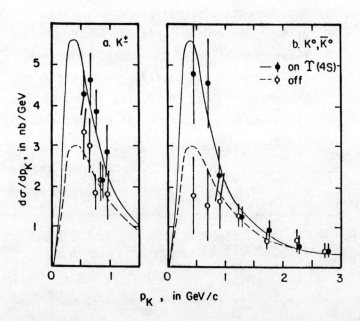

Fig. 13: Charged and neutral kaon momentum spectra observed by CLEO on the $\Upsilon(4S)$ and off resonance. The curves are from a Monte Carlo simulation assuming pure $b \to eW^-$.

b-QUARK PHYSICS

Baryons. Protons (and antiprotons) are identified by time-of-flight for momenta between 0.7 and 1.8 GeV/c, and lambdas (antilambdas) decaying into $p\pi^-$ ($p\pi^+$) are visible for momenta above 400 MeV/c. The observed baryon rates in $\Upsilon(4S)$ decays are consistent with zero (Table 6), in contrast to the rather high baryon rates in the decay of the lower upsilons.

C. Do b-Quarks Decay Independently of their Light-Quark Partners?

If the light quark acts only as a spectator in the B decay, then the lifetimes and leptonic branching ratios of charged and neutral B mesons should be equal, and observations of B decay properties can be interpreted directly as the properties of b-quark decays. But this spectator model does not work for s and c quarks; the charged K and D have much longer lifetimes and larger leptonic branching ratios than the neutrals.[39] This phenomenon has been explained[40] in terms of diagrams in which the heavy quark and light quark exchange a W boson, possible only for neutral mesons, and the diagram in which the heavy and light quarks annihilate into a W, possible only for charged mesons, as well as diagrams involving gluons.

The nonspectator amplitudes tend to be helicity suppressed, by a factor of the order of the ratio of light and heavy quark masses.[41] So they should be less prominent in B decays than in D or K decays. We have so far no measurements of lifetimes or relative branching ratios for charged and neutral B's, so we cannot test the spectator model. We will assume it for simplicity in what follows.

D. Is the b-Quark in a Left-Handed Weak Doublet?

In the standard model of Kobayashi and Maskawa[5] we have three left-handed doublets of quarks: $(ud')_L$, $(cs')_L$, and $(tb')_L$. The primed states (d's'b') are connected to the flavor eigenstates (d s b) by the K-M rotation matrix, depending on three generalized Cabibbo angles θ_1, θ_2, θ_3 and a CP violating phase δ. In this model there are no flavor changing neutral currents, and since the t is presumably very massive, the b must decay via its admixture with s and/or d:

$b \to cW^-$ or $b \to uW^-$,

$W^- \to \ell^- \bar{\nu}$ or $\bar{u}d$ or $\bar{c}s$.

From simple fermion counting (with color) one expects a leptonic branching ratio B_ℓ (here defined to be B_e or B_μ, not the sum) equal to 1/9. If we take account of phase space, which is important for c and τ final states, and of the hadronic enhancement effects of gluons, the prediction[41] is

$B_\ell = 12.2\%$ if $b \to cW^-$

$ = 10.2\%$ if $b \to uW^-$.

Either value is in good agreement with the measurements (Table 7).

There are many non-standard models, some inspired by the lack of evidence for a t-quark. Suppose that the b is in a weak doublet but has a new quantum number which forbids $b \to qW^-$ (q is any lighter quark).[42] Then either the b is stable, or all decays contain leptons. Both possibilities are ruled out by the data (Table 7).

Suppose the b is a weak singlet with a new quantum number which forbids $b \to qW^-$.[43] Then either $b \to X\tau\bar{\nu}_\tau$ or $\bar{q}\bar{q}\ell$ or $\bar{q}\bar{q}\bar{\ell}$. That is, all b decays would contain a lepton and all decays would contain either a τ or an antibaryon (baryon for $\bar{b}$ decay). The data (Tables 6,7) completely rule out this model.

Suppose the b is a weak singlet which can mix with s and/or d.[44] There is no way to suppress completely flavor changing neutral currents, and at least 25% of b decays will be through the Z°. One can derive[45] a lower limit on the dilepton branching ratio of the b:

$$(b \to X\ell^+\ell^-)/(b \to X\ell^-\bar{\nu}) > 1/8.$$

The CLEO group has seen zero dilepton events which can be attributed to flavor changing neutral current b decay.[38] The 90% confidence upper limit on the above ratio is now 10%. We are therefore almost at the point of ruling out the hypothesis that the b is a weak singlet.

Suppose that the b is one of a doublet of charge -1/3 quarks, $(b_1 b_2)_L$. Here again one can obtain the same limit[45] on dilepton decays, so that this possibility is also close to being ruled out.

Suppose that we put the b in a right-handed doublet with the c, $(c\ b)_R$.[46] This model has natural suppression of flavor changing neutral currents. Its predictions can be tailored to be very close to those of the standard six-quark model, so it will be very hard to rule out. We may have to wait for the discovery of the t-quark to know for sure.

Finally, it is possible with charged Higgs bosons of mass near 5 GeV to concoct a model in which the b decays via a Higgs instead of a W. This would show up as a preference for high mass leptons, τ and c, in the decay products. With the present data one cannot, I believe, rule this one out, but the good agreement between the standard model and the measured e and μ branching ratios makes it rather unlikely.

In short, there are no surprises. The standard model is in good shape, and most of the topless models are untenable.

E. What is the Ratio of $b \to uW^-$ and $b \to cW^-$ Rates?

The K-M matrix[5] is given in the Particle Properties Data Booklet[47] in terms of sines (s_i) and cosines (c_i) of the generalized Cabibbo angles θ_c. From strangeness conserving beta decay (proportional to c_1) we know that $s_1^2 = 0.05$. Strangeness changing decays (proportional to s_1c_3) tell us that $0 < s_3 < 0.5$.[48] If CP violation is attributed to the phase δ, then $|c_\delta| \approx 1$ ($c_\delta = \cos\delta$). Finally, from the K° mass difference we have (using the bag model) $s_2 = 0.2 \pm 0.1 + (1/2c_\delta + 1/4)s_3$. The uncertainty in the K-M angles is reduced to the uncertainty in s_3 (0 to 0.5) and in the sign of c_δ (± 1).[49]

Figs. 14 and 15 show the lifetime τ_B and the ratio of rates $(b \to uW^-)/(b \to cW^-)$ as functions of s_3 for the two signs of c_δ. The present experimental upper limit on τ_B, of the order of 10^{-10} sec,[50] is too high to be useful.

The other possibility of fixing s_3 and c_δ is a measurement of $(b \to uW^-)/(b \to cW^-)$. One indicator of the presence of charm in B decays is the average number of kaons per event. The most naive calculation goes as follows. If $b \to u W^-$, you get two kaons when W^- goes to $s\bar{c}$ (which it does 1/3 of the time by simple counting), otherwise none; so you expect an average of 2/3 kaon per B decay. If $b \to cW^-$, you always get one kaon from the c decay products and again an average of 2/3 from the W^-; so you expect 5/3 kaons per B decay. The naive calculation has to be corrected for phase space, hadronic enhancements, nonspectator contributions, and for kaons generated from the sea in the later high multiplicity hadronization stages. The predictions become very model dependent. Our best guess is that pure $b \to uW^-$ would give about 0.9 kaons per B decay and pure $b \to cW^-$ would yield about 1.6. The measured rate, combining the CLEO rates for charged and neutral kaons (Table 6) and correcting for the undetected portions of the momentum spectra, is $1.6 \pm 0.6 \pm 0.3$ per B decay. This agrees well with the expected rate for pure $b \to cW^-$, but because of the large error, it is also marginally consistent with the pure $b \to uW^-$ hypothesis.

Another charm indicator is the lepton momentum spectrum (Fig. 12). The electrons and muons above 1 GeV/c come mainly from the first generation $b \to c\ell^-\bar{\nu}$ or $b \to u\ell^-\bar{\nu}$. The end point of the

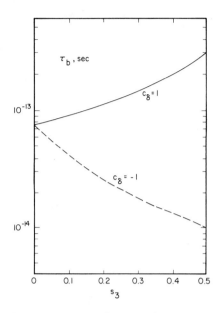

Fig. 14: Dependence of B meson lifetime on $\sin\theta_3$ for the two possible values of $\cos\delta$ in the standard model (ref. 49).

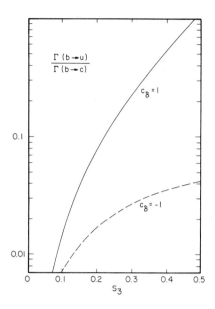

Fig. 15: Ratio of rates $(b\to uW^-)/(b\to cW^-)$ as a function of $\sin\theta_3$ for the two possible values of $\cos\delta$ in the standard model (ref. 49).

spectrum depends on the effective mass of the products of the c or u quark and the spectator quark. Since the mass is likely to be smaller in the b→u case, we expect the lepton spectrum to have a higher end point. Until we know more about the recoiling mass, though, we cannot make reliable predictions. The CUSB or CLEO data (Fig. 12, for example) are consistent so far with either b→cℓ$\bar{\nu}$ or b→uℓ$\bar{\nu}$, probably preferring b→cℓ$\bar{\nu}$.

Low momentum leptons come from later stages in the decay chain, that is, from D decays if the b decays to c. Again there are large uncertainties in the predictions; for instance, charged and neutral D's have very different leptonic branching ratios, and we don't know how many of each are made in B decays. Fig. 12 shows typical predictions. The contrast between b→cℓ$\bar{\nu}$ and b→uℓ$\bar{\nu}$ makes this potentially the best indicator of the relative rates. The decay through charm is certainly favored by the CLEO electron data, but

the model dependence prevents us from quoting a value for
$(b \to uW^-)/(b \to cW^-)$.

IV. SUMMARY

A. Conclusions

1. Masses and lepton pair widths for the first four ^3S upsilon states are well represented by a nonrelativistic flavor-independent potential which simultaneously fits psion data. The shape of the potential is consistent with gluon exchange and linear confinement. The strong coupling scale parameter $\Lambda_{\overline{ms}}$ is not very well determined by the fits, but has a rather firm lower limit at 100 MeV.

2. The purely hadronic decay of the $\Upsilon(1S)$ is consistent with the three gluon hypothesis, and the measured width implies a rather accurate and reliable determination of $\Lambda_{\overline{ms}}$ at 125 MeV.

3. The $\Upsilon(2S)$ decays hadronically to the $\Upsilon(1S)$ at a rate in good agreement with the prediction based on the multipole expansion of the gluon field.

Conclusions 1, 2, and 3 have to be counted as a triumph of QCD. I know of no other model which can simultaneously explain all these results.

4. The $\Upsilon(4S)$ is above the threshold for production of b-flavored mesons.

5. The B leptonic branching ratios and the very low rate of dilepton decays support the standard six-quark model of left-handed weak doublets. Most models not containing a t quark have been excluded.

The lepton momentum spectra and inclusive kaon rates in B decays indicate a preference of the b quark to decay through the c quark, but do not exclude a sizeable contribution of decay through the u quark.

b-QUARK PHYSICS

I confess that I am disappointed at the lack of surprises in the b quark physics. The popular orthodoxy is doing too well.

B. Unfinished Business

There are plenty of loose ends in b-quark physics. Maybe some of them will surprise us.

1. Other $b\bar{b}$ bound states. I expect that the CUSB group will establish the radiative transitions from the $\Upsilon(2S)$ to the ^3P (χ_b) states within the next six months. There is a good chance that the ^1P and ^1S (η_b) states will be seen in transitions from the $\Upsilon(3S)$. Tye and others[51] have predicted a new kind of resonance between the $\Upsilon(3S)$ and $\Upsilon(4S)$, arising from a vibrational excitation of the gluon string binding b and $\bar{b}$. With enough data we should be able to see it. Both DORIS and CESR expect, through the mini-beta trick, to have significantly higher luminosities within a year.

2. The DORIS-CESR mass calibraiton discrepancy. This will eventually be settled at CESR using the same resonant beam depolarization technique that was used at VEPP4[14] to measure the ψ and ψ' masses to 100 keV accuracy.

3. More and better decay Γ values. This requires more data and better understanding of systematic errors. It will come with time.

4. Reconstructed exclusive B decays. This will give us the mass and more information on decay modes. In CLEO it will require more events and better mass resolution. More events will come with time and increased luminosity; the resolution will improve soon when we replace the conventional solenoid with a high-field superconducting solenoid. Meanwhile, we of course have already some candidates for completely reconstructed events. Fig. 16 shows one, a possible

$$e^+e^- \to B^\circ(\to D^-(\to K^+\pi^-\pi^-)\pi^+\pi^+\pi^-) \bar{B}^\circ(\to D^\circ(\to K^-\pi^+\pi^+\pi^-)\pi^+\pi^-).$$

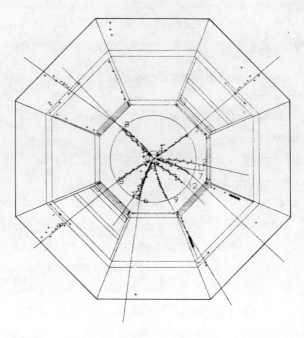

Fig. 16: Display of a CLEO event consistent with the hypothesis
$e^+e^- \to B^\circ \bar{B}^\circ$, $B^\circ \to D^-\pi^+\pi^+\pi^-$, $D^- \to K^+\pi^-\pi^-$, $\bar{B}^\circ \to D^\circ \pi^+\pi^-$, $D^\circ \to K^-\pi^+\pi^+\pi^-$.

5. The B lifetime. If it is shorter than 10^{-13} seconds, as seems likely, it will probably have to be measured at a proton machine using emulsions or some other fine-grain detector.

6. $\Gamma(b \to uW^-)/\Gamma(b \to cW^-)$. A good measurement determines the K-M mixing angles and the CP violating phase δ. It requires more $\Upsilon(4S)$ data and better understanding of the hadronization process. Significant improvement should come in the next year.

7. $B^\circ \bar{B}^\circ$ mixing. There is reason to expect mixing if θ_3 and δ are small enough. One looks for $B^\circ B^\circ$ or $\bar{B}^\circ \bar{B}^\circ$ events by looking for like-sign leptons (or D's or K's). The mixing is sensitive to the mass of the t quark. The measurement requires a large number of events.

8. CP violation in B decays. If there is $B°\bar{B}°$ mixing, one can look for a $B°B°$ vs. $\bar{B}°\bar{B}°$ asymmetry. It will require luck and a lot of data.

9. Other b-flavored hadrons. The B^*, $b\bar{s}$, and $b\bar{c}$ mesons have yet to be seen. Of the baryons, only the Λ_b has been reported.[52] Unless there are analogs of the $T(4S)$ B-factory, it may be difficult to pick these up in e^+e^- collisions.

REFERENCES

1. J.J. Aubert et al., Phys. Rev. Lett. 33, 1404 (1974); J. Augustin et al., Phys. Rev. Lett. 33, 1406 (1974).
2. S.L. Glashow, J. Iliopoulos, and L. Maiani, Phys. Rev. D2, 1285 (1970).
3. J.D. Bjorken, Proceedings of the 6th International Symposium on Electron and Photon Interactions at High Energies, H. Rollnik and W. Pfeil ed. (North-Holland, Amsterdam, 1974), p. 25.
4. S.W. Herb et al., Phys. Rev. Lett. 39, 252 (1977).
5. M. Kobayashi and T. Maskawa, Prog. Theor. Phys. 49, 652 (1973).
6. C. Berger et al (PLUTO), Phys. Lett. 76B, 243 (1978).
7. C.W. Darden et al. (DASP2), Phys. Lett. 76B, 246 (1978), and 78B, 364 (1978).
8. J.K. Bienlein et al., Phys. Lett. 78B, 360 (1978); P. Boch et al., Z. Phys. C6, 125 (1980).
9. D. Andrews et al. (CLEO), Phys. Rev. Lett. 44, 1108 (1980).
10. T. Böhringer et al. (CUSB), Phys. Rev. Lett. 44, 1111 (1980).
11. D. Andrews et al. (CLEO, Phys. Rev. Lett. 45, 219 (1980).
12. G. Finocchiaro et al. (CUSB), Phys. Rev. Lett. 45, 222 (1980).
13. M. Oreglia, preprint SLAC-PUB-2529 (1980); T.H. Burnett, High Energy Physics-1980, L. Durand and L.G. Pondrom ed. (American Institute of Physics, New York, 1981), p. 664.
14. A.A. Zholentz et al. (VEPP4), Phys. Lett. 96B, 214 (1980).
15. J.L. Richardson, Phys. Lett. 82B, 272 (1979).
16. W. Buchmüller, G. Grunberg, and S.-H.H. Tye, Phys. Rev. Lett. 45, 103 (1980).
17. A. Martin, High Energy Physics-1980, L. Durand and L.G. Pondrom ed. (American Institute of Physics, New York, 1981), p. 715.
18. R. VanRoyen and V.F. Weisskopf, Nuovo Cimento 50A, 617 (1967).
19. R. Barbieri et al., Phys. Lett. 57B, 455 (1975); W. Celmaster, Phys. D19, 1517 (1979).
20. C. Berger et al. (PLUTO), preprint DESY 80/117.
21. C. Berger et al. (PLUTO), Z. Phys. C1, 343 (1979).
22. C. Berger et al. (PLUTO), Phys. Lett. 93B, 497 (1980).

23. H. Albrecht et al. (DASP2), Phys. Lett. 93B, 500 (1980).
24. B. Niczyporuk et al.(LENA), Phys. Rev. Lett. 46, 92 (1981).
25. B. Niczyporuk et al.(LENA), Phys. Lett. 100B, 95 (1981).
26. J.J. Mueller et al.(CLEO), Phys. Rev. Lett. 46, 1181 (1981).
27. G. Mageras et al. (CUSB), Phys. Rev. Lett. 46, 1115 (1981).
28. T. Appelquist and H.D. Politzer, Phys. Rev. D12, 1404 (1975); R. Barbieri, R. Gatto, and R. Kögerler, Phys. Lett. 60B, 183 (1976); R. Barbieri, R. Gatto, and E. Remiddi, Phys. Lett. 61B, 465 (1976).
29. P. Mackenzie and G.P. Lepage (private communication).
30. H. Albrecht et al. (DASP2), preprint DESY 81-011.
31. B. Andersson, G. Gustafson, and T. Sjöstrand, Lund preprint LU TP 81-3 (1981).
32. T.-M. Yan, Phys. Rev. D22, 21 (1980); K. Gottfried, Phys. Rev. Lett., 40, 598 (1978).
33. E. Eichten et al., Phys. Rev. D17, 3090 (1978) and D21, 203 (1980).
34. E. Eichten, Harvard preprint HUTP-80/A027 (1980).
35. S. Ono, Aachen preprint PITHA 80/2 (1980).
36. L.J. Spencer et al. (CUSB), Phys. Rev. Lett. (in press).
37. C. Bebek et al. (CLEO), Phys. Rev. Lett. 46, 84 (1981).
38. K. Chadwick et al. (CLEO), Phys. Rev. Lett. 46, 88 (1981).
39. G.H. Trilling, High Energy Physics-1980, L. Durand and L.G. Pondrom ed. (American Institute of Physics, New York, 1981), p. 1139.
40. J. Ellis, M.K. Gaillard, D.V. Nanopoulos, and S. Rudaz, Nucl Phys. B131, 285 (1977).
41. J.P. Leveille, U. Mich. preprint UMHE81-18; V. Barger, J.P. Leveille, and P.M. Stevenson, High Energy Physics-1980, L. Durand and L.G. Pondrom ed. (American Institute of Physics, New York, 1981), p. 390.
42. E. Derman, Phys. Rev. D19, 133 (1979).
43. H. Georgi and S.L. Glashow, Nucl. Phys. B167, 173 (1980).
44. V. Barger and S. Pakvasa, Phys. Lett. 81B, 195 (1979).
45. M. Peskin, Cornell internal report CBX 80-65.
46. M. Peskin and S.-H.H. Tye, Cornell preprint CLNS 81-485, p. 405; V. Barger, W.Y. Keung, and R.J.N. Phillips, Wisconsin preprint DOE-ER/000881-195 (1981).
47. Particle Data Group, Rev. Mod. Phys. 52, S1 (1980).
48. R. Shrock, S.B. Treiman, and L.L. Wang, Phys. Rev. Lett. 42, 1589 (1979).
49. S.-H.H. Tye, Cornell internal report CBX 81-9. S. Pakvasa, S.F. Tuan, and J.J. Sakurai, Phys. Rev. D23, 2799 (1981).
50. W. Bartel et al. (JADE), Z. Phy. C6, 295 (1980); S. Yamada, High Energy Physics-1980, L. Durand and L.G. Pondron ed. (American Institute of Physics, New York, 1981), p. 616.
51. W. Buchmüller and S.-H.H. Tye, Phys. Rev. Lett. 44, 850 (1980).
52. M. Basile et al., preprint CERN-EP/81-38 (1981).

DISCUSSIONS

CHAIRMAN: K. BERKELMAN

Scientific Secretary: H. Yamamoto

DISCUSSION 1

- TELLER:

The potential forms you showed were for quark anti-quark state and nothing more is involved?

- BERKELMAN:

Yes, that is the potential between the quark and anti-quark that binds them to form upsilons or psions.

- ETIM:

You said that the QCD inspired potential is strikingly successful, but it is not completely true. It involves complete readjustment of the value of the mass parameter Λ. The fit was for $\Lambda = 500$ MeV, but experimentarily it goes down to 100 or 150 MeV.

- BERKELMAN:

What is claimed is that one can get acceptable fits for any Λ between 100 MeV and 500 MeV. Of course, this is not a good way to measure the value of Λ. It would be a better test if we discovered toponium states and measured the mass difference there, since that system is the one that samples shorter distances.

- WIGNER:

Why do the SPEAR data not agree with the theoretical prediction?

- BERKELMAN:

That is because one is using QCD at low energy where it probably is unreliable.

- SEIBERG:

What is the number of free parameters used in the fits of Martin and Richardson ?

- BERKELMAN:

If one does not count the quark mass, Martin has two and in Richardson no free parameter, once the value of Λ is determined.

- ETIM:

I would like to comment on the formula you have been using for the leptonic width. On this side of the Atlantic, the so-called van Royen-Weisskopf formula is considered to be wrong. The M squared in the denominator should be four times the mass of the quark squared, not the mass of the vector meson squared.

- BERKELMAN:

When one derives the three gluon decay width, the same factor occurs, so the ratio is unchanged.

DISCUSSION 2   (Scientific Secretary: H. Yamamoto)

- ETIM:

In the theoretical formula for the B meson mass, what were the values of quark masses used ?

- BERKELMAN:

The masses I used are from calculations with Coulomb plus linear potential - (see E. Eichten et al., Phys. Rev. $\underline{D\ 17}$, 3090 (1978)).

- ETIM:

Do you know the leptonic width of Υ (4S) ?

- BERKELMAN:

The width of the decay into e^+e^- is published (see D. Andrews et al., Phys. Rev. Lett. **45**,219 (1980) and G. Finocchiaro et al., Phys. Rev. Lett. **45**,222 (1980) and its value is 0.24 ± 0.4 keV.

- YAMAMOTO:

How do you eliminate γ - conversion electrons from the signal, and what is the detection efficiency of the electrons ?

- BERKELMAN:

Many cuts are done in the analysis. First of all, you have to get only one particle into any Čerenkov counter, only one electron identifier module hit, in order not to get confused by overlap. Then you have to check that the track moving toward the shower counter joins perfectly with the shower. This minimises overlap of pions with gammas which convert. Finally, you have to extract, by scanning, each one of the pair of electrons that you get by γ - conversion, say in the beam pipe or in some other part of the detector. Then the best way to convince yourself that the electrons you are seeing are genuine is to look at the inclusive electron cross-section versus beam energy. There is only a very small background below the T (4S) peak.

- KAPTANOGLU:

In the ratio of the rates of the B meson decay to the u quark and to the c quark, $\frac{\Gamma(b \to u)}{\Gamma(b \to c)}$, don't you need to know also sin δ and not only cos δ, since $|\cos \delta| = 1$ and the cos δ is changing very slowly in that region?

- BERKELMAN:

For this ratio we need only the sign of cos δ. Experimentally, sin δ is near zero, which means that δ is close to 0° or 180°.

- D'ALI:

What is the matrix element used for the simulation of the semi-leptonic B meson decay ?

- BERKELMAN:

I don't know the details, but I know that the shape of the

spectrum is dominated by Feynman and Field fragmentation technology.

- YAMAMOTO:

Do you use Čerenkov counters to identify the charged kaons ?

- BERKELMAN:

The charged kaon identification I showed you was done primarily by time of flight. In the CLEO detector, parts of the solid angle covered by time-of-flight scintillators are also covered by Čerenkov counters; dE/dx chambers cover other portions of that solid angle. In fact all the Čerenkov counters have now been replaced by dE/dx chambers. In all the cases in which one makes a redundant identification, both with time of flight and in a Čerenkov counter or dE/dx chamber, there was consistency - but the Čerenkov identification was not required.

HADRON PRODUCTION IN e^+e^- ANNIHILATION

R. Cashmore

Nuclear Physics Laboratory
University of Oxford
Oxford, OX1 3RH, England

1. INTRODUCTION

There are two major processes leading to hadronic production in e^+e^- collisions.

(i) The single photon(γ) annihilation process of Fig.1.1a in which the cross-section is given by

$$\sigma_{tot} = R\,\sigma_{\mu\mu} = R\,\frac{4}{3}\,\frac{\pi\alpha^2}{s} \qquad (1.1)$$

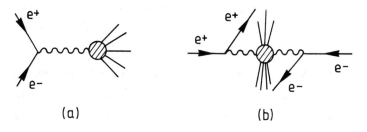

Fig.1.1(a) The single photon annihilation process
     (b) The two photon ($\gamma\gamma$) process

where s is the square of the c.m. energy i.e. $s=(2E)^2$.

(ii) The two photon($\gamma\gamma$) process of Fig.1.1b in which the cross-section is given by[1]

$$\frac{d\sigma(ee \to eex)}{dW} = \frac{dL}{dz} \sigma_{\gamma\gamma}(z) \qquad (1.2)$$

where $\frac{dL}{dz} \sim \left[\ln\left(\frac{E}{m}\right)\right]^2 \left[\ln \frac{E}{W}\right]$

and W is the $\gamma\gamma$ cm energy, $Z = \frac{W}{2E}$.

These two processes are easily separated since (i) leads to an hadronic final state with total energy ~2E and a large proportion of this is in charged particles while the energy associated with (ii) tends to be much lower. This separation can be seen in Fig.1.2. Of course as the energy is raised the $\gamma\gamma$ process will become steadily more important, particularly when LEP energies are reached. However $\gamma\gamma$ physics is only just emerging from its infancy and I only discuss the single photon annihilation, particularly at the high energies now available with the present storage rings.

A good working model exists for the description of hadron production[2],[3],[4],[5],[6]. It contains the following ingredients;

(i) the hadrons are the result of quark pair production together with the possible radiation of gluons (the quanta of QCD)
$$e^+e^- \to q\bar{q} + q\bar{q}g + \ldots$$
(ii) the quarks are spin 1/2 point like particles which have 3 colours (R,B,G) and are at present observed in 5 varieties or flavours (u,d,c,s,b). The production cross-section is given by
$$\sigma_{q_i\bar{q}_i} \propto Q_i^2$$

Thus quarks are produced "democratically" in e^+e^- collisions.

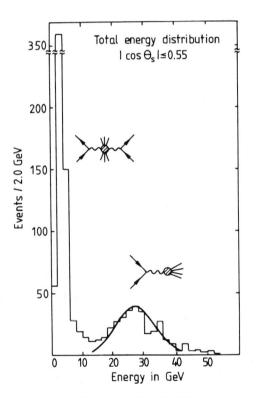

Fig.1.2 Total energy deposited in the TASSO detector for W~30 GeV data. The separation of one photon and two photon processes is clear.

(iii) the gluons have spin 1 and are radiated by the quarks with strength $\alpha_s \sim 0.1\text{-}0.2$ so that higher order radiation is progressively less likely.

(iv) α_s is the running coupling constant of QCD. However

$$\alpha_s = \frac{12\pi}{(33-2n_f)} \frac{1}{\ln Q^2/\Lambda^2} = \frac{12\pi}{23} \frac{1}{\ln Q^2/\Lambda^2}$$

and with $\Lambda \sim 0.1\text{-}0.2$ GeV it doesn't "run very far"!

(v) The quarks and gluons, which carry colour and hence are confined, fragment and result in jets of hadrons.

In the following I will review the basis of this model, particularly paying attention to the areas of difficulty, by considering

(i) The total annihilation cross-section.
(ii) General features of the data - multiplicity and neutral energy fractions.
(iii) Quark jets and their fragmentation properties.
(iv) Multijet events, gluons and QCD.

2. THE TOTAL ANNIHILATION CROSS-SECTION, σ_{TOT}, AND R.

The experimental detectors surround intersection points at e^+e^- storage rings and detect both charged and neutral particles, to different degrees depending on the different detectors. They therefore measure the rate of production of multi-hadrons and hence the total cross-section. Since this is such an important measurement I will first review the details of how σ_{tot} is extracted, paying particular attention to the sizes of errors, and then discuss its interpretation. Rather than giving σ_{tot}, the quantity R is usually displayed, where $R = \sigma_{tot}/\sigma_{\mu^+\mu^-}$, since this removes the major $1/s$ dependence of the cross section.

2.1 Measurement of σ_{tot}, R and limitations to accuracy

The number of events observed in a detector, N_{obs} is given by

$$N_{obs} = \sigma_{tot}\, L\, \varepsilon_{trig}\, \varepsilon_{acc} \qquad (2.1.1)$$

where L is the luminosity, ε_{trig} the probability of triggering the apparatus and ε_{acc} the probability of accepting the event after the cuts that have to be applied. In general the least restrictive trigger is applied eg one (or two) charged track or small energy deposit, but of course the less restrictive the trigger the higher the likelihood of backgrounds and hence the greater the need for selective cuts in defining the true annihilation sample (as distinct from $\gamma\gamma$ induced events for example). In order to have the greatest confidence in the measurements the solid angle covered by the detector should be large. This has two effects (i) more events are "seen" ie triggered on and accepted, (ii) the events recorded are better understood since more of the particles are observed, and accounts for the fact that at PETRA and PEP almost all of the detectors cover nearly the 4π solid angle.

The measurements at high energies are more easily made and hence more reliable for a variety of reasons.

(i) The multiplicities of particles are higher so that triggers based on particle multiplicity are less likely to be biassed.

(ii) The angular distribution of the hadrons at lower energies is mainly isotropic whereas at the higher energies they are associated with jets which have a well defined $1+\cos^2\theta$ distribution. A more reliable model of the production is possible making the corrections correspondingly more reliable. The production of new quarks would lead to isotropic distributions but the multiplicity would also be high.

(iii) The subtraction of backgrounds due to

$$e^+e^- \to \tau^+\tau^-$$
$$\gamma\gamma \to hadrons$$

is easier because the topologies of the events are very distinct.

The luminosity is measured from small angle Bhabha scattering ie

$$e^+e^- \to e^+e^-$$

for which the cross-section is well known and is large. Then

$$N_{ee} = L\, \sigma_{ee}$$

However this cross-section has a rather fierce dependence on angle, $\alpha\, \frac{1}{\theta^4}$, and there are systematic uncertainties which are associated with this. The geometry of the detectors and the magnetic fields transversed by the electrons have to be accurately known since a small change in the apparent θ will lead to a much larger change in the derived luminosity. Moreover radiative corrections have to be included and these require a good knowledge of the performance of the detectors.

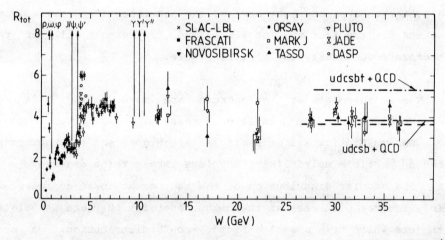

Fig.2.1 The value of $R(= \sigma_{had}/\sigma_{\mu\mu})$ as a function of W, the e^+e^- C.M. energy.

Despite all these difficulties the cross-sections have been measured and the corresponding values of $R^{(7),(8)}$ are shown in Fig.2.1 over the whole energy range. At low energies (<6-7 GeV)

HADRON PRODUCTION IN e⁺e⁻ ANNIHILATION

the various experiments (not all shown) differ by ~10-20% which reflects the problemss discussed above as well as uncertainties introduced by radiative corrections (discussed below). At high energies (>12 GeV) the agreement is much better.

As this is such an important measurement it is worth listing the factors affecting the accuracy one might expect for $\Delta\sigma_{tot}/\sigma_{tot}$.

(a) Statistics: $\dfrac{\Delta\sigma_{tot}}{\sigma_{tot}} = \dfrac{1}{\sqrt{N}}$

where N is the total number of hadronic events. Thus to obtain a 1% measurement, 10^4 events are required and it should be borne in mind that ~2 years of PETRA operation have only produced approximately that number.

(b) Luminosity: Here the statistics are satisfactory from Bhabha scattering but the systematic uncertainties dominate. Very careful design indeed is needed to reduce them to ~1-2%.

(c) Trigger and Acceptance: The variation of details of the models leads to ~(3-4)% effects in the trigger and geometrical acceptances.

If only relative measurements of the cross-sections are studied then over a reasonable energy range (b) and (c) will vary slowly and hence not be so important. However all three contribute to an absolute measurement so that we can only expect

$$\dfrac{\Delta\sigma_{tot}}{\sigma_{tot}} \geqslant 3\%$$

and this will require very careful experimentation indeed.

2.2 σ_{tot}, R and Radiative Corrections

In e⁺e⁻ annihilation the processes of Fig.2.2(a) are <u>never</u>

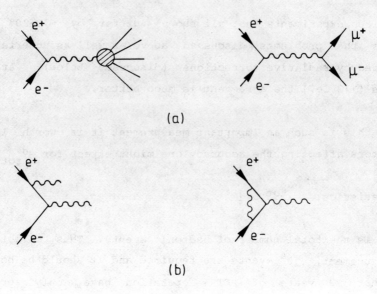

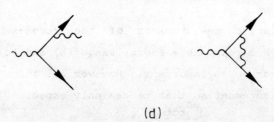

Fig.2.2(a) Lowest order em production of hadrons and $\mu^+\mu^-$
(b) Initial state radiation (real and virtual)
(c) Vacuum polarization contribution to photon propagator
(d) Final state radiation (real and virtual)

measured. The radiative effects of Figs.2.2(b), 2.2(c) and 2.2(d) are <u>always</u> present and must be accounted for in any experiment.

HADRON PRODUCTION IN e⁺e⁻ ANNIHILATION

They include

(i)   Initial state radiation

(ii) Vacuum polarization

(iii) Final state radiation

The most important of these processes is the initial state radiation, since the electrons are the lightest participants and hence the most likely to radiate. When this radiation does occur (Fig.2.2b) the effective e^+e^- centre of mass energy is lower so that the cross-section measured $\sigma(s)$ is given by an expression of the type

$$\sigma(s) = \int \sigma(s') f(s,s') ds' \qquad (2.2.1)$$

where $f(s,s')$ accounts for these "bremsstrahlung" type processes. These radiative corrections can lead to dramatic changes in the analysis of resonances ie where $\sigma(s')$ is varying rapidly close to the energy, $\sqrt{s}$, at which the measurement is being made. An example of the data at the $\psi''(3772)$ before and after radiative correction[9] is shown in Fig.2.3. Structures can easily be enhanced by this process and this must be a partial explanation of the discrepancies between experiments at low energy (4-6 GeV region).

The corrections fall into a number of categories[10].

(a) Initial state soft radiation and vacuum polarization effects:

$$\sigma_{meas} = \sigma(s) \left\{ 1 + \delta_A + \delta_\mu + \delta_\tau + \delta_{hads} \right\} \qquad (2.2.2)$$

where

$$\delta_A = \beta \ln \frac{k}{E} + \frac{2\alpha}{\pi} \left[\frac{13}{12} \ln \frac{s}{m_e^2} + \frac{\pi^2}{6} - \frac{14}{9} \right] \qquad (2.2.3)$$

$$\beta = \frac{2\alpha}{\pi} \left[\ln \frac{s}{m_e^2} - 1 \right]$$

and accounts for initial state effects including soft small angle bremstrahlung and the electron loop contribution to the vacuum polarization.

$$\delta_{\mu,\tau} = \frac{2\alpha}{\pi}\left[\frac{1}{3}\ln\frac{s}{m^2_{\mu,\tau}} - \frac{5}{9}\right] \qquad (2.2.4)$$

which accounts for μ and τ loop contributions to the vacuum polarization.

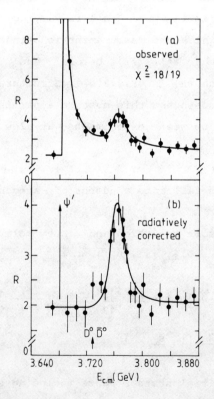

Fig.2.3 The $\psi''(3772)$ before (a) and after (b) radiative corrections.

HADRON PRODUCTION IN e⁺e⁻ ANNIHILATION

$$\delta_{hads} = \frac{s}{2\pi^2\alpha} \int_{4m_\pi^2}^{\infty} \frac{\sigma(s')}{s-s'} ds' \qquad (2.2.5)$$

which is the hadronic contribution to the vacuum polarization. These lead to 10-15% effects.

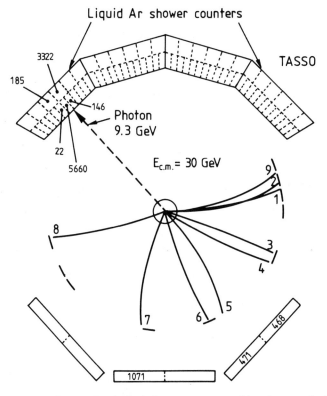

Fig.2.4 An example of initial state radiation giving a hard photon.

(b) Hard initial state radiation:

The results above (Eqn.2.2.2) only account for soft photon ($E_\gamma < k$) emission. Hard photon processes occur as shown in Fig.2.4

and must be taken into account[10]. An approximate formula is

$$\frac{d\sigma}{dxd\Omega} = \frac{2\alpha}{\pi} \sigma(\sqrt{s}) \left\{ (1-\frac{x}{2})^2 + \frac{1}{4} x^2 y^2 \right\} / x(1-y^2) \qquad (2.2.6)$$

where $x = \frac{E_\gamma}{E}$ $y = \cos\theta_\gamma$ and $\theta_\gamma \gtrsim 30°$
and leads to corrections of ~2%.

(c) Final state radiation:

The effects of final state radiation can be included in Eqn.2.2.2 by the addition of a further term

$$\delta_{FS} = \frac{3\alpha}{4\pi} \qquad (2.2.7)$$

The contribution is small $\sim \frac{1}{500}$ ie 0.2% and so can normally be ignored. However it is interesting to note that when we consider gluon radiation from quarks α is replaced by $\frac{4\alpha_s}{3}$ and the correction factor becomes $\frac{\alpha_s}{\pi}$ which is much larger ($\alpha_s \approx 0.1-0.2$) and leads to the observable effects.

In the case of Bhabha scattering large corrections also have to be made for the final state radiation and this is of course important in measuring the luminosity.

The final result quoted, R, is given by

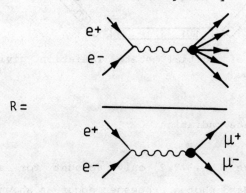

the ratio of the lowest order QED diagrams, the values for both of which have to be derived from the experimental results, since they are never directly measured.

In practice Monte Carlo programmes[10] are used to generate the reactions

$$e^+e^- \to q\bar{q}$$
$$\to q\bar{q}\gamma$$

in the correct proportions and the response of the detector evaluated. The loop corrections are then made.

Finally the reliability of these corrections might be questioned. The many theoretical calculations have converged to a final description but of course the best test is the comparison with some experimental distribution. In Fig.2.5 the acollinearity angle in

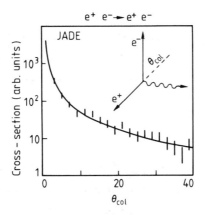

Fig.2.5 The acollinearity angle in $e^+e^- \to e^+e^-$. The curve is the expectation including radiative effects.

$$e^+e^- \to e^+e^- (+\gamma)$$

is shown. This acollinearity is due to these radiative effects. The agreement between data and theory is gratifyingly good and gives some confidence in these corrections.

2.3 Results and Interpretation

Before a detailed description of σ_{tot} (or R) is embarked upon many general comments and deductions can be drawn.

(a) General Discussion

The results for R are presented in Fig.2.1. The expectation (in lowest order) is that

$$R_o = 3 \sum_f Q_f^2 = \sum_f R_f \qquad (2.3.1)$$

where the factor 3 is due to colour and the sum is over all quark flavours. Thus we expect

(i) $E_{cm} < 3$ $R \sim 2$, due to u, d and s quark excitation
(ii) $E_{cm} > 4$ $R \sim 10/3$, due to the addition of the c quark
(iii) $E_{cm} > 11$ $R \sim 11/3$, due to the presence of the b quark.

Inspection of Fig.2.1 shows that these expectations are qualitatively correct and of course in the region of a new quark threshold we find the associated bound states, the "onia" eg ψ, ψ', ψ''[11] near charm threshold, $\Upsilon, \Upsilon', \Upsilon'', \Upsilon'''$ near b threshold[12].

From the high energy data (in Fig.2.1) it is quite clear that there is no evidence for the continuum excitation of a new quark of charge 2/3. (The same conclusion is reached by looking for the spherical topology associated with such events. More detailed investigation also rules out continuum excitation of a charge 1/3

HADRON PRODUCTION IN e⁺e⁻ ANNIHILATION

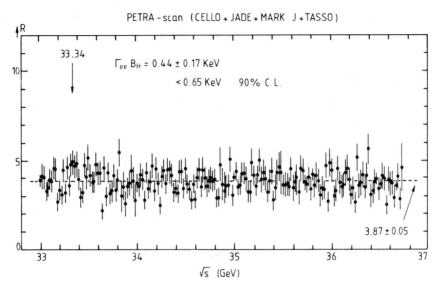

Fig.2.6 The value of R in the range 33<W<36.72. No narrow resonances are apparent in this scan.

quark). Of course it is always possible that the current highest energies might lie in the region of the bound states of new quarks so that a fine scan has been performed searching for narrow states. The results are shown in Fig.2.6 in the region up to 36.72 GeV. There is no obvious $Q\bar{Q}$ bound state and a limit can be set of

$$B_H \Gamma_{ee} < 0.65 \text{ KeV} \qquad (2.3.2)$$

which rules out charge 2/3 quarks (~5 KeV) and charge 1/3 quarks (~1.2 KeV).

Detailed inspection shows that R_o falls below the experimental results. However we do expect a 1st order QCD correction[13] of

$$R = R_o \left\{ 1 + \frac{\alpha_s}{\pi} \right\} \qquad (2.3.3)$$

and with $\alpha_s \sim 0.15$ this gives a reasonable account of the observations.

Thus we conclude that five varieties of quark (u,d,c,s,b) are necessary together with the QCD corrections.

(b) Detailed Description of R

In this section I want to discuss the many effects which can contribute to a description of R at the few percent level. We have already seen one of these, the QCD correction (2.3.3) which is ~5%. Some are more mundane while others have equally important physics content.

(i) Kinematic factors:

If new heavy quarks are produced then initially their velocity is low and eqn.2.3.1 has to be modified

$$R'_o = 3 \sum_f Q_f^2 \frac{\beta_f}{2} \{3 - \beta_f^2\} \qquad (2.3.4)$$

where β_f is the quark velocity. These effects are summarized in Table 2.3.1. Clearly the effects are appreciable and must be considered in any careful analysis of R. Often this slow rise to the asymptotic value is disguised by the presence of resonances near to a threshold.

Table 2.3.1 The Kinematic Factor $\frac{\beta}{2}(3-\beta^2)$

M_Q \ E_b	10	15	20
5	.974	.995	.998
10		.910	.97
12.5		.745	.93
15.0			.85
17.5			.67

(ii) First Order QCD:

The correction due to first order QCD is

$$R = R'_o \left\{1 + \frac{\alpha_s}{\pi}\right\} \qquad (2.3.5)$$

and for $\alpha_s \sim 0.17$ this gives a 5.4% increase. Clearly it is difficult to measure α_s, given the large error bars on the data, the resulting value being $\alpha_s = 0.22 \pm .23$. One of the major preoccupations of QCD experiments is to observe the varying nature of α_s, as expected for the running coupling constant[6]. If $\alpha_s = 0.17$ and the first order QCD formula is used (with five flavours of quark)

$$\alpha_s = \frac{12\pi}{23} \frac{1}{\ln(Q^2/\Lambda^2)}$$

Λ has a value ~ 0.24 GeV which leads to the variation of α_s with energy ($S = E^2_{cm}$) summarized in Table 2.3.2.

Table 2.3.2 Variations of α_s with S
($\alpha_s \sim 0.17$ at $S = 1300$ Gev^2)

S	α_s
100	.22
400	.19
900	.17
1600	.16
10000	.14

The changes are small, $\Delta R \lesssim 1\%$, which results in their being essentially unobservable. Thus $\alpha_s \sim .17$ is consistent with the data (which requires absolute cross-section measurement) and we have little hope of seeing any variation.

Before leaving this topic I should point out that 2.3.5 is only an approximation ignoring velocity dependent factors. In fact R_f (the value of R for a given fermion f) can be written as

$$R_f = 3 \frac{\beta_f}{2} (3-\beta_f^2) Q_f^2 \left\{1 + \frac{4}{3} \alpha_s F(\beta_f)\right\} \qquad (2.3.6)$$

where

$$F(\beta) = \frac{\pi}{2\beta} - \frac{3+\beta}{4} \left(\frac{\pi}{2} - \frac{3}{4\pi}\right) \qquad (2.3.7)$$

and this is actually only an interpolation function of a more complicated result[14]. This change can again introduce ~5% effects for different quarks in addition to those already accounted for in Eqn.2.3.5.

Clearly this whole business of kinematical factors requires great care and circumspection.

(iii) Second Order QCD:

The second order QCD formula for R, without velocity factors, is[15]

$$R = R_o \left\{1 + \frac{\alpha_s}{\pi} + C\left(\frac{\alpha_s}{\pi}\right)^2 + \right\} \qquad (2.3.8)$$

The coefficient C depends on the calculational scheme, 1.4 in the $\overline{MS}$ scheme and -1.6 in the MOM scheme[16]. The effects are small ~1/2% and are far less than even optimistic hopes for the measurement errors, thus making any tests impossible.

(iv) Weak Interactions:

At PETRA and PEP energies the weak interaction effects become appreciable and can change the total hadronic cross-section by ~ few %. The quarks are produced via the neutral weak current, the Z, as shown in Fig.2.7 and this contribution has to be added to the electromagnetic production. The resulting expression for R_f is[17]

HADRON PRODUCTION IN e⁺e⁻ ANNIHILATION

Fig.2.7 Quark production via the electromagnetic (γ) and weak (Z) currents.

$$R_f = Q_f^2 - 2sQ_f \, v_e v_f \, g \, \frac{M_Z^4(S-M_Z^2)}{(S-M_Z^2)^2+M_Z^2\Gamma^2} + s^2(v_e^2+a_e^2)(v_f^2+a_f^2)g^2 \, \frac{M_Z^4}{(S-M_Z^2)^2+M_Z^2\Gamma^2}$$

(2.3.9)

where

$$v_e = -1 + 4\sin^2\theta_W \qquad a_e = -1$$

$$v_u = v_c = +1 - \frac{8}{3}\sin^2\theta_W \qquad a_u = a_c = +1$$

$$v_d = v_s = v_b = -1 + \frac{4}{3}\sin^2\theta_W \qquad a_d = a_s = a_b = -1$$

and

$$g = \frac{G_F\sqrt{2}}{4(4\pi\alpha)}$$

The first term in 2.3.9 is due to the em production, the second the interference of em and weak amplitudes and the third due to pure weak production. If $\sin^2\theta_W \sim .25$ [18], as measured in other reactions, then the pure weak term can become important. The value of R for different s values is shown in Fig.2.8 and the effects clearly grow with s. A few % changes are apparent. Indeed, to reproduce R, $\sin^2\theta_W$ must be ~0.25 or 0.60 and certainly cannot take extreme values. A value of 0.6 is ruled out by the leptonic data ($e^+e^- \to e^+e^-$).

If α_s is allowed to run, this produces a change of ~1.5% in the cross section, approximately 1/2 of the weak interaction effect at these energies. Assuming a value for $\sin^2\theta_W$ and including all the other effects described earlier a fit for α_s leads to a value

$$\alpha_s = 0.26 \pm .17$$

However the number of small effects and the errors on the data make it a little early to conclude the presence of each and everyone of these effects (ie running α_s and $\sin^2\theta_W$). Higher energies and more data will be of great benefit.

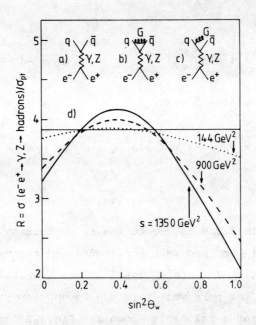

Fig.2.8 The value of R, with the inclusion of weak neutral current effects (Z), as a function of $\sin^2\theta_W$. Graphs (a), (b) and (c) contribute to the final result (d). $\sigma_{pt} = \frac{4}{3}\frac{\pi\alpha^2}{S}$.

(v) Form factors and substructure:

If the quarks have substructure then we would expect the predictions to be modified by the presence of a form factor

$$f(s) = \frac{1}{1+\frac{r^2 s}{6}} \qquad (2.3.10)$$

The approximate constancy of R implies that the quarks are indeed point like (like the μ) and thus a limit can be set to any quark radius of

$$f \lesssim 2 \times 10^{-16} \text{ cms}$$

Furthermore if constituent substructure of the quarks were to be present then at sufficiently high values of s (Q^2) we would expect dramatic variations (rises, dips, thresholds ...) before the value of R settled down to that associated with the constituents of the quarks.

(vi) The Problems of Low Energy:

As remarked earlier there are discrepancies between total cross-section measurements at low energy. Despite this all cross-sections seem to lie somewhat above the theoretical expectations. An example is given in Fig.2.9 together with the predictions from u,d,c, and s quarks together with QCD modifications. Any explanation in terms of new charge 1/3 quarks and/or heavy leptons are ruled out by PETRA data (unless they have form factors which cause suppression at the higher energies). Thus a problem remains which may have its origin in the experimental measurements.

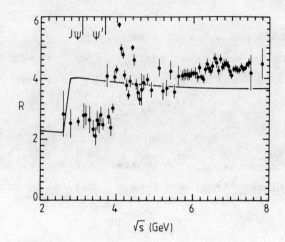

Fig.2.9 The value of R in the low energy region showing the data lying above the theoretical expectation.

2.4 Summary

The measurement of σ_{tot} is an important task but it will be difficult to reduce the error in an absolute value to <3% or <2% in relative measurements between different energies. This is a pity because many interesting phenomena appear at this level

1st order QCD	~(5-8)%
α_s running	~ 1%
2nd order QCD	~ 1/2%

and their observation would be clean tests of QCD. Even the weak interaction lead to changes ~3% when $\sin^2\theta \sim 0.25$, which would be larger for almost any other value of $\sin^2\theta$. In fact to do a detailed analysis, checking all these gauge theory predictions requires the inclusion of all effects (particularly those which are kinematic in origin) and this will be an exciting area as the energies rise and the cross-sections become more reliable and precise.

However the present value of R does demonstrate that we need the five active quarks (u,d,c,s and b) together with QCD corrections and that these quarks are point like down to distances $\sim 10^{-16}$ cms. These are all vital ingredients in the model of hadron production.

3. GENERAL FEATURES OF THE HADRONIC FINAL STATE

In this section I will briefly discuss some general features of the hadronic final state, namely the multiplicity and neutral energy fractions.

3.1 Charged Particle Multiplicity

The mean charged multiplicity observed in the hadronic final state[19] is shown in Fig.3.1. There is clearly a dramatic rise from low energies to PETRA energies. This large multiplicity means, of course, that triggers based on comparitively low charged particle multiplicities (eg $\geqslant 4$) will be very reliable. Extrapolations from low energy of the form

$$\langle n \rangle = a + b \ln s \qquad (3.1.1)$$

clearly fail. Such a form for multiplicity would imply a constant rapidity plateau, which we shall see is not observed. QCD motivated formulae of the form[20]

$$\langle n \rangle = a + b \exp\left[c(\ln s/\Lambda^2)^{1/2}\right] \qquad (3.1.2)$$

have also been fitted to the data leading to values of

$a = 2.92 \pm .04 \qquad b = 2.85 \pm .07 \qquad c = .0029 \pm .0005$

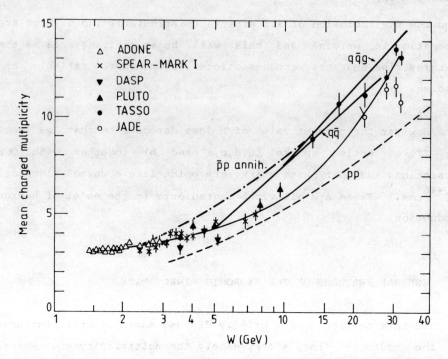

Fig.3.1 The mean charged particle multiplicity as a function of W. Expectations from $q\bar{q}$ and $q\bar{q}g$ models are shown as well as a "QCD" inspired parameterization. Also shown are the multiplicities in pp and $\bar{p}p$ interactions.

However this cannot be regarded as any test of QCD - there are no predictions of a, b or c - and the formula itself is only true at the parton level ie fragmentation is not taken into account. It is just another extrapolation function.

In Fig.3.1 are included the predictions of the Field Feynman (FF)[21] model, for $q\bar{q}$ production and $q\bar{q}g$ production. It describes the data quite well (in principle $q\bar{q}$ leads to a constant rapidity plateau!) and we must conclude that phase space effects are still important even at PETRA! Finally the results fall between the multiplicities observed in $\bar{p}p$ and pp annihilation[22]. In fact if leading protons are subtracted from the pp data, ie the cm energy of the remaining hadrons is much lower and the multiplicity reduced, then the agreement is somewhat closer.

3.2 Energy Fractions

The fraction of energy carried away by γ-rays is shown[23] in Fig.3.2. There is clearly little, if any, change from the lowest to the highest energies, even when measurements are made on the Υ and Υ'. In Fig.3.3 the energy carried by neutral particles and γ rays is compared and once again little change is seen over the whole PETRA range.

3.3 Summary

There is a large increase in multiplicity at PETRA energies. The constancy of the energy fractions and the Monte Carlo models indicate that no new dramatic physics is occurring. Rather there is just a steady development from low energies as more phase space becomes available. Thus we might expect to see the features of these models most clearly at the high energies now available.

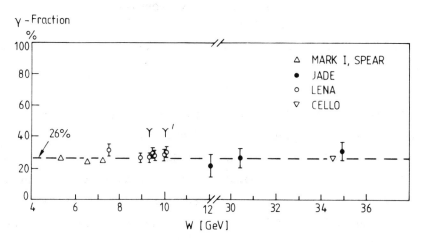

Fig.3.2 The fraction of energy carried by γ-rays as a function of cm energy (W).

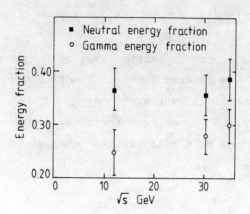

Fig.3.3 The neutral energy fraction as a function of cm energy ($\sqrt{s}$).

4. QUARK JETS AND THEIR FRAGMENTATION PROPERTIES

We expect the major production of hadrons to be via the process

$$e^+e^- \to q\bar{q}$$

since a value of $\alpha_s \sim 0.1$-0.2 implies that "hard" gluon bremsstrahlung only occurs in ~5% of events. Thus 2 jet events of the type shown in Fig.4.1 dominate the data and provide a unique set of quark jets with which to study fragmentation. In this section I will ignore, in general, the fact that gluon bremsstrahlung occurs. This smaller, though vital process, will be discussed in great detail in section 5.

HADRON PRODUCTION IN e⁺e⁻ ANNIHILATION   533

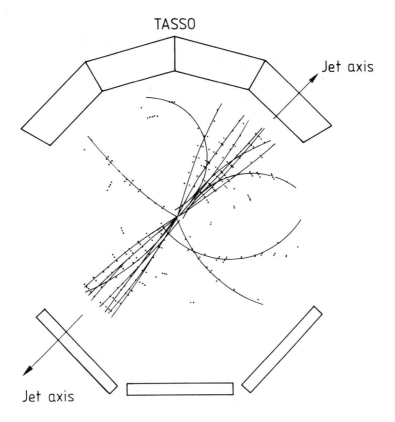

Fig.4.1   A typical 2 jet event at high energies.

4.1   Jets and Jet Measures

The supposition that hadrons were being produced in jets was first quantitatively supported at SPEAR[24]. However the jets were not as obvious as those of Fig.4.1 because the energy was low which automatically implies a low multiplicity and hence difficulty in identifying the two preferred jet directions. None the less the data was best described by a model in which the p_T was limited with respect to some direction (the quark direction). If we assume the p_T is limited as indicated in Fig.4.2 then we can calculate the mean cone angle $<\delta>$ as the e^+e^- cm energy, N, is varied

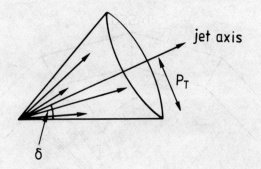

Fig.4.2 Cone angle δ and limited transverse momentum in a jet.

$$\langle\delta\rangle \sim \frac{\langle p_T\rangle}{\langle p_L\rangle} \sim \frac{\langle p_T\rangle}{W}\langle n\rangle \sim \frac{\ln W}{W} \sim \frac{1}{W} \qquad (4.1.1)$$

$\langle p_L\rangle$ is the mean longitudinal momentum and $\langle n\rangle$ is the mean multiplicity, which varies only slowly with W. Most of the momentum resides in the longitudinal component. Thus we expect the jets to become more conspicuous at higher energies, an experimental fact which is indicated in Fig.4.1

Various quantitative measures have been introduced the commonest being sphericity[2] and thrust[25]:

<u>Sphericity</u>:
$$S = \frac{3}{2}\frac{\sum_i p_{T_i}^2}{\sum_i p_i^2} \qquad 0 \leqslant S \leqslant 1 \qquad (4.1.2)$$

p_{T_i} are the transverse momenta of the particles with respect to that axis which minimizes S.

<u>Thrust</u>:
$$T = \frac{\sum |p_{Li}|}{\sum p_i} \qquad 0.5 \leqslant T \leqslant 1 \qquad (4.1.3)$$

p_{Li} are the components of the particle momenta parallel to that axis which maximizes T.

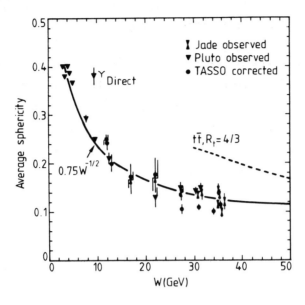

Fig.4.3 The average sphericity, S, as a function of c.m. energy W.

Very crudely we can write

$$S \sim \frac{3}{2} \langle \delta^2 \rangle \quad \text{and} \quad T \sim 1 - \frac{\langle \delta^2 \rangle}{2} \quad (4.1.4)$$

and we expect S to decrease with W and T to increase.

The variation of S is shown in Fig.4.3, this decrease being apparent. At the Υ the events are not well described by 2 jet production (the Υ is expected to decay to 3g) and there is an increase in S. However off resonance S returns to its standard behaviour. We can also see that the effect of a top quark would be dramatic and its presence easily recognised.

4.2 Jet Angular Distribution

The directions of the jet axes are vital in developing our model of e^+e^- annihilation. Up to this point there is no direct evidence that they are associated with quarks (other than the total cross-section rate). If the jets are indeed associated with partons then angular distribution of the jet axis has a specific form depending on the original parton eg

Spin 0 $\qquad\qquad\dfrac{dN}{d\cos\theta} \alpha \sin^2\theta$ \qquad\qquad (4.2.1)

Spin 1/2 $\qquad\qquad\dfrac{dN}{d\cos\theta} \alpha\ 1 + \cos^2\theta$ \qquad\qquad (4.2.2)

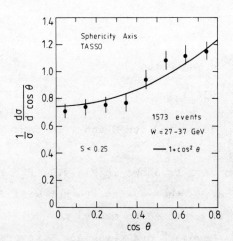

Fig.4.4 The angular distribution of the sphericity axis. The selection, S<0.25, gives predominantly 2 jet events.

HADRON PRODUCTION IN e⁺e⁻ ANNIHILATION

The latter is the angular distribution familiar to us in $\mu^+\mu^-$ production, these being spin 1/2 fermions.

In Fig.4.4 we see the experimental distribution of the jet axes at high energies, the sphericity cut having been made to ensure a clean sample of 2 jet events. It is clearly consistent with $1+\cos^2\theta$ and we are forced to conclude that the hadron production is not random but distributed in just the manner we would expect if the reaction is via primary spin 1/2 partons which subsequently fragment.

4.3 Charge Correlations and Charged Primary Partons (Quarks)

If we pursue this model of hadron production through partons (ie quark-antiquark pairs) as in Fig.4.5 then even after fragmentation we might expect there to be a correlation between the charge in one jet and in the other. After all the original quark and antiquark are oppositely charged. If we choose the jet axis then we can describe the longitudinal distribution of hadrons by their rapidity (as in Fig.4.5).

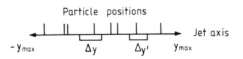

Fig.4.5 Parton (quark-antiquark) production and the rapidity distribution (y) in an event. The intervals Δy and $\Delta y'$ are indicated.

$$y = \ln \frac{E+p_L}{E-p_L} = 2 \ln(E+p_L) - \ln M_t^2 \qquad (4.3.1)$$

where $M_T^2 = M^2 + p_T^2$ and $y_{max} \sim \ln s$, and then search for these correlations by defining a charge correlation function $\Phi(y,y')$ by

$$\Phi(y,y') = -\frac{1}{\Delta y\, \Delta y'} \left\langle \frac{1}{n} \sum_{k=i}^{n} \sum_{i \neq k} e_i(y) e_k(y') \right\rangle \qquad (4.3.2)$$

$$\tilde{\Phi}(y,y') = \Phi(y,y') / \int \Phi(y,y')\, dy \qquad (4.3.3)$$

where $e_i(y)$ is the charge of particle i with rapidity y. For a given y and y' only those particles in rapidity intervals Δy and $\Delta y'$ are included in the calculation. The negative sign in 4.3.2 is included to give a positive quantity. For orientation we can consider the value of this function $\tilde{\Phi}$ for $e^+e^- \to e^+e^-$. In that case $y_1 = y_{max}$ $y_2 = -y_{max}$ and we have

$$\tilde{\Phi}(-y_{max}, y_{max}) = 1$$
$$\tilde{\Phi}(-y_{max}, y) = 0 \quad \text{otherwise}$$

The results[26] are shown in Fig.4.6, where the test particle has been chosen in the indicated y' interval and the compensation of its charge studied as a function of y. A short range correlation is immediately obvious, with a σ of ~1-1.5 units of rapidity. Similar correlations have been observed at the ISR[27] and to some extent represent the effects of resonance (eg ρ°, ω°) decays. However the important feature in Fig.4.6 is the gradual development of a statistically significant long range correlation as y' is increased indicating that the charge at high y' is compensated by particles at the opposite extreme of rapidity (eg $\int \tilde{\Phi} dy = .154 \pm .026$ for $-2.5 > y' > -5.5$ and $y > 1$). This is just what one would expect if the high rapidity particles carry some information of the primary quark (antiquark) charge, the correlation being induced because of the opposite $q\bar{q}$ charges. In Fig.4.6 are the

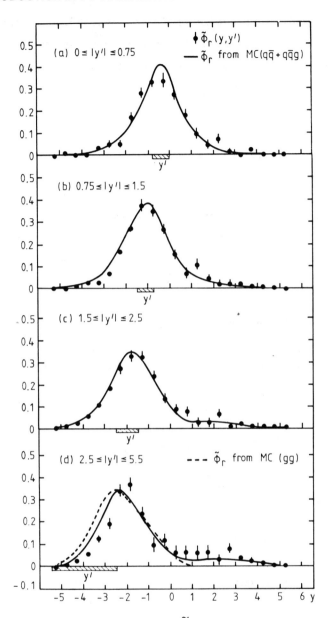

Fig.4.6 The distribution function $\tilde{\phi}$ for different y' intervals. The existence of a long tail in (d) is apparent.

predictions of the $q\bar{q}$ and $q\bar{q}g$ Monte Carlo which are in qualitative agreement with the data, as well as the predictions from an artificial model in which the partons are given zero-charge. In this case no long range correlation exists. Various other tests[26] have demonstrated that these results are not artifacts of the analyses and we are forced to conclude that the <u>primary partons are indeed charged</u>.

4.4 Quark Fragmentation

In parts (a), (b) and (c) the general properties of quark fragmentation are summarized and confronted with models of the process in 4.5. We concentrate on quantities such as rapidity, total momentum, particle composition which we hope will best reflect the process of fragmentation.

(a) Rapidity Distributions

The rapidity y has been defined in 4.3.1. It can also be written

$$y = \ln \frac{E+p_{11}}{E-p_{11}} = \frac{1}{2} \left\{ \ln \frac{E+p_{11}}{2P_b} - \ln \frac{m_T}{2P_b} \right\} \quad (4.4.1)$$

$$\sim \frac{1}{2} \left\{ \ln Z - \ln \frac{m_T}{2P_b} \right\} \quad (4.4.2)$$

where $Z = \frac{E}{E_b}$ and $E_b(P_b)$ is the energy (momentum) of the electron beam. The parallel momentum components are measured with respect to the sphericity axis.

From Eqn.4.4.2 we can deduce that

$$dy \propto \frac{dZ}{Z} \propto \frac{dE}{E} \quad (4.4.3)$$

This means that if we have a bremstrahlung like spectrum then we

HADRON PRODUCTION IN e⁺e⁻ ANNIHILATION

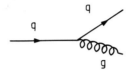

Fig.4.7 Gluon bremsstrahlung by quarks.

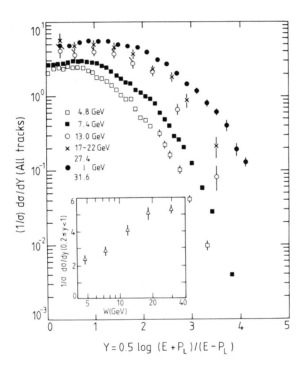

Fig.4.8 The rapidity distribution of charged tracks and the rise of the "central plateau" as a function of W.

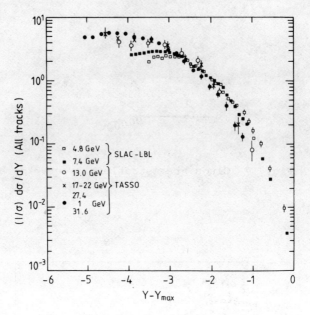

Fig.4.9 The rapidity distribution measured with respect to y_{max}.

will have a constant rapidity distribution, ie a so called rapidity plateau, and a logarithmically increasing multiplicity (since $y_{max} \propto \ln W$). A bremstrahlung like spectrum would result if quark hadronization proceeded via gluon emission as indicated in Fig.4.7.

The results are shown[28] in Fig.4.8, where distributions from a wide range of energies are compared. There is an apparent central plateau at each energy but this is certainly not constant with energy. A similar rate of rise has been observed in the central region of pp collisions at ISR energies[28]. It is also interesting to note that if the rapidity distributions are plotted as a function of $y-y_{max}$, as in Fig.4.9, then they are clearly very

HADRON PRODUCTION IN e⁺e⁻ ANNIHILATION

similar near y_{max} at all energies. This suggests that the leading particles reflect the quark properties at all energies and the increasing multiplicities are associated with production in the central region.

Thus although some of the expected properties are apparent in the rapidity distribution, the rising plateau and the correspondingly more rapid increase in multiplicity, indicate that we are still far from the asymptotic region.

(b) Inclusive Particle Spectra

In general the hadron inclusive spectrum in e^+e^- annihilation can be written[29] (in analogy with deep inelastic scattering) as

$$\frac{d^2\sigma}{dx\,d\Omega} = \frac{\alpha^2}{s}\beta\, x\{mW_1 + \frac{1}{4} x\, \beta^2\, \nu W_2\, \sin^2\theta\} \quad (4.4.4)$$

where W_1 and W_2 are in principle functions of x and s (Q^2 in deep inelastic scattering) and $x = \frac{E}{E_b}$. Thus

and

$$s\frac{d\sigma}{dx} = 4\pi\alpha^2\, \beta x\{mW_1 + \frac{1}{6} x^\nu W_2\} \quad (4.4.5)$$

$$\frac{s}{\beta}\frac{d\sigma}{dx} = 4\pi\alpha^2\{F_1(x,s) + F_2(x,s)\} \quad (4.4.6)$$

β is the velocity of the particle.

Quite frequently the charged particle is not identified (ie β not known) and an approximate expression is employed.

$$s\frac{d\sigma}{dx_p} = 4\pi\alpha^2\{F_1(x,s) + F_2(x,s)\} \quad (4.4.7)$$

where $x_p = P/E_b = P/P_b$. $F_1(x,s)$ and $F_2(x,s)$ are in general functions of x and s.

Within the quark parton model the hadron inclusive distribution is particularly simple

$$\frac{d\sigma}{dx}(e^+e^- \to h) = \sum_q \sigma_{q\bar{q}} \{D_q^h(x) + D_{\bar{q}}^h(x)\} \qquad (4.4.8)$$

while QCD would imply that the fragmentation functions are also functions of s.

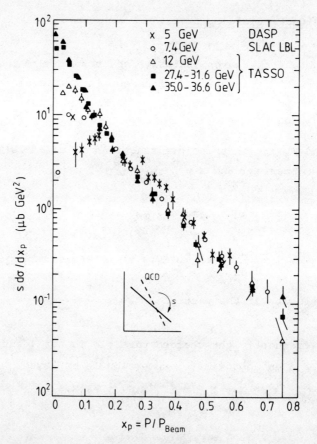

Fig.4.10 The inclusive (x_p) distribution as a function of $W(=\sqrt{s})$ showing approximate scaling. The QCD expectation.

HADRON PRODUCTION IN e⁺e⁻ ANNIHILATION

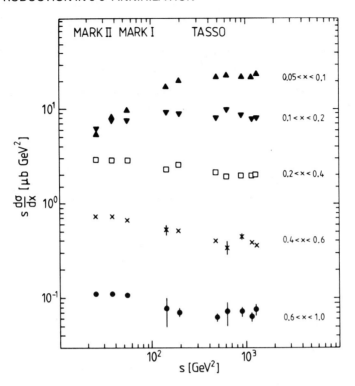

Fig.4.11 The inclusive cross-sections $\frac{d\sigma}{dx_p}$ plotted with respect to E_{cm}.

The results are shown[30] in Fig.4.10, data being used from low energies for comparison. Two features are immediately apparent. The increase in multiplicity is associated with the dramatic rise in low x_p particles and there is approximate scaling for $x \gtrsim .25$. The expectation of QCD is indicated in Fig.4.10 although the variations are small ($\lesssim$10-20% at high x). The errors clearly make any conclusions difficult and any hope of seeing scaling violation will clearly rely on the use of low energy data. In Fig.4.11 recent data[31] are plotted as $\frac{d\sigma}{dx_p}$ against E_{CM}, and some tentative indications of such scaling violations are apparent. However definitive conclusions cannot be drawn because (i) the different particle species are not considered separately and this could induce dramatic effects (ii) the observation relies on the relative normalization of low and high energy experiments. A change of 10% in one could change ones view entirely.

Thus we are left with the general impression that scaling is approximately correct and the tantalizing thought that there may be evidence for violation. Only better data will discriminate.

(c) <u>Particle Composition</u>

It is important to measure the particle composition in e^+e^- annihilation for a variety of reasons eg (i) to test QCD and scaling (ii) to see if the flavour of a quark can be identified (iii) to understand the flavour structure of the "sea". It will be the major experimental evidence necessary for a complete discription of hadronisation.

Charged particle identification is mainly accomplished at low energies by TOF (time of flight) techniques, and only now are Cerenkov and dE/dx measurements beginning to give results for higher momenta particles. Unfortunately the low data rates at high x conspire with the difficulty of identification so that large data samples are necessary before reasonable measurements can be made. Thus this interesting area of physics (although a small part of the cross-section) cannot be well discussed. We would of course like to know the particle composition at high x, which will surely reveal details of the primary quarks, and there are tentative indications that the K and proton fluxes are large[32]. However we concentrate on low x (low momentum particles) and investigate these higher momenta regions with K°'s and Λ°'s.

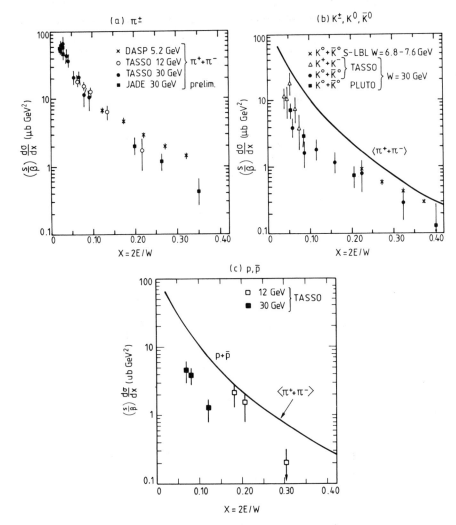

Fig.4.12 The inclusive distributions for (a) $\pi^{\pm}$ (b) $K^{\pm}$, K^0 and $\bar{K}^0$ (c) p and $\bar{p}$.

In Figs.4.12(a),(b) and (c) are the $\pi^{\pm}$, $K^{\pm}$, K^0 and proton inclusive distributions and in Fig.4.13 recent measurements of the π^0 inclusive distribution. We observe that the π distributions approximately scale, although there is marginal evidence for some violation. The $K^{\pm}$ and K^0 distributions are consistent and once again show approximate scaling, and the proton rate is surprisingly

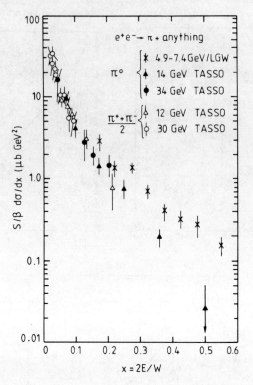

Fig.4.13 The π^0 inclusive distribution compared to $\pi^\pm$.

large. In fact these measurements were the first indication that baryons are important in quark fragmentation. In Fig.4.14 we see the Λ and $\bar{\Lambda}$ inclusive distribution, together with that for K^0's, the measurements reaching to larger values of x. In Fig.4.15 are presented the corresponding values of R_{K^0} and $R_{\Lambda,\bar{\Lambda}}$. Little evidence exists on other particle production rates although D^0 production has been estimated[33] and appears to have a somewhat harder momentum spectrum than that of the other pseudoscalar mesons. Wide states such as vector mesons, eg the ρ^0, are

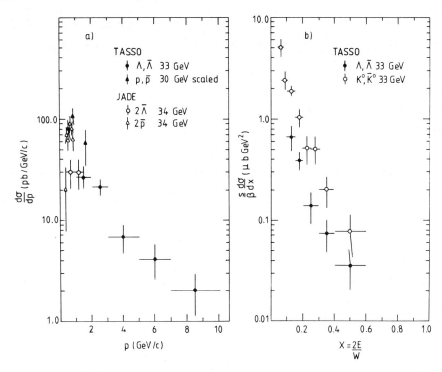

Fig.4.14 The $\Lambda,\bar{\Lambda}$ inclusive distributions compared with $p,\bar{p}$ and $K^°,\bar{K}^°$.

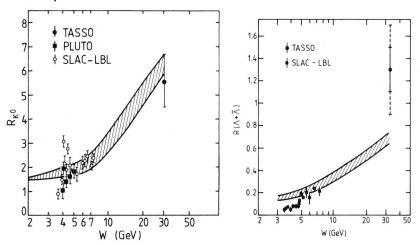

Fig.4.15 $R(K^°+\bar{K}^°)$ and $R(\Lambda+\bar{\Lambda})$ as a function of W. The shaded region is the increase that might be expected from the increased charged particle multiplicity and constant particle fractions.

difficult to observe because of large combinatorial backgrounds, with probably the K^* being the best candidate for measurement.

None the less some interesting observations can be made from these results. For a given x particles are produced in the approximate ratios

$$\pi^\pm : K^\pm : K^0, \bar{K}^0 : p, \bar{p} : \Lambda, \bar{\Lambda} = 2.5 : 1 : 1 : \sim 1 : \sim 0.3$$

and the spectra seem to be very similar in shape (see for example the Λ and K^0 distributions). Indeed the K^0 and Λ $\frac{s}{\beta}\frac{d\sigma}{dx}$ are reasonably described by a curve e^{-7x} over the region $.15 < x < .5$. The increase in K^0 multiplicity from low energies is consistent with the charged particle multiplicity increase, even though there is now b quark production. Most of the charged and neutral K's are in fact produced from the sea, rather than the primary quark. The baryon production is however greater than would be expected although this may be due partly to phase space effects. More data and even higher energies would help in this respect. The multiplicity of various particles (integrated over all particle momenta) at ~30 GeV is approximately: ~13 charged particles, ~1.4 $K^0(\bar{K}^0)$, ~1.4 $K^\pm$, ~0.3 Λ ($\bar{\Lambda}$), $\gtrsim 0.4$ p ($\bar{p}$). If protons ($\bar{p}$) from $\Lambda(\bar{\Lambda})$ decays are subtracted from the proton signal then the total baryon rate can be estimated. If we assume n production is equivalent to p production then we find that $\gtrsim 0.7$ baryon (antibaryons) are produced per event ie $\gtrsim \frac{1}{3}$ of events contain a baryon antibaryon pair.

4.5 <u>Models of Quark Fragmentation</u>

The models of quark fragmentation are at present rather crude[5],[21]. They are little more than phenomenological parametrerization of the data, although as the data improves the

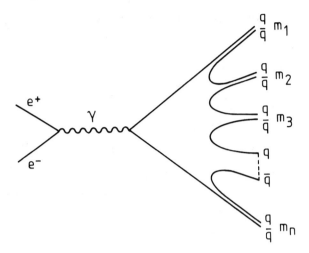

Fig.4.16 Meson production in the Field-Feynman model.

sophistication of the description does increase and the assumptions acquire some degree of support. The most popular of these parameterizations is the Field-Feynman prescription for jet formation, supplemented where necessary by QCD to describe gluon radiation. Probably the best physical insight to the process occurs through string models, the commonest version being that due to the LUND group[5].

The idea of the Field-Feynman model is indicated in Fig.4.16. Quark-antiquark pairs are produced from the vacuum at each stage of the fragmentation, the quark (or antiquark) combined with the initial antiquark (quark) to form a meson leaving the antiquark (quark) to repeat the process. At each stage the fraction of energy left to the residual quark, after the meson has been formed, must be decided. The type of quarks produced, their transverse momenta and whether vector or pseudoscalar mesons are obtained must also be chosen. Recent results also require the addition of further parameters to describe baryon production. The parameters of the model are indicated in the following:

(a) Fragmentation function:

The fraction of the initial quark energy left to the residual quark, $\eta\left(\frac{Eq_1}{Eq}\right)$, is given by

$$f(\eta) = 1 - a_F + 3a_F \eta^2 \qquad (4.5.1)$$

If the fragmenting quark is u, d or s, a_F is obtained from fits to the data, a value ~0.57 being obtained. However for c and b quarks, the fragmentation function is believed to be harder and a_F is chosen to be 0 (ie the meson on average carries more of the quark momentum). This is not a well determined parameter at present, and if $a_F(c,b) = a_F(u,d,s)$ little change results.

$$u,d,s \qquad a_F \sim 0.57$$
$$c,b \qquad a_F = 0 \qquad (4.5.2)$$

(b) Quark types

The initial quarks are produced at rates proportional to their charges (u:c:d:s:b=4:4:1:1:1). Only u,d,s quarks are obtained from the vacuum in the ratio

$$u : d : s = 2 : 2 : 1 \qquad (4.5.3)$$

Production of c and b quarks is believed to be suppressed due to their greater masses, and hence these quarks are only obtained as leading quarks. The probabilities of 4.5.3 also indicate that K mesons will be more difficult to obtain and indeed these values lead to a reasonable representation of the K yield, although a slightly lower production of $s\bar{s}$ pairs is probably more likely. It must be stressed that the majority of K's are actually produced from the sea, and not from leading strange quarks or from the decays of charm and bottom mesons.

(c) Quark Transverse Momentum:

The transverse momentum of the quarks is chosen according to an exponential distribution in p_T^2 ie $e^{-p_T^2/2\sigma_q^2}$ and

$$\sigma_q \sim 0.32 \text{ GeV} \qquad (4.5.4)$$

(d) Pseudoscalar/Vector Meson Production

When a quark and antiquark are combined a pseudoscalar or vector meson can result. The ratio, $\frac{P}{P+V}$, is a free parameter and

$$\frac{P}{P+V} \sim 0.56 \qquad (4.5.5)$$

The mesons are then allowed to decay according to the data tables or models for charm and bottom meson decay[5].

These parameters are obtained from fits to the data, principally from the sphericity distribution, the aplanarity (see section 5), the inclusive spectrum and the charged multiplicity. In Fig.4.17 are the fits to the data at W=30 GeV together with predictions at W=12 GeV. The model clearly gives a good representation. The parameters tend to be determined principally by different features of the data. σ_q by the aplanarity, a_F by the inclusive spectrum and $P/_{P+V}$ by the multiplicity.

(e) Baryon Production

The experimental data demonstrate that baryon production is important and must be included in any model[34],[35]. That baryons are produced is not so surprising as we can see in Fig.4.18. The antiquark terminates a colour field eg $\bar{R}$ antiquark is produced to terminate the field due to the R quark. However the field could equally easily be terminated by the production of a green quark and a blue quark ($\underline{3} \times \underline{3} = \underline{\bar{3}} + \underline{6}$) and we would have a baryon. Clearly we expect suppression, at least due to the extra masses involved ($M_{qq} \gtrsim M_s$), and an x distribution somewhat similar to that of the

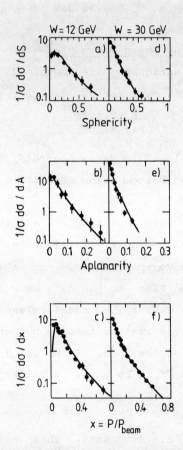

Fig.4.17 Fits using the Field-Feynman model to the sphericity, aplanarity and inclusive (x_p) distributions at 30 GeV together with predictions at 12 GeV.

HADRON PRODUCTION IN e⁺e⁻ ANNIHILATION

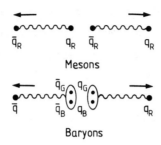

Fig.4.18 Meson and Baryon production.

K^o's, just as we observe in Fig.4.14. Baryon production can be introduced into the Feynman Field prescription[34] as indicated in Fig.4.18, PB1 and PB2, being the relative probabilities of these two processes and the meson production. Actually

$$PB1 = \frac{P(qq)}{P(q)+P(qq)} \qquad (4.5.6)$$

within the same jet. In Fig.4.20 the fits to the baryon production are shown. There is little support for production of the type of Fig.4.19(b), as we might guess. In this model the quarks from the sea, forming the baryon, are still produced in the ratio u:d:s=2:2:1. Also included are curves from the LUND model which clearly underestimate the Λ and $\bar{\Lambda}$ production rates.

It is also worth noting that the extension of the Field Feynman model[34] also predicts copious Ξ production, distinctly different from the LUND model[35] which predicts little Ξ production. Clearly baryon production will reveal further details of the fragmentation process.

Before leaving this subject of quark fragmentation models it is worth making a number of remarks:

(i) At present the u:d:s ratio is not really measured. The K° spectra will be a valuable tool in finally giving these values.

(ii) The $P/_{P+V}$ ratio is mainly determined from the multiplicity, which is totally unsatisfactory. A direct measurement of vector mesons is preferable. The ρ° is hard because of its width and the large combinatorial background in any $\pi^+\pi^-$ spectrum. The K^*,

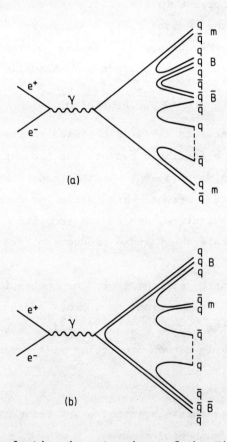

Fig.4.19 Baryon production in extensions of the Field-Feynman model.

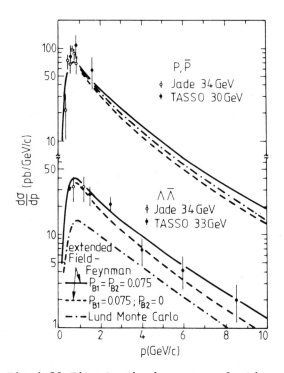

Fig.4.20 Fits to the baryon production.

particularly observed in its $K^0\pi$ decay mode, is more useful but the statistics tend to be small. Thus a combination of results on these two states will probably produce the first reliable results on this ratio.

(iii) Essentially nothing is measured of the c and b quark fragmentation functions. D and B mesons are difficult to observe at high energies and so direct measurements are unlikely. First indications will probably come from detailed study of the prompt lepton spectrum associated with the D and B meson decays.

(iv) In baryon production Σ and Ξ will give valuable new insights. Evidence for decuplet baryon production (Δ, Σ^* etc) would be the equivalent of vector meson production.

(v) $b\bar{b}$ production is important to reproduce both the sphericity distributions and the inclusive μ rates[36].

(vi) Having a good description of fragmentation is clearly important when we turn to the study of gluons and $q\bar{q}g$. After all hadronisation is presumably a soft version of the process which leads to the emission of hard gluons.

Fragmentation studies are a fascinating topic and will eventually lead to an idea of how quarks (and antiquarks) are produced in the colour fields which exist, and become confined in hadrons.

(f) Gluon production

In quantitative fits to the data gluon emission and hadronisation have to be included[5] (and are discussed at length in section 5). Briefly the gluons are produced according to QCD bremsstrahlung and then allowed to fragment. Many models for the fragmentation are used, varying from the assumption that the gluon behaves like a quark or antiquark (ie a very asymmetric splitting of energy in $g \to q\bar{q}$) to giving the gluon a fragmentation function of the form

$$z^2 + (1-z)^2 \qquad (4.5.7)$$

In fitting the data these different forms have been used, for completeness, but make little quantitative difference. After all the majority of the data (~90%) is due to just two jets.

4.6 Summary

From the discussion of this section we clearly see that

HADRON PRODUCTION IN e⁺e⁻ ANNIHILATION

(i)   The primary partons have spin 1/2

(ii) The primary partons are charged

(iii) These primary partons are the five, u,d,s,c and b, quarks produced at rates $\propto q^2$ (the quark charge squared)

(iv) The process of hadronisation produces mesons and baryons within jets, which reflect the original quark production. A detailed study of these particles must lead to a better understanding of the colour fields and the process of confinement.

All of these points are essential components of the model of hadron production in e^+e^- annihilation.

5. MULTIJETS, GLUONS AND QCD

In section 4 I have discussed quark fragmentation at some length only making vague references to the radiation of gluons and multijet events. In this section I will consider the evidence for gluon production and study in detail the properties of the gluon.

5.1 Initial Evidence for multijet events and gluon radiation

As the centre of mass energy is increased a tail becomes apparent in the p_T^2 distribution of particles, the p_T being measured with respect to the 2 jet axis[37]. This can be seen in Fig.5.1 and cannot be reproduced with the models of quark fragmentation described in the previous section (not even by increasing σ_q). The $\langle p_T \rangle$ and $\langle p_T^2 \rangle$ also increase with energy as shown in Fig.5.2 but studying these quantities is not the best way of looking for a tail in a distribution (they are heavily weighted by low p_T^2 events). The sheer existence of events with $p_T^2 \sim 5\text{-}10$ GeV^2 indicates the presence of some hard process. This was immediately interpreted as the existence of gluon bremsstrahlung[3],[4] as indicated in Fig.5.3. The strength of the radiation is expected to be proportional to α_s and on dimensional gounds

$$p_T \sim \alpha_s W \sim \frac{W}{\ln(W/\Lambda)} \qquad (5.1.1)$$

When $p_T \gg \sigma_q$ the gluon jet will be separated from the quark fragments and we have the opportunity of observing it. This clearly requires high energies, as we have at PETRA.

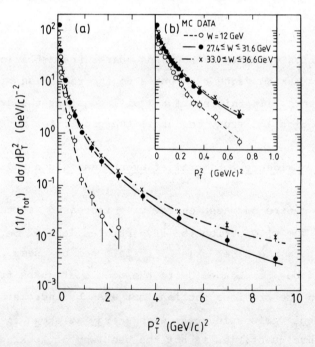

Fig.5.1 The p_T^2 distribution at different cm energies. p_T^2 is measured with respect to the sphericity axis and the development of a long tail is apparent.

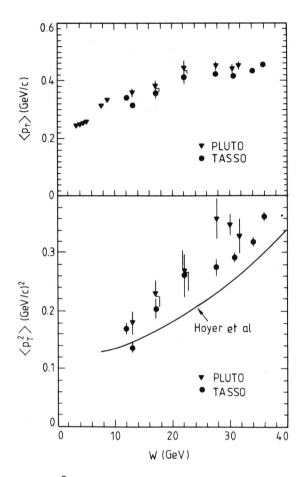

Fig.5.2 $\langle p_T \rangle$ and $\langle p_T^2 \rangle$ as a function of W. The prediction of Hoyer et al (Ref.(5)) is also shown.

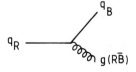

Fig.5.3 Gluon bremsstrahlung.

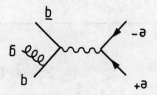

Fig.5.4 Gluon emission which leads to a 3-parton final state and hence planar events.

The emission of a gluon, as in Fig.5.4 will automatically lead to the existence of planar events, so that energy and momentum can be conserved. Thus to confirm the existence of the gluon radiation a search for these planar events has been made[38], followed by the observation of 3 jets within this plane.

To identify these events some measures have to be developed which will indicate this planar nature. Many different approaches have been followed but the most easily understood method is based on a development of sphericity measures[39]. A momentum tensor $M_{\alpha\beta}$ is defined as

$$M_{\alpha\beta} = \sum_{i,j=1}^{N} p_{i\alpha} p_{j\beta} \qquad (5.1.2)$$

where α,β are the x,y,z axes and the sum runs over all observed particles. This tensor is then diagonalized resulting in 3 eigenvectors, $\vec{n}_1, \vec{n}_2$ and $\vec{n}_3$, and three corresponding eigenvalues $\Lambda_1, \Lambda_2, \Lambda_3$. The value of Λ_i is then

$$\Lambda_i = \sum_{j=1}^{N} (\vec{p}_j \cdot \vec{n}_i)^2 \qquad (5.1.3)$$

and the eigenvalues are ordered such that $\Lambda_1 < \Lambda_2 < \Lambda_3$. $\vec{n}_3$ is then the old sphericity axis, $\vec{n}_2$ and $\vec{n}_3$ span the event plane (if this is an appropriate description) and $\vec{n}_1$ is normal to the event plane. Normalized eigenvalues can then be defined as

$$Q_i = \frac{\Lambda_i}{\Sigma_j p_j^2} \tag{5.1.4}$$

and

$$Q_1 + Q_2 + Q_3 = 1 \tag{5.1.5}$$

Roughly speaking Q_1 defines the "flatness" of the event, Q_2 the "width" of the event and Q_3 the "length" of the event. In terms of these eigenvalues the aplanarity and sphericity are

$$A = \frac{3}{2} Q_1 \tag{5.1.6}$$

$$S = \frac{3}{2}(Q_1+Q_2) = \frac{3}{2}(1-Q_3) \tag{5.1.7}$$

For orientation we expect 2 jets events to have A and S both small, 3 jet events to have S large but A small, and new quark events to have both S and A large. This is indicated in Fig.5.5 where we clearly observe small A high S events, which are incompatible with new quark production.

To further investigate the development of this planar character the distributions of $<p_T^2>_{in}$ and $<p_T^2>_{out}$ can be studied for the events, where these quantities are defined as

$$<p_T^2>_{out} = \frac{1}{N} \Sigma_j (\vec{p}_j \cdot \vec{n}_1)^2 = Q_1 <p^2> \tag{5.1.8}$$

$$<p_T^2>_{in} = \frac{1}{N} \Sigma_j (\vec{p}_j \cdot \vec{n}_2)^2 = Q_2 <p^2> \tag{5.1.9}$$

These distributions at low and high W are shown in Fig.5.6. Both quantities increase as W increases, but a long tail develops in the $<p_T^2>_{in}$ distribution. Furthermore $\frac{<p_T^2>_{out}}{<p_T^2>_{in}}$ shows a slight increase, whereas a decrease would be expected if the events remained cylindrical in topology (ie in 2 jet events at infinite energy $<p_T^2>_{out} \sim <p_T^2>_{in}$ because the problems of limited phase space would have been overcome). If a $q\bar{q}$ model is used in an attempt to "explain" the data σ_q has to be arbitrarily increased (~.45 GeV) to fit the p_T^2 distribution and it then fails to reproduce the long tail in the $\langle p_T^2 \rangle_{in}$ distribution, as shown in Fig.5.6.

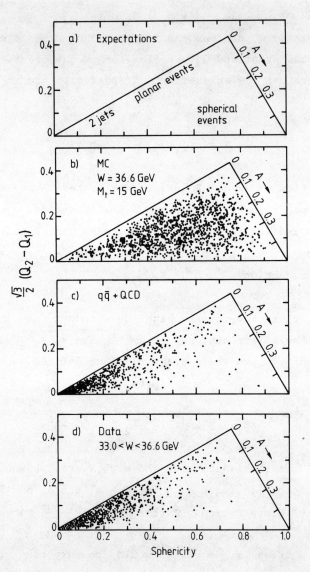

Fig.5.5 The triangle plots for data (d) and MC models of heavy quark production (b) and $q\bar{q}g$(c).

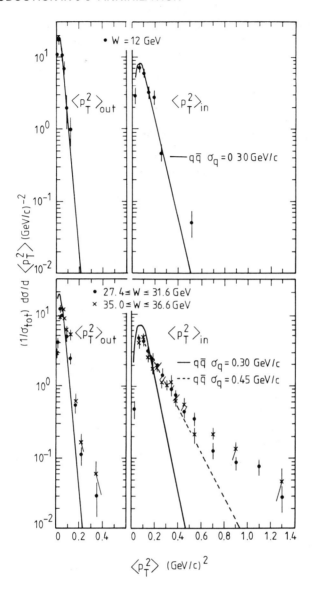

Fig.5.6 $\langle p_T^2 \rangle_{in}$ and $\langle p_T^2 \rangle_{out}$ distributions at different W together with $q\bar{q}$ MC predictions.

Finally if the topology within the event plane is studied there is clear eveidence for the existence of 3 jet events, as shown in Fig.5.7. In order to finally confirm this interpretation the more quantitative studies described in the following sections must be performed.

5.2 A Gluon Radiation Model

To first order in α_s the three graphs contributing to hadron production are indicated in Fig.5.8. The infra red divergences of Fig.5.8(b) are cancelled by the interference between the diagrams of Fig.5.8(a) and Fig.5.8(c). Straightforward calculation of the bremsstrahlung process for gluons lead to the following cross-sections[3],[4]

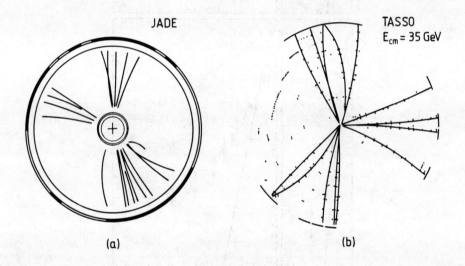

Fig.5.7 Examples of three-jet events observed in the JADE and TASSO detectors as viewed along the beam axis.

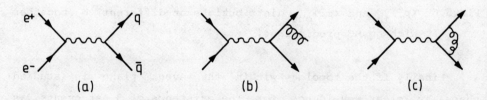

Fig.5.8 Graphs which lead to hadron production at 1st order in α_s.

Scalar:
$$\frac{d^2\sigma}{dx_1 dx_2}(q\bar{q}g) = \frac{\tilde{\alpha}_s}{3\pi} \sigma_0 \frac{x_3^2}{(1-x_1)(1-x_2)} \qquad (5.2.1)$$

Vector:
$$\frac{d^2\sigma}{dx_1 dx_2}(q\bar{q}g) = \frac{2\alpha_s}{3\pi} \sigma_0 \frac{x_1^2+x_2^2}{(1-x_1)(1-x_2)} \qquad (5.2.2)$$

leading to

$$\frac{d\sigma}{dT} = \frac{2\alpha_s}{3\pi} \sigma_0 \left\{ \frac{2(3T^2-3T-2)}{T(1-T)} \ln\left(\frac{2T-1}{1-T}\right) - \frac{3(3T-2)(2-T)}{1-T} \right\} \qquad (5.2.3)$$

The divergences in these formulae are apparent as $x_1, x_2 \to 1$ or T (the thrust) $\to 1$.

Quark and gluon fragmentation clearly complicates the issue, at some point the gluon radiation becoming indistinguishable from the hadronisation of quarks. Furthermore as we can see

$$\int_{2/3}^{1} \frac{d\sigma(q\bar{q}g)}{dT} dT \qquad (5.2.4)$$

diverges and thus we have to choose a cut off, T_0. In Fig.5.9 we show the thrust distribution for $q\bar{q}$ events, including fragmentation, a peak occuring at T ~0.96[40]. T_0 is thus chosen to be ~0.95 and beyond this value $q\bar{q}g$ events are indistinguishable from $q\bar{q}$ events.

The gluon fragmentation must also be included in any model, but this does not have any dramatic consequences. The gluon has

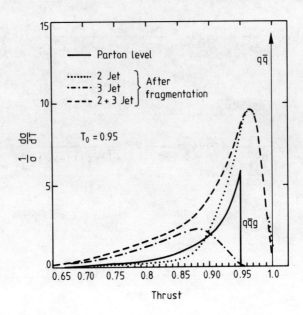

Fig.5.9 The thrust distribution from $q\bar{q}$ and $q\bar{q}g$ before and after parton fragmentation.

been allowed to hadronize either purely as a quark (or antiquark) or as a $q\bar{q}$ pair with the fractional energies chosen from a distribution

$$f(z) = z^2 + (1-z)^2 \qquad (5.2.5)$$

This is the QCD model, to first order in α_s, which has been used to describe the hadronic data, with α_s and the quark fragmentation properties (a_F, σ_q, $P/_{P+V}$) as parameters. The curves in Fig.5.1, the distribution in Fig.5.5 and the excellent fits to the $\langle p_T^2 \rangle$ distribution in Fig.5.10 are all obtained with this model.

A major component of all these arguments is the existence of tails in the p_T^2 distributions, when compared with models in which

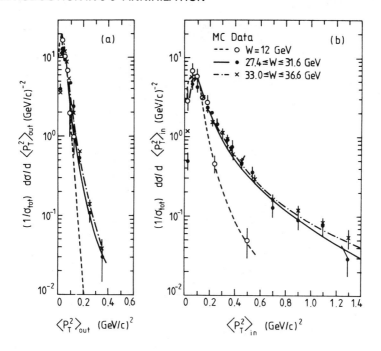

Fig.5.10 $\langle p_T^2 \rangle_{in}$ and $\langle p_T^2 \rangle_{out}$ distributions and the results of $q\bar{q}+q\bar{q}g$ Monte Carlo calculations.

the p_T^2 distribution is generated according to $e^{-p_T^2/2\sigma_q^2}$. An exponential in p_T would have a longer tail but fortunately it does not reproduce the data, as will become apparent in the next section.

5.3 Planar events, Identification of 3 jets and their properties

In Fig.5.7 we have seen topologies consistent with the existence of 3 jet events. Before identifying the three jets we can study the energy flow within the event plane. The Mark J data[41], for events with oblateness >0.3 (planar events) are shown in Fig.5.11 and the topology characteristic of 3 jets is once again

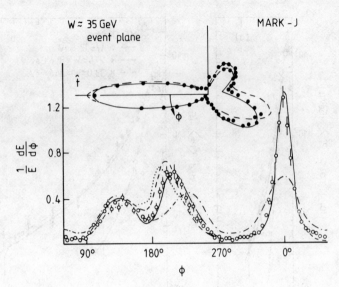

Fig.5.11 Energy flow within the plane for planar events defined by oblateness O>0.3. The curves show fits with a variety of models

(i) QCD calculation $\alpha_s = 0.18$ (———)
(ii) Two jet $q\bar{q}$ with Gaussian p_T (- - - -)
(iii) Two jet $q\bar{q}$ with Exponential p_T (·····)
(iv) Pure phase space (-·-·-·-)

$$O = \frac{\Sigma |\vec{E}_{\perp in}| - \Sigma |E_{\perp out}|}{E_{tot}}$$

apparent. The QCD model is clearly the most successful and the data is in disagreement with models based on $q\bar{q}$, phase space or $q\bar{q}$ with an exponential p_T ($\exp(-p_T/0.65)$) fragmentation function.

To identify the 3 jets a selection is first made to reject 2 jet events eg S>0.25. The three jets, of Fig.5.12, and their associated particles can then be determined by a variety of methods
(i)   Minimize p_T^2 with respect to 3 axes [39]
(ii) Maximize p_L with respect to 3 axes [25]
(iii) Arrange particles in clusters by virtue of cone angles and require the cones to be separated [42].

All these techniques have been used and the properties of the jets can then be studied.

(a) p_T^2:

In Fig.5.13 the p_T^2 of the particles with respect to their jet axes (in 3 jet events) are plotted and compared with p_T^2 in predominantly 2 jet events at 12 Gev [43]. The jets obviously have similar properties and emphasise that the procedure used does result in jets (in 3 jet events) of sensible properties.

(b) Multiplicity:

In Fig.5.14 we see the multiplicity of these jets [43], the jets being ordered according to energy. The multiplicity is always

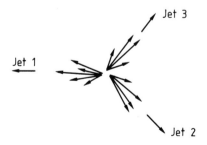

Fig.5.12 A 3-jet event and the jet axes.

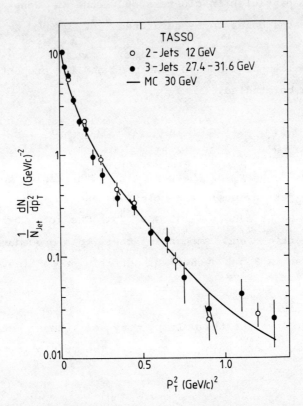

Fig.5.13 p_T^2 of particles with respect to their jet axes in 3 jet events at high energy compared with p_T^2 in events at W = 12 GeV (which are predominantly 2 jets).

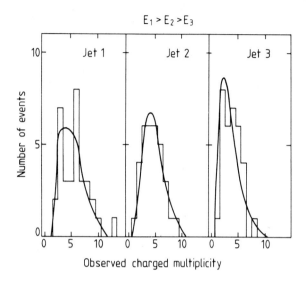

Fig.5.14 The observed charged multiplicity of each jet in 3 jet events.

appreciable, ruling out the possibility that the planar structure is created by single hard particles.

(c) <u>Thrust in "qg" system</u>:

If the 3 jet events are due to gluon bremsstrahlung then the two lowest energy jets are probably the radiating quark and the gluon and will be contained in the broader jet in a 2 jet classification. A Lorentz boost can then be made to the "quark gluon" centre of mass system and studies of the two jet properties made. In Fig.5.15 the thrust distribution[38] in this qg system is again compared with that of 2 jet events ($q\bar{q}$) at a C.M. energy of 12 GeV. The distributions are remarkably similar.

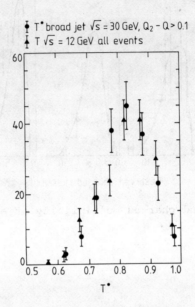

Fig.5.15 Distribution of thrust of the broad jet in planar events at W = 30 GeV compared with the two jet thrust distribution at 12 GeV as measured by the JADE group.

All of this analysis indicates that we are indeed dealing with genuine 3 jet systems and that the model $q\bar{q}g$ is indeed a reasonable representation of the data.

HADRON PRODUCTION IN e⁺e⁻ ANNIHILATION

(d) Gluon spin:

The distribution of $\tilde{\theta}$, defined in Fig.5.15, is valuable in demonstrating that the spin of the gluon is most likely 1. The expected angular distributions[44] are

spin 0, scalar: $\dfrac{d\sigma}{d\cos\tilde{\theta}} \propto 1 + 0.2 \cos^2 \tilde{\theta}$ (5.3.1)

spin 1, vector: $\dfrac{d\sigma}{d\cos\tilde{\theta}} \propto 1 + 2.0 \cos^2 \tilde{\theta}$ (5.3.2)

which are modified by the acceptance and selection criteria for the events. In Fig.5.16 the observed distribution[45] is compared with the predictions for spin 0 and spin 1 gluons, the latter clearly being favoured.

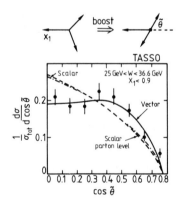

Fig.5.16 Distribution of $\cos\tilde{\theta}$ demonstrating that spin 1, vector, gluons are favoured.

From this discussion we conclude that three jet events exist, the jets have "sensible" properties and that a gluon spin of 1 is highly favoured.

5.4 Quantitative Discription and the value of α_s

Providing α_s is small then the major contributions to the cross-section are 2 and 3 jet production. Indeed measurements of

$$\frac{\sigma_{3\text{-jet}}}{\sigma_{2\text{-jet}}} \sim \frac{\sigma_{4\text{-jet}}}{\sigma_{3\text{-jet}}} \sim \alpha_s \qquad (5.4.1)$$

would lead directly to values of α_s. However, as remarked earlier, fragmentation confuses the identification of the jets and necessitates quantitative models being made.

Analyses of the data have been made in different manners by all the groups at PETRA, the analyses being based on the formalism of Hoyer et al, the Lund group or Ali et al[5]. The first two include only terms to order α_s ie 2- and 3-jet events only, while the model of Ali does include some order α_s^2 terms (but not all) and hence some 4-jet events.

The relative sizes are normalized such that

$$\sigma_2 + \sigma_3 + \sigma_4 = \sigma_0 \left[1 + \frac{\alpha_s}{\pi}\right] \qquad (5.4.2)$$

where σ_n is the n-jet cross-section (σ_4 only being present in the model of Ali et al). The value of α_s together with the fragmentation parameters are then varied to obtain good fits to the data.

Each group fits its own "favourite" distributions to arrive at a value for α_s. Studies are of course made of the interdependence of the various parameters, an example being given in Table 5.1

drawn from the TASSO analyses. Good fits to the data are always obtained (see earlier figures in this section and Fig.5.17) and result in the values of α_s indicated in Table 5.2. The resulting value is

$$\alpha_s \sim 0.17 \qquad (5.4.3)$$

Correlations certainly exist eg α_s and σ_q. From the MARK J analyses, Fig.5.11, it is clear an exponential does not fit the data. None the less any broadening of the fragmentation p_T distribution will clearly lead to a reduction in α_s.

Table 5.1

Fitted Values of α_s and σ_q for different input values of a_F and $P/_{P+V}$

a_F		P/(P + V)				
		0.1	0.3	0.5	0.7	0.9
0.1	α_s	0.17±0.03	0.17±0.03	0.17±0.03	0.16±0.03	0.16±0.03
	σ_q	0.44±0.11	0.46±0.10	0.47±0.10	0.48±0.09	0.48±0.08
0.3	α_s	0.17±0.04	0.16±0.04	0.16±0.04	0.15±0.04	0.15±0.04
	σ_q	0.42±0.12	0.44±0.11	0.46±0.10	0.47±0.09	0.48±0.08
0.5	α_s	0.17±0.04	0.16±0.04	0.16±0.04	0.15±0.04	0.14±0.04
	σ_q	0.35±0.12	0.38±0.12	0.41±0.12	0.43±0.10	0.44±0.09
0.7	α_s	0.17±0.03	0.17±0.04	0.16±0.04	0.15±0.05	0.14±0.05
	σ_q	0.28±0.09	0.30±0.10	0.33±0.10	0.36±0.10	0.39±0.09
0.9	α_s	0.17±0.03	0.17±0.04	0.16±0.04	0.15±0.04	0.14±0.05
	σ_q	0.21±0.08	0.23±0.08	0.26±0.08	0.30±0.09	0.33±0.08

Table 5.2

Values for α_s

Expt.	Reference	α_s
JADE	(46)	0.18 ± .03 ± .03
MARKJ	(47)	0.19 ± .02 ± .04
PLUTO	(48)	0.15 ± .03 ± .02
TASSO	(43)	0.17 ± .02 ± .03
First error	statistical	
Second error	systematic	

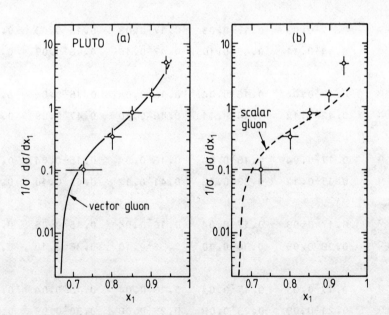

Fig.5.17 Distribution of the energy, x_1, of the fastest parton ($x_1 = \frac{2E_{max}}{W}$). Vector gluons are again favoured.

Thus there are large uncertainties in α_s due to the models employed at this order ($\Delta\alpha_s \sim 0.02-0.03$) and this should be borne in mind when making further deductions from α_s (eg in the value of Λ).

5.5 Higher Order Corrections and interpreting α_s

Only in the analyis based on the programme of Ali are higher order effects included. The contributions to the aplanarity distribution from 4-jet structures are shown in Fig.5.18 and an example of an apparent 4 jet event indicated in Fig.5.19. These higher order processes should be included.

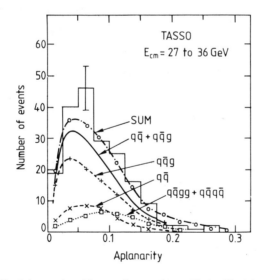

Fig.5.18 Contributions to the aplanarity distribution in the model of Ali. Multijet events are clearly important.

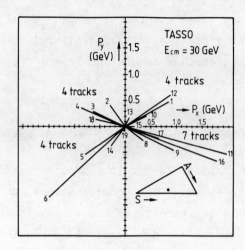

Fig.5.19 A possible 4-jet event.

A number of groups have calculated the contribution of all diagrams to order α_s^2 (49),(50),(51). These are shown in Fig.5.20. The groups agree on their analytic formulae but disagree in their application to the data. The CALTEC/CERN group imply that there are large effects in the thrust distribution

$$\sigma(\alpha_s) + \sigma(\alpha_s^2) \sim \sigma(\alpha_s) \left[1 + 15 \frac{\alpha_s(Q^2)}{\pi}\right] \quad (5.5.1)$$

but that the shape of $\frac{d\sigma}{dT}$ is essentially the same. The DESY group however believe that there is a small effect, which means that $\frac{d\sigma}{dT}$ automatically has essentially the same shape as at $0(\alpha_s)$. The concesus of opinion seems to indicate that the effect is large (eqn.5.5.1) and a recent analysis by Ali[52] has led to a value of $\alpha_s \sim 0.12$. This is just the sort of value one would obtain by rescaling the values of $\alpha_s \sim .17$ by the correction factor of eqn.5.5.1.

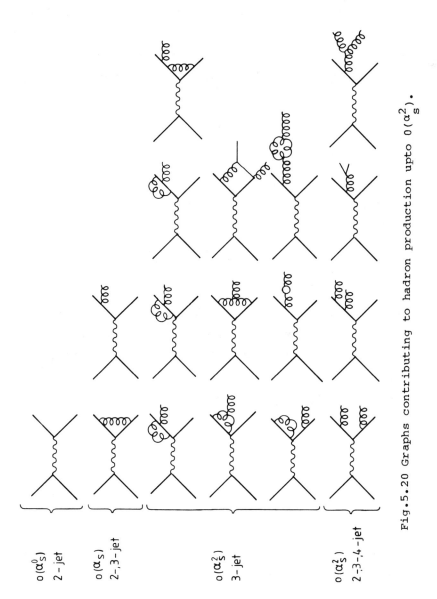

Fig. 5.20 Graphs contributing to hadron production upto $O(\alpha_s^2)$.

Two remaining comments are important

(i) if the $O(\alpha_s^2)$ correction is so large (~60%) one should be sceptical of the results. After all $O(\alpha_s^3)$ corrections might be even larger. We can only await the titanic efforts of devoted theorists to reassure us on this point.

(ii) The actual value of α_s is important in calculating Λ eg a change of α_s from .17 to .11 will result in a change in Λ from .385 GeV to .025 GeV if the first order equation

$$\alpha_s = \frac{12\pi}{23 \ln(W^2/\Lambda^2)} \qquad (5.5.2)$$

is used. If the second order equation is used for $\alpha_s = .125$ then the value of $\Lambda_{\overline{MS}}$ is 110^{+70}_{-50} MeV.

Needless to say these changes have a dramatic effect on quantities like the proton lifetime ($\propto \Lambda^4$).

5.6 The Gluon and its properties

Up to this point we have mainly been concerned with the 3 jet events observed in e^+e^- annihilation and their interpretation as gluon bremsstrahlung by the quarks. This has necessarily involved some discussion of the gluon properties eg its coupling α_s at high energies and its spin. In this section we will gather together the various pieces of information on the gluon.

(a) Spin

We have already seen that the high energy data favours a spin 1 assignment. Further evidence can be obtained in Υ decay which is hypothesised to occur via the 3 gluon process of Fig.5.21. The gluon momenta are given by[4]

$$\frac{1}{\Gamma}\frac{d^2\Gamma}{dx_1 dx_2} = \frac{1}{\pi^2-9}\left\{\left(\frac{1-x_3}{x_1 x_2}\right)^2 + \left(\frac{1-x_2}{x_3 x_1}\right)^2 + \left(\frac{1-x_1}{x_2 x_3}\right)^2\right\} \qquad (5.6.1)$$

from which the thrust distribution can be derived. In Fig.5.22 the

Fig.5.21 The three gluon decay process in Υ decay.

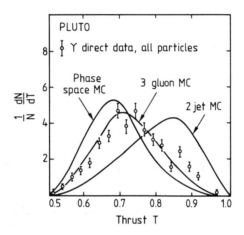

Fig.5.22 The thrust distribution in Υ decay. A 3 gluon process is favoured.

thrust distribution is well represented by such a three gluon decay. Furthermore the angular distribution of the thrust axis can be predicted to be

spin 0 gluons $\quad \dfrac{d\sigma}{d\cos\theta} = 1 - 0.995 \cos^2\theta \quad$ (5.6.2)

spin 1 gluons $\quad \dfrac{d\sigma}{d\cos\theta} = 1 + 0.39 \cos^2\theta \quad$ (5.6.3)

In Fig.5.23 the agreement with a spin 1 hypothesis is clear.

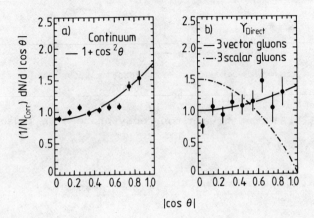

Fig.5.23 The angular distribution of the thrust axis on an off resonance at the Υ. Vector gluons are clearly favoured.

Thus all evidence points towards spin 1 for the gluon.

(b) Coupling α_s

At high energy we obtained a value of $\alpha_s \sim 0.12$ (which may change depending on higher order corrections and their interpretation). A value for α_s can also be obtained from Υ decay via the formula[54]

$$\frac{\Gamma(\Upsilon \to 3g, 4g, q\bar{q}gg)}{\Gamma(\Upsilon \to \mu^+\mu^-)} = \frac{10(\pi^2-9)}{81\pi} \frac{\alpha_{\overline{MS}}^3(M_\Upsilon)}{Q_b^2 \alpha^2} \left[1 + 9.1 \frac{\alpha_{\overline{MS}}(M_\Upsilon)}{\pi}\right] \quad (5.6.4)$$

yielding a value of $\alpha_{\overline{MS}}(M_\Upsilon) = 0.14 \pm .01$ or $\Lambda_{\overline{MS}} \sim 121^{+44}_{-31}$ MeV. The higher order corrections can be minimized by using as the argument of $\alpha_{\overline{MS}}$ the value of $\sqrt{s} = (.48M_\Upsilon)$ which results in a value of $\alpha_{\overline{MS}}(.48M_\Upsilon) = 0.16$ and implies $\Lambda_{\overline{MS}} = 100^{+34}_{-25}$ MeV.

Unfortunately there is no evidence that the running coupling constant α_s does indeed run. Its variation is very small.

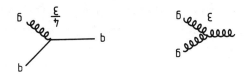

Fig.5.24 The relative strengths of the coupling at ggg and q$\bar{q}$g vertices.

(c) Fragmentation

The non-abelian nature of the gluon is indicated in Fig.5.24, the strength of the coupling being greater for g→gg than q→qg. At very high energies this would result in a higher multiplicity in gluon jets than in quark jets

$$\text{eg} \quad \frac{n_{ch}(g)}{n_{ch}(q)} = \frac{9}{4} \qquad (5.6.5)$$

Unfortunately we are not at infinite energy.

(i) <u>Υ decay</u>:

The difference in charged multiplicities on and off Υ resonance[55] is

$$\Delta n = n_\Upsilon - n_{continuum} = 0.55 \pm 0.19 \qquad (5.6.6)$$

$$n_\Upsilon = (8.12 \pm .11), \quad n_{cont} = (7.57 \pm .15)$$

If account is taken of the fact that the 3 gluon events would necessarily have jets of lower energy, then even if quarks and gluon fragmentation were identical we would expect $\Delta n=0.4$. Thus there is no evidence for a difference in jet multiplicity.

However if the species of particles are studied there are differences. K mesons, Λ's and protons appear to be more prolific in Υ decay although there are differences in quoted results[56],[57] As the data at high energies improves we may expect to see similar effects.

(ii) $q\bar{q}g$

There are a few very tentative pieces of evidence for quark and gluon differences. If the distribution of particles is studied as a function of θ, the angle between the quark and gluon jets, then JADE find differences[46] between the quark quark jets and the quark gluon jets, as shown in Fig.5.25, which they ascribe to differences in quark and gluon fragmentation. Furthermore[58] if they order the three jets such that

$$E_1 > E_2 > E_3$$

and then plot the $\langle p_\perp \rangle$ with respect to the jet axes, as in Fig.5.26, they find that the lowest energy fit appears broader. From the Lund Monte Carlo they find that the gluon will be the lowest energy jet 55% of the time, the middle jet 27% of the time and the highest energy jet 12% of the time. (The remaining 11% of 3 jet events are actually $q\bar{q}$ events which are misclassified). Thus there are some indications that the gluon jets are indeed broader than the quark jets.

It is quite obvious that these studies at high energy are quite primitive but will improve as both the quality and the quantity of data increases. There are already interesting suggestions.

(d) Non Abelian Coupling

This is the single most important feature of QCD, which results in the specific running nature of the coupling constant α_s. There is unfortunately no direct evidence for this ggg coupling. perhaps the best indications would be the existence of a glueball, a bound state of glue, although it would again be difficult to deduce that the coupling was of exactly the correct magnitude.

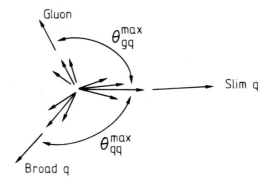

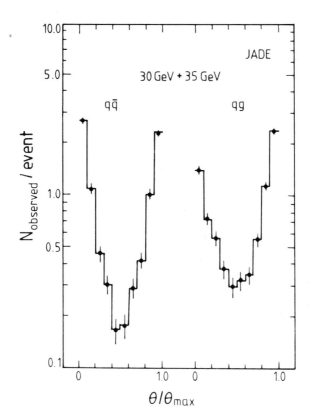

Fig.5.25 The angular distribution, θ/θ_{max}, of particles between the jet axes.

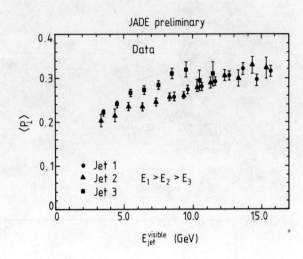

Fig.5.26 The $\langle p_\perp \rangle$ distribution for the three jets as a function of the visible energy of the three jets.

All gluon studies would, of course, be greatly benefitted by the discovery of a toponium system. The gluon jets would be far more characteristic ($E \gtrsim 10$ GeV), a good measurement of α_s could be obtained from Υgg compared with ggg decays, and perhaps even some evidence for the non abelian coupling found in the presence of 4 jet events compared to 3 jet events. We can only hope toponium turns up soon, preferably in the next PETRA energy range 36→40 GeV.

5.7 Summary

From this discussion there is no doubt that gluon jets are observed, that they are radiated at a rate corresponding to a coupling $\alpha_s \sim 0.12-.17$ (depending on higher order corrections) by high energy quarks, and that they have spin 1. What would be really nice to observe would be a dramatic difference in quark and gluon jets, even though some differences are already indicated, and direct evidence for the non abelian gluon coupling.

Improving our knowledge on all these topics will be a major goal of e^+e^- annihilation experiments in the next few years.

6. CONCLUSIONS

Over the last few sections we have been substantiating our working model for e^+e^- annihilation to hadrons via the one photon process. There is no doubt that the major process is quark-antiquark pair production where the quarks are point like, have spin 1/2, 3 colours and 5 flavours and are produced at a rate proportional to the squares of their charges. These quarks radiate gluons, which have spin 1, with a coupling constant $\alpha_s \sim 0.12$-0.17. These statements represent a tremendous triumph for QCD. In fact as far as QCD is concerned the only open questions are the direct demonstration of the non abelian ggg coupling and the running nature of α_s. Unfortunately it is hard to see how to get direct evidence bearing on these questions. Indeed the queries surrounding the true value of $\alpha_s(Q^2)$ have still to be resolved but we probably can expect progress on this topic in the near future.

Of course the area where we have most information and least understanding is in fragmentation of quarks and gluons (the bound state or confinement problem of QCD). This will be a major area of progress in the next few years. The data will get progressively more detailed and hopefully drive a better theoretical understanding. We can look forward to some excitement as this picture steadily unfolds, a picture which would be dramatically enhanced by the discovery of a toponium system.

REFERENCES

(1) S.J.Brodsky, T.Kinoshita, H.Terazawa, P.R. D4 1532 (1971); PRL 25, 972 (1970).

(2) S.D.Drell, D.J.Levy, T.M.Yan, PR 187, 2159 (1969), PR D1, 1617 (1970)
N.Cabibbo, G.Parisi, M.Testa, Lett. N.C. 4, 35 (1970)
J.D.Bjorken, S.J.Brodsky, P.R. D1, 1416 (1970)
R.P.Feynman, Photon-Hadron Interactions (Benjamin, Reading Mass.) p.166 (1972).

(3) J.Ellis, M.K.Gaillard, G.G.Ross, N.P. B111, 253 (1976), erratum B130, 516 (1977).

(4) A.de Rujula, J.Ellis, E.G.Floratos, M.K.Gaillard, N.P. B138, 387 (1978).

(5) P.Hoyer, P.Osland, H.G.Sander, T.F.Walsh, P.M.Zerwas, N.P. B161, 349 (1979)
R.Field and R.P.Feynman, N.P. B136, 1 (1978)
A.Ali et al PL 93B, 155 (1980)
J.B.Anderson et al, ZP C3, 223 (1980)
J.B.Anderson et al, NP B135, 273 (1980)
J.Anderson et al, PL 94B, 211 (1980)

(6) H.Fitzsch, M.Grell-Mann, H.Leutwyler, PL 47B, 365 (1973)
D.J.Gross, F.Wilczek, PRL 30, 1343 (1973)
H.D.Politzer, PRL 30, 1346 (1973)
S.Weinberg, PRL 31, 31 (1973)
H.D.Politzer, Phys.Rep. 19, 129 (1979)

(7) R.Felst, 1981 Inst.Symp.on Lepton and Photon Interactions at High Energies (Bonn 1981).

(8) P.Soding, G.Wolf, Ann.Rev.Nucl.Part.Sci. 31, 231 (1981).

(9) P.A.Rapidis et al, PRL 39, 526 (1977)
W.Bacino et al, PRL 40, 671 (1978)

(10) F.A.Berends, K.J.F.Gaemers, R.Gastmans, NP B63, 381 (1973); B66, 541 (1974)

F.A.Berends, G.J.Komen, P.L. 63B, 432 (1976)

F.A.Berends, R.Gastmans, NP B61, 414 (1973)

F.A.Berends, R.Kleiss, DESY 80/66

T.F.Walsh, P.Zerwas, DESY 72/77

(11) R.J.Cashmore, Progress in Particle and Nuclear Physics 7, 225 (1981).

(12) K.Berbelman, Proceedings this Erice School.

(13) T.Appelquist and H.Georgi, P.R. D8, 4000 (1973)

A.Zee, P.R. D8, 4038 (1973).

(14) J.Schwinger, Particles, Sources and Fields, Vol.II (1973) (Addison-Wesley, New York).

(15) M.Dine et al., PRL 43, 668 (1979)

K.G.Chetyrkin, PL 85B, 277 (1979)

W.Celmaster, R.J.Gonsalves, PRL 44, 560 (1979)

(16) G.L.Hooft, M.Veltman, NP B44, 189 (1972)

W.A.Bardeen, A.J.Buras, D.W.Dukes, T.Muta, PR D18, 3998 (1978)

W.Celmaster, R.J.Gonsalves, PRL 42, 1435 (1979); PR D20, 1420 (1979)

W.Celmaster, D.Sivers, PR D23, 227 (1981)

R.Barbieri et al, PRL 81B, 207 (1979)

(17) Proceedings of LEP Summer Study CERN 79-01.

(18) Particle Data Group Compilation RMP 52 (1980).

(19) R.Brandelik et al, PL 89B, 418 (1980)

W.Bartel et al, PL 88B, 171 (1979)

Ch.Berger et al, DESY 80/69

C.Bacci et al, PL 86B, 234 (1979)

R.Brandelik et al, NP B148, 189 (1979)

(20) W.Furmanski et al, NP B155, 253 (1979)

A.Bassetto et al, PL 83B, 207 (1978)

(21) T.Meyer, Private communication.

(22) W.Thomé et al, NP B129, 365 (1977)
R.Stenbacka et al, NC 51A, 63 (1979)

(23) W.Bartel et al, DESY 81/025.

(24) G.G.Hanson et al, PRL 35, 1609 (1975)
R.F.Schwitters et al, PRL 35, 1320 (1975).

(25) S.Brandt et al, PL 12, 57 (1969)
E.Fahri, PRL 39, 1587 (1977)
S.Brandt, H.D.Dahmen, ZP C1, 61 (1979).

(26) R.Brandelik et al, PL 100B, 357 (1981).

(27) D.Drijard et al, NP B155, 269 (1979)
D.Drijard et al, NP B166, 233 (1980).

(28) M.Holder, Rapporteur talk at the EPS conference on High Energy Physics, Lisbon (1981).

(29) S.D.Drell, D.Levy, T.M.Yan, PR 187, 2559 (1969); D1 1035, 1617, 2402 (1970).

(30) R.Brandelik et al, PL 89B, 418 (1980).

(31) P.Soding, Private communication.

(32) B.H.Wiik, XXth International Conference on High Energy Physics 1980 (Madison, Wisconsin)
R.Bandelik et al, to be published.

(33) M.W.Coles, LBL-11513.

(34) T.Meyer, DESY 81/046.

(35) B.Anders et al, LU TP 81-3.

(36) Mark J Collaboration, to be published.

(37) R.Cashmore, Proc.Int.Conf.on High Energy Physics, Geneva, p.330 (1979).

(38) P.Söding, Proc.Int.Conf.on High Energy Physics, Geneva, p.271 (1979)
R.Brandelik et al, PL 86B, 243 (1979)
D.P.Barber et al, PRL 43, 830 (1979)
Ch.Berger et al, PL 86B, 418 (1979)
W.Bartel et al, PL 91B, 142 (1980).

(39) S.L.Wu, G.Zobernig, ZP C2, 107 (1979).

(40) J.Freeman, Ph.D.Thesis, Dept.of Physics, University of Wisconsin (1981).

(41) D.P.Barber et al, MIT-LNS report no.115 (1981).

(42) J.Dorfan, ZP C7, 349 (1981)
K.Lanius, H.E.Roloff, H.Schiller, ZP C8, 251 (1981)
H.J.Daum, H.Meyer, H.Burger, ZP C8, 167 (1981).

(43) R.Brandelik et al, PL 94B, 437 (1980).

(44) J.Ellis, I.Karliner, NP B148, 141 (1979).

(45) R.Brandelik et al, PL 97B, 453 (1980).

(46) S.Yamada, Proc.XX Int.Conf.on High Energy Physics, Madison Wisconsin (1980).

(47) D.P.Barber et al, PL 89B, 139 (1979).

(48) Ch.Berger et al, PL 97B, 459 (1980).

(49) R.K.Ellis, D.A.Ross, A.E.Terrano, PRL 45, 1226 (1980); NP B178, 421 (1981)
D.A.Ross, NP B188, 109 (1981)
Z.Kunszt, PL 99B, 429 (1981) and CERN TH-3141 (1981).

(50) K.Fabricus, I.Schmitt, G.Schierholz, G.Kramer, PL 97B, 431 (1981).

(51) J.A.M.Vermaseren, K.J.F.Gaemers, S.J.Oldham, NP B187, 301 (1981).

(52) A.Ali, DESY 81-059.

(53) Ch.Berger et al, ZP C8, 101 (1981).

(54) P.B.Mackenzie, G.P.Lepage, CLNS/81-498.

(55) K.Berkelman, Proceedings of this 1981 Erice School.

(56) S.Stone, CLNS 81-514.

(57) H.Albrecht et al, DESY 81/011.

(58) JADE collaboration, to be published.

DISCUSSIONS

CHAIRMAN: R. CASHMORE

Scientific Secretary: M. J. Duncan

DISCUSSION 1

— *CASTELLANI*:

How would quark substructure manifest itself in the variation of R with Q^2 ?

— *CASHMORE*:

If you look at high enough Q^2 you will see the sum of the squares of the constituent charges and hence expect R to change. This change is model dependent and at present the data shows no substructure.

— *BERTANI*:

In investigating the $\gamma\gamma$ interaction, Frascati used an e^- in the beam pipe as a trigger. Will this be used at LEP ?

— *CASHMORE*:

PEP and PETRA have this trigger but the efficiency of tagging is small due to the detectors covering a small solid angle about the beam pipe. In LEP you have higher energy for the electrons and hence for constant Q^2 of the photons, the electrons will have smaller angle and you will have to go further down the beam pipe or bend the electrons out more. If you want to look at these events you can do it in this way. I was more interested in whether they form a background to one photon annihilation.

- DUNCAN:

How much information about the hadronisation of quarks do you need to compare experiment and theory close to threshold for new quarks?

- CASHMORE:

Not a lot if you just want to see the effect of the new quarks as they sit in the centre of mass, and you expect the weak decays to be isotropic. Nera isotropy is also ensured by the spin, J=1, of the photon. The details of the decay, however, would be very model dependent.

- TELLER:

Can you think of a three jet event as the decay of an excited meson which emits a gluon in addition to a quark and an antiquark ?

- CASHMORE:

You could call it a heavy excited photon, with J=1 controlling the angular distribution of quarks.

- COLEMAN:

It exists at too small a distance scale for one to properly think of it as a meson since the chromodynamic forces are weak at short distances.

- YAMAMOTO:

How important is hard radiation ?

- CASHMORE:

The initial state radiative corrections lower the centre of mass energy at which the e^+e^- appear to be colliding and the final state corrections are small as you are dealing with more massive particles.

If you see a hard photon effect then it is not a radiative correction so it is not what I am interested in.

- YAMAMOTO:

Is the hard photon asymmetry in e^+e^- experiments due to the interference of initial and final state photons ?

- CASHMORE:

No, the muon pair asymmetry of purely electromagnetic origin that you were alluding to is due to the interference of the following graphs:

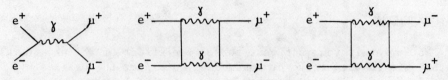

The effect is important when one looks at the weak electromagnetic interference where there is also the graph

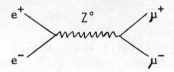

- SEIBERG:

Is there any experimental evidence for proving that there are eight different gluons without QCD ?

- CASHMORE:

No, except that if there were only one gluon which was colourless then the photon could decay through it giving two jet events.

- BERKELMAN:

Suppose the Upsilon had not been discovered at Fermilab or seen in other experiments, would you be able to tell from PETRA data alone that there had to be a fifth quark, and if so, how ?

- CASHMORE:

If one had no b-quark then $R = 10/3$ and we would need a bigger correction to take us up to $R = 4$. This would imply that α_s is larger, giving more gluon radiation, and hence even more planar events. People at PETRA need a fifth quark to get good fits to sphericity data. Without it you would get a mismatch between the cross-section and the type of events within that cross-section. Also recent measurements of inclusive cross-sections would be too large. A b-quark contribution is needed.

- SPIEGELGLAS:

How certain is $r = 10^{-16}$ cm and can it be traded for QCD corrections ?

- CASHMORE:

I have been asked before if you could have a quark form factor and another quark together making R constant. This would be remarkable, but you would have difficulty fitting the detailed topology of the event.

DISCUSSION 2   (Scientific Secretary: M.J. Duncan)

- WITTEN:

You showed us some beautiful examples of two and three jet events. How typical were these events and what fraction of all events are obviously "jet-like" to the naked eye ?

- CASHMORE:

At least three quarters of all events are clearly jet-like. This is particularly so of the two-jet events. The high momentum tracks usually define clear jet directions, although the low momentum tracks may not. The visibility of three-jet events depends on the orientation of the plane defined by the jets relative to the direction of observation and is more delicate.

- NOVERO:

What is the possibility that the mean multiplicity might grow like some power of the logarithm of the centre of mass energy ?

- CASHMORE:

You could try a fit of the form: $a + b(\log s) + c(\log s)^2$ and get values for a,b,c but that is only curve fitting.

- HERTEN:

You showed a randomised model where you looked at charge correlation along the jet axis. Ought one to include the possibility of fragmentation of charge zero quarks in your model ?

- CASHMORE:

If the quarks had zero charge then they would not couple to the photon. However we can pretend that they could be produced, would hadronise and then look for a remnant contribution over large distances. This is to check if there is any accidental long range correlation.

- ROBERTS:

You showed rapidity distributions defined with respect to the jet axis and also how for higher beam energies the distribution lengthens in rapidity space. I seem to recall that the plot of rapidity with respect to the beam axis in pp collisions is roughly the same and I would like to know if there is a fundamental connection between the two.

- CASHMORE:

In pp everything is essentially going along the beam axis in contrast to e^+e^- where the angular distribution is $1 + \cos^2\theta$.

A significant number comes out at right angles and so you would get a large difference if you measure rapidity with respect to the beam direction.

- LUIT:

Could the long range correlation be produced by events with low charge multiplicity ?

- CASHMORE:

No, at least five charged tracks are required.

- KARLINER:

Have you performed any charge correlation for three-jet events to verify that the third jet originates from a gluon ?

- CASHMORE:

We have not performed this analysis since the bulk of the PETRA data is rather small for three-jet events although it would be a nice test of the chargelessness of the gluon.

- D'HOKER:

When you extend the Feynman-Field quark fragmentation model to baryons, why is it a good approximation to include only one quark line ?

- CASHMORE:

I put in a double quark line so that I conserve baryon number. The model can be extended to give a baryon and an antibaryon in opposite jets. There is not a good prescription for putting baryons into the fragmentation of quarks as yet, although PEP and PETRA will give us information in the next few years. Also there is as yet no reasonable information on baryon-antibaryon correlations.

- YAMAGISHI:

Is there any possibility in the near future of determining whether the leading partons are fractionally charged ?

- CASHMORE:

It is a very difficult thing to show, as ultimately you measure integral charges in the experiment and you have to decide whether a track comes from one jet or another quark. You are very susceptible to the particles at small rapidity which can often come from either jet.

- LUCHA:

You mentioned that the number of three-jet events is about 5% of the number of two-jet events. Can one distinguish four-jet events and is there any rule about the occurrence of n-jet events ?

- CASHMORE:

As you go further down in energy it becomes more difficult to define what you call a jet and it can be confused with bottom particle decays. You would have to fit some distribution expected for two, three, four jets and then get some estimate of the contribution of four-jet events, although it would be low.

- YAMAMOTO:

When you measure the multiplicity, presumably you have to make corrections for the acceptance of low momentum particles. How sensitive is the multiplicity to these corrections ?

- *CASHMORE:*

We simulate the detector and from a given multiplicity in the model see what we would get in the detector.

- *PETERSEN:*

Why do you use the fragmentation function: $1 - a_F + 3a_F n^2$?

- *CASHMORE:*

You can choose any distribution you like and fit the parameters. The motivation for this particular distribution is that it fits the data well.

SEARCH FOR NEW PARTICLES AND ELECTROWEAK

INTERFERENCE EFFECTS IN e^+e^- INTERACTIONS

        P. Duinker

        National Institute for Nuclear and
        High Energy Physics
        Amsterdam, the Netherlands

1. INTRODUCTION

This talk will review some of the experimental results up to the middle of 1981 obtained with electron-positron colliding beam storage rings. An important result has been obtained at PETRA where the luminosity has been increased considerably due to the installation of mini-beta quadrupoles in the four intersection regions[1].

The time integrated luminosity per month obtained since the beginning of 1979 with the MARK-J detector is shown in Fig. 1. The months without any luminosity are the periods of the major machine shutdowns. The striking increase, a factor $\sim$ 3.5 in 1981, is due to the decrease in value of the vertical beta function at the intersection point using the mini-beta scheme.

This talk will concentrate on the searches for new particles which have been performed over the last years. The evidence for the non existance of the top quarks at the highest e^+e^- center of mass energies at PETRA will be presented. In addition, the searches for

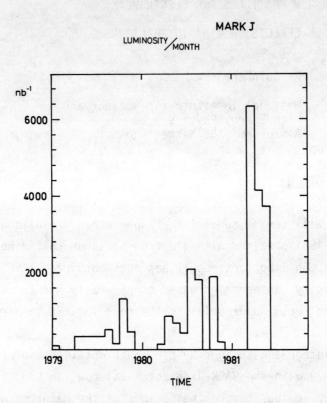

Figure 1   The integrated luminosity per month obtained with the MARK-J detector since the beginning of 1979 as function of time.

free quarks, heavy leptons, axions and numerous other particles will be reviewed. Finally, the evidence for weak interaction effects in e^+e^- interactions will be presented.

2. SEARCH FOR TOP QUARKS

The total cross section for $e^+e^- \to$ hadrons was measured over a wide range of center of mass energies from 12 to 36.7 GeV, including results obtained by extended periods of running at a fixed beam energy and by fine energy scans covering the ranges of 29.92 to 31.46 GeV and 33.00 to 36.70 GeV.[2] The results are expressed in terms of R:

$$R = \sigma(e^+e^- \to \text{hadrons})/\sigma(e^+e^- \to \mu^+\mu^-).$$

In the naive quark-parton model, the cross section for the hadron production process is simply given by the sum over flavors of the pointlike $q\bar{q}$ pair cross sections. Using this picture with spin 1/2 massless quarks and with three colors gives

$$R_o = 3\Sigma\, e_q^2$$

where e_q is the charge of the quark with flavor q. Considering the five known quarks (u,d,s,c and b), and correcting the naive model for gluon emission as predicted by QCD, one expects $R \simeq 4$, over the entire PETRA energy range, with only a slight decrease in R with increasing beam energy.

The experimental R values[2] have to be corrected for initial state radiative corrections, for contamination of the sample by hadronic events produced by the two-photon process $e^+e^- \to e^+e^- +$ hadrons, and for the contribution of $e^+e^- \to \tau^+\tau^- \to$ hadrons (+ leptons). In addition to the statistical errors, there is additional systematic uncertainty due to the model dependence of the acceptance calculations of ~ 10%. The measurements of R from four PETRA groups are averaged and plotted in Fig. 2. Some recent results reported at this conference from CELLO and MARK-J at PETRA and MARK II at PEP are also included. Above $\sqrt{s}$ = 12 GeV there is a large range up to 36 GeV where the R-values vary only very little. The error weighted average of the combined PETRA data above $\sqrt{s}$ = 12 GeV yields a value R = 3.84 ± 0.04 (solid line in Fig. 2). This value rules out the production of $t\bar{t}$ quarks as the expected value for R would be ~ 5.

In addition to the $t\bar{t}$ contribution to the hadronic continuum, the toponium system should form one or more bound states, with the number of such states depending on the shape of the binding potential.[3] Interpretation of the vector mesons ρ, ω, ϕ, J, ψ', T, T', and T'', as non relativistic $q\bar{q}$ bound states, or "quarkonia" leads to the prediction that the gap between the lowest bound state and the continuum is probably ~ 1 GeV, and very likely $\leq$ 2 GeV.

In order to check for the excistence of $t\bar{t}$ bound states lying below 36.7 GeV, the energy scans mentioned earlier were performed in 20 MeV center of mass energy steps (matching the r.m.s. energy spread of PETRA), with an average of ~ 25 nb^{-1} to 50 nb^{-1} per point and per experiment. The combined result of the highest energy scan is shown in Fig. 3. The figure shows that the data are entirely consistent with the predictions of QCD for u, d, s, c, and b quarks, that is, with constant values of R over the regions of the scans. There is no indication of an upward slope with increasing energy signaling the onset of a new contribution to the continuum. The value of R averaged over the energy range of the scan is 3.87 ± 0.05.

e⁺e⁻ INTERACTIONS

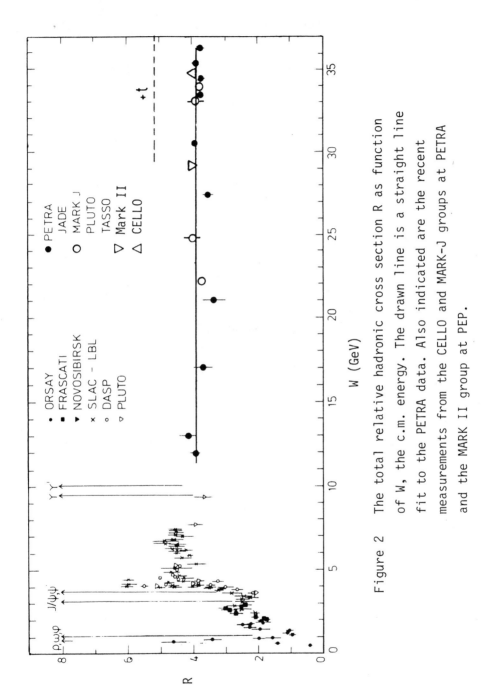

Figure 2 The total relative hadronic cross section R as function of W, the c.m. energy. The drawn line is a straight line fit to the PETRA data. Also indicated are the recent measurements from the CELLO and MARK-J groups at PETRA and the MARK II group at PEP.

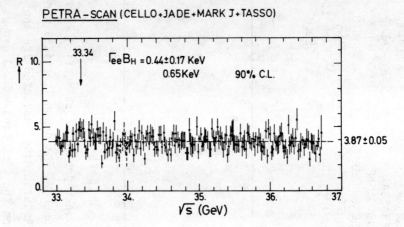

Figure 3   Average R values as measured by the PETRA experiments during the energy scan between 33.00 and 36.70 GeV.

In order to set a quantitative upper bound on the possible production of a narrow resonance, the data in Fig. 3 were fitted by a constant plus a gaussian. Using the relation between the resonance strength, the width into e^+e^- (Γ_{ee}), the hadronic width (Γ_h), the total width (Γ) and the hadronic branching ratio ($B_h \equiv \Gamma_h/\Gamma$) one obtains as an upper limit for

$$B_h \Gamma_{ee} < 0.65 \text{ KeV} \qquad (90\% \text{ C.L.})$$

From the experimental fact that the reduced width Γ_{ee}/e_q^2 is approximately constant for the vector meson ground states ρ, ω, ϕ, J, and Υ (predicted also by duality arguments[3]) and assuming the branching ratio for the lowest mass meson in the toponium family to hadrons to be $B_h \geq 0.7$, upper limits for $q\bar{q}$ bound states for quark charges 2/3e and 1/3e can be derived. This is illustrated in Fig. 4 where the reduced width Γ_{ee}/e_q^2 is plotted as function of the mass of the known vector mesons. The upper limit derived for $q\bar{q}$ bound states where the quark charge is 2/3e is clearly excluded. The upper limit for the production of a state with a quark charge q = 1/3e is also indicated.

Another way to search for the appearance of a new quark is to investigate the general properties of the hadronic events.

The TASSO Collaboration uses in the analysis of their hadronic events[4] the sphericity tensor:[5]

$$T_{\alpha\beta} = \Sigma_i P_{i\alpha} P_{i\beta} \qquad \alpha, \beta = x, y, z$$
$$\text{and } i = 1, \ldots, N \text{ particles.}$$

The summation is over the charged particles, which have their momenta measured with the central detector. The tensor $T_{\alpha\beta}$ has eigen-vectors $\vec{n}_1$, $\vec{n}_2$, $\vec{n}_3$ and the normalized eigen-values

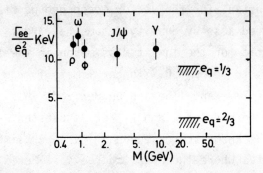

Figure 4 The reduced width Γ_{ee}/e^2 as function of mass. The 90% C.L. upper limits for the production of a $t\bar{t}$ bound state with either 2/3e or 1/3e charge are indicated.

$$Q_k = \frac{\Sigma_i (\vec{p}_i \cdot \vec{n}_k)^2}{\Sigma_i \vec{p}_i^{\,2}}$$

which satisfy $Q_1 + Q_2 + Q_3 = 1$ and $0 \leq Q_1 \leq Q_2 \leq Q_3$. The quantity sphericity S and aplanarity A can be expressed in the eigen values

$S = 3/2 \, (Q_1 + Q_2)$

$A = 3/2 \, Q_1$

and a third quantity is defined as

$Y = \frac{\sqrt{3}}{2} (Q_2 - Q_1).$

The event plane is defined by the eigen-vectors $\vec{n}_2$ and $\vec{n}_3$, the sphericity axis is given by $\vec{n}_1$. Plotting the events in the plane spanned by the quantities S and A, the two jet events, the planar and the spherical events can be separated as indicated in Fig. 5a. Spherical events arising from the production of heavy quark states would populate the center part of the triangle (see Fig. 5b for a Monte Carlo calculation for the production of top quarks with a mass of 15 GeV at W = 36.6 GeV). The model calculation for five quarks plus gluon bremsstrahlung corrections would populate the triangle as shown in Fig. 5c and the data at the highest energies is shown in Fig. 5d.

For A > 0.15 and S > 0.25, 12 events are found. From the $q\bar{q}$ (five different quarks) + QCD corrections one expects 11 ± 1 events, indeed a good agreement. The addition of a six quark with charge 2/3e or 1/3e would give rise to 138 or 43 events, respectively. Consequently the production of such quarks are excluded by the TASSO data[6].

In the framework of the six quark model for the weak decays of heavy quarks (c, b, and t) copious muon production is expected

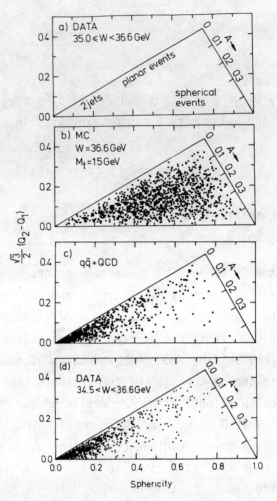

Figure 5   Triangle plots for hadronic events of spericity versus aplanarity as calculated and measured by the TASSO collaboration.

(a) Schematic representation of different event types.

(b) Monte Carlo calculation for the production of a heavy quark M_t = 15 GeV at W = 36.6 GeV.

(c) Monte Carlo calculation for five quark plus bremsstrahlung corrections.

(d) The data at 34.5 < W < 36.6 GeV.

from the cascade decays $t \to b \to c^7$. The onset of the production of a new heavy lepton would also lead to an increase in muon production. Thus, in addition to indications based on R measurements and event shapes, a measurement of inclusive muon production in hadronic final states should provide a clear indication of the formation of top quarks or new leptons. All the hadron data for $\sqrt{s}$ from 12 to 36.7 GeV have therfore been analyzed by the MARK-J group and scanned in search for muons. Fig. 6 summarizes the result of the MARK-J group for the relative production rate of hadronic events containing muons as a function of the c.m. energy. The figure demonstrates once again the absence of new heavy mesons containing t-quarks up to 36.7 GeV. The observed rate agrees with the Monte Carlo predictions for five quark flavors but is more than 5 standard deviations away from the prediction which includes the top quark. Fig. 7 shows the thrust distribution of the hadronic events containing muons compared with all the hadronic events. The scarcity of events at low thrust in the figure also rules out the existance of the top quark. The agreement between the data and the Monte Carlo containing five quarks is good. With this distribution the production of a new quark with charge $q = 1/3e$ is also ruled out. Below a thrust value of 0.75, 14 events are observed and 13.8 are predicted by the QCD model containing 5 quarks and gluon bremsstrahlung. The calculation including an additional quark predicts 163 or 60 events for the charge of 2/3e or 1/3e respectively.

3. WEAK DECAYS OF BOTTOM QUARKS

The main source of prompt muons in the hadronic events are decay products of bottom and charm quarks. Background contributions to the muon signal, arising from hadron punch through and decays in flight of pions and kaons have been calculated using a Monte Carlo simulation[7] to be $\sim 2\%$ of the hadronic events. The contribution

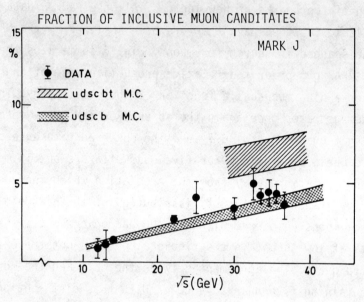

Figure 6 Relative production rate of hadronic events containing muons as a function of the c.m. energy E_{CM}. The hatched areas are the predictions containing five and six quarks.

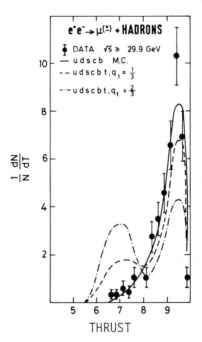

Figure 7 Thrust distributions of inclusive muon events compared with all hadronic events from MARK-J (solid line).

of $\tau^+\tau^-$ events to the μ+ hadron sample becomes negligible when the total energy cuts and energy balance cuts are applied. The question can be asked whether it is possible to obtain event samples where the secondaries originate from the fragmentation of either $c\bar{c}$ or $b\bar{b}$ quark pairs. Fig. 8 shows a calculation for the transverse momentum of the muons, P_t (transverse with respect to the thrust axis of the jet not containing the muon).

The P_t distribution for muons from c quark decays is peaked at relative low momenta, while the distribution for muons from b decay is stretched out to higher values. Also indicated in this graph is the P_t distribution of the muons from π or K decay. By making a cut at 1 GeV one obtains a sample of events dominated by muons from the decay of the b-quark.

Fig. 9 shows the muon P_t distribution for 234 events as measured by the MARK-J collaboration. A cut at 1.2 GeV leaves 48 events of which $\sim$ 45% are estimated to be due to b-quark decay. The preliminary result for the semi-leptonic branching ratio for mesons with a b-quark denoted by B is found to be:

$$\frac{(B \to \mu + x)}{(B \to all)} = 8.0\% \pm 2.8\% \text{ (STAT)} \pm 2.0\% \text{ (SYST)}.$$

The large systematic error is due to the uncertainty of the punch through contribution. This result compares well with the results obtained at CESR[8] which are:

$$\frac{(B \to \mu + x)}{(B \to all)} = 9.4\% \pm 3.6\% \quad \text{and} \quad \frac{(B \to e + x)}{(B \to all)} = 13.0\% \pm 4.0\%$$

The JADE group using their jet chambers has searched for secondary decay vertices near the primary vertex of the hadronic events containing a prompt muon.[9] By demanding that the muon momentum is greater than 1.8 GeV/c an enriched sample of events from $b\bar{b}$ production is obtained.

e⁺e⁻ INTERACTIONS

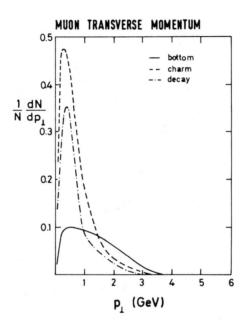

Figure 8   Muon transverse momentum distribution as calculated by a Monte Carlo program for different sources of muons as indicated.

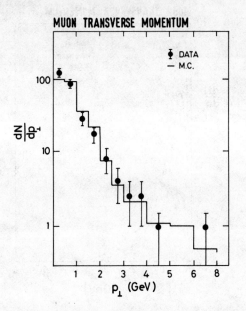

Figure 9　The muon transverse momentum distribution as measured by the MARK-J group. The drawn histogram is a M.C. calculation

The shortest distance d of the muon track to the primary vertex is a sensitive parameter to find a possible secondary vertex. In Fig. 10 the quantity $x_\mu = d/\sigma$ is plotted where σ is the resolution of the method derived by studying the tracks of pure hadron events. No change in slope is seen in this distribution and $<x_\mu>$ is found to be 0.80. From this average value of x_μ an upper limit for the lenght $d_\mu < 790$ µm at 90% C.L. is derived. This result can be converted into an upper limit of the life-time of mesons containing a b-quark. The JADE-group finds:

$$\tau_b < 5. \ 10^{-12} \text{sec} \quad (90\% \text{ C.L.})$$

In the standard six quark model of Kobayashi and Maskawa[7] the mixing of the three left handed quark doublets

$$\begin{pmatrix} u \\ d \end{pmatrix} \quad \begin{pmatrix} c \\ s \end{pmatrix} \text{ and } \begin{pmatrix} t \\ b \end{pmatrix}$$

is related to three angles α, β and γ, where α is the normal Cabibbo angle. The transition rate of b → u quarks is proportional to $\sin\beta$ in this model and the transition b → c to $\sin\gamma\cos\beta$. The life time of the b-quark can be expressed as a function of the angles β and γ according to:[9]

$$\tau_b = \frac{10^{-14}}{2.74 \sin^2\gamma + 7.7 \sin^2\beta} \text{ sec.}$$

This relation will be represented in the $\sin\beta$-$\sin\gamma$ plane by an ellips as is shown in Fig. 11. The bound on $\sin\beta$ in this figure is derived from the existing knowledge of the Cabibbo angle. The bound on $|\sin\gamma|$ is determined by the error on the value of the mass-difference between the K_S^0 and the K_L^0. Also indicated in the plot are the regions where the transition b → c or b → u would be dominant. The hatched area at small angles is the bound found by the JADE group.

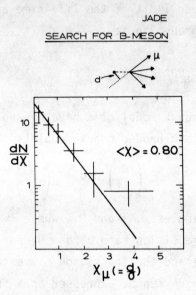

Figure 10 Event distribution of the quantity χ_μ for the events containing a prompt muon.

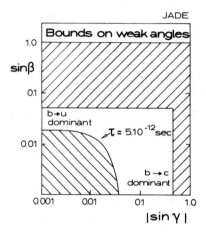

Figure 11 The sin β - sin γ plane with the existing bounds on these angles. The ellips $\tau_b = 5.10^{-12}$ sec is also indicated.

4 FREE QUARK SEARCHES

The MARK II group[10] at SPEAR has searched in their 1.4×10^6 e^+e^- annihilation events for particles with a mass in the range 1 to 3 GeV/c^2 and charge $q = -1e$ and $-2/3e$. Particles originating from the interaction point have their time of flight measured by a system of trigger counters. Combining the time measurement with the apparent momentum measurement P_q one can determine a mass:

$$m_T^2 = \frac{P_q^2}{q^2}(1/\beta^2 - 1) = \frac{m_q^2}{q^2}$$

where $\beta = v/c$. Figures 12a and 12b display the observed spectrum of time of flight masses m_T^2 for positive and negative charged particles. In the positive particle spectrum deuterons and tritons are observed produced in beam-gas collisions. This spectrum can serve as a calibration for the detection of negative charged particles ($\bar{d}$ and $\bar{t}$ are not observed). Upper limits can be derived in the mass regions of 1 - 3 GeV/c^2 for the production of $q = -1$ particles, assuming an exponential momentum distribution. Expressing the limits in terms of $R_q = \frac{\sigma(e^+e^- \to q\bar{q})}{\sigma(e^+e^- \to \mu\mu)}$ the MARK II group finds for

m_q = 1.7 - 2.3 GeV/c^2 $R_q \leq 3.1 \times 10^{-4}$ and
m_q = 2.3 - 3.0 GeV/c^2 $R_q \leq 1.9 \times 10^{-3}$

at the 90% confidence limit.

In the search for $q = 2/3e$ particle use is made of the pulse height measurements of the trigger counters. The calibration is done with muon pairs and electrons from Bhabha scattering. Two sets of limits are found for the production of quarks: (1) the exclusive production of quark-antiquark pairs, giving rise to two nearly collinear tracks in the detector and (2) the inclusive production of quarks, in which case a quark within a hadron jet is searched for. The limits found, expressed as the fraction of the point-like muon pair cross section

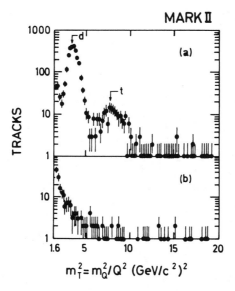

Figure 12 The apparent mass squared m_T^2 for (a) positive and (b) negative particles as measured by the MARK II detector.

R_q are shown in Fig. 13 as function of mass.

The JADE group, using their jet chamber, a cylindrical drift chamber filled with an argon, methane, isobuthane mixture at a pressure of 4 atm, is able to identify particles by a simultaneous measurement of the mean energy loss and the apparent momentum[11]. The inclusive production of quarks with q = 2/3e and 1/3e is studied. The limits found by this group are shown in Fig. 13 for the two cases. For each case two limits are presented, one for an exponential and another for a flat momentum distribution for the produced quarks.

At this conference, the first results of an experiment at PEP specially devised for quark searches, were presented by R. Ely[12] There limits are also shown in Fig. 13.

5 SEARCHES FOR OTHER PARTICLES THAN QUARKS[13]

(a) Heavy Leptons

Following the initial searches[14] for the τ-lepton, the discovery[15] and the further study of its properties[16], there has been great interest in searching for a new heavy lepton, which would extend the series e, μ, τ. Analogous to the τ-lepton one assumes that a new heavy lepton HL couples universally to leptons and quarks according to the standard V-A weak interaction theory and has the following decay modes: (in %), $\tau^- \bar{\nu}_\tau \nu_{HL}$ (9.2), $\mu^- \bar{\nu}_\mu \nu_{HL}$ (10.6), $e^- \bar{\nu}_e \nu_{HL}$ (10.6) and hadrons $+\nu_{HL}$ (69.6). The branching ratios[17] which are mass dependent, are given here for a heavy lepton mass of 14 GeV. Owing to their large mass and low velocity, the decay products would be expected to have large angles with respect to the HL line of flight. This contrasts to the decay products of the τ at PETRA energies which are tightly collimated.

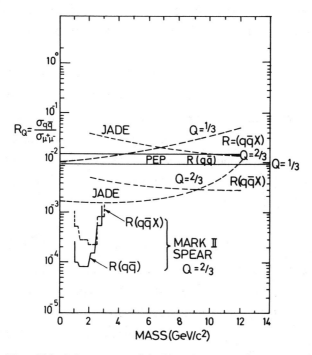

Figure 13 The 90° C.L. upper limits from various quark searches expressed in R_q as function of mass.

Heavy lepton production is recognized in the MARK-J detector for events in which one lepton decays into a muon and neutrinos and the other lepton decays into hadrons and neutrinos. Hadrons are detected by their energy deposit, E_{vis}, in the calorimeter[18].

Heavy lepton candidates with masses greater than 6 GeV are selected by applying criteria involving cuts on the total deposited energy, the acoplanarity, the charge multiplicity of the events and the total amount of hadronic energy and its direction.

No events were observed in the data with $33 < \sqrt{s} < 36.7$ GeV, corresponding to a time integrated luminosity of 6.9 pb^{-1}. The number of events predicted by a Monte Carlo calculation as a function of the heavy lepton mass M_{HL} is shown in Fig. 14 along with the 95% confidence level upper limit from the data. Fig. 14 demonstrates that the existence of a sequential heavy lepton with a mass between 6 and 16 GeV is excluded.

Heavy leptons with $M_{HL} < 6$ GeV would decay into final states similar in appearance to those from τ decay, and would tend to be included in the sample of $e^+e^- \rightarrow \tau^+\tau^-$ events. The inset in Fig. 14 shows the Monte Carlo prediction for the total number of events in the sample from $e^+e^- \rightarrow \tau^+\tau^-$ [19] and from heavy lepton production, as a function of M_{HL}. The inset demonstrates that the predicted number of heavy lepton events exceeds by more than 2σ the number of observed τ events. Therefore, with more than 95% confidence they exclude the excistence of the heavy lepton with mass $M_{HL} < 6$ GeV. One is thus able to rule out the existanse of a new heavy lepton for a $M_{HL} < 16$ GeV.

The criteria employed by the PLUTO group[20] for finding heavy leptons are very similar to the ones mentioned above and they find a limit for M_{HL} of 14.5 GeV.

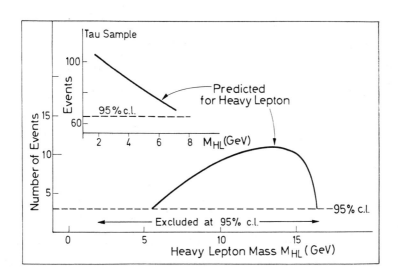

Figure 14 Number of events expected by the MARK-J group for the production of a new (sequential) heavy lepton as a function of mass. The inset shows the number of events expected in the τ sample from tau and heavy lepton production. A total of 52 τ events is observed. The dashed line corresponds to the 95% confidence upper limits for τ events.

The TASSO group[21] searches for a single charged particle recoiling against many hadrons. The angle between the single or "lone" track and the nearest neighbour is plotted in Fig. 15. A new heavy lepton would give an excess of events in this plot at large angles. From the actual observed number, taking into account the background from hadron production, an upper limit on the mass is found to be $M_{HL} < 15.5$ GeV at 90% C.L.

The JADE group concentrates on the possible hadronic decays of the heavy leptons. They calculate the direction of the two hadron jets and demand that the jets are very acoplanar, excluding by such criteria the events produced by quark-antiquark pairs which give rise to two back-to-back jets. They expect for a 17 GeV lepton $\sim$ 10 events, and they find none[13].

(b) Heavy Neutral Electron

The JADE collaboration has searched for a heavy neutral partner of the electron. These particles appear in non-standard models of the weak interaction[22]. When the neutral current contains non-diagonal terms they can be produced in neutrino and e^+e^- interactions. In models where only diagonal terms are present to describe the neutral current the heavy neutral leptons can only be produced in e^+e^- interactions according to:

$$e^+e^- \to \bar{E}^\circ + \nu_e \quad \text{and} \quad e^+e^- \to \bar{E}^\circ + E^\circ$$

Depending on the coupling of the charged current to these particles which can be either V-A or V+A, the differential cross sections for the case of an exchange of a charged weak boson in the t channel can be written as:

$$\frac{d\sigma^{V-A}}{d\Omega_{\bar{E}^\circ}(E^\circ)} = \frac{G_F^2}{32\pi^2}\left(1 - \frac{M^2}{S}\right)\left(\frac{1}{1 - \frac{q^2}{M_W^2}}\right)\{2s(1 \pm \cos\theta) - (s - M^2)\sin^2\theta\}$$

e^+e^- INTERACTIONS

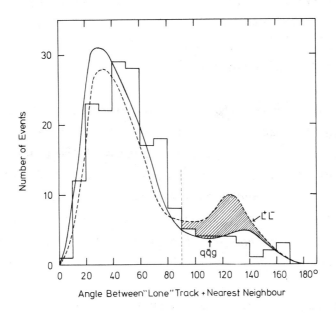

Figure 15 The distribution of the angle between the "lone" track and the nearest neighbour as measured by the TASSO collaboration. The prediction of the production of a heavy lepton is indicated by the dashed line. The QCD expectation is given by the solid curve.

$$\frac{d\sigma^{V+a}}{d\Omega \, \overline{E}^\circ} = \frac{G_f^2}{32\pi^2} \left(1 - \frac{M^2}{s}\right)\left(\frac{4s}{1 - \frac{q^2}{M_W^2}}\right)$$

where G_f is the Fermi compling constant, M the mass of $\overline{E}^\circ$ or E°, M_W the mass of the exchanged charged weak boson, q^2 the momentum transfer in the t-channel and θ the angle between the beam direction and the produced $\overline{E}^\circ$. The posible decay channels for the E° are $E^\circ \to e^+e^-\nu_e$ (10%), $e^-\mu^+\nu_\mu$ (10%), $e^-\tau^+\nu_\tau$ (10%), $e^-u\bar{d}$ (35%) and $e^-c\bar{s}$ (35%) respectively.

The JADE group[23] searched for the events produced via t-channel exchange and for hadronic decays of the lepton. As a ν_e is always produced in such a reaction, the signature of the events is the imbalance of energy and momentum along the line of flight of the neutrino. The experimental cuts were made such that the event had an empty cone (half opening angle 50°) containing no charged particles or neutral energy, opposite to a hadronic jet containing one electron.

The expect number of events as a function of the mass are presented in Fig. 16 for the V + A and V - A case. One event was found which could be due to $\tau^+\tau^-$ production. From the 95% C.L. limit it can be concluded that the neutral heavy leptons do not exist below 18 GeV if the coupling is V-A and not below 20 GeV if the coupling is V+A[23].

(c) Scalar Electrons and Muons

In the framework of supersymetric theories[24], spin zero partners of the electron and the muon are expected to decay only according to the reactions:

$s^- \to e^- \, (\mu^-) +$ photino (goldstino)
$t^- \to e^- \, (\mu^-) +$ antiphotino (antigoldstino)

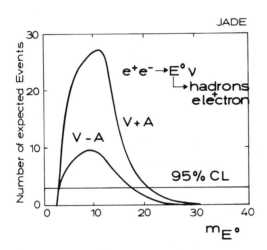

Figure 16 The number of expected events for heavy neutral lepton production as function of the mass. The 95% C.L. line is also indicated.

where s and t are the spin zero partners of the electron (muon) associated with the left and right handed parts of the electron (muon) field respectively, and the photino and goldstino are the spin 1/2 partners of the photon and the goldstone boson. Since s and t carry unit electric charge they may be produced in pairs in e^+e^- annihilation according to the cross section:

$$\frac{d(e^+e^- \to s^-s^+ \text{ or } t^-t^+)}{d(\cos\theta)} = \frac{\pi\alpha^2\beta^3\sin^2\theta}{4s} \; ; \; \beta = \{1-(\frac{m}{E})^2\}^{\frac{1}{2}}$$

which is characteristic of spin zero particle production. m is the mass of s or t, E is the beam energy, and θ is the scattering angle.

Because of the uniqueness of the decay reactions and the extremely short lifetimes of s and t and the prediction that the interaction cross section of photino and goldstino are expected to be very small, only electron and muon pairs are observed in the final state. Near threshold production of s and t, the two residual electrons or muons would be produced isotropically in space. Data from SPEAR place a lower limit of 3.5 GeV[25] on the mass of s and t. Thus, over the PETRA energy range of 12 and 36.7 GeV an increase in the production of acoplanar electron-positron or muon pairs should be observed if a new threshold is passed.

The event selection criteria used by the PETRA groups to obtain upper limits for the production of these particles are based on the considerations mentioned above. I will describe the procedure followed by the CELLO group in some detail[26].
At a average beam energy $<E_b>$ = 17.3 GeV the events triggered by at least two charged particles were subjected to the following cuts;

1. The angle θ between the tracks and the beam direction is constrained to be 35° < θ < 145°.
2. The acoplanarity is less than 30°.
3. The charged track multiplicity is less than 5.
4. Two tracks have a momentum greater than 0.8 GeV.
5. For e^+e^- pairs the sum of the energies must exceed 5 GeV and for $\mu^+\mu^-$ pairs the momentum of the muon must be greater than 20% of the beam energy.

In the CELLO detector the electrons and photons are identified with a lead-liquid argon calorimeter and the muons with proportional chambers behind 80 cm of iron. No candidates were found. Fig. 17 shows the expected event rate for the e scalar leptons under the above mentioned conditions. At the 95% C.L. the mass range $2 < M_s < 16.6$ GeV is excluded. Similar limits were obtained by the other PETRA experiments.[27,28]

(d) Search for Hyperpions

In order to avoid the necessity of the existance of Higgs particles, which were introduced to give masses to the gauge bosons, Weinberg and Susskind[29] independently demonstrated that it is possible to generate the $W^{\pm}$ and $Z^°$ boson masses by introducing a hypercolour. Heavy quarks are introduced, e.g. U and D, which can form so called hyperpions. These hyperpions could be produced in e^+e^- interactions[30] at relatively low energies as their masses could be quite small.

The cross section behaviour of their production is the same as other spin 0 particles (see section on scalar electrons and muons). Possible decay modes are:

$$\pi' \to \tau\nu_\tau$$
or
$$\pi' \to c\bar{s} \to \text{hadrons}$$

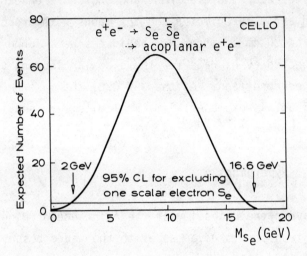

Figure 17 Number of events expected for the production of a spin zero partner s_e or t_e of the electron as a function of mass. The upper limit of events (95% confidence) and the mass range excluded is also indicated.

As the branching ratio for these two decays is not known the JADE group[31] searched for these particles as function of the branching ratio to $\tau \nu_\tau$ (B_τ). The events looked for are the ones where one hyperpion decays to a τ lepton which subsequently decays to one charged particle plus neutrino's. Opposite to the charged particle but acollineair with it one searches for a hadronic jet originating from the decay of the π' to $c\bar{s}$ quarks. This method of searching breaks down if B_τ is close to one or zero. Fig. 18 shows the result of the JADE group. The mass of the hyperpion is plotted as function of B_τ. The two-dimensional region which is excluded at 90% C.L. is indicated. Hyperpion masses from 5 to 15 GeV are excluded by this search.

(e) Search for Axions.

In order to give an explanation of the experimentally observed CP conservation in the strong interactions[32], the existance of a very light boson called the axion has been postulated. These particles could be produced in beam dump experiments[33] and also in the decay of a vector meson like: $J/\psi \to \gamma + a$.

F. Porter has reported the search for this decay of the J/ψ resonance using the Crystal Ball detector at Spear. This detector is specially constructed to detect γ rays with high accuracy.

The axion behaves like a neutrino so it escapes direct detection. As the mass is supposed to be very small, one can look for photons at $\sim$ 1.5 GeV without anything else observed in the detector. The photon has of course to be in time with the time of crossing of the electron and positron bunches at the intersection point. The total number of J/ψ decays for this search were $1.43 \cdot 10^6$ events.

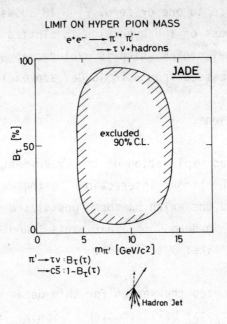

Figure 18 Limit on the mass of the hyperpions as found by the JADE group. Plotted is the mass as function of B_τ. The contour gives the 90% C.L. in the plane. The inside of the contour is the region which is ruled out by the data.

The branching ratio of $J/\psi \to \gamma + a$ was estimated by Peccei[32] to be $\sim 5 \cdot 10^{-5}$. Five candidate events were found, but they could all be attributed to cosmic ray events hitting the detector from the sides. The upper limit for the branching ratio $J/\psi \to \gamma + a$ was found to be $< 3 \cdot 10^{-5}$ at 90% C.L.

(f) Excited State of the Muon.

Heavy muons, excited states of the muon could be in pinciple produced in e^+e^- interactions. Their production could proceed according to:
$$e^+e^- \to \mu^{*\pm} + \mu^{\mp} \quad \text{or} \quad e^+e^- \to \mu^{*-} + \mu^{*+}$$
The μ^* subsequently decays according to $\mu^{*\pm} \to \mu^{\pm} + \gamma$.

For the simplest case of spin ½ the differential cross section can be written as
$$\frac{d\sigma}{d\Omega} = \lambda^2 \alpha^2 \frac{(s-M^2)^2}{s^3} \left[(s + M^2) - (s - M^2)\cos^2\theta\right]$$
where λ is a new coupling constant, α the fine structure constant, M the mass of the μ^* and s the square of the centre of mass energy. The MARK-J group has looked for events with two muons and 1 photon in their detector using the following criteria:

1. Two identified and momentum analyzed muon tracks.
2. One hard photon with an energy greater than 20% of the beam energy.
3. The angle between the photon and the beam direction must be greater than 20°.
4. The event must satisfy a kinematical fit in order to insure energy and momentum conservation within errors.
5. The acoplanarity of the muon tracks must be greater than 20°.

The data used for the search were 12.34 pb^{-1} obtained at energies ranging from $27.4 < \sqrt{s} < 36.7$ GeV.

Possible background comes from radiative processes like:
$$e^+e^- \to \mu^+\mu^-\gamma$$
and from two photon processes
$$e^+e^- \to \mu^+\mu^- e(e)$$
where one of the electrons stays within the beam pipe and the other is scattered into the detector and cannot be distinguished from a photon. Using Monte Carlo programs developped by Berends, Kleiss[34] and Vermaseren[35] the background was estimated to be 12 events, in good agreement with the number of observed events of 11. This result can be expressed as the maximum allowed cross section for $\mu^*\mu$ production, $\sigma_{\mu^*\mu}^{max}$ at the 90% C.L. limit. In Fig. 19 the mass of the μ^* is plotted versus the cross section ratio $\sigma_{\mu^*\mu}/\sigma_{\mu\mu}$ (right hand side vertical scale). The 90% C.L. line is drawn as solid dotted curve.

A bound on the coupling λ can be found using the relation:
$$\sigma_{\mu^*\mu}^{max}/\lambda^2 = 4.5\ 10^{-4}\ (1 - \frac{M^2}{s})\ (1 + 2\frac{M^2}{s})$$
The solid line (left hand vertical scale) shows the maximum allowed coupling for μ^* production. At the lower masses the bound on μ^* production is obtained by searching for $\mu^*\mu^*$ production in events with two muons and two photons in the final state. Fig. 19 demonstrates that the existance of μ^* particles is excluded up to masses of 30 GeV if the coupling is not too small.

(g) Excited State of the Electron.

To test the exchange of a hypothetical heavy electron in the two photon annihilation channel $e^+e^- \to \gamma\gamma$ the differential cross section can be written as[36]:
$$\frac{d\sigma_{e^*}}{d\Omega} = \frac{\alpha^2}{2s}\ \{\frac{q'^2}{q^2} + \frac{q^2}{q'^2} \pm \frac{2s^2 - 4q^2q'^2}{\Lambda_{e_\pm}^{4*}}\}\ \{1 + C(\theta)\}$$

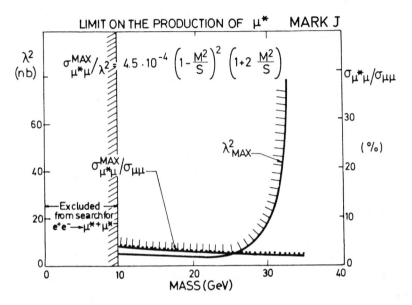

Figure 19 Limit on the production of the μ* from the MARK-J group. The right hand scale is the cross section for μ*μ production over the μμ cross section in percentage versus the mass of the μ*. The left hand verticale scale is the coupling strength.

The solid-dotted curve is the upper limit at 90% C.L. for $\sigma^*_{\mu\mu}/\sigma_{\mu\mu}$. The solid curve is the 90% C.L. limit for the coupling denoted with λ^2_{max}. The region excluded below 10 GeV mass of the μ* is also indicated.

The value of Λ_{e*} can be interpreted as the mass of a heavy electron assuming its coupling strength is the same as that of the electron. In the above expression $q^2 = -s\cos^2\theta/2$, $q'^2 = -s\sin^2\theta/2$ and $C(\theta)$ is the radiative correction term as a function of θ, the angle between the outgoing photon and the beam-axis.

Fig. 20 shows the angular distribution normalized with the predicted QED cross sections for $e^+e^- \to \gamma\gamma$ obtained by the five PETRA groups[37] in the energy range $27.4 < \sqrt{s} < 36.7$ GeV.

The agreement between the data and the QED expectation (dashed line in the figure) is good. The data was fitted to the above expression of the differential cross section. The results of these fits for the cut-off parameters are collected in Table I. The lower limits for the cut-off parameters are about 50 GeV (95% C.L.). At the highest PETRA energies the lower bounds are raised by a factor 4 to 5 over the values obtained with the lower energy e^+e^- colliding beam machines[38]

A heavy electron with a mass smaller than ~ 50 GeV is ruled out by the measurements. It is interesting to note that the ratio of the excited electron mass over the electron mass is greater than 10^5 while the excited proton, the Δ for example is very close in mass to the proton. This indicates that there is no structure inside the electron down to a size of 10^{-16} cm.

6. TESTS OF ELECTRO-WEAK INTERACTION MODELS

The Fermion-pair production processes $e^+e^- \to F + \bar{F}$ where the $F\bar{F}$ pair stands for either a lepton pair (e^+e^-, $\mu^+\mu^-$ and $\tau^+\tau^-$) or quark-antiquark pairs, proceeds via the exchange of a virtual photon. At sufficient high energies, however, the Z° exchange effects have to be taken into account. One of the aims of the e^+e^- experiments at the highest machine energies is to detect the weak effects due to this exchange.

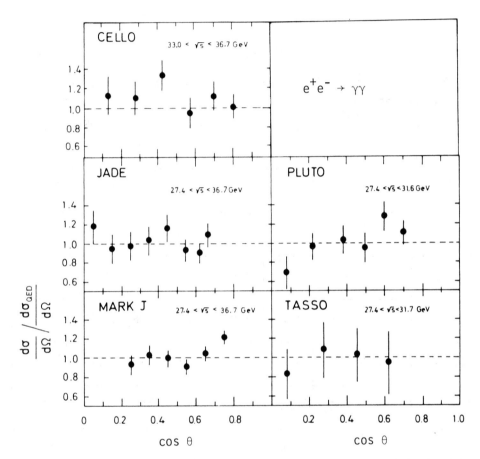

Figure 20 The differential cross section normalized with the QED cross section as function of cos θ for the process $e^+e^- \to \gamma\gamma$.

TABLE I

The cut-off parameters obtained by the various experiments for the QED processes $e^+e^- \to \gamma\gamma$, e^+e^-, $\mu^+\mu^-$ and $\tau^+\tau^-$. The values given for Λ are the lower limits in GeV at 95% C.L.

	$\gamma\gamma$		e^+e^-		$\mu^+\mu^-$		$\tau^+\tau^-$	
	Λ_+	Λ_-	Λ_+	Λ_-	Λ_+	Λ_-	Λ_+	Λ_-
CELLO	43	48	83	155	-	-	-	-
JADE	50	52	118	138	146	126	111	93
MARK-J	51	49	106	193	194	153	126	116
PLUTO	46	46	80	234	107	101	79	63
TASSO	34	42	150	136	80	118	88	103

The analysis of the weak effects is performed in two ways:

a. The $\sin^2\theta_W$ is determined within the context of the standard G.W.S. model[39] using all the available data of the production of Fermion - anti-Fermion pairs.

b. A model independant analysis is presented for the determination of the weak neutral current coupling constants using the data of the lepton pair production processes alone.

In this section the cross section determinations and the asymmetry measurements for muon and tau pairs are presented. In addition to the differential Bhabha cross sections, the analysis of R in terms of $\sin^2\theta_W$ is discussed.

6.1. Muon and Tau Pair Production

The detectors are able to distinguish muons from electrons and hadrons and to eliminate cosmic ray muons as a source of background for the back-to-back muon pairs.

I will describe in detail the criteria used by the MARK-J collaboration for the muon and tau pair production identification.

Single muons are identified as particles which:

(i) Are reconstructed in the inner drift chambers to come from the interaction region;

(ii) Leave minimum ionizing pulse heights in the seven layers of counters of the calorimeter;

(iii) Leave a track in the outer drift chamber.

In addition back-to-back muon pairs are distinguished from cosmic rays by requirement that:

(i) The D counter timing signals are coincident with one another (and not relatively off time as in the case for cosmic rays traversing the detector);

(ii) The muons should be collinear and coplanar, and they should pass through the intersection region.

Tau leptons are identified with the MARK-J detector by detecting μ-hadron and μ-electron final states. The cross section is determined using the known branching ratio of $\tau \to \mu \nu\nu$ (16%) and $\tau \to$ (e, hadron or multihadrons) + ν (84%). The measured muon momentum and hadron energy for the $\tau^+\tau^-$ candidates is in agreement with calculations based on the known decay properties of the τ lepton. The acceptance is calculated using a Monte Carlo method to generate $\tau^+\tau^-$ production including radiative corrections. A detection efficiency of $\sim$ 10% for τ pairs at various energies, when requiring one decay muon to be detected in association with a single electron, or one or more hadrons, is obtained with the MARK-J detector. The selection criteria for the tau leptons for the PLUTO [40] and TASSO [41] groups are different from the ones mentioned above. PLUTO identifies one charged particle and neutrals recoiling against one or more charged particles for the decay of the tau, and TASSO requires at least three charged and neutral particles, recoiling against one charged particle.

The CELLO group uses essentially the same criteria as the ones used by the PLUTO collaboration including one charged particle recoiling against one other charged particle. The muon pairs and the Bhabha events are removed from their τ sample.

The resultant $e^+e^- \to \mu^+\mu^-$ and $\tau^+\tau^-$ cross sections as a function of s are plotted in Fig. 21 and 22. We see that from $q^2 = s = 169$ to $q^2 = 1225$ GeV^2 the data agree well with the predictions of QED for the production of a pair of point-like particles. In particular, Fig. 22 represents the evidence that the τ lepton is a point-like particle over a large range of q^2, and demonstrates that it belongs in the same family as the electron and muon. To parametrize the maximum permissible size (radius) of the particles, one uses the form factor:

$$F(s) = 1 \mp s/(s - \Lambda_\pm^2)$$

By comparing the data with the cross sections including this form factor, lower limits on the cut-off parameters, at the 95% confidence level are calculated and they are summarized in Table I.

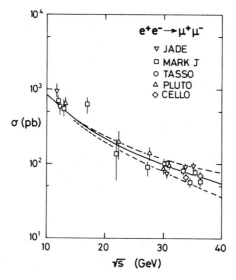

Figure 21 Observed cross section for the reaction $e^+e^- \to \mu^+\mu^-$ compared to the predictions of QED (solid curve). The dotted curves are the QED cross sections modified by cut-off parameter $\Lambda_\pm = 100$ GeV.

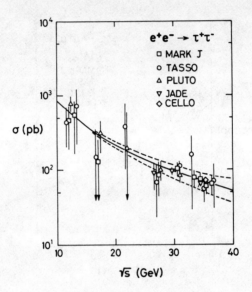

Figure 22 Observed cross section for the reaction $e^+e^- \to \tau^+\tau^-$ compared to the predictions of QED (solid curve). The dotted curves are the QED cross sections modified by cut-off parameters $\Lambda_\pm$ = 100 GeV.

e⁺e⁻ INTERACTIONS

The values obtained for the cut-off parameters demonstrate that the muon and the tau are point like particles in their electromagnetic interaction with characteristic ratio $\leq 10^{-16}$ cm. In Fig. 23 the $\cos\theta$ distribution of $\tau^+\tau^-$ production is shown obtained by the CELLO collaboration[42] at the highest PETRA energies. The agreement with the expected QED distribution of $1 + \cos^2\theta$ is good.

6.2. Asymmetry measurements in muon pair production.

Within the frame work of the standard GWS model the differential muon pair cross sections can be written as:

$$\frac{d\sigma}{d\Omega} = \frac{\alpha^2}{4s} \left[(B + 1)(1 + \cos^2\theta) + A \cos\theta\right]$$

where

$$B = 8 \, s g g_V^2 \, \frac{M_Z^2}{s - M_Z^2} + 16 \, s^2 g^2 (g_V^2 + g_A^2) \left(\frac{M_Z^2}{s - M_Z^2}\right)^2$$

$$A = 16 \, s g g_A^2 \, \frac{M_Z^2}{s - M_Z^2} + 128 \, s^2 g^2 g_V^2 g_A^2 \left(\frac{M_Z^2}{s - M_Z^2}\right)^2$$

$$g = \frac{G_F}{8\sqrt{2} \, \pi \alpha} = 4.49 \, 10^{-5} \, \text{GeV}^{-2}$$

and g_A and g_V are the axial end vector coupling constants of the neutral weak current.

In the standard model $g_A = 0.5$ and g_V is related to $\sin^2\theta_W$ according to: $g_V = (1 - 4 \sin^2\theta_W)$.
The term B is for a value of $\sin^2\theta_W = 0.23$ at $\sqrt{s} = 35$ GeV very small (B ∼ 3.5 10^{-3}). As a result the change in the total muon pair cross section due to weak effects is not noticeable at PETRA energies. As the term $A = -.26$ at $\sqrt{s} = 35$ GeV the asymmetry effect becomes detectable.

Defining θ as the angle between the incident e⁻ and the outgoing

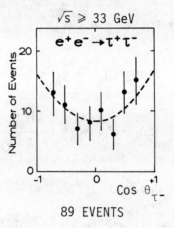

Figure 23 The cos θ distribution for $\tau^+\tau^-$ production. The dotted line is the $1 + \cos^2\theta$ expectation.

muon in the forward hemisphere, the asymmetry is:

$$A_{\mu\mu} = \frac{N_- - N_+}{N_- + N_+}$$

where N_- (N_+) is the corrected number of events with the negatively (positively) charged muon in the forward hemisphere, integrated over the accessable angular range.

In first order $A_{\mu\mu}$ is directly related to g_A^2 according to

$$A_{\mu\mu} = 6\, g\, g_A^2\, s\, \frac{M_Z^2}{s - M_Z^2}$$

Due to the propagator term $\frac{M_Z^2}{s - M_Z^2}$ the asymmetry is negative below the Z°. A positive asymmetry is, however, introduced by the interference between QED loop diagrams (C = +1) and the first order annihilation amplitude (C = -1).

In Fig. 24 the asymmetries due to the QED and weak effects are plotted[43] separately as function of the scattering angle θ. Due to bremsstrahlung effects, cuts are introduced on the energy of the muons and the acollineanity angle (ζ) between them. The energy variation for the QED asymmetry is very small (within the thickness of the line). The weak effects show a considerable energy dependance and this effect is lineair in s. In Fig. 25 the sum of QED and weak asymmetry is plotted as function of θ.

It has to be noted that in all practical cases the experiments do not measure muons below a scattering angle θ of 30°. The statistics so far do not allow to measure the asymmetry as function of θ at different energies and the measurements for $A_{\mu\mu}$ are integrated over the available θ range.

Fig. 26 shows the $\cos\theta$ distribution for muon pairs obtained by the MARK II collaboration[44] at PEP for $\sqrt{s}$ = 29 GeV. The dashed line gives the QED expectation. The measured asymmetry is found to be $A_{\mu\mu} = (-4.7 \pm 4.2)\%$.

For the standard model the expected value is -4.0% at this energy and taking the angular range of the experiment into account.

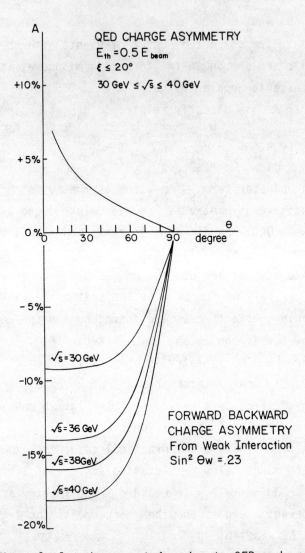

Figure 24 The calculated asymmetries due to QED and weak effects as function of the scattering angle θ for various energies. The weak effects are calculated within the standard model using $\sin^2\theta_W = 0.23$.

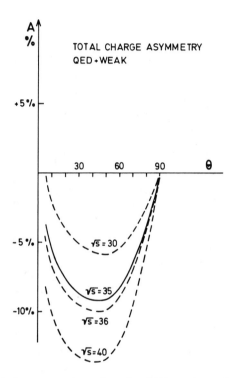

Figure 25 The total asymmetry due to QED and weak effects as function of θ for different energies.

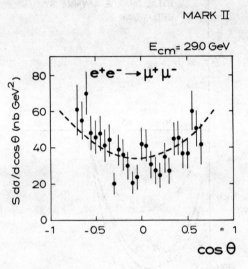

Figure 26 The cos θ distribution for muon pair events obtained by MARK II at PEP. The dashed line is the $1 + \cos^2\theta$ expectation.

The value for the axial vector coupling obtained from this distribution is

$$|g_A| = 0.60 ^{+.18}_{-.25}$$

The PETRA results for the cos θ distribution[45] are summarized in fig. 27 and Table II. The asymmetry averaged over the four experiments is:

$A_{\mu\mu}$ = (-2.8 ± 3.4)%, while -6.7% is expected.

Also listed in Table II are the preliminary results for the asymmetry obtained from the cosθ distribution of the $\tau^+\tau^-$ pairs. The average value is $A_{\tau\tau}$ = (-3 ± 8)%, while -6% is expected.
From the values of $A_{\mu\mu}$ and $A_{\tau\tau}$ one can derive upper limits for the axial coupling constants of the muon and tau without assuming lepton universality. If one takes the axial coupling for the electron to be $g_A(e) = 0.5$ one finds:[45]

$g_A(\mu) \leq 0.72$ } at 95% C.L.
$g_A(\tau) \leq 1.62$

6.3. Tests of Electro-weak Interaction Models

a) Measurements of $\sin^2\theta_W$.

The theory of Glashow-Weinberg-Salam (GWS)[39] is characterized by a parameter denoted $\sin^2\theta_W$.
Neutrino-nucleon scattering experiments[46], which yield a value of $\sin^2\theta_W = 0.234 \pm 0.011$, are characterized by the following experimental conditions:
1. the relatively low c.m. energy, namely s ∼ 300 GeV²
2. the relatively low momentum transfer q² ∼ -100 GeV² in the space-like region
3. the use of nuclear targets.

At e^+e^- machines complementary experiments have been performed measuring electro-weak parameters from the reactions:

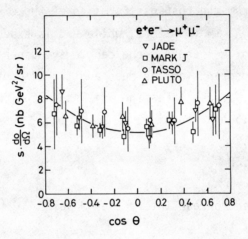

Figure 27 The cos θ distributions as measured by the PETRA experiments for muon pair production. The solid line is the 1 + $\cos^2\theta$ expectation.

TABLE II

The values for the muon and tau asymmetry for the various experiments. The expected values from the GWS model are also given.

	JADE	MARK II	MARK-J	PLUTO	TASSO
$A_{\mu\mu}$%	-5 ± 6	-4.7 ± 4.2	-1 ± 6	7 ± 10	-6 ± 8
Expected	-6	-4.0	-7.7	-5.8	-6
$A_{\mu\mu}$%			-6 ± 12 prel.		0 ± 11 prel.
Expected			-5		-7

$$e^+ + e^- \rightarrow (\gamma + Z^0) \rightarrow \ell^+ \ell^- \qquad (\ell = e, \mu, \tau) \qquad (1)$$

and

$$e^+ + e^- \rightarrow (\gamma + Z^0) \rightarrow \text{hadrons} \qquad (2)$$

where the production of final state leptons and hadrons is via both virtual photons (γ) and neutral intermediate vector bosons (Z^0).

In comparison with neutrino experiments, these experiments have the following unique properties:

a. high c.m. energy, namely $s \sim 1300$ GeV²,

b. most data is at the large momentum transfer $q^2 \sim 1300$ GeV², and in both the space-like and the time-like region.

c. purely point-like particles in the initial state.

In the framework of the parton model, the present neutrino experiments are viewed as lepton scattering on u and d quarks whereas at PETRA one uses either purely leptonic reactions such as in reaction (1) or a mixture of 5 quarks with substantial contributions from the heavy c and b quarks as in reaction (2). First we will describe the derivation of $\sin^2\theta_W$ with the lepton pair data.

The measurements consist of the already discussed data on muon and tau pair production. In addition the measurements of the angular distribution of the Bhabha scattering events are used. In Fig. 28 the available $e^+e^- \rightarrow e^+e^-$ data [37] at the highest PETRA energies are shown. The differential cross section $d\sigma/d\Omega$ is normalized with the first order QED cross section $d\sigma_{QED}/d\Omega$ and is plotted as function of $\cos\theta$.

The luminosity is measured by Bhabha scattering in the central detectors and independently in monitoring stations centered at small scattering angles (~ 30 mrad). The two luminosity measurements generally agree within 3% once radiative corrections have been taken into account [47,48], and one can take 3% as a conservative estimate of the error in the luminosity for reaction (1).

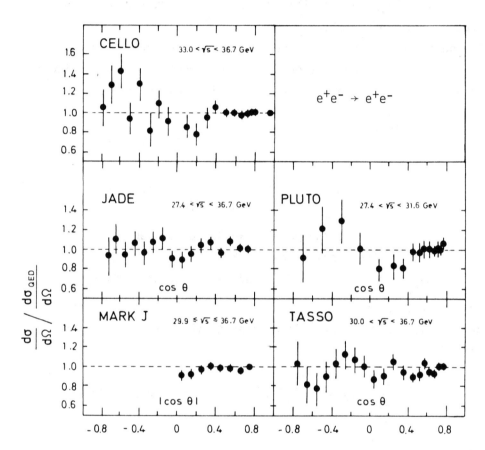

Figure 28 The differential Bhabha cross sections $d\sigma/d\Omega$ normalized with the point-like QED cross section. The dashed lines are the expected distribution from QED.

The results of the determination of $\sin^2\theta_W$ are summarized in Table III. The values obtained range from 0.23 - 0.25 for the different experiments. As no deviations are seen from QED so far (see Table I for the cut-off parameters for $e^+e^- \to e^+e^-$) the values for $\sin^2\theta_W$ naturally fall in this region as the smallest effects are expected at $\sin^2\theta_W = 0.25$.

The next step is to use the hadron data to determine $\sin^2\theta_W$. In reaction (2) above, if one takes weak neutral current effects into account, one can write the hadronic cross sections as[49]

$$R_f = \frac{\sigma(e^+e^- \to \gamma, Z^0 \to f\bar{f})}{\sigma_p} = Q_f^2 - 8s\, Q_f \cdot g_V \cdot g_{V_f} \cdot p(s)$$

$$+ 16s^2 g^2 (g_V^2 + g_A^2)(g_{V_f}^2 + g_{A_f}^2) \cdot p'(s),$$

where $\sigma_p = 4\pi\alpha^2/3s$ is the point-like QED cross section,
Q_f = charge of the final state quark f,
g_{V_f} and g_{A_f} are the weak vector and axial vector coupling constants of quark f,
and $g = 4.47 \times 10^{-5}$ GeV^{-2} is related to the Fermi coupling constant.

$$p(s) = \left(\left(\frac{s}{m_Z^2} - 1\right) + \frac{\Gamma_Z^2}{s - m_Z^2} \right)^{-1} \quad \text{for } \gamma \text{ and } Z^0 \text{ interference,}$$

and

$$p'(s) = \left(\left(\frac{s}{m_Z^2} - 1\right)^2 + \frac{\Gamma_Z^2}{m_Z^2} \right)^{-1} \quad \text{for pure } Z^0 \text{ exchange.}$$

In the framework of the GWS model, assuming e-μ universality, the weak coupling constants are given by:

$$g_{V_f} = T_{3_f}^L - 2Q_f \sin^2\theta_W$$

$$g_{A_f} = T_{3_f}^L$$

TABLE III

The values for $\sin^2\theta_W$ obtained by the various experiments.

	$\sin^2\theta_W$	leptons
CELLO	$0.22 ^{+0.15}_{-0.10}$	e^+e^-, $\tau^+\tau^-$, $A_{\mu\mu}$
JADE	0.25 ± 0.15	e^+e^-, $\mu^+\mu^-$, $A_{\mu\mu}$
MARK-J	0.25 ± 0.11	e^+e^-, $\mu^+\mu^-$, $\tau^+\tau^-$, $A_{\mu\mu}$
PLUTO	0.23 ± 0.17	e^+e^-, $\mu^+\mu^-$
TASSO	0.25 ± 0.10	e^+e^-, $\mu^+\mu^-$, $A_{\mu\mu}$

where T_{3f}^L is the weak isospin of the left handed quark. The mass of the Z^0 can be expressed as $m_Z = 37.28/\sin\theta_W\cos\theta_W$ GeV and the width is taken to be constant: $\Gamma_Z = 2.3$ GeV.

The electroweak hadronic cross section R_T is given in the quark model by the incoherent sum over all final state quarks including a color factor of three and QCD corrections[50].

$$R_T = 3 \cdot \Sigma_f R_f \left(1 + \frac{\alpha_s(s)}{\pi} + (1.98 - 0.115 \cdot N_f) \frac{\alpha_s^2(s)}{\pi^2} \right)$$

where α_s is the coupling constant of the strong interaction, which has been measured at PETRA energies to be[51] $\alpha_s = 0.18$ for $\sqrt{s}$ in the 30 to 36 GeV range, and which changes slowly with energy. $N_f = 5$ is the number of different quark flavors.

Fig. 29 shows[52] the dependance of R on $\sin^2\theta_W$ for different values of the centre of mass energy squared, s.

The measurement of R of the JADE collaboration is ~ 4. At the highest PETRA energy (s = 1350 GeV²) large deviations are expected for values of $\sin^2\theta_W$ at ~ 0 and ~ 1. The solid curve intersects the value of R at $\sin^2\theta_W \sim 0.2$ and ~ 0.55. As the systematic error in the determination of R is $\sim 7\%$, a normalisation parameter is introduced in a fit to the data. The result of the fit is displayed in Fig. 30 where the 1σ contours are plotted in the $\sin^2\theta_W$ - normalization plane - (dashed contours). Two values for $\sin^2\theta_W$ are obtained namely

$$\sin^2\theta_W = \begin{matrix} 0.22 \pm 0.03 \text{ (stat)} \\ 0.56 \pm 0.03 \text{ (stat)} \end{matrix} \pm 0.10 \text{ (syst)}$$

Also indicated in the figure is the result if one combines the hadron data with the lepton pair results (solid contour). One finds:

$$\sin^2\theta_W = 0.22 \pm 0.08.$$

The MARK J collaboration[53] makes a quantitative comparison by using a χ^2-method, where χ^2 is defined as:

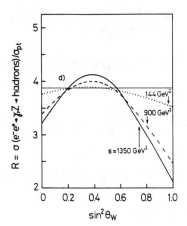

Figure 29   The variation of the total cross-section R with s and $\sin^2\theta_W$ as predicted by the Glashow-Salam-Weinberg model. The horizontal line at R = 3.87 corresponds to γ exchange only.

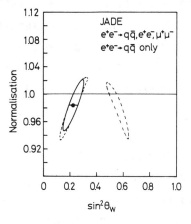

Figure 30   The log-likelihood function contour (68% C.L.) in the $\sin^2\theta_W$-normalization plane. The dashed curves show the solution from the fit to the hadron data, the solid curve is the solution for the combined data.

$$\chi^2 = \frac{(F-1)^2}{\sigma_{syst}^2} + \sum_i \frac{(F R^i - R_T^i)^2}{\sigma_i^2}$$

where the sum runs over all measurements at different c.m energies, and where σ_i and σ_{syst} are the statistical and systematic errors respectively. The systematic error σ_{syst} is estimated to be 10%. The scaling factor F is used in order to take into account the overall normalization uncertainties.

The theoretical cross section R_T is compared with the experimental result R for different values of $\sin^2\theta_W$ [54] in Fig. 31.
The results of the fit are:

$$0.12 \leq \sin^2\theta_W \leq 0.65 \qquad (95\% \text{ C.L.})$$

and

$$\sin^2\theta_W = 0.27 {}^{+0.34}_{-0.08}$$

with a minimum χ^2 per degree of freedom = 1.14.

The limits on $\sin^2\theta_W$ change by $\lesssim 0.02$ for relatively large changes in α_s (0 to 0.24) and σ_{syst} (8% to 12%). The values for the coupling constant α_s and the scaling factor F are correlated; i.e. changes in α_s are compensated by an appropiate change in F. This demonstrates that the method applied is sensitive to the behaviour of R as a function of energy and not to the absolute R values.
A combined analysis of MARK J data for leptonic (reaction (1)) and hadronic final states (reaction (2)) yields:

$$0.13 \leq \sin^2\theta_W \leq 0.42 \qquad (95\% \text{ C.L.})$$

and

$$\sin^2\theta_W = 0.27 \pm 0.08.$$

In conclusion, the determination of $\sin^2\theta_W$ is in good agreement with the value obtained from neutrino scattering experiments. The two types of experiments are done in entirely different kinematic regions. The good agreement in $\sin^2\theta_W$ obtained in two types of experiments gives important support to the validity of the GWS theory, and its applicability in the timelike region up to $q^2 = 1300 \text{ GeV}^2$.

e⁺e⁻ INTERACTIONS

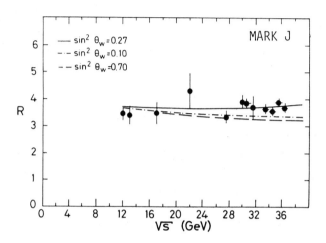

Figure 31 Experimental results on R with statistical errors as a function of c.m. energy $\sqrt{s}$. The solid line shows the theoretical prediction R_T for the best fit to our data ($\sin^2\theta_W = 0.27$ using $\alpha_s = 0.18$ at $\sqrt{s} = 30$ GeV). The dependence of R_T on $\sin^2\theta_W$ is also shown for two extreme values ($\sin^2\theta_W = 0.70$ and 0.10). The best fitted results require that the normalization of the curves be scaled down by 7.5% as shown (the normalization uncertainty is 10%).

b) Determination of g_A and g_V [45,53].

In comparing the data on $e^+e^- \to \ell^+\ell^-$ ($\ell = e, \mu, \tau$) with the predictions of electroweak theories containing one Z^o, the yields for reactions (1) are calculated according to QED using Monte Carlo programs which include order α^3 radiative corrections [48] The cross section for $e^+e^- \to e^+e^-$ has been calculated to second order in the weak interaction coupling strength by Budny [55]. The MARK J experiment has a total of 2×10^5 events for an integrated luminosity of 10,818 nb^{-1}, which represents a 4-fold increase in statistics from a previous result [56]

The modification of the Bhabha scattering angular distribution by weak effects is not immediately apparent in the distribution of $d\sigma(e^+e^- \to e^+e^-)/d\Omega$ commonly used in tests of pure QED. This is shown in Fig. 32a where the combined cross section from 29.9 to 36.7 GeV is compared with the predictions of QED. The sensitivity to weak effects is, however, demonstrated in Fig. 32b, which shows the deviation of the combined data from the QED predictions as a function of $\cos\theta$. The errors shown in Fig. 32b include both statistical errors and an estimated point to point 3% systematic uncertainty. Also shown in Fig. 32b are the predictions of the electroweak theory in terms of the values of vector and axial vector coupling constants g_A and g_V. The best fitted curve obtained from Bhabha scattering as well as $\mu^+\mu^-$, $\tau^+\tau^-$ production and $\mu^+\mu^-$ asymmetry is also included. It has to be noted that it is the Bhabha data that is mainly sensitive to the vector coupling. The constraint on the axial coupling is mainly obtained from the asymmetry measurement.

The results for the vector and axial vector couplings squared h_{VV} and h_{AA} are shown in Fig. 33.

In the $h_{VV} - h_{AA}$ plane the GWS model is represented by a line parallel to the h_{VV} axis and intersecting the h_{AA} axis at 0.25. The MARK J group converted these results including correlations

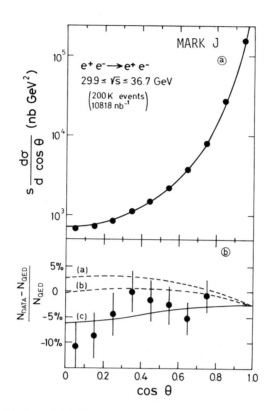

Figure 32a The differential cross sections $d\sigma/d\cos\theta$ for $e^+e^- \to e^+e^-$.
b The deviation of the QED predictions from the combined data is plotted as a function of $\cos\theta$. The error includes both statistical and 3% point to point systematic uncertainties. The curves a, b and c represent the prediction of the electroweak theory with ($g_A^2 = 1/4$, $g_V^2 = 0.16$), ($g_A^2 = 1/4$, $g_V^2 = 1/4$) and the best fitted result ($g_A^2 = 0.21$, $g_V^2 = -0.05$) respectively. The best fitted results require that the normalization of the curves a, b and c be scaled down by 2.5% as shown (the normalization uncertainty is 3%).

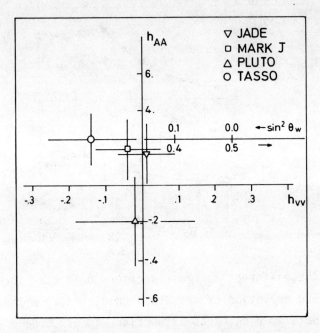

Figure 33 Best fit values for h_{VV} and h_{AA} and their uncorrelated 1σ errors from the PETRA experiments.

into an allowed region in the $g_V - g_A$ plane, as shown in Fig. 34. Also shown in the figure are neutrino electron scattering data $\nu_\mu e^- \to \nu_\mu e^-$, $\bar{\nu}_\mu e^- \to \bar{\nu}_\mu e^-$, $\bar{\nu}_e e^- \to \bar{\nu}_e e^-$, which limit the possible value of g_A and g_V to two regions in the $g_V - g_A$ plane around $(g_V = 0, g_A = -\frac{1}{2})$ and $(g_V = -\frac{1}{2}, g_A = 0)$ [57]. Combining the MARK J and the neutrino scattering data the second solution is ruled out with more than 95% confidence.

This confirms the conclusion drawn on the basis of deep inelastic neutrino nuclear scattering and polarized electron deuterium scattering data. The allowed region from these experiments plus the additional requirement of factorization relations is shown in the figure as the hatched area [58]. It has to be noted, however, that the MARK J result is obtained without having recourse to models of hadron production by the weak neutral current.

7. CONCLUSIONS.

The main conclusions of this talk can be summarized as follows:

a. No new particles have been found in e^+e^- interactions.

b. The preliminary leptonic branching ratio for mesons containing a b-quark found by the MARK J collaboration is:

$$\frac{B \to \mu + X}{B \to all} = 8.0 \pm 2.7 \text{ (stat)} \pm 2.0 \text{ (syst)}$$

in good agreement with results obtained at CESR. The JADE group obtained as an upper limit for mesons containing a b-quark $\tau_B < 5 \cdot 10^{-12}$ at 90% C.L.

c. The PETRA experiments obtained an average value for the muon asymmetry $A_{\mu\mu} = (-2.8 \pm 3.4)\%$, while -6.7% is expected from the GWS model.

MARK II at PEP found for the axial coupling constant
$$|g_A| = 0.60 ^{+0.18}_{-0.25}$$

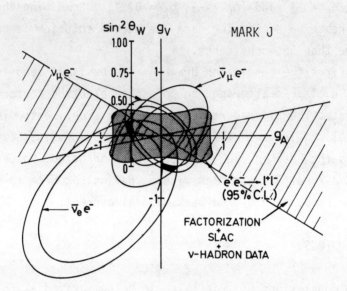

Figure 34 Results obtained from neutrino experiments and the
MARJ-J experiment expressed in terms of limits on g_V
and g_A. The region in between the concentric ellipses
corresponds to 68% C.L. limits from the neutrino electron scattering experiments. The two black areas indicate the two allowed regions for g_V and g_A from the
combined neutrino data. The shaded area represents the
95% confidence limit contour from the MARK -J experiment. The hatched area is the allowed region obtained
from deep inealstic neutriono nuclear scattering and
polarized electron deuterium scattering data.

d. Combining the results from MARK J and JADE on their measurement of $\sin^2\theta_W$ using fermion-antifermion production one finds
$\sin^2\theta_W = 0.245 \pm 0.057$.

e. The coupling constants of the weak neutral current are uniquely determined by lepton - lepton scattering alone and they are in good agreement with the predictions of the standard $SU(2) \otimes U(1)$ theory.

ACKNOWLEDGEMENTS

I want to thank my colleagues of the MARK-J collaboration and acknowledge helpful discussions with Drs. H. de Groot, E. Eichler, G. Grindhammer, J.P. Revol, H. Takeda, J.A.M. Vermaseren and G. Wolf. I thank Prof. S.C.C. Ting for his continuous support and encouragement.

REFERENCES

(1) K. Steffen, DESY Internal Report DESY M-79/07 (1979).

(2) W. Bartel el al., (JADE-Collaboration), Phys. Lett. 88B, 171 (1979)
S. Orito, Proc. Int. Symp. on Lepton and Hadron Interactions 1979, p.52
D.P. Barber et al., (MARK-J Collaboration), Phys. Rep. 63, 7, 337 (1980)
Ch. Berger et al., (PLUTO-Collaboration), Phys. Lett. 81B, 410 (1979); 86B, 413 (1979)
R. Brandelik et al., (TASSO-Collaboration), Phys. Lett. 83B, 261 (1979); Z. Phys. C4, 87 (1980)
H. Ryckaczewski, Ph. D. Thesis, Aachen, 1981 (unpublished).

(3) J.L. Rosner, C. Quigg, H.B. Thacker, Phys. Lett. 74B, 350 (1978)
C. Quigg, contribution to the 1979 Intern. Symp. on Lepton and Photon Interactions, Fermilab
M. Greco, Phys. Lett. 77B, 84 (1978)
T. Applequist and H. Georgi, Phys. Rev. D8, 4000 (1973)
A. Zee, Phys. Rev. D8, 4038 (1973)
M. Dine, J. Sapirstein, Phys. Lett. 43, 152 (1979).

(4) R. Brandelik et al., (TASSO-Collaboration), Phys. Lett. 94B, 437 (1980).

(5) J.D. Bjorken and S. Brodsky, Phys. Rev. D1, 1416 (1970).

(6) G. Wolf, (TASSO-Collaboration), private communication.

(7) M. Kobayashi and T. Maskawa, Progr. Theor. Phys. <u>49</u>, 652 (1973)
A. Ali et al., "Heavy Quarks in e^+e^- Annihilation", DESY Report 79/63 (1979)
A. Ali, Z. Physik C, Particles and Fields <u>1</u>, 25 (1979).

(8) C. Bebek et al., Phys. Rev. Lett. <u>46</u>, 84 (1981)
K. Chadwick et al., Phys. Rev. Lett. <u>46</u>, 88 (1981).

(9) R. Eichler and D. Haidt, (JADE-Collaboration), private communication.

(10) J.M Weiss et al., (MARK II Collaboration), Phys. Lett. <u>101B</u>, 439 (1979).

(11) W. Bartel et al., (JADE-Collaboration), Z. Physik C, Particles and Fields <u>6</u>, 295 (1980).

(12) A. Marini et al., "A free quark search at PEP", paper contributed to this conference.

(13) D. Cords, Invited talk at the International High Energy Conference 1980 (Wisconsin).

(14) ADONE Proposal INFN/AE-67/3 (March 1967), ADONE-Frascati (unpublished) and M. Bernardini et al., Nuovo Cimento <u>17A</u>, 383 (1973)
S. Orito et al., Phys. Lett. <u>48B</u>, 165 (1974).

(15) M. Perl et al., Phys. Rev. Lett. <u>35</u>, 1489 (1975)
G. Feldman et al., Phys. Rev. Lett. <u>38</u>, 117 (1977).

(16) J. Burmester et al., Phys. Lett. <u>68B</u>, 297 (1977)

J. Burmester et al., Phys. Lett. **68B**, 301 (1977)
For a review of our present knowledge of the τ lepton, see
G. Fluegge, Zeitschr. f. Physik C1, Particles and Fields,
121 (1979), and the references therein
R. Brandelik et al., Phys. Lett. **92B**, 199 (1980).

(17) Y.S. Tsai, SLAC-Preprint, SLAC-PUB-2450, Dec. 1979.

(18) D.P. Barber et al., Phys. Rep. **63**, 7 (1980), 337-391.

(19) D.P. Barber et al., Phys. Rev. Lett. **45**, 1904 (1980).

(20) Ch. Berger et al., (PLUTO-Collaboration), Phys. Lett. **99B**, 489 (1981).

(21) R. Brandelik et al., (TASSO-Collaboration), Phys. Lett. **99B**, 163 (1981).

(22) F. Bletzacker and H.T. Nieh, Phys. Rev. **D16**, 2115 (1977).

(23) H. Takeda, (JADE-Collaboration), private communication.

(24) Yu. A. Gol'fand and E.P. Likthman, JETP Letters **13**, 323 (1971)
J. Wess and B. Zumino, Nucl. Phys. **B70**, 39 (1974)
For a review article, see:
P. Fayet and S. Ferrara, Phys. Reports **32C**, 249 (1977)
P. Fayet, Phys. Lett. **69B**, 489 (1977)
G.R. Farrar and P. Fayet, Phys. Lett. **76B**, 575 (1978); **79B**, 442 (1978)
P. Fayet, Phys. Lett. **84B**, 421 (1979)
P. Fayet, Phys. Lett. **86B**, 272 (1979)
G.R. Farrar and P. Fayet, Phys. Lett. **89B**, 191 (1980).

(25)   F.B. Heiel et al., Nucl. Phys. B138, 189 (1978).

(26)   "Search for Scalar Leptons at PETRA" (CELLO-Collaboration), contributed paper to the "International Conference on High Energy Physics" at Lisbon (1981).

(27)   D.P. Barber et al., Phys. Rev. Lett. 44, 1722 (1980).

(28)   Ch. Berger et al., (PLUTO-Collaboration), DESY Report 80/42 (1980).

(29)   S. Weinberg, Phys. Rev. D19, 1277 (1979)
        L. Susskind, Phys. Rev. D20, 2619 (1979).

(30)   A. Ali and M.A.B. Beg, DESY Report 80/90
        A. Ali, H.B. Newman and R.Y. Zhu, DESY Report 80/110 (1980) to be published in Nucl. Phys. B
        A. Ali, DESY Report 81/032 (1981).

(31)   H. Takeda, (JADE-Collaboration), private communication.

(32)   R.D. Peccei and H.R. Quinn, Phys. Rev. Lett. 38, 1440 (1977); Phys. Rev. D16, 1791 (1977)
        S. Weinberg, Phys. Rev. Lett. 40, 223 (1978)
        F. Wilczek, Phys. Rev. Lett. 40, 279 (1978).

(33)   H. Faissner et al., Phys. Lett. 103B, 234 (1981).

(34)   F.A. Berends and R. Kleiss, DESY Report 80/66.

(35)   All Monte Carlo calculations concerning $e^+e^- \to e^+e^-\mu^+\mu^-$ were made using J.A.M Vermaseren's event generator, and a detailed simulation of the MARK-J detector.

(36)   A. Litke, Harvard University, Ph. D. Thesis (1970), unpublished.

(37)   P. Dittman and V. Hepp, DESY Report 81/030 (1981).

(38)   C. Bacai et al., Lett. Nuovo Cim. $\underline{2}$, 73 (1971)
       G. Hanson et al., Lett. Nuovo Cim. $\underline{7}$, 587 (1973)
       B.L. Beron et al., Phys. Rev. Lett. $\underline{33}$, 663 (1974)
       E. Hilger et al., Phys. Rev. $\underline{D15}$, 1809 (1977).

(39)   S.L. Glashow, Nucl. Phys. $\underline{22}$, 579 (1961)
       S. Weinberg, Phys. Rev. Lett. $\underline{19}$, 1264 (1967)
       S. Weinberg, Phys. Rev. $\underline{D5}$, 1412 (1972)
       A. Salam, Proc. 8th Nobel Symposium, Aspenäsgaden, 1968 Almqvist and Wiksell, Stockholm 1968, 367.

(40)   PLUTO-Collaboration, G. Flügge, talk presented to Intern. Conf. on High Energy Physics (Geneva, 1979), DESY Report 79/37 (1979).

(41)   R. Brandelik et al. (TASSO-Collaboration), Phys. Lett. $\underline{92B}$, 199 (1980).

(42)   "Measurement of the Reaction $e^+e^- \to \tau^+\tau^-$ at PETRA" (CELLO-Collaboration), contributed paper to the International Conference on High Energy Physics at Lisbon, 1981.

(43)   J.P. Revol, MIT Ph. D. Thesis (unpublished).

(44)   Contribution to the LISBON conference (1981).

(45)   M. Pohl, Review talk given at the XVI-th Recontre du Moriond (1981).

(46)   P. Langacker et al., Proc. Neutrino-79, Bergen Vol. 1, 276 (1979)
         J.E. Kim et al., Pennsylvania Report UPR-158T (1980)
         I. Liede and M. Roos, Proc. Neutrino-79, Bergen Vol. 1, 309 (1979)
         J.J. Sakurai, ibid, 267
         L.M. Sehgal, Proc. Symposium on Lepton and Hadron Interactions, Visegard (1979)
         Ed.F. Csikor et al., (Budapest), Aachen Report PITHA-79/34, 29.

(47)   G. Ripken, private communication.

(48)   F.A. Berends, K.J.F. Gaemers and R. Gastmans, Nucl. Phys. $\underline{B57}$, 381 (1973); $\underline{B63}$, 381 (1973)
         F.A. Berends et al., Phys. Lett. $\underline{63B}$, 432 (1976).

(49)   J. Ellis and M.K. Gaillard, Physics With Very High Energy e^+e^- Colliding Beams, CERN 76-18, 21 (1976).

(50)   K.G. Chetyrkin et al., Phys. Lett. $\underline{85B}$, 277 (1979)
         M. Dine & J. Sapierstein, Phys. Rev. Lett. $\underline{43}$, 668 (1979)
         W. Celmaster and R. Gousalves, Phys. Rev. Lett. $\underline{44}$, 560 (1980).

(51)   D.P. Barber et al., Phys. Rev. Lett. $\underline{43}$, 830 (1979)
         D.P. Barber et al., Phys. Lett. $\underline{89B}$, 139 (1979)
         W. Bartel et al., Phys. Lett. $\underline{91B}$, 142 (1980)
         Ch. Berger et al., Phys. Lett. $\underline{86B}$, 418 (1979)
         R. Brandelik et al., Phys. Lett. $\underline{86B}$, 243 (1979)
         R. Brandelik et al., DESY-Report 80/40 (1980
         H.B. Newman, XXthe Int. Conf. on High Energy Physics, Madison (1980).

(52) W. Bartel et al., (JADE-Collaboration), Phys. Lett. 101B, 361 (1981).

(53) D.P. Barber et al., Phys. Rev. Lett. 46, 1663 (1981).

(54) D.P. Barber et al., to be published
H. Rykaczewski, Aachen thesis (1981), unpublished.

(55) R. Budny, Phys. Lett. 55B, 227 (1975)
E.H. De Groot et al., University of Bielefeld Report Bi-TP 80/08

(56) A. Böhm, Technische Hochschule, Aachen, Report No. PITHA 80/9, 1980, (Invited Talk at the XXth Intern. Conf. on High Energy Physics, Madison, 1980), to be published
D.P. Barber et al., Phys. Lett. 95B, 149 (1980)
P. Duinker, DESY-Report 81/012 (1981)

(57) H. Faissner, New Phenomena in Lepton-Hadron Physics, eds. D.E. Fries and J. Wess (Plenum, New York, 1979), p. 371.
H. Reithler, Phys. Blätter 35, 630 (1979)
R.H. Heisterberg et al., Phys. Rev. Lett. 44, 635 (1980)
L.W. Mo, Contribution to Neutrino 80 (Erice, 1980)
H. Faissner and H. Reithler, private communication.

(58) P.Q. Hung and J.J. Sakurai, Phys. Lett. 88B, 91 (1979).

DISCUSSION

CHAIRMAN: P. DUINKER

Scientific Secretary: R. Clare

DISCUSSION

- KARLINER:

Can you give us a lower limit on the Z^0 mass from the corrections to QED ?

- DUINKER:

From the μ-charge asymmetry the lower limit is 60 GeV.

- WITTEN:

Your bounds on g_V and g_A are hard to evaluate because they depend on several experiments each with large errors. If you had favoured the other solution with the same confidence, how confident would you have been ?

- DUINKER:

The bounds on g_V and g_A obtained with $e^+e^- \to e^+e^-$ and $\mu^+\mu^-$ are straightforward and come from one experiment. They single out the solution that is in agreement with the standard model. The Hung-Sakurai analysis which uses various sources of data also singles out that solution. I do not know how to answer the second part of your question.

UPDATE ON CP-VIOLATION

Val L. Fitch

Princeton University
Princeton, N.J., U.S.A.

In the surroundings of a lovely, medieval, walled town high on a mountain, it is perhaps not inappropriate to begin with a few historical observations. First, it has been slightly more than 50 years since Dirac proposed his famous equation which has contained in it, intrinsic in its structure, the notion of antimatter. And completely independent, without knowledge of the theoretical developments, Carl Anderson discovered the positron, the antimatter twin of the electron. Now, of course, we are very sophisticated and recognize that the existence of antimatter - symmetric with matter - is a natural consequence of the joining of relativity and quantum mechanics.[1]

Second, it has been almost precisely 25 years since Lee and Yang proposed[2] a series of tests to see if the puzzle posed by the tau and theta particles was indeed a manifestation of a failure of parity conservation in the weak interactions. This was immediately followed by that series of definitive experiments which showed with startling clarity that parity was indeed violated; not only that, but charge conjugation also. The complete symmetry between matter and antimatter, supposed for the previous 25 years, was not true. In an elegant paper written between the time of the Lee-Yang paper and the experimental discovery, Landau observed[3] that parity nonconservation would create very difficult problems in physics unless at the same time charge conjugation noninvariance was invoked, unless the notion of space inversion symmetry (parity conservation) was replaced by space inversion (P) and particle-antiparticle interchange (C). Indeed, this was the path that nature appeared to take. The failure of parity was compensated by a failure of charge conjugation. All interactions appeared invariant under CP.

Third, it has been 17 years since it was discovered that the physical world was not completely symmetric under CP, a small but unmistakable violation appeared in the system of neutral K mesons. Subsequently, it has been found that the CP violation appears to be compensated by a failure of time reversal invariance, T, preserving the general idea of TCP invariance. Clearly, a failure of time-reversal invariance touches on our understanding of nature at its deepest level. It has ramifications which extend from particle physics to cosmology. And it serves to remind us that at the present time we are still far from really understanding the nature of the physical world.

Status of CP Violation in the Neutral K-meson System

The CP violation in the neutral K system could have occurred either in the mass-decay matrix, which determines the time dependence of the K^0 and $\overline{K}^0$ amplitudes, or in the amplitudes for decay of the K^0-$\overline{K}^0$ system to the observed 2π states. From the measurements which have been performed to date we know that most, if not all, the violation comes from the mass-decay matrix, viz:

$$-\frac{d}{dt}\begin{pmatrix} K^0 \\ \overline{K^0} \end{pmatrix} = \begin{pmatrix} A & p^2 \\ q^2 & B \end{pmatrix} \begin{pmatrix} K^0 \\ \overline{K^0} \end{pmatrix}$$

where we have let the particle symbol stand for the amplitude of the corresponding state. With TCP invariance A=B, and CP invariance also, $p^2=q^2$. The eigenstates are

$$K_S^0 = (p|K^0\rangle + q|\overline{K^0}\rangle)/\sqrt{p^2+q^2} \; .$$

and

$$K_L^0 = (p|K^0\rangle - q|\overline{K^0}\rangle)/\sqrt{p^2+q^2} \; .$$

with eigenvalues

$$\Gamma_{\substack{S \\ L}} = \frac{A+B}{2} \pm \sqrt{\left(\frac{A-B}{2}\right)^2 + p^2 q^2} \; .$$

The relative amplitudes for the decay of the K_L^0 to $\pi^+\pi^-$ mesons compared to K_S^0 to the same state is

$$\frac{A(K_L \to \pi^+\pi^-)}{A(K_S \to \pi^+\pi^-)} = \eta_{+-} = \varepsilon + \varepsilon'$$

where ε is the contribution from the mass-decay matrix and ε' is from the decay amplitudes A_0 and A_2, where the subscript refers to the corresponding isotopic spin states of the 2 pions. In particular

$$\varepsilon = \frac{p-q}{p+q} \quad \text{and} \quad \varepsilon' = \frac{ie^{i(\delta_2-\delta_0)}}{\sqrt{2}} \operatorname{Im}\frac{A_2}{A_0}$$

where δ_2 and δ_0 are the π-π scattering phase shifts. The corresponding parameter for $2\pi^0$ decay is $\eta_{00}=\varepsilon-2\varepsilon'$. The notation is that established by Wu and Yang[4] who took A_0 to be real and allowed for the possibility of A_2 to be complex. This is all well known but we repeat it because in a later discussion the natural choice will be to make A_2 real.

Current experiments show

$$\eta_{+-} = (2.27 \pm 0.02) \times 10^{-3} \, e^{i\phi_{+-}}$$

$$\phi_{+-} = 44.7 \pm 1.2^0$$

and $\eta_{00} = 2.31 \pm 0.1 \times 10^{-3} \, e^{i\phi_{00}}$

$$\phi_{00} = 54.7 \pm 6^0$$

The experiments on the charge asymmetry in K_L^0 semi-leptonic decay, assuming the $\Delta S = \Delta Q$ rule, are sensitive to the real part of ε, viz.,

$$\delta_e = \frac{\Gamma(K_L^0 \to \pi^+ e^- \nu) - \Gamma(K_L^0 \to \pi^+ e^- \bar{\nu})}{\Gamma(K_L^0 \to \pi^- e^+ \nu) + \Gamma(K_L^0 \to \pi^+ e^- \bar{\nu})} = 2\text{Re}(\varepsilon)$$

and is measured to be

$$= 3.33 \pm 0.14 \times 10^{-13} \ .$$

Correspondingly, for $K_L^0 \to \pi^\pm \mu^\mp \nu$

$$\delta_\mu = 3.19 \pm 0.24 \times 10^{-3} \ .$$

From the experimental data it is derived that

$$\left| \frac{\varepsilon'}{\varepsilon} \right| \cong \frac{1}{3} \left(1 - \left| \frac{\eta_{00}}{\eta_{+-}} \right| \right) = .007 \pm 0.014 \ .$$

That is, to within about 2%, the violation of CP is in the mass-decay matrix, in ε.

In the most qualitative terms the CP violation appears as a difference in the transition amplitude of $K^0 \to \overline{K}^0$ (q^2) and $\overline{K}^0 \to K^0$ (p^2). The transition amplitude from antiparticle to particle differs in both amplitude and phase for the reverse process.

In these few numbers we have summarized a very great effort devoted to many elegant and ingenious experiments performed in the past 17 years on the neutral K system.

Additional conclusions require accessory data. For example, in testing whether the violation is in TCP instead of CP it is necessary to know the sign of the K_L, K_S mass difference. If TCP were violated and CP conserved the predicted phase of η_{+-} would be 90^0 removed from the measured value. Clearly, the data admits only a very small violation of TCP[5] and is completely consistent with CP and T violation.

Time Reversal Violation in other Systems

Concurrent with the elucidation of CP violation in the neutral K system there has been an exhaustive search for effects in other systems. Initially it was thought likely that the effect could be very large in some interactions and only happened to be small in the neutral K's. This optimism has not proven justified and the effect remains measured only where it was discovered. As we will see, several theoretical models predict milliweak effects everywhere but thus far none of the experiments outside the K system has been sufficiently sensitive to measure CP or T violating amplitudes at the 10^{-3} level. The electric dipole moment of the neutron is a

Table I

	Measured Parameter	Result	
$K_{\mu 3} \to \pi\mu\nu$	$\vec{\sigma}_\mu \cdot \vec{P}_\pi \times \vec{P}_\mu$	$-2.1 \pm 4.8 \times 10^{-3}$	(15)
$n \to pe\nu$	$\vec{\sigma}_n \cdot \vec{P}_e \times \vec{P}_\nu$	$-1.1 \pm 1.7 \times 10^{-3}$	(16)
Ne^{19} decay	$\vec{\sigma}_{Ne} \cdot \vec{P}_e \times \vec{P}_\nu$	$-0.5 \pm 1.0 \times 10^{-3}$	(17)
Mixed transitions in decay of polarized nuclei		$4.7 \pm 0.3 \times 10^{-3}$	(18)
(expected from final state int.)		4.3×10^{-3}	(19)
Detailed balance	$\dfrac{A_{mpl} \text{ T violating}}{A_{mpl} \text{ T conserving}}$	$<3 \times 10^{-3}$	(20)
Electric dipole moment of neutron		$0.4 \pm 1.5 \times 10^{-24} e$ cm	(21)
		$0.4 \pm 0.75 \times 10^{-24} e$	(22)

special case to which we will return later. The most precise data is given in Table I. As seen most of the experimental errors remain at the few times 10^{-3} level which means that T or CP violation is currently well tested only at the centiweak level. Another factor of ten improvement in the experiments is clearly needed.

For the neutron to have an EDM both parity and time reversal must be violated. A number of search experiments of exquisite sensitivity have been performed. Most of these have used the magnetic resonance split field technique originally devised by Ramsay - for minimum line width. The basic idea is to do the magnetic resonance experiment with an electric field superimposed

UPDATE OF CP-VIOLATION

on the magnetic and measure the change in the resonant frequency on reversal of the electric field. The magnetic moment is $\sim 10^{-14}$ e cm. The current limit on the electric moment is $<10^{-24}$ e cm or $\sim 10^{-10}$ of the value for the magnetic moment. Further improvements on this important number are to be obtained by narrowing the resonance - using slower neutrons. It is hoped that by increasing the neutron holding time within the resonance apparatus the limit can be extended in the near future to 10^{-26} e cm.

Matter-Antimatter Asymmetry

As we have noted above the CP violation in the neutral K system shows that the transition rate from particle (K^0) to antiparticle ($\bar{K}^0$) does not equal the inverse. This fact was seized on very early by A. Sakharov[6] to provide a mechanism for explaining the observed matter-antimatter asymmetry in the universe. This subject is treated in detail by Professor Wilczek in his lectures and we only summarize here. Required is a lepto-quark, L_q, which couples leptons and quarks, viz.,

$$L_q \leftrightarrow e^+ + \bar{u}$$

and

$$L_q \leftrightarrow u + d \; .$$

With these postulated couplings of the quarks to the leptons, proton decay occurs via

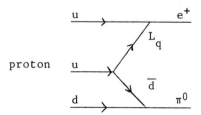

As is well-known grand unification specifies the leptiquark to have a mass of $\sim 10^{14}$ GeV which gives the proton an appropriately long life. But we should keep an open mind and be prepared for leptoquarks of any smaller mass with other mechanisms invoked to inhibit proton decay.

We have seen already that with CP violation the K^0_L decays preferentially to positrons. Indeed, with 50-50 production of K^0's and $\bar{K}^0$'s the net result is an excess of positrons from K^0_L and K^0_S decay. Through the leptoquark a positron is equivalent to a proton

$$e^+ \to L_q + u$$
$$\to u+d + u = p \ .$$

So an analog of the neutral K meson which also shows CP violation (the D^0 and B^0 for example) could decay, through the leptoquark, to an excess of protons.

The Neutral K and Gravity

The question is of strong universality, that is whether different objects, particle and antiparticle with the same inertial mass behave the same in a gravitational field. M.L. Good first observed[7] that if the K^0 and $\bar{K}^0$ had opposite gravitational potential energy the particles would mix so quickly the long-lived neutral K would never have been observed. Within the quark model the problem goes to a new level because mesons are composites of quarks and antiquarks. However, in the case of the K mesons, which contain the strange and down quarks ($K^0 \equiv d\bar{s}$, $K^0 \equiv \bar{d}s$), the thrust of Good's argument still holds because the strange quark is much more massive than the down quark. The presence of the down quark dilutes the argument somewhat but does not destroy it. Using modern data one can demonstrate that the K^0 and $\overline{K}^0$ must have the same gravitational potential to within one part in 10^{10} if just the earth's potential is considered. For the galaxy the limit is less than one part in 10^{13}. Within the framework of the quark model we use $m_s = m_\phi/2 = 510$ MeV and $m_d = m_\rho/2 = 385$ MeV and assume the gravitational-gluon interaction to be color independent. With these assumptions the limits above, applied to quark matter and antimatter, are larger by a factor of four. It is still an enormously sensitive system for testing strong universality.

CP Violation and Gauge Theories

This has been extensively discussed by Wolfenstein.[8] Within the framework of gauge models CP violation may arise within the six quark Kabayashi-Maskawa[9], KM, model or it could be due to the exchange of Higg's particles.[10] These are not necessarily orthogonal ideas and both could be present. While the six quark model does not explicitly "explain" CP violation any more than the Cabibbo angle explains weak interactions it does, in fact, provide a natural home for the effect. Higgs exchange may account for the effect found at home in the KM six quark model.

The current high interest in these models lies in their imminent testability. In particular, if CP violation lies in the Higgs sector, milliweak effects should be seen rather uniformly in the weak interactions, e.g. in $K_{\mu 3}$ and neutron decay. In addition, it is predicted that the electric dipole moment of the neutron should lie in the range 10^{-24} to 10^{-25} e cm - just beyond the present experimental limits. On the other hand, if Higgs' exchange is not the answer but the effect is at home in the KM model then the EDM of the neutron will be in the range of 10^{-30} e cm - unobservable in any of the currently planned experiments. But with the KM model it is predicted that ε'/ε is in the range of .003-.02. New experiments at Fermilab and Brookhaven are addressing precisely this question. As we noted before the natural choice of phase in the K-M model is to make A_2 real whereupon the parameters ε and ε' become (to a good approximation)

$$\varepsilon = \frac{p-q}{p+q} + i \frac{I_m A_0}{R_e A_0}$$

$$\varepsilon' = \frac{i}{\sqrt{2}} \frac{R_e A_2}{R_e A_0} \frac{I_{mag} A_0}{R_e A_0} e^{i(\delta_2 - \delta_0)}$$

Recent theoretical results suggest that the answer to the question of Higgs exchange may already be known. Deshpande and independently, Sanda[11] have shown that Higgs exchange should lead to $\frac{\varepsilon'}{\varepsilon}$ equal to .05, a number already excluded by experiment. The calculation assumes little more than natural flavor conservation. Of course, the calculation also deals with quark amplitudes and the extrapolation from quarks to the real world of hadrons is still an uncomfortable transition.

CP Violation in $D^0 \overline{D}^0$ and $B^0 \overline{B}^0$ Systems

When charmed particles were discovered and the general features of the charmed physics were found to be in accord with the prescient article of Gaillard, Lee, and Rosner[12], GLR, the possible opportunities for studying CP violation in the neutral charmed meson system were quickly noted. Unfortunately, as observed by GLR, the common modes of decay of the D^0 and $\overline{D}^0$ are Cabibbo suppressed so the possible mixing is correspondingly small. Therefore, the expected lifetime difference between D^0_L and D^0_S is small and the highly favorable situation of the $K^0-\bar{K}^0$ system is missing. This is especially regretted since the intrinsic decay length of these particles is comparable to the achievable spatial resolution in the best detectors available. Because of this circumstance it is feasible to examine only integrated intensities (for example, of charge asymmetries) which dilutes the magnitude of CP-violating effects.[13]

A number of theoretical treatments of the magnitude of CP violating effects in $D\bar{D}$, $B\bar{B}$, and $T\bar{T}$ systems using the KM model.[14] In general, quite large effects in the $B\bar{B}$ and $T\bar{T}$ complexes might be expected though these theoretical conclusions are by no means universally accepted. Clearly experiments are in order. It is a pity that the luminosities of CESR and DORIS are not larger. As we have heard from Professor Berkelman at this school the rate at which semi-leptonic decay data is acquired is excruciatingly low.

REFERENCES

1. Weinberg, Steven, Gravitation and Cosmology, John Wiley & Sons, New York, p.61.
2. Lee, T.D., and Yang, C.N., 1956, Phys. Rev. 104, 254.
3. Landau, L., 1957, Nucl. Phys. 3, 254.
4. Wu, T.T., and Yang, C.N., 1964, Phys. Rev. Lett. 13, 380.
5. Cronin, J.W., 1981, Rev. Mod. Phys. 53, 373.
6. Sakharov, A., 1967, JETP Lett. 5, 24.
7. Good, M.L., 1961, Phys. Rev. 121, 311.
8. Wolfenstein, L., 1979, Nucl. Phys. B160, 501.
9. Kobayashi, M., and Maskawa, K., 1973, Prog. Theor. Phys. 49, 652.
10. Lee, T.D., 1973, Phys. Rev. D8, 1226, and Weinberg, S., 1976, Phys. Rev. Lett. 31, 657.
11. Deshpande, N., 1981, Phys. Rev. D23, 2654.
    Sanda, A., 1981, Phys. Rev. D23, 2647.

12. Gaillard, M.K., Lee, B.W., and Rosner, J.L., 1975, Rev. Mod. Phys. 47, 277.
13. Okun, L.B., Zakharov, V.I., and Pontecrovo, B.M., 1975, Nuovo Cim. Lett. 13, 218.
    Pais, A., and Treiman, S.B., 1975, Phys. Rev. D12, 2744.
14. Gilman, F.J., and Wise, M.B., 1979, Phys. Rev. D20, 2392, Guberina, B., and Peccei, R.D., 1980, Nucl. Phys. B163, 289, and references contained therein.
15. Schmidt, M.P., Blatt, S.R., Campbell, M.K., Grannau, D.M., Lauterbach, M.J., Kasha, H., Adair, R.K., Morse, W.M., Leipuner, L.B., and Larsen, R.C., 1980, Phys. Rev. D21, 1750.
16. Steinberg, R.I., Liaud, P., Vignon, B., and Hughes, V.W., 1974, Phys. Rev. Lett. 33, 41.
17. Baltrusaitis, R.M., and Calaprice, R.P., 1977, Phys. Rev. Lett. 38, 464.
18. Gimlett, J.L., 1979, Phys. Rev. Lett. 42, 354.
19. Davis, B.R., Koonin, S.E., and Vogel, P., 1980, Phys. Rev. C22, 1233.
20. Driller, H., Blanke, E., Genz, H., Richter, A., Schrieder, G., and Pearson, J.M., 1979, Nucl. Phys. A317, 300.
21. Dress, W.B., Miller, P.D., Pendlebury, J.M., Perrin, P., and Ramsey, N.F., 1977, Phys. rev. D15, 9.
22. Altarev, I.S., Borisov, Y.V., Brandin, A.B., Egorov, A.I., Ezhov, V.F., Ivanov, S.N., Lobashov, V.M., Nazarenko, V.A., Porsev, G.D., Rjabov, V.L., Serebrov, A.P., and Taldaev, R.R., 1978, in Proceedings of the Third International Symposium on Neutron Capture Gamma-Ray Spectroscopy and Related Topics, BNL and SUNY Stony Brook, edited by Robert E. Chrien and Walter R. Kane, Plenum, New York, p.541.

DISCUSSIONS

CHAIRMAN: V. FITCH

Scientific Secretary: S. Gottlieb

DISCUSSION 1

- CASTELLANI:

You mentioned this morning the possibility of matter and antimatter being gravitationally repelled. Is this a serious suggestion?

- FITCH:

As recently as ten years ago I recall there were some serious experiments attempting to determine if positrons would fall up or down in a gravitational field: these experiments failed, to my knowledge.

- CASTELLANI:

One should also construct a theory of gravity which predicts this. I suppose it should be a quantum theory since the concept of antiparticle is essentially quantum mechanical. Has such a theory been constructed?

- FITCH:

Not to my knowledge. Perhaps there is a greater expert here.

- WILCZEK:

The exchange of an even spin object is attractive between particle and antiparticle. Since the graviton is spin 2, it would result in an attractive force.

- CASTELLANI:

What about the gravitino ?

- WILCZEK:

Simple gravitino exchange does not introduce a long range force because it is not coherent. Exchange of two will introduce a force which I believe falls faster than $1/r$.

- FITCH:

To the extent that quarks are similar to leptons, we may draw a conclusion from the neutral kaon system. There the forces on K^0 and $\bar{K}^0$ from the earth's gravitational field are equal to within one part of 10^{10}.

- BERTANI:

Why doesn't the $D^0 \bar{D}^0$ system have the same behaviour as the neutral kaon system ?

- FITCH:

One can expect the mixing to be very small simply because they decay primarily into strange particles and the channels common to the D^0 and $\bar{D}^0$ are the nonstrange channels. Hence, the mixing would be very small. In the case of the B mesons, the mixing can be quite large. We are all looking forward to experiments on that system.

- BERKELMAN:

One must have the right competition between the decay rate and the mixing rate. The D mesons have no Cabbibo suppression of the decay to strange particles, so their lifetimes are quite short.

DISCUSSION 2   (Scientific Secretary: S. Gottlieb)

- YAMAMOTO:

How can we look for CP violation using the $\bar{p}p$ collider ?

- FITCH:

We simply look at the momentum spectrum of positive and negative pions and kaons at $90°$ in the centre of mass. If there is a difference between positive and negative particles, one has explicit evidence of violation of charge conjugation. This kind of experiment is tricky because there is a natural charge asymmetry off 90° through the leading particle effect.

- WILCZEK:

What kind of accuracy can one expect ?

- FITCH:

If the energy and luminosity are as expected, one might achieve an accuracy of one part in 10^4 for $p_\perp$ of 2 GeV. That is a purely statistical limit.

- WITTEN:

To what extent can the determination of CPT symmetry in neutral kaons be improved ?

- FITCH:

I said earlier that the masses of K^0 and $\bar{K}^0$ are equal to within one part in 10^{17}. One can, at this moment, improve on that information. The parameter η_{+-}, which we take to be equivalent to ε has its phase largely determined by the denominator,

$$i\,(m_s - m_L) + \tfrac{1}{2}\,(\Gamma_s - \Gamma_L)$$

The value of Δm determines the phase to be nearly 45°. That is the measured value. One can allow a deviation of perhaps 0.1 radian due to a mass difference in the diagonal elements of the mass matrix. This corresponds to an energy of 1.17×10^{-9} eV and since the mass of the kaon is about 500 MeV, we have equality of K^0 and $\bar{K}^0$ masses to about two parts in 10^{18}.

One can improve this using a bit of theoretical prejudice for example, that 0.1 radian allows a maximum contamination due to $\Delta S = \Delta Q$ amplitudes. If you say they are absent, there is a factor of 2 or 5 improvement.

- KARLINER:

Is it possible to measure very accurately the magnetic moments of electrons or muons of opposite signs ? Since the mass enters into the calculation of magnetic moments, could one compare the moments of positive and negative muons to reach a quantitative conclusion about CPT ?

- FITCH:

I see no difficulty with that whatsoever. It should be possible at the famous machine at CERN.

- ZICHICHI:

This has already been done. The magnetic moments of μ^+ and μ^- have already been measured, and they provide a very good test of CPT invariance in electromagnetic interactions.

- WIGNER:

I want to ask a crazy question. Is it possible that some of the symmetry violation comes from the environment, that is from the fact that the sun and planets are made of particles, not anti-particles ? Is it possible that this would have an effect ?

- FITCH:

So far as we know, all of the interactions we are dealing with are extraordinarily short in range. The photon is the only massless quantum included here.

- WIGNER:

The origin of the forces may be a field which is surrounding us. Perhaps without that field some of the reactions would not take place at all.

- FITCH:

Such interactions have been considered in the past. In the neutral kaon system they have been lumped under the general name of "superweak". Those interactions of a vector or tensor nature would result in a branching ratio which varies with laboratory velocity. Variations in the branching ratio into two pions have not been seen. Some kind of scalar or pseudoscalar interaction might be possible, but it is equivalent essentially to the superweak interaction.

- WIGNER:

In other words, such a possibility is terribly unlikely.

- FITCH:

I would say it is terribly unlikely, and that it would have to be a pseudoscalar interaction.

- KARLINER:

Since we are considering rather unlikely possibilities, how well has Lorentz covariance been checked?

- FITCH:

I can't give you a precise answer. At Cornell, DESY and PEP, where electrons are accelerated to 20,000 times their rest mass and stored, everything works well. If there was any departure from the usual relations between mass, velocity and energy, it would be unlikely that such machines would work. However, quantifying these observations takes a specific model for violation.

- ZICHICHI:

The storage time of particles probably provides a good test. Since the particles are pumped with radio-frequency electromagnetic waves, we would probably detect any violation greater than one part in 10^4.

WHAT WE CAN LEARN FROM HIGH-ENERGY,

SOFT (pp) INTERACTIONS

M. Basile, G. Bonvicini, G. Cara Romeo, L. Cifarelli,
A. Contin, M. Curatolo, G. D'Ali, B. Esposito, P. Giusti,
T. Massam, R. Nania, F. Palmonari, A. Petrosino, V. Rossi,
G. Sartorelli, M. Spinetti, G. Susinno, G. Valenti,
L. Votano and A. Zichichi

CERN, Geneva, Switzerland
Istituto di Fisica dell'Università di Bologna, Italy
Istituto Nazionale di Fisica Nucleare, Laboratori
 Nazionali di Frascati, Italy
Istituto Nazionale di Fisica Nucleare, Sezione di Bologna
 Italy
Istituto di Fisica dell'Università di Perugia, Italy
Istituto di Fisica dell'Università di Roma, Italy

Presented by A. Zichichi

1. INTRODUCTION

The purpose of this review is to report on a series of similarities between multiparticle hadronic systems produced in (pp) interactions and in (e^+e^-) annihilation[1-12]. The new feature of this work is the fact that, in order to establish these similarities, the basic principle is to evaluate, for each (pp) interaction, the correct energy available for particle production. The old trend was to select high-p_T processes.

One can produce multiparticle hadronic states in basically three ways: i) (hadron-hadron) interactions; ii) (e^+e^-) annihilation; iii) (lepton-hadron) deep inelastic scattering (DIS). As mentioned above, the only way, so far, to correlate the hadronic systems produced in these different ways was to compare the properties, measured in (e^+e^-) and DIS, with high-p_T (hadron-hadron) data. The low-p_T data were considered as examples of typically hadronic phenomena, not suitable for comparison with (e^+e^-) and DIS physics. Our new way of analysing (pp) interactions[1-12] brings into the exciting field of physics the large amount of, so far abandoned, low-p_T (pp) data.

The study of the properties of multihadronic systems produced in low-p_T proton-proton interactions, once the energy for particle production in a (pp) interaction has been correctly calculated, shows that a striking series of analogies can be established between (pp) and (e^+e^-) processes[1-12]. I will not discuss the case of multiparticle production in DIS. The way in which DIS data can be compared with (pp) data will be mentioned at the very end.

The comparison of the multiparticle hadronic systems produced in (pp) interactions and (e^+e^-) annihilation will be discussed in terms of the following five quantities:

i) the inclusive, single-particle, fractional momentum distribution of the produced particles;
ii) the inclusive, single particle, transverse momentum distribution of the particles produced;
iii) the average charged particle multiplicity;
iv) the ratio of "charged" to "total" energy of the multiparticle hadronic systems produced;
v) the planarity of the multiparticle hadronic systems produced.

2. EXPERIMENTAL (pp) DATA AND COMPARISON WITH (e^+e^-) DATA

2.1 The Main Points

The experimental data for the (pp) interactions have been taken at the CERN Intersecting Storage Rings (ISR) using a large-volume magnetic field, the so-called Split-Field Magnet (SFM), coupled with a powerful system of multiwire proportional chambers (MWPCs)[13]. It is the nearest system to a bubble-chamber-like instrument at the ISR. The data[1-12] have been collected at three different ISR energies, i.e. $(\sqrt{s})_{pp}$ = 30, 44, and 62 GeV, either using the "minimum" bias mode of running, or triggering on fast particles to enrich the sample of events with "leading" protons.

The comparison of our (pp) data with (e^+e^-) is made using (e^+e^-) data from PETRA, from SPEAR, and from ADONE[14-21].

The key point of our analysis can be illustrated as follows: when you have a (pp) collision, the total energy available for particle production is not $(\sqrt{s})_{pp} = 2E_{inc}$, where E_{inc} is the incident energy of each colliding proton. In fact, quite often a large fraction of the primary energy is carried away, by the incoming proton, into the final state. This is the "leading" proton effect; it is there, and not only in the case of a proton interacting with another proton. As we will see later, it is a very general phenomenon, which is present when a hadron interacts, no matter if strongly, electromagnetically or weakly[22,23]. The "hadron-in-the-final-state" which is coming from the initial state has an energy sharing, with all other particles produced, which is highly priviliged: this "leading" hadron effect must be accounted for correctly in order to compare the properties of the multiparticle hadronic system produced in the interaction. For example, in the (pp) case, if you study the interaction of two protons, each with 31 GeV, the total energy for particle production is not going to be 62 GeV, but a fraction of this. This fraction depends on the energies taken away by the two leading protons in the final state.

The first evidence for analogies between (pp) and (e^+e^-) data[1] was found using (pp) interactions at the nominal ISR energy, $(\sqrt{s})_{pp}$ = 62 GeV. As mentioned above, this "fixed" nominal energy corresponds to a set of "effective" energies, available for particle production, which range from few GeV up to about 60 GeV. In Fig. 1 the relation between the incoming proton beam energy E_{inc} and the effective hadronic energy available for particle production E_{had} is shown, once the experimental cuts have been taken into account. In addition to E_{inc} = 31 GeV, two other cases, E_{inc} = 22 GeV and E_{inc} = 15 GeV, are also shown. A large "leading" proton effect produces a small effective hadronic energy; conversely, a small "leading" proton effect produces a large effective hadronic energy. It is the variation of the "leading" proton effect that allows different nominal fixed proton beam energies E_{inc} to produce the same effective hadronic energy $2E_{had}$ available for particle production.

The reason why we have collected data at three ISR energies, $(\sqrt{s})_{pp}$ = 30, 44, and 62 GeV, is as follows. It was a crucial point for our study to show that the multiparticle hadronic systems produced in (pp) interactions, with the same values of $2E_{had}$ but with different values of E_{inc}, had the same properties in terms of the five quantities mentioned above, i.e. fractional momentum distribution, transverse momentum properties, average charged particle multiplicity, ratio of charged to total energy, and planarity.

2.2 The Leading Hadron Effect

As anticipated earlier, the "leading" proton effect is an example of a very general phenomenon: the leading hadron effect[22,23].

Let F(x) be the inclusive single-particle cross-section integrated over p_T:

$$F(x) = \frac{1}{\pi} \int \frac{2E}{\sqrt{s}} \frac{d\sigma^2}{dx dp_T^2} dp_T^2 ,$$

where $x = 2p_L/\sqrt{s}$, with p_L the longitudinal momentum of the hadron under consideration.

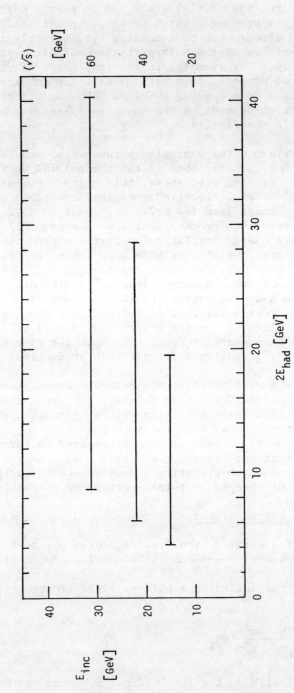

Fig. 1 Ranges of effective hadronic energy available for particle production ($2E_{had}$) for a given ISR incident energy E_{inc}. These ranges depend on the proton x_F range selected (0.35–0.86).

The range $0 \lesssim x < 0.2$ is typical of the so-called "central" production; the range $x > 0.8$ is near to the region where diffractive and quasi-elastic processes start to appear. There is a range of x which is far from the above two well-known physical processes [the so-called "central" production ($x \simeq 0$) and the peripheral production ($x \simeq 1$)]: $0.2 \leq x \leq 0.8$. If we divide this range of x into two parts, $0.2 \leq x \leq 0.4$ and $0.4 < x \leq 0.8$, and if we evaluate the integral of $F(x)$ in these two intervals, the ratio of these two quantities will be a good parameter for expressing, in a quantitative way, the "leading" effect:

$$L = \int_{0.4}^{0.8} F(x)dx \bigg/ \int_{0.2}^{0.4} F(x)dx .$$

The quantity L tells us how strong the "leading" effect is in a given process. Figure 2 shows the values of L for different types of particles produced in (pp) interactions at the ISR. The quantity L scales, in the ISR energy range investigated, from $(\sqrt{s})_{pp} = 25$ up to $(\sqrt{s})_{pp} = 62$ GeV. The important point is: the lower the number of propagating quarks the lower the value of L. Notice that the value of L for zero propagating quarks is below 0.5, as in (e^+e^-) annihilation.

A very interesting result is shown in Fig. 3 where the analysis of the $(\bar{p}p)$ data, obtained at Fermilab, is given. The values of L show that the leading effect holds true for antiprotons as well as for protons.

Finally, Fig. 4 shows a spectacular result: the leading effect is present in electromagnetic and in weak interactions. In fact, in this figure we report the analysis of Λ^0 production in (ep) and $(\bar{\nu}p)$ interactions. The values of L, calculated for the Λ^0, in the electromagnetic production (ep) and in the weak production $(\bar{\nu}p)$ processes, are the proof that the Λ^0 is leading.

This synthetic analysis shows that the "leading" proton effect, which is at the basis of our discovery of the similarities between the two ways of producing multiparticle hadronic states, (pp) and (e^+e^-), is a very general phenomenon. It holds true in strong, electromagnetic, and weak interactions, provided a hadron is present in the initial state. The leading effect in the final state is present, even if the initial hadron has transformed itself into another hadron. More transformation implies less leading effect.

We will now discuss the five quantities studied in order to establish the similarities between (pp) and (e^+e^-) processes.

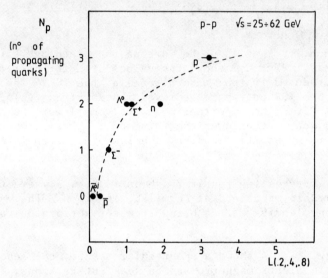

Fig. 2 The leading quantity L(0.2, 0.4, 0.8), for various final-state hadrons in pp collisions at ISR energies (25-62 GeV), is plotted versus the number of propagating quarks from the incoming into the final-state hadrons. The dashed line is obtained by using a parametrization of the single-particle inclusive cross-section, as described in Refs. 22 and 23.

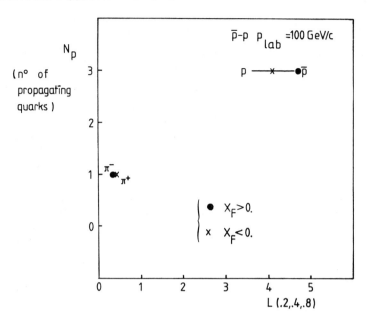

Fig. 3 L(0.2, 0.4, 0.8) for final-state hadrons produced in ($\bar{p}$-p) collisions at p_{lab} = 100 GeV/c (Refs. 22, 23).

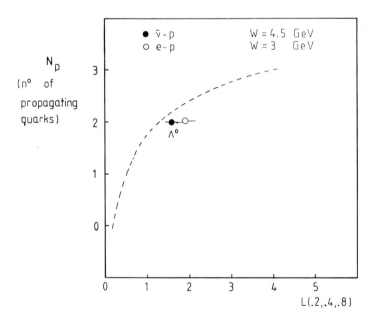

Fig. 4 L(0.2, 0.4, 0.8) for Λ^0 production in ($\bar{\nu}$p) and (ep) reactions. The dashed line is the same as described for Fig. 2 (Refs. 22, 23).

2.3 The Inclusive Fractional Momentum Distribution

The inclusive, single-particle, fractional momentum distributions $d\sigma/dx_R^*$, of the particles produced in (pp) interactions at the nominal total c.m. energy, $(\sqrt{s})_{pp}$ = 62 GeV[1], are shown in Figs. 5a-5c for different values of the effective hadronic energy E_{had} available for particle production, i.e. once the leading proton effect is correctly accounted for. The fractional momentum x_R^* is evaluated with respect to the effective hadronic energy E_{had}, according to the definition $x_R^* = p/E_{had}$, p being the momentum of the particles. As mentioned above from a sample of (pp) interactions at fixed (pp) energy, $(\sqrt{s})_{pp}$ = = 62 GeV, we have been able to obtain three samples of events at different E_{had} by selecting the leading proton in different energy ranges. The data in Fig. 5a refer to hadronic energies in the range E_{had} = 5-8 GeV; in Fig. 5b the data with E_{had} = 8-11 GeV are reported; in Fig. 5c the data at the highest hadronic energy range, E_{had} = 14-16 GeV, are given. High values of E_{had} means less leading proton effect. At ISR energies it is not easy to identify a "proton" via the Čerenkov effect or via the time-of-flight technique. Our method of identifying a proton in the final state is the simplest: i.e. by its leading property. Experimental measurements at the ISR[24] show that when the proton is more leading, there is less π contamination. For example, if a proton is leading with 80% of its initial momentum, the probability of having a π in the same momentum configuration is ∿ 2%. If the proton has 50% of its primary momentum, the π contamination can be as high as ∿ 15%. In Fig. 5a the π contamination is expected to be at the 2% level; in Fig. 5c we expect a π contamination of the order of ∿ 15%.

In Figs. 5a-5c the (e^+e^-) data[14] at equivalent energies are shown. The agreement is remarkable in Figs. 5a and 5b. The (pp) data in Fig. 5c are systematically below the (e^+e^-) data; this can be understood, as mentioned above, in terms of the unavoidable π contamination present in the (pp) data, in this E_{had} range.

Now we come to the proof that E_{had} is a good variable, independent of the nominal c.m. (pp) energy $(\sqrt{s})_{pp}$: this is shown in Figs. 6a-6c. The first of this set of figures, Fig. 6a, shows the data obtained at three nominal c.m. (pp) energies, namely

$$(\sqrt{s})_{pp} = 30, 44, 62 \text{ GeV},$$

but with the same effective hadronic energy range

$$(10 \leq 2E_{had} \leq 16) \text{ GeV}.$$

Irrespective of the primary proton energies, the data on $d\sigma/dx_R^*$ follow the same line. In Fig. 6b, again the same set of (pp) interactions at three values for $(\sqrt{s})_{pp}$ are used, but with a higher hadronic energy range

HIGH-ENERGY, SOFT (pp) INTERACTIONS

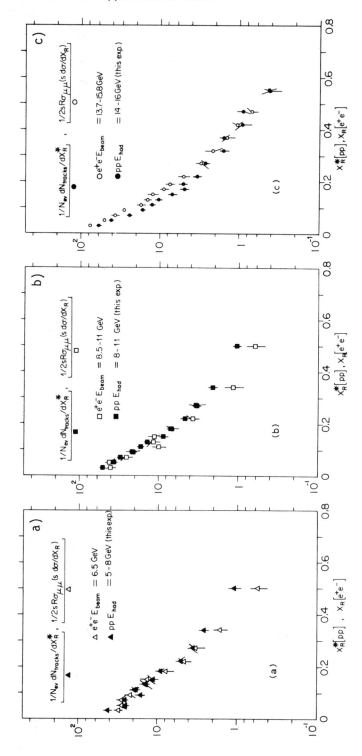

Fig. 5 The (pp) single-particle inclusive distributions in terms of the variable x_R^* (Ref. 1), are compared to the single-particle inclusive distributions obtained in (e^+e^-) annihilation (Ref. 14): a) E_{had} = 5.8 GeV; b) E_{had} = 8-11 GeV; c) E_{had} = 14-16 GeV.

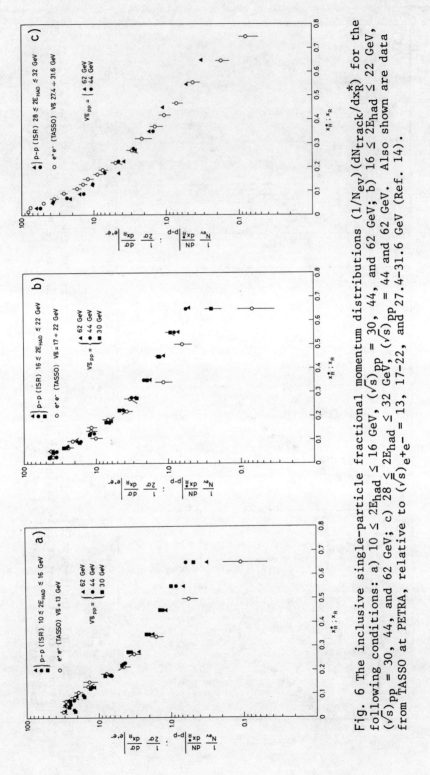

Fig. 6 The inclusive single-particle fractional momentum distributions $(1/N_{ev})(dN_{track}/dx_R^*)$ for the following conditions: a) $10 \leq 2E_{HAD} \leq 16$ GeV, $(\sqrt{s})_{pp} = 30$, 44, and 62 GeV; b) $16 \leq 2E_{had} \leq 22$ GeV, $(\sqrt{s})_{pp} = 30$, 44, and 62 GeV; c) $28 \leq 2E_{had} \leq 32$ GeV, $(\sqrt{s})_{pp} = 44$ and 62 GeV. Also shown are data from TASSO at PETRA, relative to $(\sqrt{s})_{e^+e^-} = 13$, 17-22, and 27.4-31.6 GeV (Ref. 14).

$(16 \leq 2E_{had} \leq 22)$ GeV .

Finally, in Fig. 6c the lower energy data at $(\sqrt{s})_{pp} = 30$ GeV cannot be used because the hadronic energy range is too high, i.e.

$(28 \leq 2E_{had} \leq 32)$ GeV .

In the same graphs the (e^+e^-) data[14] at equivalent energies, i.e.

$$(\sqrt{s})_{e^+e^-} = 2E_{had} ,$$

are shown for comparison with our data. The agreement is very good.

Having seen that $(d\sigma/dx_R^*)$ does not change with $(\sqrt{s})_{pp}$, we can limit ourselves to the $(\sqrt{s})_{pp} = 62$ GeV data in order to avoid too many points in a graph. This is done in Fig. 7 in order to show a very interesting fact. If we had had the idea of subtracting the leading proton effect in a (pp) collision earlier, we could have discovered the dramatic rise in $(d\sigma/dx_R^*)$ at low x_R^* values. This dramatic rise in the inclusive fractional momentum distributions has been discovered with (e^+e^-) at PETRA, and it is one of the most important (e^+e^-) results obtained in this energy range. As shown by the SPEAR data at lower (e^+e^-) energy[15], the shape of $(d\sigma/dx_R)$ was flattening at low x_R values.

Now there is a key problem: Who tells us that, going to the equivalent low E_{had} values, $d\sigma/dx_R^*$ follows the same trend as the one observed in the SPEAR energy range? When we first observed the similarity between (pp) and (e^+e^-) data in the quantity $d\sigma/dx_R^{*1}$, we had at our disposal ISR data at

$(\sqrt{s})_{pp} = 62$ GeV.

This is why we could not answer the above question. Using our new ISR data at

$(\sqrt{s})_{pp} = 30$ GeV

we are able to reach very low hadronic energies[11]. Figures 8a-8c show our data at

$(3 \leq 2E_{had} \leq 4)$ GeV ,

$(4 < 2E_{had} \leq 6)$ GeV ,

$(6 < 2E_{had} \leq 9)$ GeV ,

compared with SPEAR data[15] at equivalent (e^+e^-) energies: $(\sqrt{s})_{e^+e^-} \simeq 2E_{had}$.

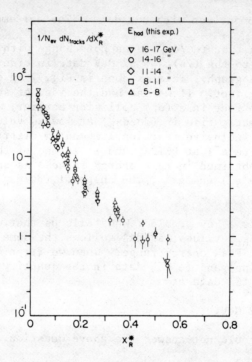

Fig. 7 The inclusive single-particle distributions $(1/N_{ev})(dN_{track}/dx_R^*)$ versus x_R^* for various bands of E_{had}, obtained from (pp) collisions (Ref. 1).

HIGH-ENERGY, SOFT (pp) INTERACTIONS

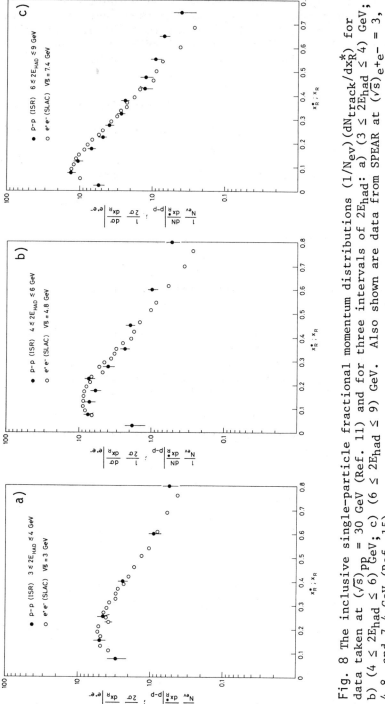

Fig. 8 The inclusive single-particle fractional momentum distributions $(1/N_{ev})(dN_{track}/dx_R^*)$ for data taken at $(\sqrt{s})_{pp} = 30$ GeV (Ref. 11) and for three intervals of $2E_{had}$: a) $(3 \leq 2E_{had} \leq 4)$ GeV; b) $(4 \leq 2E_{had} \leq 6)$ GeV; c) $(6 \leq 2E_{had} \leq 9)$ GeV. Also shown are data from SPEAR at $(\sqrt{s})_{e+e-} = 3$, 4.8, and 7.4 GeV (Ref. 15).

The data reported in Figs. 8a-8c show that, also in the very low hadronic energy range, the agreement between our data and (e^+e^-) annihilation is very good. This proves that the dramatic rise in the single-particle inclusive momentum distribution could have been observed in (pp) interactions at the ISR before being discovered in (e^+e^-) annihilation at PETRA.

2.4 The Inclusive Transverse Momentum Distribution

Let me now go to another important property of the multiparticle hadronic systems produced in (pp) interactions; namely, the inclusive transverse momentum distribution of the particles produced in a (pp) interaction where the effective hadronic energy has been correctly calculated. Figure 9 shows our results, obtained with $(\sqrt{s})_{pp} =$ = 30 GeV data, where a sample of events in the hadronic energy range

$$(11 \leq 2E_{had} \leq 13) \text{ GeV}$$

has been selected.

In Fig. 9 this is compared with (e^+e^-) data at $(\sqrt{s})_{e^+e^-} =$ = 12 GeV[16]. The agreement is excellent.

A new way of analysing the transverse momentum distribution of the particles produced in (e^+e^-) annihilation has recently been introduced[17]. It is based on the so-called reduced variable $p_T/\langle p_T \rangle$, where p_T is the transverse momentum of the single particle observed and $\langle p_T \rangle$ is the average value of the same quantity for all particles produced.

The results[7] are shown in Fig. 10, where we compare two ranges of hadron energies,

$$(8 \leq 2E_{had} \leq 16) \text{ GeV}$$
$$(24 \leq 2E_{had} \leq 32) \text{ GeV} ,$$

with the (e^+e^-) equivalent energy data obtained by the PLUTO group[17]. The data overlap very well. So even the transverse momentum distributions of the particles produced follow the same trend in (pp) and (e^+e^-).

2.5 The Average Charged Particle Multiplicity

Another important quantity for comparing (pp) and (e^+e^-) is the average number of charged particles produced at a given energy.

The data shown in Fig. 11 refer to the results obtained using three samples of data at $(\sqrt{s})_{pp} =$ 30, 44, and 62 GeV[8]. These data have been analysed, on an event-by-event basis, with the method of

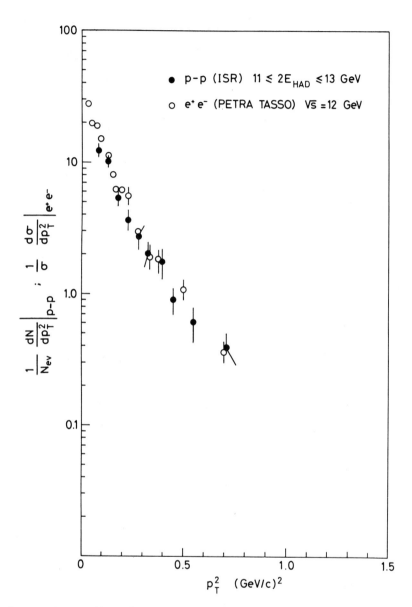

Fig. 9 The inclusive single-particle transverse-momentum distribution $(1/N_{ev})(dN/dp_T^2)$ for $(11 \leq 2E_{had} \leq 13)$ GeV (Ref. 10) compared with $(1/\sigma)/(d\sigma/dp_T^2)$ measured at PETRA (TASSO) at $(\sqrt{s})_{e^+e^-} = 12$ GeV (Ref. 16).

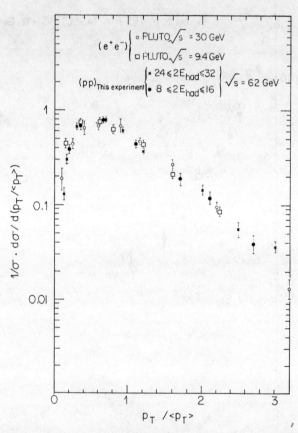

Fig. 10 Renormalized differential cross-section $(1/\sigma)[d\sigma/d(p_T/\langle p_T \rangle)]$ versus the "reduced" variable $p_T/\langle p_T \rangle$ (Ref. 7). These distributions allow a comparison of the multiparticle systems produced in e^+e^- annihilation (Ref. 17) and in pp interactions in terms of the "reduced" transverse momentum properties.

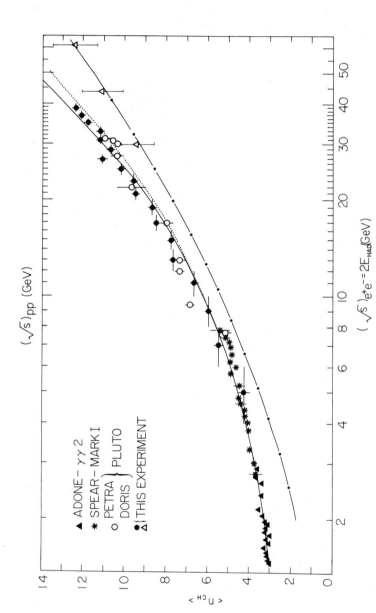

Fig. 11 Mean charged particle multiplicity, averaged over different $(\sqrt{s})$pp, versus $(2E_{had})$ (Ref. 8), compared with (e^+e^-) data (Refs. 15, 18, 19). The contribution of $K_S^0 \to \pi^+\pi^-$ has been subtracted. Each value is an average over 2 GeV. The quoted errors are statistical only. The systematic uncertainty is less than 8%. The continuous line is the best fit to our data according to the formula $\langle n_{ch} \rangle = a + b \exp\left[c\sqrt{\ln(s/\Lambda^2)}\right]$. The dashed-dotted line is the best fit using PLUTO data (Ref. 19). The dotted line is the (pp) total charged particle multiplicity (Ref. 25). For this curve, the abscissa is $(\sqrt{s})$pp. The open triangular points in this curve are our data using the standard way (i.e. without the subtraction of the leading proton effects), to calculate $\langle n_{ch} \rangle$ at fixed $(\sqrt{s})$pp.

calculating the effective hadronic energy available in the (pp) interaction, E_{had}, after the leading proton effects are subtracted. Once again, the use of different $(\sqrt{s})_{pp}$ to produce the same E_{had} is crucial in order to prove that the quantity E_{had} is a good variable. In the same figure (Fig. 11), the results from (e^+e^-) experiments are also reported[15,18,19]. The agreement between (pp) and (e^+e^-) data is very satisfactory.

It should be remarked that the "standard" way of analysing (pp) data, i.e. without the subtraction of the "leading" proton effect, would produce the dashed-dotted curve plotted in Fig. 11[25], where the (e^+e^-) data clearly stay above the standard (pp) ones. An important cross-check is given by the three points (open triangles) plotted in Fig. 11. These points are the results of our analysis at $(\sqrt{s})_{pp}$ = 30, 44, and 62 GeV, once we proceed in the "standard" way.

So, even with an important quantity such as the number of charged particles produced on the average in (e^+e^-) annihilation and in (pp) interactions at equivalent energies, we find an excellent agreement, once the energy available for particle production in a (pp) interaction is correctly calculated.

2.6 The Ratio of Charged to Total Energy

In a (pp) interaction, studied in the standard way, the total energy is $(\sqrt{s})_{pp}$. If we now try to calculate the energy carried away by the charged particles, and make the ratio

$$\frac{\langle \text{Energy carried by the charged particles} \rangle}{\text{Total energy}},$$

we find no agreement between this value and the ratio

$$\frac{\langle E_{charged} \rangle}{E_{total}}$$

measured in (e^+e^-) annihilation.

The method of removing the leading proton effects makes it possible, in (pp) interactions, to calculate on an event-by-event basis the effective hadronic energy E_{had}, i.e. the total hadronic energy which is really useful for that given (pp) interaction. On the other hand, the energy carried by the charged particles, $E_{charged}$, is directly measured. From these two quantities, measured for each event, the average value $\langle E_{charged} \rangle$ can be calculated, and the ratio $\langle E_{charged} \rangle / E_{had} = \alpha_{pp}$ can be compared with $\langle E_{charged} \rangle / E_{total}$ measured in (e^+e^-) annihilation.

The (pp) results[5] are shown in Fig. 12 together with the (e^+e^-) data[15,20]. Once again we use three sets of data at $(\sqrt{s})_{pp}$ = 30, 44, and 62 GeV to produce events with equal E_{had} values.

The results of Fig. 12 show that the average fractional energy available for charged particle production in a (pp) interaction is, within the experimental uncertainty, the same as for the (e^+e^-) annihilation.

2.7 The Event Planarity

The problem under investigation is now the following: Do the events produced in a (pp) interaction show a planarity structure? To answer this question we proceed[4] in the same way as our colleagues working on (e^+e^-), adopting the same terminology and the same symbols. Let me remind you about a few basic points.

For each event we

i) construct a quantity (see Fig. 13)

$$M_{\alpha\beta} = \sum_{j=1}^{N} P_{j\alpha} P_{j\beta} \qquad (\alpha,\beta = 1,2) ,$$

where N = number of particles in the event;

ii) determine the eigenvectors $\vec{n}_1$, $\vec{n}_2$, and eigenvalues Λ_1, Λ_2 ($\Lambda_1 < \Lambda_2$);

$\vec{n}_1$ is the direction in which the sum of the square of the momentum components is minimized;

$\vec{n}_2$ together with the (pp) line of flight, defines the event plane.

The next step is to define the average p_T^2 "in" and "out" of the event plane, according to

$$\langle p_T^2 \rangle_{out} = \frac{\Lambda_1}{N} = \frac{1}{N} \sum_{j=1}^{N} (\vec{p}_j \cdot \vec{n}_1)^2 ,$$

$$\langle p_T^2 \rangle_{in} = \frac{\Lambda_2}{N} = \frac{1}{N} \sum_{j=1}^{N} (\vec{p}_j \cdot \vec{n}_2)^2 .$$

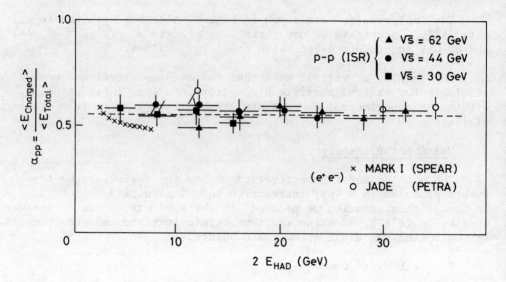

Fig. 12 The charged-to-total energy ratio obtained in pp collisions (Ref. 5), α_{pp}, plotted versus $2E_{had}$ and compared with e^+e^- data obtained at SPEAR (Ref. 15) and PETRA (Ref. 20).

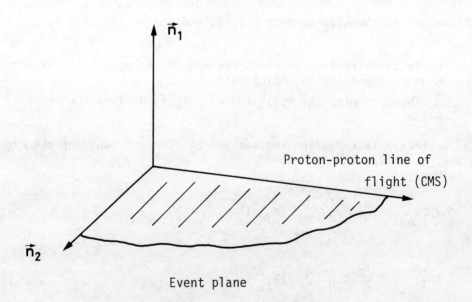

Fig. 13 Event plane definition.

Figure 14 shows $\langle p_T^2 \rangle_{in}$ and $\langle p_T^2 \rangle_{out}$ distributions in the low-energy range $(10 \leq 2E_{had} \leq 16)$ GeV^4. The equivalent (e^+e^-) energies are $[13 \leq (\sqrt{s})_{e^+e^-} \leq 17]$ GeV. The agreement between (pp) and (e^+e^-) data[21] is very good.

Figure 15 shows the results[4] at higher energy, i.e. $(28 \leq 2E_{had} \leq 34)$ GeV. Notice that the $\langle p_T^2 \rangle_{out}$ distribution follows the Monte Carlo prediction. The deviation of the $\langle p_T^2 \rangle_{in}$ distribution from Monte Carlo predictions is the proof that the events produced in (pp) interactions do show a planarity effect. Notice that the (e^+e^-) data[21] stay above the (pp) data for the "in" and the "out" distributions. If this effect is going to be confirmed by more accurate (pp) and (e^+e^-) data, the interpretation could be in terms of "heavy" quarks being produced in (e^+e^-) more easily than in (pp) interactions. In fact, once the energy is over the threshold, the production of a quark depends on its charge squared in (e^+e^-) annihilation. However, the phenomenological trend in (pp) interactions, also expected from QCD, is in terms of the inverse mass squared of the quark to be produced. This implies that in (e^+e^-) the production of heavy quarks competes well with the production of light quarks, but in (pp) interactions the light quark production dominates the "heavy" quark one. The results at different E_{had} values, reported in Fig. 14 and 15, are all from the same (pp) nominal energy $(\sqrt{s})_{pp} = 62$ GeV.

We now come to a very interesting point; namely, the $(\sqrt{s})_{pp}$ independence of the planarity effect. You can see in Fig. 16 and 17 that the data obtained from different values of $(\sqrt{s})_{pp}$, i.e. $(\sqrt{s})_{pp} = 30, 44, 62$ GeV, overlap each other, and show the same planarity structure.

2.8 Conclusions

Let me summarize the most important points.

The hadronic production in (e^+e^-) annihilation takes place in such a way that no hadron has a privileged energy sharing. The inclusive fractional momentum distribution of the particles produced in (e^+e^-) annihilation follows the same shape as for hadronic systems produced in (pp) interactions[1]. This similarity can be established only if the "leading" proton effects are removed[1]. Thus the "leading" effect plays an important role in the understanding of the similarities and therefore of the connection between these two different ways of producing multiparticle hadronic systems[1-12]. The leading effect is a very general phenomenon[22,23]. It shows up whenever there is a hadron in the initial state of an interaction -- no matter whether the interaction is strong, electromagnetic, or weak -- no matter whether or not the initial hadron remains exactly as it is. For example, the leading effect is present if a proton becomes a neutron, or a Λ_s^0, or even a Λ_c^+. Even if the quantum numbers of the hadron in the initial state are not fully carried out in the final-state hadron,

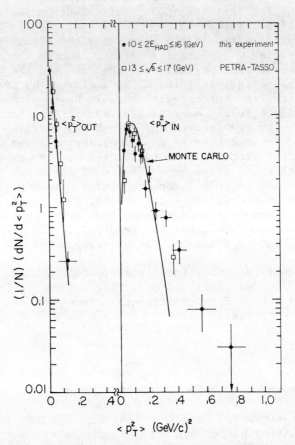

Fig. 14 The average transverse-momentum squared distributions for the "out" and the "in" cases, in the "low-energy" range ($10 \leq 2E_{had} \leq 16$) GeV (Ref. 4), compared with (e^+e^-) data at $[13 \leq (\sqrt{s})_{e^+e^-} \leq 17]$ GeV (Ref. 21). The full lines are the results of a p_T-limited phase-space Monte Carlo.

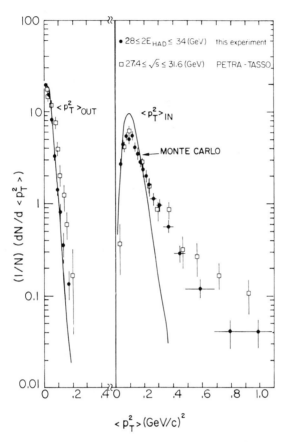

Fig. 15 The average transverse-momentum squared distributions for the "out" and the "in" cases, in the "high-energy" range ($28 \leq 2E_{had} \leq 34$) GeV (Ref. 4), compared with (e^+e^-) data at $[27.4 \leq (\sqrt{s})_{e^+e^-} \leq 31.6]$ GeV (Ref. 21). The full lines are the results of a p_T-limited phase-space Monte Carlo.

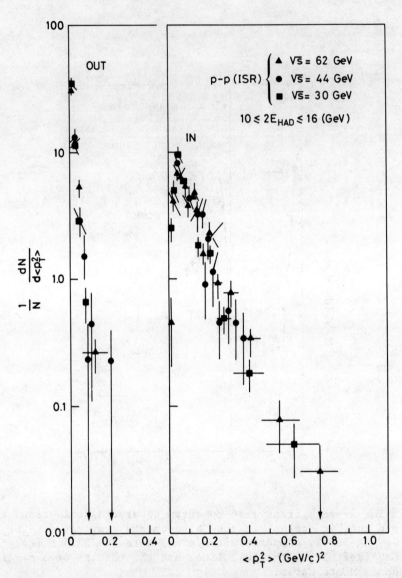

Fig. 16 The average transverse-momentum squared distributions for the "out" and the "in" cases. Data relative to the effective hadronic energy interval $(10 \leq 2E_{had} \leq 16)$ GeV obtained from (pp) collisions at $(\sqrt{s})_{pp}$ = 30, 44, and 62 GeV are shown.

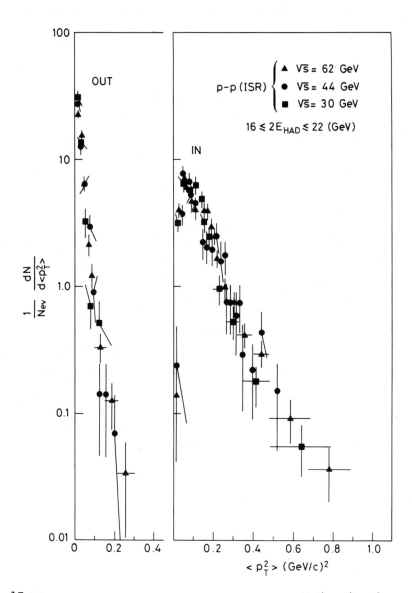

Fig. 17 The average transverse-momentum squared distributions for the "out" and the "in" cases. Data relative to the effective hadronic energy interval $(16 \leq 2E_{had} \leq 22)$ GeV obtained from (pp) collisions at $(\sqrt{s})_{pp}$ = 30, 44, and 62 GeV are shown.

the leading effect is present. It is the quantum number flow, be it colour, flavour, J^{PC}, etc., which gives to the initial-state hadron a highly privileged energy sharing, when compared with the particles produced.

It should therefore be clear that all reactions where an initial-state hadron is present should be analysed by taking into account the "leading" phenomenon. For example, the final-state hadronic system produced in DIS should also be analysed in such a way as to identify the leading hadron and to subtract its effects from the global interaction. In the present study we could not report on the comparison between (pp) and deep inelastic lepton-hadron scattering (DIS) because the DIS data have so far never been analysed by subtracting the presence of a leading hadron in the final state.

The use of the leading hadron effect in (pp) interactions has brought the vast amount of low-p_T physics into the very interesting domain of comparison with (e^+e^-) physics[1-12]. Let me emphasize, once again, that the (pp) data presented to you are obtained with the simplest possible logic: in a proton-proton collision try to do nothing, as would be the case with a bubble chamber, and look at the (pp) event. This means: study the lowest possible p_T interactions. This has been done in terms of the five quantities investigated: i) the inclusive fractional momentum distribution; ii) the inclusive transverse momentum distribution; iii) the average charged particle multiplicity; iv) the ratio of charged-to-total energy; v) the planarity. The low-p_T physics appears to be as interesting as the high-p_T physics. So, the old belief that only high-p_T (pp) phenomena could be the basis for comparisons with (e^+e^-) seems to be over. In order to understand the way in which the multiparticle hadronic systems are produced in strong, electromagnetic, and weak interactions, all p_T physics needs to be investigated.

REFERENCES

1. M. Basile, G. Cara Romeo, L. Cifarelli, A. Contin, G. D'Ali, P. Di Cesare, B. Esposito, P. Giusti, T. Massam, F. Palmonari, G. Sartorelli, G. Valenti and A. Zichichi, Phys. Lett. 92B 367 (1980).
2. M. Basile, G. Cara Romeo, L. Cifarelli, A. Contin, G. D'Ali, P. Di Cesare, B. Esposito, P. Giusti, T. Massam, R. Nania, F. Palmonari, G. Sartorelli, G. Valenti and A. Zichichi, Nuovo Cimento 58A 193 (1980).
3. M. Basile, G. Cara Romeo, L. Cifarelli, A. Contin, G. D'Ali, P. Di Cesare, B. Esposito, P. Giusti, T. Massam, R. Nania, F. Palmonari, G. Sartorelli, G. Valenti and A. Zichichi, Phys. Lett. 95B 311 (1980).

4. M. Basile, G. Cara Romeo, L. Cifarelli, A. Contin. G. D'Ali, P. Di Cesare, B. Esposito, P. Giusti, T. Massam, R. Nania, F. Palmonari, G. Sartorelli, G. Valenti and A. Zichichi, Nuovo Cimento Lett. 29 491 (1980).
5. M. Basile, G. Cara Romeo, L. Cifarelli, A. Contin, G. D'Ali, P. Di Cesare, B. Esposito, P. Giusti, T. Massam, R. Nania, F. Palmonari, G. Sartorelli, M. Spinetti, G. Susinno, G. Valenti and A. Zichichi, Phys. Lett. 99B 247 (1981).
6. M. Basile, G. Cara Romeo, L. Cifarelli, A. Contin, G. D'Ali, P. Di Cesare, B. Esposito, P. Giusti, T. Massam, R. Nania, F. Palmonari, G. Sartorelli, M. Spinetti, G. Susinno, G. Valenti and A. Zichichi, Nuovo Cimento Lett. 30 389 (1981).
7. M. Basile, G. Cara Romeo, L. Cifarelli, A. Contin, G. D'Ali, P. Di Cesare, B. Esposito, P. Giusti, T. Massam, R. Nania, F. Palmonari, G. Sartorelli, M. Spinetti, G. Susinno, G. Valenti, L. Votano and A. Zichichi, Nuovo Cimento Lett. 31 273 (1981).
8. M. Basile, G. Cara Romeo, L. Cifarelli, A. Contin, G. D'Ali, P. Di Cesare, B. Esposito, P. Giusti, T. Massam, R. Nania, F. Palmonari, V. Rossi, G. Sartorelli, M. Spinetti, G. Susinno, G. Valenti, L. Votano and A. Zichichi, Nuovo Cimento 65A 400 (1981).
9. M. Basile, G. Cara Romeo, L. Cifarelli, A. Contin, G. D'Ali, P. Di Cesare, B. Esposito, P. Giusti, T. Massam, R. Nania, F. Palmonari, V. Rossi, G. Sartorelli, M. Spinetti, G. Susinno, G. Valenti, L. Votano and A. Zichichi, Nuovo Cimento 65A 414 (1981).
10. M. Basile, G. Cara Romeo, L. Cifarelli, A. Contin, G. D'Ali, P. Di Cesare, B. Esposito, P. Giusti, T. Massam, R. Nania, F. Palmonari, V. Rossi, G. Sartorelli, M. Spinetti, G. Susinno, G. Valenti, L. Votano and A. Zichichi, Nuovo Cimento Lett. 32 210 (1981).
11. M. Basile, G. Cara Romeo, L. Cifarelli, A. Contin, G. D'Ali, P. Di Cesare, B. Esposito, P. Giusti, T. Massam, R. Nania, F. Palmonari, V. Rossi, F. Rohrback, G. Sartorelli, M. Spinetti, G. Susinno, G. Valenti, L. Votano and A. Zichichi, Nuovo Cimento 67A 53 (1982).
12. M. Basile, G. Bonvicini, G. Cara Romeo, L. Cifarelli, A. Contin, M. Curatolo, G. D'Ali, P. Di Cesare, B. Esposito, P. Giusti, T. Massam, R. Nania, F. Palmonari, A. Petrosino, V. Rossi, G. Sartorelli, M. Spinetti, G. Susinno, G. Valenti, L. Votano and A. Zichichi, preprint CERN-EP/81-146 (1981), submitted to Nuovo Cimento.
13. R. Bouclier, R.C.A. Brown, E. Chesi, L. Dumps, H.G. Fischer, P.G. Innocenti, G. Maurin, A. Minten, L. Nauman, F. Piuz and O. Ullaland, Nucl. Instrum. Methods 125 19 (1975).
14. TASSO Collaboration (R. Brandelik et al.), Phys. Lett. 89B 418 (1980).

15. J.L. Siegrist, Ph.D. Thesis, SLAC Report No. 225 (1979).
16. G. Wolf, DESY Report 80/85 (1980).
17. PLUTO Collaboration (Ch. Berger et al.), DESY Report 80/111 (1980).
18. C. Bacci, G. de Zorzi, G. Penso, B. Stella, D. Bollini, R. Baldini Celio, G. Battistoni, G. Capon, R. Del Fabbro, E. Iarocci, M.M. Massai, S. Moriggi, G.P. Murtas, M. Spinetti and L. Trasatti, Phys. Lett. $\underline{86B}$ 234 (1979).
19. PLUTO Collaboration (Ch. Berger et al.), Phys. Lett. $\underline{95B}$ 313 (1980).
20. JADE Collaboration (W. Bartel et al.) DESY Report 80/46 (1980).
21. TASSO Collaboration (R. Brandelik et al.), Phys. Lett. $\underline{86B}$ 243 (1979).
22. M. Basile, G. Cara Romeo, L. Cifarelli, A. Contin, G. D'Ali, P. Di Cesare, B. Esposito, P. Giusti, T. Massam, R. Nania, F. Palmonari, V. Rossi, G. Sartorelli, M. Spinetti, G. Susinno, G. Valenti, L. Votano and A. Zichichi, Nuovo Cimento $\underline{66A}$ 129 (1981).
23. M. Basile, G. Cara Romeo, L. Cifarelli, A. Contin, G. D'Ali, P. Di Cesare, B. Esposito, P. Giusti, T. Massam, R. Nania, F. Palmonari, V. Rossi, G. Sartorelli, M. Spinetti, G. Susinno, G. Valenti, L. Votano and A. Zichichi, Nuovo Cimento Lett. $\underline{32}$ 321 (1981).
24. J.W. Chapman, J.W. Cooper, N. Green, A.A. Seidl, J.C. Vender Velde, C.M. Bromberg, D. Cohen, T. Ferbel and P. Slattery, Phys. Rev. Lett. $\underline{32}$ 257 (1974).
    P. Capiluppi, G. Giacomelli, A.M. Rossi, G. Vannini and A. Bussière, Nucl. Phys. $\underline{B70}$ 1 (1974).
25. W. Thomé, K. Eggert, K. Giboni, H. Lisken, P. Darriulat, P. Dittman, M. Holder, K.T. McDonald, H. Albrecht, T. Modis, K. Tittel, H. Preissner, P. Allen, J. Derado, V. Eckardt, H.-J. Gebauer, R. Meinke, P. Seyboth and S. Uhlig, Nucl. Phys. $\underline{B129}$ 365 (1977).
    A. Albini, P. Capiluppi, G. Giacomelli and A.M. Rossi, Nuovo Cimento $\underline{32A}$ 101 (1976).

HIGH-ENERGY, SOFT (pp) INTERACTIONS

DISCUSSIONS

CHAIRMAN : A. Zichichi

Scientific Secretaries: G. D'Ali, G. Sartorelli

DISCUSSION 1

- *SEIBERG:*

You described a method for analysing hadronic processes, defining an "effective energy" in terms of which proton-proton interactions behave like e^+e^- processes. What is the maximum effective energy you can obtain in this way and what are the applications of this energy?

- *ZICHICHI:*

The range of hadronic effective energies available varies according to a) the x_F cut one applies to the leading proton; b) the total energy ($\sqrt{s}$) available in the pp collision. For example, one can select the leading proton in the range

$$0.35 \leq x_F \leq 0.86 .$$

In this case, the effective hadronic energy available for various values of $\sqrt{s}$ in the proton-proton centre-of-mass system is as follows (see Fig. 1 of the lecture):

$\sqrt{s}$ = 30 GeV   4.2-19 GeV ,
$\sqrt{s}$ = 44 GeV   6.2-28 GeV ,
$\sqrt{s}$ = 62 GeV   8.7-40 GeV .

When you have two protons colliding with energy E_1 and E_2, $E_1 = E_2 = E_{inc}$, the total c.m. energy available in this collision is generally taken to be
$$\sqrt{s} = E_1 + E_2 = 2E_{inc}.$$

However, we say that this is not the right energy to use in analysing the process. After the collision, the two incoming protons keep, on the average, a large fraction of the available energy and play a privileged role in the energy-momentum sharing among the particles in the final state. Our statement is that you have to subtract, from the total energy, the energy carried by the two leading protons in order to obtain the effective energy available for particle production. We call it E^{had}, with $E^{had} = E_1^{had} + E_2^{had}$. (The reason for this splitting into two terms will soon be clear).

There are many ways of identifying this energy experimentally. The best one is obtained when you identify both protons in the final state. In this case, and with the definitions

$$\text{four momentum of incoming proton 1} \equiv p_1^{inc}$$
$$\text{four momentum of incoming proton 2} \equiv p_2^{inc}$$
$$\text{four momentum of leading proton 1} \equiv p_1^{lead}$$
$$\text{four momentum of leading proton 2} \equiv p_2^{lead}$$

$$p_1^{had} = p_1^{inc} - p_1^{lead} \quad \text{and} \quad p_2^{had} = p_2^{inc} - p_2^{lead}$$

you can construct the relativistic invariant quantity

$$Q_{1,2}^2 = \left(p_1^{had} + p_2^{had}\right)^2 \quad \text{and} \quad E_{1,2}^{had} = \sqrt{Q_{1,2}^2} \ .$$

However, to a good approximation the analysis can be done quite satisfactorily in terms of the simpler energy variable,

$$\begin{cases} E_1^{had} = E_1^{inc} - E_1^{lead} \\ E_2^{had} = E_2^{inc} - E_2^{lead} \end{cases} \quad \text{and} \quad E_{1,2}^{had} \cong E_1^{had} + E_2^{had} \ .$$

Let me come now to the important point, namely the limits of this analysis. In proton-proton collisions at ISR energies you do not have Čerenkov counters which can discriminate 25 GeV/c protons, nor any other system such as time of flight, so that you can only rely on nature. Nature has been kind enough to give us a very powerful tool in terms of the behaviour of the proton in this kind of interaction.

The ratio p/π in pp collisions at the ISR behaves exponentially (see Fig. 18) as a function of x_F, and scales with $\sqrt{s}$ and p_T. So, if you take, for example, particles with $x_F = 0.5$ you will find a sample of $\sim 85\%$ protons with $\sim 15\%$ π contamination.

So the proton identification can be done by finding the fastest positive particle in an event and applying to it a lower x_F cut. We

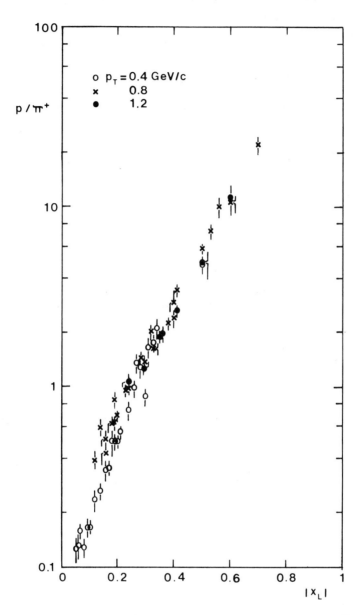

Fig. 18 The inclusive p/π^+ ratio measured at the ISR ($\sqrt{s}$ = 23, 31, 45, 53 GeV).

choose $x_F > 0.4$ as a safe cut. Notice that where the proton has $x_F \simeq 0.8$ you have only a few percent of π contamination, and only 20% of the initial energy is used for particle production.

In this case a high-energy pp collision at, say, 62 GeV gives an effective energy of only 12 GeV, and this is the e^+e^- energy you have to take for a comparison.

DISCUSSION 2   (Scientific Secretaries: G. D'Ali, G. Sartorelli)

- *HERTEN:*

Could you describe the event selection. What cuts do you use in the analysis? and how are the leading protons separated from all other hadrons?

- *ZICHICHI:*

A first sample of data has been collected using the so-called minimum bias trigger -- where you require nothing but a real proton-proton interaction (two tracks in the apparatus) -- in such a way as to minimize beam-gas background and interactions with the beam-pipe wall. In this sample the analysis was performed looking at the jet on one side only and making assumptions on the other side.

There was in fact only one proton per event identified in most of the triggered events. The idea was that this could be an approximation that would not have destroyed the meaning of the method and the results.

A second run has been done, selecting events rich in leading protons. It has been checked that this trigger did not bias the event; in other words, one has only to make sure one does not lose, for acceptance reasons, the leading proton, without which one cannot define the right energy available for particle production.

- *HERTEN:*

Is the energy of the leading proton always so high that it is possible to distinguish it from all the other hadrons produced?

- *ZICHICHI:*

This is an important question. As I have already explained, we limit our analysis to the x_F region

$$0.35 \leq x_F \leq 0.86$$

and this restricts our effective hadronic energy range. To go below this lower limit you would need some kind of particle identification system (Čerenkov counters or a TOF system). We are seriously thinking of the possibility of introducing such a system and of extending our study to the low x_F (high E_{had}) region.

HIGH-ENERGY, SOFT (pp) INTERACTIONS

I am personally convinced that pp machines are really very good: it is very difficult to work with them, but, once you disentangle the right physics you want to do, they compete very well with other sources of particles such as e^+e^- machines.

DISCUSSION 3   (Scientific Secretaries: G. D'Ali, G. Sartorelli)

- *KARLINER:*

As you mentioned in your talk, in pp scattering (unlike in e^+e^- scattering) gluon-gluon scattering is almost unavoidable. Is there any chance of seeing the signature of the glueball pole in that kind of scattering?

- *ZICHICHI:*

If we find this effect, I will invite you to participate in a future "Ettore Majorana" School, free of admission fee! It would be extremely difficult to isolate this kind of effect. We have been looking at all possible sources of differences which could show up in charged particles produced -- higher transverse momenta, charge-correlations etc. -- but, up to now, we have not been able to find any.

- *KARLINER:*

As you know, some theorists have estimated the glueball mass based on J/ψ decay. Has anyone seriously tried to see what its effects would be in pp scattering?

- *ZICHICHI:*

I think that the trouble with the glueball mass is that there has been no adequate theoretical investigation of the problem. However, it is not easy to work with glueballs. Sam Lindenbaum worked a lot on this problem and he claims he has evidence for glueballs.

DISCUSSION 4   (Scientific Secretaries: G. D'Ali, G. Sartorelli)

- *WIGNER:*

You reported on a interesting similarity in jets and particle production between pp and e^+e^- collisions, when processes at proper energies are compared. Are there similar results for ep collision products?

- *ZICHICHI:*

This is a field which the previous speaker is working on, and the interaction being studied is not ep but μp. The EMC Collaboration has been investigating the production of hadronic systems (jets) in the collision between space-like virtual photons (emitted by the muon) and the proton. The recoiling proton should be "leading" in the γp c.m. system.

My statement is based on the analysis we performed on data for the process

$$\bar{\nu}_\mu + p \to \Lambda^0 + X .$$

In this case we found that the Λ^0 is "leading" in the sense defined previously. The reaction is

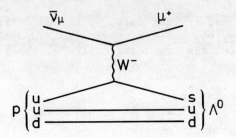

The W^- interacts with the u quark, converting it into an s quark which, with the remaining u and d quarks of the incoming proton, forms the Λ^0. The latter then goes on in the same direction as the incoming proton with a significant fraction of the incoming energy. In terms of our variable L the value of L_{Λ^0} in this case is

$$L_{\Lambda^0} \simeq 2 .$$

in agreement with what I showed you for the values of L for leading particles. This proves that the Λ^0 produced in weak interactions is leading.

So when the EMC Collaboration will analyse their data, identifying the hadrons, they will find a "leading" baryon, and if they analyse the final state in terms of the effective energy we have introduced, I think they will discover the same structures we claim are common to e^+e^- and pp final states.

- *GABATHULER:*

The leading proton effect in μp collisions corresponds to the target proton in the backward hemisphere in the c.m. system. In the EMC experiment I reported, the detector collected only events in the forward hemisphere. At present we have added a vertex detector consisting of magnet and streamer chamber to study the backward hemisphere, i.e. the leading proton.

- *YAMAMOTO:*

The data you showed seem to suggest that leading protons have nothing to do with jet formation. Is it true?

– ZICHICHI:

To some extent we can say that leading protons have nothing to do with the formation of jets. We have been looking for some correlations and we are still studying the problem since it is an interesting one, but so far we can only say that there are no correlations but kinematics.

– YAMAMOTO:

What is the connection with the Drell-Yan process which suggests that there are forward and backward jets and two away-side jets?

– ZICHICHI:

The Drell-Yan process itself is an annihilation process. Proton-proton collision is different. But if you are referring to the qq hard scattering I can tell you that the analysis I presented is on low-p_T and not high-p_T physics. In order to study the Drell-Yan process in a clear way you need to look at final states with leptons of opposite charges produced in strong interactions. That kind of physics is completely different from the one we are doing.

DISCUSSION 5   (Scientific Secretaries: G. D'Ali, G. Sartorelli)

– D'HOKER:

I would like to make a comment on the problem of the observation of the glueball. The glueball is bound to be very difficult to observe, for two reasons. Firstly, the glueball state mixes with many states of bound quarks. Secondly, at very high energy, a deconfining phase transition occurs and gluons are effectively free and the glueball does not exist. This means that it will always be confused with radiative corrections.

– KARLINER:

It is not clear that gluons arrive at thermal equilibrium during the scattering process, so I am not sure how meaningful the concept of temperature is here. However, even if we were to assume that the temperature is so high that it causes a deconfining phase transition, this would probably have some other consequences which would not agree with what one calculates in QCD, assuming zero temperature, and we know that these zero temperature predictions have been experimentally confirmed.

– ZICHICHI:

I must remind you of the important message given to us by Professors Dirac, Wigner and Teller: we should be more modest and not pretend to understand everything at once. Now, the reason why people have taken asymptotic freedom very seriously is due to the fact that, firstly, one knows that the theory is non-Abelian. This is the starting point. The discovery that in non-Abelian gauge theories the coupling constant tends to zero when the energy tends to infinity was very useful in understanding deep-inelastic scattering phenomena. These rather simple considerations form the basis of hundreds of papers written by theorists. From this moment on, the confusion increases because none of these papers has really contributed towards a better or deeper understanding of these two very simple facts. I do believe in asymptotic freedom as an important property of non-Abelian gauge forces. It is also corroborated by experiment. We should be very happy that at least we understood this point. The glueball is something else; it is much more complicated.

- *KARLINER*:

I absolutely agree -- we should believe in asymptotic freedom. What I said is that if we take the idea seriously that the temperature is not zero, then we might get into trouble and we know that experimentally we are not in trouble with asymptotic freedom.

- *D'HOKER*:

The final temperature of deconfining phase transition is not in contradiction with asymptotic freedom. It is just the same thing.

- *SPIEGELGLAS*:

Are there any "next to leading" particles? If so, what good reasons do you have for neglecting them?

- *ZICHICHI*:

Once you take out the leading proton, you find that the $d\sigma/dx$ distribution (this is the first quantity measured by us) collapses into the $d\sigma/dx$ distribution of particles produced in e^+e^- interactions. Consequently there is no reason to believe that there is another leading particle. To be more clear: there is not a second leading particle.

So, physically speaking, the question is as follows. You have some energy E which you want to transform into particles. If this energy E is not associated with any colour singlet state, then the energy is shared between the particles according to the e^+e^- $d\sigma/dx$ distribution.

If, on the contrary, as is the case in pp collisions, the energy E is associated with a colour singlet state, then the proton will not share its energy equally between the produced particles. This is the "leading particle phenomenon". So, once you have taken out the leading particle there is no other leading effect left; in other words, there is no second leading particle. If such a second leading particle existed it would have been seen in our analysis.

HIGH-ENERGY, SOFT (pp) INTERACTIONS

— *PETERSON:*

How do you determine the jet axis?

— *ZICHICHI:*

There are many ways in which this can be done. One can use the thrust or the spericity axis in exactly the same way as for e^+e^- data. In some cases we use the missing momentum axis because we do tet the leading proton, i.e. $\vec{P}_M = \vec{P}_{inc} - \vec{P}_{lead}$, where $\vec{P}_{inc}$ is the momentum of the incoming proton and $\vec{P}_{lead}$ is the momentum of the leading proton. Anyway there are no big differences in the results using different axes; the effects we find do not depend on a particular choice of the jet axis.

— *GABATHULER:*

QCD is a theory of strong interactions. Can we understand the very good agreement between the e^+e^- and pp data within the framework of QCD?

— *ZICHICHI:*

I think that QCD is going to be the common tool with which to explain the data. In fact, when one analyses pp data using the same criteria as in e^+e^- physics, one finds the same results. This is shown by the comparison between the set of quantities, $d\sigma/dx$, $d\sigma/dp_T$, $\langle n_{ch} \rangle$, planarity, etc., measured in (pp) and (e^+e^-) interactions. So, if e^+e^- data are explained in terms of QCD, pp data must also be explained in terms of QCD.

MY LIFE AS A PHYSICIST

Paul A.M. Dirac

The Florida State University
Department of Physics
TALLAHASSEE, Florida 32306, USA

I would like to talk to you about my life as a physicist and try to explain to you why I look upon a good many things rather differently from other physicists. I went to a school which specialized in mathematics, physics, chemistry, and modern languages. I was very lucky to go to this school; it gave me just the subjects I wanted, and there were good teachers there. My interest was in mathematics and physics right from the earliest time that I can remember.

There was one thing about my school days which was especially lucky for me. While I was at school, there was a war going on, the first World War. That resulted in all the older boys being taken away to serve in the army, and then the younger boys were pushed on as fast as they could absorb the knowledge to fill up the empty places that arose in the higher forms of the school. In that way, I got pushed on to the higher forms and was introduced to the main ideas in mathematics and science at an earlier age than I would otherwise have been. I think this was an advantage to me. I did not have to wait in order to satisfy age requirements, but all the time I was moved ahead as fast as I was able to absorb the new knowledge.

and I thus learned such things as the foundations of mechanics, the calculus and the atomic theory of chemistry at an earlier age than most children would.

On leaving the school, I proceeded to the Engineering faculty of the University of Bristol, which was held in the same buildings as the school, and it had excellent labs. And again, I was pushed ahead as much as possible, because of the war, and I started my University course at the age of sixteen. The war ended soon after I went to the University.

The Engineering course influenced me very strongly. Previously, I was interested only in exact mathematical relations. Any approximations, I found disagreable, and I wanted to avoid them, but as an engineer, I had to learn to tolerate them. An engineer is dealing with the real world and is doing calculations referring to it, and he has to make approximations all the time. A good engineer is one who has a feeling for what approximations are good and which ones will lead to trouble. I've learned that, in the description of nature, one has to tolerate approximations, and that even work with approximations can be interesting and can sometimes be beautiful.

Another important thing that I learned during my Engineering training, was in connection with Relativity. I started at the University of Bristol in 1918, and it was just in November 1918 that the war ended, and then we suddenly heard about Relativity and Einstein. Previously, nobody knew anything about either Relativity or Einstein except for a very few specialists in the universities. There was no one in the Faculty of Engineering at Bristol who knew anything about these things. Then suddenly, everyone was talking about them. It just burst upon the world, a world that was sick of the war and was craving for new ideas. Relativity provided just what was wanted.

But no one had any accurate information. Among the students and even among the professors, no one had accurate information about what the theory of Relativity was. We just had to discuss it in terms of the information that was given in the popular press, information that was largely philosophical and did not contain any of the equations which one really needed. As engineers, we had been basing all our work on Newton, and then we heard that Newton was wrong, in some strange way. That was a great surprise to us. No one understood why it was wrong, but we had to accept that Newton's laws were inaccurate. That led me to think that any laws of nature were possibly only approximations and might have to be modified with increases of our knowledge.

In England, we had Eddington, the one man who really understood Einstein and Relativity. The way that came about was that Eddington had a friend in Holland, de Sitter. Holland was neutral and so Eddington and de Sitter could communicate with each other, and de Sitter could also communicate with Einstein in Germany. In that indirect way, Eddington learned about the Einstein theory, and he understood it completely and explained it to the public. He was very good at popular exposition, but still we had to wait a few years for Eddington to publish his book on the Mathematical Theory of Relativity before we really had an adequate basis for studying the equations. When this book did appear, people like engineering students had enough mathematics to be able to read it through and get the real foundations of the theory.

This engineering period led me to be content with approximations when one is studying nature. Maybe all our laws of nature are only approximate, and our knowledge of nature should be looked upon as having nothing final in it. Anything may be subject to modification with the increasing knowledge we shall obtain in the future. That I feel is something we should bear in mind all the time. Anything we have in Science is not a final truth but is something which is liable to be altered with increasing knowledge.

I finished my engineering course after three years, that was in 1921, and then I found I could not get a job as an engineer. There was a depression on at the time. I continued at Bristol University studying mathematics, for two years.

The thing which interested me most in this mathematics was Projective Geometry. This is something which physicists don't use very much, but I found it really very useful to have the basic ideas of Projective Geometry. They formed a powerful generalization of the ideas of Euclidean Geometry. I am continually using Projective Geometry in my research work, even though you won't see any reference to it in my published papers. I use it to picture relationships between vector and tensor quantities in Euclidean and also in Minkowski space. One can get a clearer picture of the relationship in this way. Then, when one has got the picture one can, of course, explain the relationship without using Projective Geometry. In all my published work, I avoided Projective Geometry, because I thought that most physicists would not appreciate it, and it wasn't essential that they should learn it. But still, it is a great help in research. You can't very well picture things in four dimensions in any direct way, but the relationships between vectors can be pictured very well in three dimensions in terms of the ideas of Projective Geometry. There was a very good teacher of Projective Geometry at Bristol, Mr. Fraser, a man who never did any research but who was an extremely stimulating teacher.

After these two years doing mathematics at Bristol, I proceeded to Cambridge as a research student, with a small government grant which helped me. There, I was put under the supervision of R. H. Fowler. Fowler was an expert in Statistical Mechanics and in Quantum Theory. Then, I learned for the first time about Bohr orbits. Nobody in Bristol seemed to know about them, at least they never mentioned them to me. The Faculty of Engineering was not concerned with them and the mathematics I studied also was not concerned about them.

It was a complete surprise to me to find that one could actually use equations of motion to describe conditions within the atom, equations rather like Newton's equations which have to be applied with certain special conditions. I became very much excited with these Bohr orbits and studied them intensively. One needed to know more about them. One only had very primitive applications for them. The big problem was how do the different Bohr orbits in an atom interact with one another ? In dealing with the Bohr orbits, it was sometimes useful to use the Hamiltonian methods of dynamics, so I studied the Hamiltonian theory of dynamics a good deal. There is an extensive transformation theory connected with it, which I learned.

I worked very hard on the question of how to extend the use of Bohr orbits and get some understanding of the interaction between the Bohr orbits in an atom. This work was essentially futile. It led to nothing.

The sudden change came in 1925, when Heisenberg introduced his matrix mechanics. Heisenberg got away altogether from the idea of Bohr orbits. He concentrated on quantities which are more closely connected with observation and not with Bohr orbits. These quantities are each connected with two stationary states of an atom, and they form a matrix. Now it was quite an effort for me to get away from Bohr orbits. It took quite a while to do so, because I was so much attracted by them previously. I first got a copy of Heisenberg's work a few weeks before publication. A copy had been sent to Fowler, and he'd passed it on to me. I remember when I first read this work, I didn't like it very much. I didn't see any point in it, and put it aside. I went back to it about two weeks later, and then suddenly realized that it provided the key to the whole question of getting an atom mechanics and took it up excitedly.

Heisenberg had introduced matrix elements, each associated with two energy levels. At that time, I was just full of Relativity, and the first idea I had was that each matrix element should be associated with two momentum values as well as two energy values. It's the natural thing that one would think of under those conditions. Then it seemed to me that one could not very well allow the two momentum values to be completely arbitrary. Maybe one should restrict them in some way. The natural restriction would be to have the jump in momentum equal to c^{-1} times the jump in energy. That was the first bit of work I did on Quantum Mechanics. I never published this idea, because I soon got caught up in more fundamental ideas, but this original idea was incorporated in my later work on the Compton effect.

The first main advance I made with Quantum Theory was realizing that there is an analogy between the commutator of two dynamical variables uv - vu, which is not zero, with Heisenberg's Matrix Mechanics, and the Poisson bracket of classical mechanics. I remember this idea occurred to me when I was taking a walk out in the country, and I was very excited about it. I could not check on it immediately, because I wasn't sure what a Poisson bracket was. It was something I'd read about in my studies of Hamiltonian methods, but my memory was not very clear about it. I had to wait till I could go to a library and check out whether my ideas were really right or not, and I found then that everything was O.K.

As a question of notation, I might mention that in the early writings on Hamiltonian methods and the invariants associated with them, there are two bracket expressions occurring, the Poisson bracket and the Lagrange bracket. And the standard notation in those days was to use round brackets for the Poisson bracket and square brackets for the Lagrange bracket. Now the Lagrange bracket is of no importance for Quantum Mechanics. The Poisson bracket is all important. It seemed to me that to use the round bracket for the Poisson bracket was not very suitable, because one was not used to having something which is anti-symmetrical in its two members represented by a round bracket. It is quite different from the notation which one uses in Vector analysis, for instance. So I boldly introduced the new notation of square brackets for Poisson brackets, and I'm glad to be able to say that people copied me, and it is now standard practice. No one refers anymore to the Lagrange bracket. I must say that the appearance of non-communicative algebra was a tremendous surprise at that time. It was something quite foreign to anything that physicists had ever thought of previously. I heard that Heisenberg was extremely perturbed about it and wondered whether his whole theory would have to be abandoned on account of it, but Heisenberg received a lot of encouragement from his professor, Max Born, and did proceed with it. It turns out that it is the really fundamental difference between the new mechanics and the previous mechanics of Newton, as developed by Lagrange, Hamilton, and others.

After I got the idea of the connection between the commutator of two variables in Heisenberg's new theory and the Poisson bracket, the foundation was provided for making a rapid advance, and I was quite enthusiastic and devoted my whole energy towards developing this new theory. Then Schrödinger came along with a new theory, and at first, I quite resented Schrödinger. I was quite happy with the Heisenberg mechanics, just wanted to go on developing it, and did not want to have to get used to new ideas in the way that Schrödinger required. So it was just a rather unwanted diversion. Then it turned out that Schrödinger's theory was not at all contradictory to Heisenberg's but just supplemented it.

Schrödinger's theory was a great help in providing a physical interpretation for the new mechanics which Heisenberg had originated.

We then had a quite unprecedented situation in Physics in that we were working with equations and did not know precisely what their interpretation was. We had some simple examples where we were able to guess. We chose a certain energy function, and we were able to work out its eigenvalues. By the way, with regard to notation, I may say that English people at that time didn't like to use the word eigenvalue; they used the word proper value instead. That seemed to me to be a bit unsuitable, and so I emphasized the eigenvalue. I was able to get it standardized. Schrödinger's theory was a great help in providing a physical interpretation for the new mechanics. This physical interpretation involved probabilities. One had a Schrödinger wave function. One took the square of its modulus and interpreted that as the probability of certain variables having certain values. I was able to develop a general transformation theory of the new mechanics that enabled one to express the Schrödinger function in terms of any commuting variables having specified values. I think the development of this transformation theory was the piece of work that gave me most satisfaction in all my work in physics, because I was proceeding definitely along a chain of ideas and not just lucky guesses. This definite, logical development led to the complete transformation theory. If you read the early papers of Schrödinger, you will see that they are all non-relativistic. Now Schrödinger's starting idea was the waves associated with particles, the waves which had been introduced by de Broglie, and de Broglie waves were connected with the particles in a relativistic way. Thus, his starting idea was a relativistic one, but still Schrödinger's early papers were all non-relativistic. How did that come about ? Schrödinger explained it to me, many years later, during a little conversation we were having, and he said that he first got his wave equation as a generalization of the de Broglie equation, referring to an electron moving in an electro-magnetic field. This equation which he first got was a relativistic one.

And then, of course, he applied it to the electron in the hydrogen atom. He got a result not in agreement with observation. The reason why it did not agree with observation was that there was no reference in it to the spin of the electron. People had just begun to think about the spin of the electron in those days, and had not set up any detailed theory involving it. Schrödinger's original wave equation had no reference to the spin at all and therefore gave the wrong result. And when Schrödinger found this wrong result, he was excessively disappointed. He thought his whole idea was no good at all, and he just abandoned it. It was some months later, when he went back to this work and noticed that, in non-relativistic approximation, the result of his calculations did agree with the observation, that he published his work therefore as a non-relativistic theory.

The original Schrödinger equation was rediscovered later on by Klein and Gordon, and they published it. So it is now known as the Klein-Gordon equation, even though it was discovered a good while earlier by Schrödinger. Schrödinger just did not have the courage, I might say, to publish a paper which was definitely in disagreement with observation. People nowadays don't have this kind of lack of courage. It seems that Schrödinger was too timid, and the development of Quantum theory was held up a little bit by his timidity.

The Klein-Gordon equation is an equation of the second order in d/dt. It is quite relativistic, but it does not conform to the standard kind of equations which we have in quantum mechanics, to which the general transformation theory would apply. That provided me with a problem, to get a relativistic quantum theory of the electron, and I was studying this for quite a while. During this time I remember Bohr coming and asking me, at the meeting of the Solvay conference in 1927, "what are you working on now?" I said, "I'm trying to get a relativistic theory for the motion of the electron." Then Bohr said, "but that problem has already been solved by Klein

and Gordon." I wanted to explain to Bohr that this solution by
Klein and Gordon was not satisfactory, because it did not agree with
the standard transformation theory of quantum mechanics. The
standard transformation theory required that the wave equation should
be linear in d/dt and then one could use the square of the modulus
of the wave function to determine the probability of any commutating
variables having any particular values. I wanted to explain my
dissatisfaction to Bohr at that time, but I wasn't able to because
the start of another lecture put an end to this private discussion
we were having between lectures.

However, I continued to work on this problem. The problem was
to get a wave equation which would be compatible with the foundations
of quantum mechanics and thus with transformation theory which I had
worked out, which I considered to be essential. The transformation
theory demanded a wave equation which should be linear in d/dt. Well,
from a sort of a lucky guess I suppose, I ultimately thought of a
new wave equation, which is now accepted as the relativistic wave
equation of the electron. It gave us probabilities for finding the
electron in any place.

With the introduction of this wave equation, one had the problem
of negative energies for the electron becoming of a paramount
importance. Previously, the negative energies were always present
in the equation but had been obscured by the fact that the equation
had more serious difficulties. When the more serious difficulties
were removed, the negative energy difficulty remained as the next
most serious one to be disposed of. I was able to dispose of it by
introducing a vacuum with all the negative energy states occupied
and then discussing what a few unoccupied negative energy states
would appear like. I first of all wrote up my work with the holes
in the negative energy states considered as protons, because they
had a positive charge. Protons were the only things known at that

time which had a positive charge. At that time, the whole climate of opinion was against postulating new particles. One just had two particles, electron and proton, associated with the two charges, negative and positive, and there was nothing else. However, the mathematicians insisted the holes had to have the same mass as the electron, and therefore they had to be a new particle. It was discovered a little later by Anderson. It was also discovered independently and somewhat earlier by Blackett, but Blackett was slow in publishing his work, because it was of a statistical nature, and it did not have the same definiteness that Anderson's had, and Blackett wanted confirmation. Blackett was a bit too cautious, and Anderson scooped that discovery.

At this stage we had a satisfactory theory for an electron moving in a given electro-magnetic field. In this work one had proceeded step by step, accounting for the difficulties one by one, taking the most serious difficulty first, getting rid of that, and then the next most serious difficulty and so on. That has led me to the view that the whole of the development of physics will be like this. We see a lot of difficulties in front of us. We must find a way of explaining those difficulties one by one. A good many people are trying to find an ultimate theory which will explain all the difficulties, maybe a grand unified theory. I believe that is hopeless. It is quite beyond human ingenuity to think of such a theory. I believe that many new ideas will be needed in the future, and that these new ideas must be obtained, one by one, each of them explaining just one of the difficulties which lie ahead. One should be quite satisfied if one can explain one little difficulty and get a theory which is an improvement in one little respect, and not be too ambitious and hope to be able to explain all the difficulties together.

Well, this work which led up to the theory of the electron and the positron applied only for particles in a specified electro-magnetic field. The next step was to try to generalize it to apply to an electro-magnetic field which is subject to the laws of Quantum mechanics, a quantized field. Here we ran into a difficulty. One can set up the equations for electrons interacting with the electro-magnetic field. One gets quite a definite Hamiltonian function, leading to a definite Schrödinger equation. One can proceed to solve this Schrödinger equation by a perturbation method, starting off with a zero order approximation where there is no interaction between the electrons and the electro-magnetic field and then getting successive terms in the approximation. That method appears to be a good method because the coupling constant is so small. However, when we get to the second order of approximation, we find infinities appearing in the solution of the Schrödinger equation. That is the way the infinities appear in quantum mechanics in the first place. Subsequent work on setting up relativistic theories of particles and fields of interaction always leads to such infinities except in trivial cases. The proper conclusion to draw, I will say, is that the basis of our theory is not correct. But we don't have any better basis. So people are just working with these equations containing infinities. They have developed a way of discarding the infinities, turning a blind eye to them, and they calculate the residual effects which are left when one adopts a suitable procedure for discarding the infinities, involving renormalization.

But with an infinite renormalization factor, you are neglecting quantities in your equations which are infinitely large. You are neglecting them without any logical reason for doing so. You are neglecting them just because you don't want to have them in the theory. Now I find this intolerable. It is especially intolerable for someone with an engineering training. You're just going against all principles which you have been taught, those that are really fundamental in your theory.

What can you do if you refuse to accept the discarding of infinities ? You might try to do quantum electro-dynamics without discarding infinities at all, by introducing a cutoff in the interaction between the electron and the field, a cutoff at certain frequencies of the field. You then say that the interaction terms from the frequencies above the cutoff don't occur in the equations. If you do that, of course, you're making the theory non-relativistic; you are making it ugly. But even so, I think that a theory which is non-relativistic and which is ugly is preferable to one which goes so much against logic as to require discarding infinite terms.

I did a good deal of work with this quantum electro-dynamics with a cutoff, and I found that, thanks to the smallness of the coupling constant, it's possible to introduce the cutoff at a certain place, so as to make the first order correction in the Lamb shift and the first order correction in the anomalous magnetic moment of the electron come out correctly. But still, it is too complicated to proceed with the higher order corrections in that way. There is a good deal of experimental evidence that these higher order corrections, as provided by the method of discarding infinities, are in agreement with observation, so that this method with the cutoff is not a method to be recommended, and I don't think there's any future in it.

What is the situation now ? I still cannot tolerate discarding infinite quantities, and I have spent a long time looking for a way of reformulating relativistic quantum mechanics so that the infinities do not occur, but I have not had any success. I've tried several ideas. I've been working for a good while just with streams of matter interacting with each other instead of individual particles, and with these streams one gets equations that do not contain infinities, but they do not contain the detailed interaction which is needed for getting specific results. I will say that it is an unsolved problem, but still a very necessary problem to get an improved quantum mechanics where the infinities do not occur at all. Even though I have failed, it is something that one should continue to strive for.

One should keep one's eyes open for any possibilities in that direction. Another question that has interested me very much concerns the numerical value for the fine structure constant, $\frac{hc}{e^2}$, the reciprocal of the fine structure constant α. It has the value of about 137. The question is why does it have this value ? Why did Nature choose this particular number ? That is a problem which people have considered very much from 1930 onward. Eddington, in particular, devoted a lot of work to it, and set up an argument involving counting degrees of freedom which required this number to be exactly 136. Then he improved the argument to get 137. I have studied Eddington's arguments intensively but was unable to understand them. I don't think anybody else really understood them. Eddington's arguments would require this number to be exactly 137, and at the time that Eddington brought them out the number was not known experimentally very accurately. But with more recent observations, it appears that this number is not accurately 137, not accurately an integer, and thus Eddington cannot be right. So I don't think there's any logical basis behind Eddington's arguments; at any rate, I could not find it. I would like to refer to some other work on this problem by Euler and Kockel. Before I can tell you what this work is, I must explain a modification of electro-dynamics, which was made by Born, a modification of the classical Maxwell field theory. Born supposed that the Maxwell equations that we all work with are valid only for weak fields, and that when you get to strong fields, fields comparable to the electric field at the surface of the electron, you have to change the Maxwell equations. Now Born thought out a simple, neat way of making such a change and adjusted the coefficients so that the total Coulomb energy of the electron would just have the correct value, mc^2. This was entirely a classical theory. Now with this theory, the propagation of light is no longer governed by linear equations and there is a possibility of scattering of light by light. One can work out what this probability is.

Now there is also a scattering of light by light that can occur in standard quantum theory. If you have two photons coming together, they may interact and form an electron-positron pair. The electron-positron pair can then annihilate each other and form two photons going in different directions. Now even if the two original photons do not have enough energy to form a real electron-positron pair, they may form a virtual pair which is sufficient to give a scattering of light by light.

So you have two theories in which light is scattered by light, Born's theory and the standard quantum electro-dynamics. You can compare the coefficients in these two theories; one of them involves Planck's constant; the other doesn't. So in comparing them you get an equation which fixes Planck's constant in terms of the electric charge. You get a value for the fine-structure constant. Now Euler and Kockel did this comparison, and they obtained a value for α^{-1} of about 82, 82 when it ought to have been 137. But still that's not too bad a discrepancy.

Soon after Euler and Kockel published this work, Infeld took up the problem and he pointed out that there are various possible ways in which one could modify the standard classical electro-dynamics in accordance with the ideas of Born. Infeld proposed an alternative to Born, also based on a simple involves assumption, and again made the comparison of the scattering of light by light according to <u>his</u> modified Maxwell theory and the standard quantum theory. He got for α^{-1}, 130. That was much better; I might mention that if you are interested in this work, you could look up Infeld's paper which is published in Nature, volume 137.

These ideas which involved a connection between two discordant theories and fitting them together are interesting. There may be some truth in them. But still I don't feel that any final solution of the problem will be obtained in such a way. What we really want is a quantum electro-dynamics which shall demand that α^{-1} has just this one value close to 137. We want fundamental equations which are consistent only provided α^{-1} has this one value. Can one find such equations?

I spent a good deal of time thinking it over, and at one time I got an idea which did suggest that we could get quantized singularities occurring in the electro-magnetic field. But further study of this idea led to the appearance of magnetic monopoles in the equations. The monopole has a quantized strength in terms of the strength of the charge of the electron. We then had a theory which connected the quantized monopole and the quantized electron, but we got no information about the value of the quantized electron by itself. When I realized this, it was a great disappointment to me. Still I had to make the best of it, so I published my work as the theory of the monopole. That work has led to quite a lot of development some of which we've heard of in this school. But to me it remains just a disappointment. It is a disappointment because it gives no help in solving the fundamental problem of why α^{-1} has just this one value.

Well, at the present time, I feel that we have two fundamental problems lying ahead in Physics. One of them is the problem of getting a quantum mechanics which works without infinities, and the other is getting a quantum mechanics which fixes the value of the constant quantity "α", i.e., of the fundamental electric charge. I've spent years just examining these problems, and there has been no significant progress. I feel that we need some new ideas to answer these problems, and should not just continue with the calculations based on the old ideas. It is, of course, very useful to continue these calculations and see what information we can get by fitting in experimental results. That is what most of the world's physicists are doing now. But we shall not make any substantial progress till we get another Einstein or another Heisenberg who will bring in some entirely new idea to help us understand these two difficulties. That is the main part of what I wanted to tell you.

MY LIFE AS A PHYSICIST*

Edward Teller

University of California
Lawrence Livermore National Laboratory
Livermore, California, U.S.A.

Originally I was asked to talk about how it was in physics a long time ago when I was young. Indeed, physics was a little different from what it is now. I have decided to begin by telling you not about what I did in physics, which was not much, but about how I learned physics. I will finish just at the beginning of the second world war. Once a physicist, always a physicist. But after the beginning of the war, the world and physics never again were the same for me.

At the time I was a student, physics was quite a remarkable subject. Many people today have almost forgotten the many ways in which it was changing. When I was still in high school, I had a terrible habit which I have since abandoned. Occasionally I wrote in a diary. The first thing that I know I wrote in my diary was connected with physics. Shortly before the time that deBroglie wrote his paper, I wrote something that in a strange way I still believe. I wrote:

---

*This work was performed under the auspices of the U.S. Department of Energy by the Lawrence Livermore National Laboratory under contract No. W7405-Eng-48.

What is called understanding is often no more than a state where one has become familiar with what one does not understand.

I don't want to talk about where I went to school and what I studied. I went somewhere, and then I went somewhere else, and in the end, I wound up with Heisenberg. At that time quantum mechanics existed, and I even knew a little bit about it. Heisenberg showed me a paper from the United States. It showed that a hydrogen molecular ion could not exist because the Schroedinger equation for that case had no solution. Heisenberg thought that this was improbable and suggested that I should find the mistake. I found it in a day. The mistake was small and obvious. I won't even tell you what it was. I went back to Heisenberg, very proud that I had found the error. He said, "If you say that this is mistaken, why don't you now calculate all the energy levels of the hydrogen molecular ion?" And that is how I got my doctor's thesis.

However, a very peculiar trick was involved in actually getting my doctor's degree which everyone, including the ladies, will understand. You see, it was too difficult to find all the solutions explicitly, so I had to use a computing machine. At that time in Leipzig where I was studying, there was a computing machine. But it did not work by means of horsepower or electricity. It worked by hand power. It needed a lot of hand power because the friction was very great and much of the power went into acoustic waves. Heisenberg, still unmarried, lived on the third floor which was two floors above the place where I calculated.

I did not like to work in the morning. I got up at noon, and then I calculated until midnight. One night around 10:30 or 11:00, Heisenberg came in and asked, "How does it go?" I said, "It goes well. I have calculated a few states, but I haven't calculated them all." (There are infinitely many.) Heisenberg asked, "How much longer do you think you will need?" I replied, "Oh, one or two more years." Heisenberg asked, "Which states have you calculated?" So I showed him. He said, "Maybe this is enough. Please just write them down, and we will give you a doctor's degree." These were not his precise words (we spoke in German), but this was the implication. So I wrote what I had done, and I got my doctor's degree.

At that time, quantum mechanics was not yet very far removed from the days of the correspondence principle. It was still good practice to assess a new solution in terms of how it fit in both with the new theory and with the viewpoint held before all the strange ideas had occurred. This attracted me. Starting from the hydrogen molecular ion, I went into the study of vibrations of more and more complicated molecules. These vibrations behaved almost, but not quite, according to classical mechanics. Had I not been rescued by nuclear physics, I might still be calculating ever bigger molecules.

Heisenberg looked after his students very well. Once a year or thereabouts, all of us went up to Berlin to hear a lecture by a more advanced physicist. I want to tell you a story about the first time I went to such a lecture because it has become a very famous story in our family. We went to a lecture by Einstein. At that time, Einstein was working on a very modest idea which he called the unified field theory. Needless to say, it was very much more modest than those things which you are working on today. It was the first time I had heard Einstein. I listened. I listened with all my ears and all my eyes. I didn't understand anything.

That is an exaggeration. I understood that there were four coordinates, and I understood that they were different coordinates in every point of space. This much I understood. But then I heard that this was somehow different from Riemannian geometry, and at that point I was completely lost.

It was a beautiful afternoon, and after the lecture a few people, including a slightly advanced physicist by the name of Eugene Wigner, went for a walk to the zoo. Everybody except me was in excellent spirits. Eugene had a peculiar interest in other people. Seeing that I was not enjoying myself, he came to me and asked what was the matter. I answered him in a very short and complete manner: "I am very stupid." Of course, I meant it.

Eugene is polite, and, perhaps for that reason, he did not contradict me. If he had, I would not have believed him, and it would not have helped a bit. But Eugene told me the truth. When I told him that I was very stupid, he said, "Yes, yes. Stupidity is a very general human property." You see, that made sense. That I was stupid, I knew. But that others also were stupid, that was a certain amount of consolation. That I could believe, and that cheered me up. So my children know that when one talks about a general human property in the Teller family, that means stupidity. In fact, it may be the most general human property.

Heisenberg, a really excellent teacher, did not imagine that he was the only one from whom we could learn. Not only did he take us to Einstein's lecture, but some of us got invited to Bohr's institute. This was a short time after Bohr and Heisenberg had worked on the uncertainty relation and on complementarity. I want to dwell on these exceedingly strange ideas because I suspect that many of the younger people here have forgotten about complementarity, or at least about how big a shock complementarity was to everyone including Bohr and Heisenberg. I can illustrate the shock in the case of Bohr by a very peculiar story which happened about two or three years later when I was back in Copenhagen for a longer period with a Rockefeller fellowship.

We had a great conference of philosophers. Those in attendance were mostly positivists, whatever that means. Everybody agreed that waves and particles are complementary, and that one had to approach many problems from two mutually exclusive directions in order to attain a complete comprehension. I have never seen a conference that was so unanimous. The morning after the conference was over, Bohr came in extremely depressed and very unhappy. Eugene was not there, but somebody asked him what was the matter. He gave a very different answer from Eugene's. He said, "If somebody talks about Planck's constant and does not feel at least a little dizzy, then he does not know what he is talking about."

Planck invented an equation that stated that energy equals a constant times frequency. Then for years physicists tried to explain it. Bohr's first great contribution was the statement that this equation is unexplainable. I believe that what he meant was that in order to explain something one must derive it from a simpler statement. However, an equation which states that one quantity is the product of two other quantities--one of which is a constant--is so simple that it cannot be derived from something simpler. One should look at that equation not as something to be explained, not as something to be derived, but as something to be accepted as a paradox. It is a paradox that energy, which we originally associate with ideas of particles, should have something to do with frequency, which we associate with a vibration. Therefore, there must be some relation between particles and vibrations. We knew almost nothing else except this simple statement and that whatever new laws would come out of Planck's equation should not be in some special cases in direct contradiction to what we knew of the world of big objects. From this incredibly slender foundation, Bohr then started to build atomic physics.

Until 1913, science was characterized by certainty. The

systematics of science was connected by an understanding that proceeded along a straight line from one or more assumptions (which could not be contradictory) to conclusions which one could verify. I believe that Bohr's initiative marked the first time that science based its ideas, not on this kind of understanding, but on a contradiction. This does not mark the first time that human thought had based its knowledge on a contradiction, however. Thousands of years ago, as long ago as any serious religion began, the question was asked: Am I a body or a soul? Now that is a very peculiar question because, in reality, I do not know what a body is because I haven't understood life. And while I may understand what a body is to a slight extent, I understand even less what a soul is. Nevertheless, people long ago said that we must be considered as both body and soul. The scientists in the nineteenth century contradicted this idea. They argued that describing people as both body and soul evades the question. They objected that such a statement is not an answer, but is dishonest and certainly not scientific.

I am quite sure that Bohr had been influenced by this age-old dualism. The remarkable point is that, as you all know, the outcome of Bohr's initiative was dualism in physics. One formulation is that we must describe matter both as particles and as waves. To say this more accurately, we must describe matter sometimes as particles and sometimes as waves, and we must understand the relation between the two. To imagine that the particle is somewhere without describing precisely how that particle was found did not make sense. That was a very peculiar idea. We have become so used to it that perhaps we have forgotten how peculiar it was.

There is a further peculiarity in the interpretation of quantum mechanics which has been accepted by people close to Bohr but not by everyone. The structure of quantum mechanics itself is peculiar.

There is one portion which obeys the laws of causality and another portion which does not. It is possible to talk about particles in a way of causality. It is possible to talk about waves in terms of causality. But when you switch from a particle description to a wave description, you have to make a measurement. The result of this measurement is unpredictable and cannot be described by a differential equation. This means in common language: it is not subject to the laws of causality. You can say something about probabilities, and you can say nothing more.

All of you are familiar with the fact that Einstein--a most imaginative scientist who contributed a great deal to the development of quantum mechanics--never accepted this statement. He felt that if you deny any part of causality, you have broken the link between an observed object and your observation. In fact you have. Bohr pointed out that in every observation there is a necessary uncertainty which becomes very important and very obvious at atomic dimensions. There are many situations where the uncertainty is there but not in an obvious manner.

This is the end of the shortest formulation of the story I can give you. This transformation of science from something that dealt with certainties into something that dealt with uncertainties was a very considerable shock.

Bohr went much farther with the idea of complementarity. He suspected, and he never proved, that in every part of life that is of any importance, this complementary approach is essential. He believed that every young person by the age of eighteen should understand complementarity because (and this I add myself, I hope not wrongly) it is the strongest antidote to any unquestioning conviction. A related statement is that doubt applied at the right place is the most essential part of knowledge.

What Bohr advocated was not compromise. His suggestion was not to take a bit of this truth and a bit from that, mix them well, and come up with a better truth. Bohr advocated complementarity, a way of making contradictory ideas illuminate the same situation. Without bringing both to bear, the situation remains incompletely described.

Bohr was a strong proponent of uncompromising statements. Perhaps the story I will tell to illustrate this seems funnier to me than to others because it happened to me. I had the experience once, and only once, of contradicting Bohr. I have forgotten the details of our discussion which was about a molecule, but I remember that in this particular unimportant little case, Bohr made an incorrect statement. I was very young, but it was the style in science, then as now, to contradict anyone if it was necessary. So I did. That wasn't my error. But Bohr began his reply by saying, "I know Teller knows a hundred times as much about this as I do." I was young and inexperienced, and what I said next may have sounded rude although I certainly did not intend it to be. I said, "This is an exaggeration."

Bohr glared at me as though he were really angry. He said, "Teller says I am exaggerating. Teller does not want me to exaggerate. If I cannot exaggerate, I cannot speak. All right. You are right. You know only 99 times as much as I do."

Of course, to this there was no answer. But I appreciate the truth in Bohr's remarks, and this is really why I told the story. Bohr's point--that if he could not exaggerate, he could not speak-- was sincerely meant. Bohr loved paradoxes. When he saw the possibility of a paradox, he made it even sharper. He tried to formulate it in as strong a fashion as was possible. Many years later, in about 1953, at a meeting in the Pocono Mountains in the United States, we ran into some of the same complications that we are encountering today. (I know that our first speaker today said that all problems are almost solved, but perhaps this is true and perhaps not.) At that time, Bohr said, "The trouble with high energy physics is that there are no contradictions, and without contradictions, you cannot make any progress."

I did not participate in the development of quantum mechanics. I only saw it when it was complete, and then I played with it a little. But I like to think about the development of quantum mechanics, because from my point of view, it was the golden age of physics. And I believe this period was so rich because it was the time of the greatest contradiction that ever occurred in science. During these years a contradiction occurred not between this fact and that but between this science and that science.

A science is an organized system of facts which are so strongly linked that none of them can fall without all of them falling--or at least so it would appear. Atomic theory involved a seeming contradiction between the science of physics and the science of chemistry. These sciences dealt with the same subjects from opposite points of view. Anybody who has any knowledge of the stable, well-defined formulae that chemists write, and who also knows the variable composition and configurations of which atoms consisting of electrons and nuclei are capable, can see that these two refuse to fit together. Something very radical had to happen in order for these two to become compatible. The result was that now, instead of two sciences, we have one. But we had to pay a price for this, and the price is complementarity. The price is the limited applicability of causality.

While I was in Copenhagen (and more so shortly after I left), I began to get interested in nuclei. A young Danish physicist, Kalchar, and I worked on a strange problem which is not very interesting--the transformation between orthohydrogen and parahydrogen. However, the transformation is not the point of this further terrible story I will tell about myself. Bohr wanted us to report on our work, and he fixed the time for a colloquium. Kalchar and I came, but Bohr did not, so we could not have the colloquium. That was not unusual. The colloquium was scheduled a second time. Again Bohr did not come.

As I mentioned, I liked to sleep late, and one day I came into Bohr's institute very late indeed. There was no plan for the colloquium to be held, but when I arrived, it was already starting. Bohr was there, and Kalchar was talking. The institute was like this conference, an international gathering in steady session. There was a rule that no one should lecture in his native language, perhaps with the exception of Professor Dirac, because if somebody spoke his native language, he talked too fast for those translating to understand him. I arrived, and Kalchar was just starting to explain our work in German.

Bohr was not always logical. For one reason or another, perhaps in order to be polite to me as a guest, Bohr wanted me to talk rather than Kalchar. But he also did not want to be impolite to Kalchar. He said, "It would be better to have this talk in English. I am sure that Teller could give us the talk in English." So Kalchar had to sit down, and I had to give the first lecture I ever delivered in English. I did speak English after a fashion in those days, but I had never before given a lecture in that language. Bohr asked me to talk in English because I could speak that better than German.

At any rate, I learned English, I came to the United States, and I became involved with atomic nuclei and with the way the sun produces energy. All of you have heard a lot about that and about all those parts of nuclear physics which were of interest then but which are now so old that I dare not mention them. My lecture will end with the coming of the second world war. That was the end of a wonderful period in physics. When the war was over and physics came back, it was not quite the same.

During the war applications of physics were born including the subject on which we worked in Los Alamos. That was a lot of physics, and that was also a lot of other ingredients. We all agreed then as now that an atomic war must be avoided. The real question is how to do so. I want to answer this question in the spirit of Niels Bohr. While no two people are apt to agree on all points, and some do not even agree on the general approach, I claim that the question of how to avoid war is one of the best examples of the application of complementarity.

You are not going to avoid war without considering two opposites which appear mutually exclusive. One is preparation for peace and for the increasing interdependence of nations; the other is preparation for war, in particular, for defense. Remarkably enough, nuclear weapons which first seemed to be exclusively destructive can be used for defense against conventional and against nuclear attack.

Today it is more necessary than ever before that every young man and woman be familiar with the principle of complementarity. However, complementarity is much less accepted today than some of us had hoped when complementarity was new. We should learn to debate without polarizing our opinions. We should realize that the answers to difficult questions are not simple, not fully expected, and often involve seeming contradictions.

QUESTIONS AND ANSWERS:

Q: We all heard that the start of the Manhattan Project was motivated by the fear of the German atomic bomb. Could you tell us at the time you were working on the hydrogen bomb what you knew about what the Soviets were doing?

A: It is true that when we worked on the atomic bomb we were afraid that the Nazis would get the bomb first. It is also true, without any question, that the man who was in charge of the atomic bomb project in Germany, Heisenberg, did not succeed. It is also true that one of the reasons, in my opinion the most important reason, he did not succeed was that when he found difficulties, he was happy about them. He had a very different approach than we had. We did not know any of this at the time, but it is a fact. My feelings about Heisenberg have a lot to do with that fact.

Wigner: Perhaps I should mention that in the middle of our efforts, we received a telegram from Switzerland, from Dr. Houtermans, a German, saying: "Hurry up. We are on the track." That alarmed us a great deal. Excuse me.

Teller: Since you mention Houtermans, I may tell you a story about Houtermans. Houtermans was one of the Germans who fled from Hitler, not to the United States, but to Soviet Russia. Very soon, he was in jail. At the time of the Hitler-Stalin pact, he was handed over to the Nazis. It was a practical certainty that he would go into a concentration camp. Heisenberg saved him. Heisenberg said he needed Houtermans for his atomic bomb project. A further fact is that Houtermans suggested to Heisenberg that if they made a reactor, they could get an element out of it from which one could possibly make an atomic bomb. It is a fact that Heisenberg disregarded this statement to a peculiarly thorough extent.

Finally Germany surrendered, and the bomb was dropped, unfortunately on Hiroshima. I say unfortunately because I think that we ought to have demonstrated it and thereby finished the war. However, when it was dropped, Heisenberg and maybe nine or ten other physicists were in a British castle under observation. When they were told the news, they did not believe it. A little while later, Heisenberg said, "Perhaps we can believe it. Perhaps it is the kind of thing. . ." He then described Houtermans' idea. That to me looks very much like forgetting on purpose, or halfway on purpose at least.

The question was what did we know about the Russian effort on the hydrogen bomb. I can tell you that I was interested in the hydrogen bomb already during the second world war and worked on it. I wanted to work on it for many reasons. One was that I considered it something new, something we did not know about. Then as now, I believe that we should go to the limits of what one can know and also go to the limits of applying knowledge. That was one of my motives, and my answer to you would not be complete if I did not mention it.

Shortly after the end of the second world war, Stalin made a public statement which many people did not notice. However, I did, for obvious reasons. Stalin said, "We soon will have the atomic bomb, and we soon will have much more." Stalin said this explicitly. So in addition to wanting to go ahead, I was also worried that Stalin might have meant precisely the hydrogen bomb when he said this. It turned out that he did. The Russians developed their atomic bomb four years after ours. However, on the record, only nine months elapsed from our first thermonuclear reaction to their first thermonuclear reaction. I could talk about this in even greater detail, but these are the bare facts.

Zichichi: Any more questions? If not, we thank you very much

Professor Teller, and we invite Professor Teller and all of you to a drinking party on the terrace.

Teller (inaudible): Drinking and thinking are complementary.

DISCLAIMER

This document was prepared as an account of work sponsored by an agency of the United States Government. Neither the United States Government nor the University of California nor any of their employees, makes any warranty, express or implied, or assumes any legal liability or responsibility for the accuracy, completeness, or usefulness of any information, apparatus, product, or process disclosed, or represents that its use would not infringe privately owned rights. Reference herein to any specific commercial products, process, or service by trade name, trademark, manufacturer, or otherwise, does not necessarily constitute or imply its endorsement, recommendation, or favoring by the United States Government or the University of California. The views and opinions of authors expressed herein do not necessarily state or reflect those of the United States Government thereof, and shall not be used for advertising or product endorsement purposes.

THE GLORIOUS DAYS OF PHYSICS

Eugene P. Wigner

Princeton University
Department of Physics, P.O. Box 708
Princeton, NJ 08540, USA

INTRODUCTION

The title "Glorious Days of Physics" implies that our science's present days are less glorious. If we accept this, we are induced to think of the changes both in the development of physics, and in the attitude of physicists toward science, which have taken place in the last 40 to 50 years and these will be the subject of my talk.

I'll discuss the subject under three headings. The first of these will refer to the more modest nature of our earlier discipline, that it did not imagine that it can explain all phenomena - in fact, the explanations of an increasing number of phenomena often came as a pleasant surprise - thus contributing to the glorious nature of those days. My second topic will be the change in the subjects of physics research - partly the result of our increasing confidence in the generality of our theories' validity. Lastly, I will speak about the changed relation of physics and of physicists to society and also the somewhat changed relation of ourselves to science.

1. THE INCREASE IN THE AREA OF PHYSICS - OUR DECREASED MODESTY.

It is surely unnecessary to recall to this audience that, ever since Newton, physics makes a sharp distinction between initial conditions and laws of nature, and that theoretical physics deals almost exclusively with the latter, that is, not with isolated events but with correlations between events, called laws of nature. However, the area for which such laws of nature were found has increased enormously in the course of time - Newton's theory enables us to calculate the future positions and velocities of planets if their present positions and velocities are given, but even before the "glorious days" our science was extended to a wealth of other phenomena, electromagnetic, thermodynamic and some other. And that was wonderful.

Yet, while I was studying chemical engineering and visited the Berlin University's physics colloquium every week, I gained the impression that many of the participants had doubts whether the human mind is strong enough to extend physics to the microscopic domain, to the behavior of atoms, molecules, but particularly also to the constituents of these, nuclei and electrons. Perhaps I should mention that the first physics book I read said "atoms and molecules may exist but this is irrelevant from the point of view of physics". And this was true - physics dealt at that time only with macroscopic phenomena. But by the time of my studies of chemical engineering the interest in microscopic phenomena was very alive - partly as a result of N. Bohr's work - even though it was doubted by many, if not most, that we can cope with them.

The first big change in this pessimistic attitude came with Heisenberg's paper, modest though the aim of that was. He adopted a positivistic philosophy and felt that the theory should not deal with the positions and motions of the electrons in the atom - these are not observable - but only with the energy levels, and transition probabilities between these. And he proposed a method for calculating these which, interestingly enough, turned out to be correct.

His ideas were almost immediately adopted by M. Born and P. Jordan and the joint work of these authors fundamentally changed the pessimistic attitude toward the future of microscopic physics.

An equally large change came with Schrödinger's paper which proposed not only a method to calculate the energies of the energy levels - he proposed an equation which turned out to be identical to that of Born, Heisenberg and Jordan - but also described the time development of all states - those corresponding to the energy levels were stationary. His theory was, essentially, an extension of Einstein's idea of the "guiding field " but guided not the individual particles but the whole configuration ("Schrödinger's second equation"). This resulted in the fact that his equation is consistent with the conservation laws of energy and momentum - which laws cannot be preserved if each particle is guided by a separate field. It also resulted in the very important fact that his equation permitted the description of collision phenomena the importance of which became increasingly realised. It may be worth mentioning at this point that Einstein never published his idea of the guiding field (Führungsfeld) because he realised that it is in conflict with the conservation laws of energy and momentum.

The glory of the aforementioned articles introduced the delightful atmosphere of the following years. It was a pleasure to apply their theory to a vast number of phenomena and produce an understanding of these. This will be discussed in the next section. We did realise the remaining inconsistencies of the theory but were glad and surprised that we had any theory which described a great variety of phenomena, phenomena with which physicists were familiar for a long time but which could not be incorporated into physics, for which their theory provided no description.

The first and most obvious weakness of the theory just acquired by physics was the non-relativistic nature of Schrödinger's equation, its enormous distance from the theory of general relativity. The Klein-Gordon equation, and even more the Dirac equation of the electron, were admired. But they applied to single-particle systems and did not encompass the greatest miracle which was inherent in Schrödinger's equation: this described also many-particle systems. It was able, therefore, to describe also collision phenomena - the most important phenomena according to a later article of Heisenberg.

Another limitation of the theory which we accepted, and which persists also to the present, was also recognised: the impossibility to describe the process of observation, the "measurement", by means of the equation of our theory, by means of the equations of quantum mechanics which were just praised. This was recognised already by Von Neumann, the founder of the mathematical formulation of the theory of measurement. He realised that the process of observation cannot be described by the Schrödinger equation - it is in fact conflicting with it. This eventually led to a set of new questions and ideas but, at that time, we accepted the existence of this difficulty.

In spite of recognising the limitations of "our theory" just mentioned - some of which are still with us - we were proud of the development we witnessed, most of us considered it a miracle. It was a development which occurred even though it was not necessary for the assurance of the survival of the human race, as Darwin postulated to be the reason of developments.

To repeat it: we were much more modest, we accepted and worked with a theory which we knew to be incomplete, the limitations of which we recognised. We were both proud and surprised when we could extend it to a new set of phenomena. Now, instead, the work of theoretical physicists is directed principally toward establishing a perfect theory, in particular a relativistic one, and both the older generation, in particular Dirac, and the younger one have significantly contributed to this effort.

II. THE CHANGE IN THE SUBJECT AND SPIRIT OF PHYSICS RESEARCH.

As was mentioned before, one of the changes that have taken place since the Glorious Days is our increased confidence in the validity of quantum mechanics and, even more importantly, the increased confidence that we'll make it "perfect". The second and equally significant change concerns the phenomena to which we wish to apply it.

That quantum mechanics will reproduce the atomic spectra was largely taken for granted, particularly when Hylleras' calculation of the ground state energy of helium was found to reproduce the experimental value of this quantity with an amazing accuracy. Hylleras' calculation referred to a two-electron system - a system for which no equation could be formulated before the advent of quantum mechanics.

We rather expected that the theory will describe the molecular spectra also. But, perhaps strangely, when the article of Heitler and London on the binding energy of two hydrogen atoms into a hydrogen molecule became known, this was a surprise to most of us.* This started the extension of physics to the area of chemistry - a wonderful extension which results in our belief that we possess the basic information from which all chemistry could be deduced. This does, of course, not mean that chemistry became uninteresting - just as the moon's position and state of illumination remain interesting even though we could calculate it. But chemistry is also immensely useful - and it continues to develop on its own, not on the basis of quantum mechanical calculations. Just the same, it is truly wonderful that two parts of science, physics and chemistry, acquired a common basis.

---

*Dr. Teller tells me that he was not surprised.

Other phenomena which were described by means of the equations of quantum mechanics include the electrical, and also the thermal, conductivity of metals, even superconductivity, the Van der Waals forces between molecules, the cohesion of solids, their strength. In fact, all the properties of normal matter were expected to be derivable from the basic equations.

All this indicates that the interest was centered on well known phenomena and we were pleased at the ease these could be described. I will admit that I do not consider all these descriptions fully satisfactory - in particular I consider the phenomenon of the permanence of the currents in superconductors a remaining mystery. But the success was great and pleased us greatly. In contrast, the present principal interest of physics concerns the prediction of new phenomena, the formulation of new theories, and produces new and interesting but often hardly acceptable ideas. This is largely due to the fact that it is believed that the common phenomena are "understood", and that the uncommon phenomena, about most of which we know very little, if anything, require new basic approaches.

The difficulty to describe the measurement process and hence "life" and "consciousness" was realised but not much emphasized. This remains a real difficulty also for the present, affecting the foundation of quantum mechanics. In classical physics we describe the state of the system by giving the positions and velocities of the objects - these are natural quantities, acceptable also from a positivistic point of view, at least for macroscopic objects. The wave function or state vector has no similar observable meaning and an interpretation of the quantum mechanical state, by a description of the observation, is a necessity - an unsatisfactory one.

I cannot resist at this point to mention the impossibility to describe the cosmological events by quantum theory. The collision of two stars leads to a branching of their wave function, i.e. to different probabilities of different directions; what has happened to these branches? Only one branch remained and it is not reasonable to say that the others were eliminated by "observations". I'll try to give an explanation of this kind of phenomena in a couple of weeks at a meeting in Bad Windsheim, but am not sure it will be correct.

Perhaps I should reiterate that the greater modesty of the "Glorious Days" was much fostered by the earlier fear, and the realisation of the possibility, that man may not be able to describe atomic phenomena in an interesting way, that he is not gifted enough to do that. And the pleasure and glory of the time resulted from the recognition that we are able to do so. The concentration of the theorists on showing that the theory does describe the bulk of the well known phenomena of inanimate nature was largely the result of the induced modesty. Of course, it can be claimed that the present physics is even more glorious - we have overcome our fear, perhaps to a too large extent, and are willing to speculate wildly. Let me now turn to my last subject:

III. THE INCREASED NUMBER OF PHYSICISTS AND THEIR CHANGED ATTITUDE TOWARD SCIENCE.

I often tell a story about my childhood when my father asked me what I'd like to become as a grown-up. I said that I'd like to become a scientist and, if possible, a physicist. But when he asked me about the number of physics jobs in Hungary, I had to admit that there were only four. (We decided that I'd study chemical engineering.) At present there are about 400 physics jobs in Hungary. And even in the United States, the number of the members of the Physical Society quadrupled since I joined it. In the "glorious days" we knew a much larger fraction of our colleagues than we can hope to know now. And the same holds for articles - we can read only a small fraction of those we should read and, in fact, the Physical Review of the earlier days has split into Physical Review Letters, and Physical Review A, B, C and D journals.

All this influenced the attitude of the physicist greatly. To be a scientist was, in the glorious days, "crazy". It was difficult to find a job, and the salary was low. There was, of course, a wonderful compensation: the pleasure we derived from our work from both research and teaching. To-day we have good salaries and we also have public influence - we are no longer considered to be "crazy". As a result, we also have obligations, in particular to inform the public on technical problems, and we try to satisfy these.

Another reason which justifies our calling the old times "glorious" is that we did know physics. We knew a much larger fraction of the accumulated information than we know now. Our present knowledge of physics as a whole is very similar to the chemists' knowledge of their science in our glorious days. I have listened to many lectures at this meeting, on a great variety of subjects, including the origin of our universe, queer particles unknown so far, including magnetic monopoles, new symmetry postulates and others. And even though I am convinced they were well presented, quite a few of them I did not really understand. And I do not believe there was only a small minority to which this applies.

It must be admitted that this constitutes a danger to science which was not really present in the glorious days. It is much more attractive to be a physicist than to be a solid state physicist or a low energy nuclear physicist or anything equally specialized. The need for specialization decreases the attractiveness of our discipline - as it decreases the attractiveness of all science, including chemistry. Yet it is very important for the future of man that attractive interests be available to him which can divert him from the search for power even if his physical needs are satisfied. And now the physical needs can be satisfied for practically everybody, particularly in the developed nations. We must do all we can to keep science interesting and truly attractive - if all people could see only power as a goal our societies would be demoralized. And art has a problem similar to that of science - we have enough wonderful pictures and beautiful music to give pleasure to everyone - there is no need for new ones to ornament all houses and satisfy all tastes. It remains wonderful to hear about new ones, and of producers of new ones, yet not as wonderful as it was in Michelangelo's times.

What I just touched upon is a serious problem for our societies and a serious problem for physics. Yet, I believe that at least the present days of physics are also glorious. After all, it is good that we can have so many more colleagues than before and perhaps also good that so many of us acquired social influence. And that the interest now extends to almost unreasonably large energy ranges, to the origin of our world and to the many other questions which were discussed at our meeting. To some degree this kind of diversification does counteract the problems I mentioned. It was considered somewhat unreasonable when I proposed the consideration of the SU(4) symmetry in nuclear physics - now we heard about the consequences and breakup of an SU(7) symmetry! And it is very good also that we have conferences like this one or the one of last year!

Let me, finally, support the more optimistic side of the preceding discussion by telling you that I believe, and am convinced, that our basic concepts need, and will receive, as basic modifications as were those furnished by Planck, Einstein, Heisenberg, Dirac (and perhaps some others) in the glorious days. It is not likely that they will be attributed to a single physicist, but they will be basic, interesting and will contribute to the beauty of our discipline. To give an indication, I mention that, rather recently, H.D. Zeh has demonstrated that the principle of determinism cannot be maintained for the microscopic (that is quantum mechanical) description of macroscopic bodies. This means that Newton's greatest accomplishment, the theorem that the initial conditions and the laws of nature are separate concepts, will have to be modified (I'll propose such a modification at the conference in Bad Windsheim). Another example is the observation of Fleming and of Hegerfeldt (actually based on an article of Roger Newton and myself) that the idea of space-time points should be modified - there is no state vector which corresponds to the situation that a particle occupies at a definite time a definite point in space. Clearly, the corresponding set of state vectors would be invariant with respect to Lorentz transformations about that point of space-time. There are some proposals in the literature for the modification of the concept of space-time points but it is not clear whether they are final. The solutions of the aforementioned two problems, and probably of others, are probably in the near future but I believe that they will be the result of the ideas and collaboration of several people, not of single individuals. They will become of fundamental importance for physics - physics' glorious days are not over !

And, as I said before, we _should_ maintain glorious days for science. We should love science and not power and extend its area not only to high energy phenomena but also to life and psychology, of ourselves and of animals, also to the social problems of man and his happiness !

WHAT HAVE WE LEARNED?*

Edward Teller

University of California
Lawrence Livermore National Laboratory
Livermore, California U.S.A.

I will not summarize the unsummarizable. I will start with a quote from Kepler, and if I don't give it in Latin but almost in Hungarian, please forgive me. When Kepler published his great discoveries, he made a statement in the introduction which was incredibly presumptuous but at the same time even a little more modest than presumptuous. He said, this book 'may wait 100 years for a reader.' (Incidentally, he was right. Newton read it a century later and used it as a basis of classical physics.) I don't mind, he went on, 'since God has also waited six thousand years for a witness.'

My impression of this conference is that of an assembly of Keplers, where everybody knows (or almost knows) the secrets of 15 billion years. However, some seem to lack Kepler's patience.

There is an alternation of periods in the development of physics--periods where more material is gathered, often in chaotic

---

*This work was performed under the auspices of the U.S. Department of Energy by the Lawrence Livermore National Laboratory under contract No. W7405-Eng-48.

form, and periods in which these chaotic facts become surprisingly simple. In the beginning of this century, we experienced one of the greatest simplifying periods. We are now in the opposite phase. Listening to everything that has been said here, I feel that in many cases more explanation would have been healthy.

Let me give you one example, perhaps ill-chosen. One of the most important and wonderful accomplishments which promises to unify what we know, is the gauge invariance applied to non-commuting variables. I had trouble understanding this gauge invariance. Because I had trouble, I went to a number of people and asked them. Everybody understood it, but almost everybody understood it differently.

The one answer that I came closest to understanding myself was this. If the proton and neutron had the same charge and varied only in iso-spin which has no influence on their interaction, then everything would be fine. But they differ in charge, and therefore the symmetry is spontaneously broken. So we can use the symmetry only as an approximation. This much I understand. We realize, of course, that we are talking about a gauge transformation that can change a proton into a neutron, and therefore produces an observable change which is not permitted in a gauge transformation.

According to some others with whom I talked, that is not a complete gauge transformation and actually the proton has to be turned back into a neutron. But if you do that, you have to describe the results in greater detail. I think that it is a healthy development to spell out points as fundamental as these, even if most people know the answers.

Of course, in the case of quantumchromodynamics, the symmetry

WHAT HAVE WE LEARNED?

may be perfect. Then there is no trouble. But in the case of the neutron-proton approximate symmetry, the Yang-Mills gauge transformation leads to an observable change, and that point should be definitely clarified.

The problem in quantum mechanics of how to reconcile the uncertainty principle with the meaning of a classical measurement had to be discussed again and again, not with unanimity, but with clarity or at least a conscientious effort at clarity. I think in some phases of current work that same type of effort would be useful. All this is a side remark.

Let me come to the main point. It seems to me that we have arrived at a number of particles that includes many more than existed not so many years ago. The list includes leptons, hadrons, quarks, and other particles, some of which have the peculiar, almost exceptional property of existing, and others lacking this quality. Some people say they don't exist because they are not really particles but states that can decay or can interact. Or they are not particles because they are only imagined for the purpose of introducing further changes in the theory.

In a most disturbing way, there is a third class of particles which are not quite particles. I will mention the quarks which have difficulty in getting a visa to leave the hadron or meson. If the particle cannot be isolated (although perhaps Fairbanks has done so), then it is a little different from other particles. We ought to understand that in much greater detail. Dr. Hendricks in Livermore is now producing incredibly uniform droplets whose behavior bears out Millikan with extreme precision but seems to contradict Fairbanks. In a year, in this or some other manner, the question will be decided.

I hesitate even to mention the gluon ball. I don't know whether
or not a gluon ball can get an exit permit, nor do I know how a gluon
ball would behave if it had an exit permit. These are questions to
which I would like to have answers, and I know you share my feelings.

But even though we all wish to have that answer, remarkably
enough we have more answers than we have questions. I remember in
about 1953, at the beginning of this era of revolutionary change
in physics, Bohr participated in a conference in the hills called
the Poconos not far from New York. He was very unhappy, and he
muttered a short, rather obscure comment which I will paraphrase
more lengthily in hopes of catching his meaning. He said, "You
know what the trouble is with you? The trouble with you is that
you do not have any clear contradictions. Without a clear,
sharply formulated contradiction, you will never get a new idea
because we get new ideas only as a last resort--when all other
means have failed."

A new idea is not a new particle. A new idea is not SU7, not
even SU137, about which I will consult Eugene tomorrow. A new
idea is something that I will not name, because neither I nor
probably any one of you has thought about it.

We have these particles (or potential particles), and we have
interactions between them. These interactions are studied with a
diligence by a number of physicists with the aid of a number of
monetary units (I think currently the only monetary unit to which
I dare refer is the Swiss franc) which surpasses the imagination.
So we get more and more data, and we do get contradictions. But
they are small contradictions. In many cases, in fact in most
cases, we are not even sure whether they are real contradictions
or whether they will go away tomorrow. This is an exciting part
of the story because this state of accumulating evidence is the

prelude to a state of clarification. At least this is a firm article of belief shared by all physicists.

One of my younger Hungarian friends, not Eugene, came to me with a sad face recently and asked, "Did I come too late?" He did not use these exact words, but I understood him. I assured him that he did not come too late. Perhaps he came much too early; I hope not, because I hope we are approaching a state of clarification. Here and there, there are some little signs of it. That a proton/anti-proton collision behaves the same way (if you neglect what you should neglect) as an electron/positron collision is a little encouraging. That bombardment by muons and bombardment by neutrinos have far reaching analogies is a little more encouraging. So the confusion is by no means complete.

Now let me say a few words about a topic which as I heard (and I hope I am not misinformed) had a great importance at an Erice conference for the first time this year. That is to bring the universe into the arena of elementary particles. The connection is not obvious. It is a connection between elementary particles, gravitation, and the Big Bang.

Is it proper to quantize the gravitational field? If you want to measure the gravitational interaction between an electron and a proton in a hydrogen atom, you have to measure the distance between the two to an accuracy approximately of 10^{-49} centimeters because the effect is so small that the change in the coulomb field becomes as great as the gravitational action if you change the distance by this small amount. Yet 10^{-33} centimeters seems to be the absolute limit of localizing a particle because the energy of localization suffices to produce a black hole.

If instead you want to get away from the horribly big force

due to electricity and take two neutrons, the Bohr orbit will be half as big as the universe. While they are still practically moving on straight lines in the Bohr orbit (if I am allowed to talk about motion in a Bohr orbit), they will disintegrate before a tiny fraction of the orbit is completed. I don't deny, of course, the necessary existence of gravitational waves, but they contain so many quanta that to look at them from the quantum point of view might pose some difficulties.

Will you please forgive me if I show you that I can talk unadulterated nonsense. Can I talk about gravitation in the same way as I talk about other phenomena? Is gravitation perhaps not an elementary phenomenon but a statistical phenomenon which you can experience only if you have many particles. This need not conflict with the geometric interpretation of gravitation. I dare to ask this question because, while I don't believe that gravitation is statistical, it's highly probable that I will not be disproved in my lifetime. I just wanted to say something that is very different from what we have been talking about, although this talk, of course, should be a summary.

Now I want to discuss some topics that we did talk about--one point where I am very happy with a possible result and another one where I am unhappy. For years we were talking about the universe having the duty to have an equal number of protons and anti-protons in a very early stage. It makes sense to go back to these early stages and calculate neutrino absorptions and emissions as events which occurred in great abundance in every pico-second. If one forgets about the anti-protons, one eventually gets the right relation between hydrogen and helium. This makes the Big Bang theory one step more real. Here is a concrete bridge between high energy physics and cosmology.

Similarly, in an earlier stage, there may have been lots of pions, and their collision should have been even more frequent a phenomenon. We also know that the behavior of the linear superposition of the K_0 and the anti-K_0 mesons violates time reversal, and therefore also violates the symmetry between protons and antiprotons. In the end, only the protons are left over. Thus we may obtain the universe as we know it. While all this is not proven, it appears very hopeful.

There is another thought that the neutrino mass may change the model of the universe. I have not the slightest doubt that this is a possibility, but please don't hold the galaxies together by concentrating neutrinos in the galactic plane. If that should happen in an early stage, then the neutrinos have light velocity, or almost light velocity, and cannot be retained. If it happens at a late stage, they still retain high velocities because of their small rest mass. Therefore they will escape. Furthermore, on account of their small interaction, it will be difficult to concentrate them in the galactic plane. There probably is extra matter in the galactic plane, for instance, black dwarfs or planetoids which we can't see.

I will try to say a few words about novel instruments. I shall not comment on accelerators because I know too little about their development. But I should mention computing machines which will play an increasingly important role. They developed very fast in the 1950s and 1960s. For some strange reason, they stagnated in the 1970s. Our best computing machines are now better than they were in 1970, but not enormously better. I know of new developments which make it absolutely certain that the development of computing machines will resume their fast progress, and in ten years, for $10 million you will get a computing machine that is one-thousand times better than what that amount of money will buy today. They will not only be one thousand times faster, but, perhaps more importantly, their memory will be comparable to the human memory in extent.

Now that does not mean that the computing machines will take over. It means that we should begin to give a lot of thought to the question of how we can use these computing machines in our field. I want to talk about that in a little detail. I will talk about it in connection with the place where I am working--the Lawrence Livermore National Laboratory--because I am familiar with that work. I am specifically speaking of the S-1 effort under the leadership of Lowell Wood. There may be similar efforts elsewhere.

I have a watch which is made out of quite complicated integrated circuits. There are hand-held computers which are doing similar things. These large circuits on a single chip have in general not yet been integrated into really big machines. I can tell you why. To integrate elements with simple individual functions, to make them work together, is a relatively easy task. The logic that is needed to couple really complicated elements usefully, to integrate uncounted thousands of such elements, is a much bigger intellectual effort. That intellectual effort is now being performed.

The machine we are planning has sixteen arithmetic units with reasonably big, but not very big, memories in them. There are sixteen memory units, whose memory is comparable to the memory of a man, and there is a switchyard which can couple any of these elements to any of the other elements. Now, it is impossible to program such a machine. No human can do it. The human can only write a program in a standard language, and then ask the machine to figure out which elements to couple to which elements in order to execute this program in the best way.

This is nothing new. Machines in use today do this. However, they are much simpler, and if they do it, they do it by sacrificing speed. They then don't work as fast as they would if a really clever programmer had planned the program. They fall short of the optimum by a factor of two or three. Our planned new big machines will do the programming in a much more ingenious manner and fall short of ideal functioning by 20 percent. This is the kind of development that is coming. If the functioning of gluons is calculated today by too small a lattice, in ten years you probably can increase this lattice to 50 elements, or maybe 200. You thereby will have the means to make the connection between assumptions and results in a more complete and reliable way. Using this machine may itself become an art.

There is another proposal which is entirely different although it is also an international proposal. We have, as you all know, succeeded in sending up a space shuttle. The important thing about the space shuttle is that it can lift the same weight at less expense. The United States, as President Reagan has told all of us, is broke. We can send up these space shuttles at a cost of maybe $20 million apiece. We are not so broke that we can't afford that. But the apparatus that one could put into that space shuttle may be much more expensive. Even if it isn't, it may be of interest to all the 400 Hungarian physicists and perhaps a few more outside of Hungary, for instance from Sicily, as well.

I don't know what to put on the space shuttle. I won't try to enumerate all the kinds of things that might be of interest. But one thing we certainly want to put on the space shuttle (and this is already being planned although not in the best form) is a telescope. This telescope is not one of those clumsy beasts where there is one piece of reflecting surface which has to be polished by an almost perfect polishing mechanism. Instead, you have small plane mirrors which are electronically guided so that if somebody sneezes in the shuttle, these flimsy mirrors will be electronically readjusted and will give the image of a distant star as a point.

The possibility which this telescope provides, of course, is to look farther. It may be possible to look all the way to the equator of the universe (if you know what I mean by the equator) and see not stars but galaxies. In this way it may be possible to find out to what extent the Hubble constant has changed, to what extent the expansion of the universe has slowed down, and to answer the question of whether we will eventually come to the crunch or whether we live in a world of ever-expanding horizons.

I believe this will not happen without a lot of international cooperation. I don't think it should or can happen without complete openness in executing these experiments. In spite of all the troubles in the world, this has been demonstrated as possible in one field in which I have had an interest--controlled fusion. Today there is worldwide cooperation and open cooperation.

I am afraid I may have talked too long. I want to mention one more unconnected subject which I consider extremely important. In quantum mechanics, during the golden days of physics, we learned that all we needed to do was to solve the two-body problem. Once the two-body problem was solved, the rest came with incredible rapidity. The many-body problem, polyatomic molecules, the conductivity of solids, and a number of ways to look at chemistry all fell into line. Although we did not solve thousand-dimensional equations, we learned the essential facts about semiconductors, which in turn had practical applications. It is clear that once the two-body problem was solved, the principles were solved.

I don't think this will be so in high energy physics. Here you are working with the devotion of devils and the good will of angels on solving the two-body problem--be it electron/positron collision or whatever. But it isn't a two-body problem, it is a twelve-body problem, because no sooner do the two touch than there are a few more. And one has to know all these interactions before one gets to quantitative results.

If you can't solve your problems exclusively by working with very few particles, you might try to solve them by working with a great many particles and using statistics. After all, thermodynamics and its statistical interpretation were among the sharpest tools that gave us information about the structure of matter and led us to Planck's law.

It was statistical mechanics that gave us sharp enough formulations to force us into contradictions which, a quarter of a century later, brought about in turn the most amazing new ideas. Yet the many-body problem in high energy physics is neglected by almost everybody except by us in the land famous for fruits and nuts--in California. It is not our monopoly, but we are doing it, and I want to report on a continuing experiment.

So far we have accelerated iron nuclei to an energy of 2 GeV per nucleon. When this slams into a similar nucleus, one gets a many-body problem. It might be most useful to develop colliding beam machines for this very purpose. This would increase our ability to compress the nucleus. Of course, the nuclei are already compressed in neutron stars, but experience in the interior of neutron stars is a little inaccessible. If you compress nuclei, do the baryons remain intact? It is to be expected that baryons at short distances should repel like the closed shells of helium repel.

If baryons get together and exchange a red quark, nobody is

likely to be in despair. But if they exchange a red quark for a blue quark, then there will be some trouble. I would love to see what would happen under those conditions, although the process is too short and the disintegration that follows too fast. Thus a concrete experiment will be most difficult.

But there is an experiment by a group of people led by Friedlander, and they have discovered what they call the anomalon.[1] These experiments are continuing, and the results remain consistent. An iron nucleus with a 2 GeV per nucleon falls on an emulsion. The emulsion is good and thick so that you can see tracks on it. What one finds is that an iron nucleus bumps into another heavy nucleus, a silver or a bromine, and makes a star. In the laboratory system, the star moves forward. Fragments having values of Z between 3 and 26 have been investigated. The ionization density in the emulsion does not change. The fragments are hardly slowed down. Each one of them ends in another star. The mean free path is 25 centimeters for $Z = 26$, which approximately corresponds to nuclear cross sections. These are not the anomalons.

But when you subtract these, what is left over are more similar collisions whose mean free path is considerably shorter. They seem to have considerably bigger cross sections. The actual cross sections cannot be well estimated because the result depends on the previous subtraction process. But the effect appears to be real.

Now I am simpleminded. I believe that these are fragments whose radius is appreciably bigger than that of a nucleus or corresponding z-value. How can that be? Now that is the point which I know least and will tell you last.

When a 2 GeV per nucleon particle falls on another similar particle, the average energy per nucleon in the center of gravity system is 1 GeV, and there are two times as many nucleons. So you really have only 500 MeV, and probably somewhat less because the z-values have not been equal. Why does the fragment not just fall apart into protons and neutrons? Apparently in many cases it does not. Indeed, in a less systematic way, this has been observed earlier in cosmic rays. Now, of course, this nucleus will evaporate particles and lose energy. It will emit mesons, not virtual mesons, but there is enough energy to emit real mesons.

I shall propose an imprudently simplified model of a nucleus with a lot of kinetic energy and a few mesons. I will say that the nucleus should hang together. There will be radiation emitted with consequent cooling. The pi-mesons which obey Bose statistics might condense. One may think of an expanded nucleus where all neutrons have turned into protons, and a corresponding number of negative pi-mesons have been created. This would give the chance for coherence between a maximum number of pi-mesons.

Single emission or absorption of a pi-meson from an N-fold occupied state would have matrix elements proportional to $N^{\frac{1}{2}}$. But because of the parity of these mesons, single emission would transfer angular momentum to nucleons which will cost more energy. Pairwise emission and absorption might be favored with matrix elements proportional to N and negative contributions to energy proportional to N^2. A relative minimum, which should not be lower than the energy of a normal nucleus, might be established at high N-values.

The resultant fragment may well have a lower density than a normal nucleus. This is so partly because of the exclusion principle in a nucleus in which all nucleons are protons, and partly because the dependence of the distances on the Compton wavelength of pions may lead to a less close packing. It should be reemphasized, however, that the experiments are consistent with a variety of cross sections depending on the assumption of the fraction of anomalons present. It is highly interesting that the effect seems enhanced in consecutive star-producing collisions.

Returning to questions of the greatest generality, the conference emphasized events occurring at distances of 10^{-14} to 10^{-15} centimeters. Since 10^{-33} centimeters plays the role of a minimum distance, a big field exists for phenomena unknown to us. This region has been called on some occasions a <u>desert</u> in which no new phenomena will be encountered and no new ideas will grow. I prefer the attitude of Newton. He wrote:

> I do not know what I may appear to the world, but to myself I seem to have been only like a boy playing on the seashore and diverting myself in now and then finding a smoother pebble or a prettier shell than ordinary, whilst the great ocean of truth lay all undiscovered before me.

I think we should remember that we should be perhaps a little selective in looking for pebbles and shells. I also think that those we have found, we should closely examine. Perhaps we should also be aware of something of which no one can be certain but which I believe: that the ocean of knowledge is immeasurable.

Reference

1. E. Friedlander, et al., Evidence for Anomalous Nuclei Among Relativistic Projectile Fragments from Heavy Ion Collisions at 2 GeV/Nucleon, <u>Physical Review Letters</u> 45:1084 (1980).

DISCLAIMER

This document was prepared as an account of work sponsored by an agency of the United States Government. Neither the United States Government nor the University of California nor any of their employees, makes any warranty, express or implied, or assumes any legal liability or responsibility for the accuracy, completeness, or usefulness of any information, apparatus, product, or process disclosed, or represents that its use would not infringe privately owned rights. Reference herein to any specific commercial products, process, or service by trade name, trademark, manufacturer, or otherwise, does not necessarily constitute or imply its endorsement, recommendation, or favoring by the United States Government or the University of California. The views and opinions of authors expressed herein do not necessarily state or reflect those of the United States Government thereof, and shall not be used for advertising or product endorsement purposes.

CLOSING CEREMONY

The Closing Ceremony took place on Monday, 10 August 1981. The Director of the School presented the prizes and scholarships to the winners as specified below.

PRIZES AND SCHOLARSHIPS

Prize for Best Student - awarded to Marek KARLINER, Tel-Aviv University, Israel.

Fourteen Scholarships were open for competition among the participants. They were awarded as follows:

Patrick M.S. Blackett Scholarship - awarded to Jay L. BANKS, Ohio State University, Columbus, USA.

James Chadwick Scholarship - awarded to Callum DUNCAN, University of Oxford, UK.

Amos-de-Shalit Scholarship - awarded to Marek KARLINER, Tel-Aviv University, Israel.

Gunnar Källen Scholarship - awarded to Hidenaga YAMAGISHI, Princeton University, USA.

André Lagarrigue Scholarship - awarded to Simon KAPTANOĞLU-METU, Turkey.

Giulio Racah Scholarship - awarded to Nathan SEIBERG, Weizmann Institute of Science, Israel.

Giorgio Ghigo Scholarship - awarded to Eric D'HOKER, Princeton University, USA.

Enrico Persico Scholarship - awarded to Mordechai SPIEGELGLAS, Tel-Aviv University, Israel.

Peter Preiswerk Scholarship - awarded to Gary Lynn DUERKSEN, University of Chicago, USA.

Gianni Quareni Scholarship - awarded to Peter FORGÁCS, Hungarian Academy of Sciences, Hungary.

Antonio Stanghellini Scholarship - awarded to Steven GOTTLIEB, FERMILAB, USA.

Alberto Tomasini Scholarship - awarded to JOHN M. GIPSON, Yale University, USA.

Ettore Majorana Scholarship - awarded to Gregor HERTEN, DESY, FRG.

Benjamin W. Lee Scholarship - awarded to Leonardo CASTELLANI, INFN, Florence, Italy.

Prize for Best Scientific Secretary - awarded to Jay L. BANKS, Ohio State University, USA.

The following students received *honorary mentions* for their contributions to the activity of the School:

Elcio ABDALLA, Universidade de São Paulo, Brazil.

Rose Mary BALTRUSAITIS, CALTECH, USA.

Harry W. BRADEN, University of Cambridge, UK.

Isaac COHEN, Weizmann Institute of Science, Israel.

Henrik FLYVBJERG, Niels Bohr Institute, Denmark.

G.C. JOSHI, University of Melbourne, Australia.

Robert D.C. MILLER, University of Melbourne, Australia.

Caroline MILSTENE, Tel-Aviv University, Israel.

Emil MOTTOLA, Institute for Advanced Study, USA.

Hitoshi YAMAMOTO, SLAC, USA.

CLOSING CEREMONY

The following participants gave their collaboration in the scientific secretarial work:

Jay L. BANKS	Zvonimir HLOUŠEK
Robert CLARE	Marek KARLINER
Isaac COHEN	Robert D.C. MILLER
Giacomo D'ALI'	Caroline MILSTENE
Eric D'HOKER	Gabriella SARTORELLI
Gary Lynn DUERKSEN	Nathan SEIBERG
Callum DUNCAN	Mordechai SPIEGELGLAS
John M. GIPSON	Hidenaga YAMAGISHI
Steven GOTTLIEB	Hitoshi YAMAMOTO

The following EPS Scholarships were awarded to:

Jànos BALOG, Hungary
Peter FORGÁCS, Hungary
Sinan KAPTANOĞLU, Turkey
Stephen MIKHOV, Bulgaria
Krysztof REDLICH, Poland

PARTICIPANTS

Elcio ABDALLA   Universidade de São Paulo
                               Istituto di Fisica
                               Caixa Postal 20516
                               01000 SAO PAULO, SP, Brazil

Simon ANTHONY   University of Oxford
                               Department of Theoretical Physics
                               1 Keble Road
                               OXFORD OX1 3NP, UK

Francesco ANTONELLI   Istituto di Fisica dell'Università
                               Piazzale Aldo Moro, 5
                               00185 ROMA, Italy

Augusto ARGENTO   Istituto di Fisica dell'Università
                               Via Irnerio, 46
                               40126 BOLOGNA, Italy

Maged ATIYA   Columbia University
                               Nevis Laboratories
                               P.O. Box 137
                               IRVINGTON, NY 10533, USA

János BALOG   Eötvös Lorand University
                               Department of Atomic Physics
                               VIII Puskin u. 5-7
                               1088 BUDAPEST 8, Hungary

Rose Mary BALTRUSAITIS   California Institute of Technology
                               High Energy Physics Department (256-48)
                               PASADENA, CA 91125, USA

Jay L. BANKS	Ohio State University Department of Physics 174 West 18th Avenue COLUMBUS, OH 43210, USA
Giovanni BATIGNANI	Istituto di Fisica dell'Università Via A. Valerio, 2 34100 TRIESTE, Italy
Anders BENGTSSON	University of Göteborg Chalmers University of Technology Institute of Theoretical Physics 41296 GÖTEBORG, Sweden
Ertugrul BERKCAN	Ecole Normale Supérieure Laboratoire de Physique Théorique 24 Rue de Lhomond 75231 PARIS CEDEX 05, France
Karl BERKELMAN	Cornell University Newman Laboratory of Nuclear Studies ITHACA, NY 14853, USA
Ruggero BERTANI	Istituto Nazionale di Fisica Nucleare Via Livornese 56010 S. PIERO A GRADO (PISA), Italy
Gisèle BORDES	College de France Laboratoire de Physique Corpusculaire 11, Place Marcelin-Berthelot 75231 PARIS CEDEX 05, France
Harry W. BRADEN	University of Cambridge Department of Applied Mathematics and Theoretical Physics Silver Street CAMBRIDGE CB3 9EW, UK
Roger CASHMORE	University of Oxford Nuclear Physics Laboratory Keble Road, OXFORD OX1 3RH, UK
Leonardo CASTELLANI	Institute for Theor. Physics SUNY, STONY BROOK, USA

PARTICIPANTS

Yoonsuhn CHUNG	Universität Heidelberg Institut für Theoretische Physik Philosophenweg 16 6900 HEIDELBERG, FRG
Robert CLARE	DESY - Group F13 Notkestrasse 85 2000 HAMBURG 52, FRG
Isaac COHEN	Weizmann Institute of Science Department of Physics REHOVOT, Israel
Yuval COHEN	Hebrew University Racah Institute of Physics JERUSALEM, Israel
Sidney COLEMAN	Harvard University Department of Physics Lyman Laboratory of Physics CAMBRIDGE, MA 02138, USA
Michael J. CREUTZ	Brookhaven National Laboratory Physics Department Associated Universities Inc. UPTON, NY 11973, USA
Luigi DADDA	Politecnico di Milano Piazza Leonardo da Vinci, 32 20133 MILANO, Italy
Giacomo D'ALI´	Istituto di Fisica, Università di Bologna Via Irnerio, 46 40126 BOLOGNA, Italy
P. DE CAUSMAECKER	Katholieke Universiteit Leuven Instituut voor Theoretische Fysica Department Natuurkunde Celestijnenlaan 200D 3030 LEUVEN, Belgium
Eric D'HOKER	Princeton University Physics Department P.O. Box 708 PRINCETON, NJ 08540, USA

Sovas DIMOPOULOS   Stanford University
                          Department of Physics
                          STANFORD, CA 94305, USA

Paul A.M. DIRAC   The Florida State University
                          Department of Physics
                          TALLAHASSEE, Florida 32306, USA

Simon DUANE   University of Cambridge
                          Department of Applied Mathematics and
                          Theoretical Physics
                          Silver Street
                          CAMBRIDGE CB3 9EW, UK

Gary Lynn DUERKSEN   University of Chicago
                          Enrico Fermi Institute
                          5630 Ellis Avenue
                          CHICAGO, IL 60637, USA

Pieter DUINKER   NIKHEF-H
                          Kruislaan 409
                          P.O. Box 41882
                          1009 DB AMSTERDAM, The Netherlands

                          DESY
                          Notkestrasse 85
                          2000 HAMBURG 52, FRG

Callum DUNCAN   University of Oxford
                          Department of Theoretical Physics
                          1 Keble Road
                          OXFORD OX1 3NP, UK

Etim ETIM   Laboratory Nazionali di Frascati
                          C.P. 13
                          00044 FRASCATI, Italy

Tom FEARNLEY   CERN
                          EP Division
                          1211 GENEVA 23, Switzerland

Val L. FITCH   Princeton University
                          Department of Physics
                          PRINCETON, NJ 08544, USA

Henrik FLYVBJERG   Niels Bohr Institute
                          Blegdamsvej 17
                          2100 COPENHAGEN Ø, Denmark

PARTICIPANTS

Peter FORGÁCS	Hungarian Academy of Sciences Central Research Institute for Physics P.O. Box 49 1525 BUDAPEST 114, Hungary
Francesco FUCITO	Istituto di Fisica dell'Università Piazzale Aldo Moro, 5 00185 ROMA, Italy
Erwin GABATHULER	CERN EP Division 1211 GENEVA 23, Switzerland
James G.M. GATHERAL	University of Cambridge Department of Applied Mathematics and Theoretical Physics Silver Street CAMBRIDGE, CB3 9EW, UK
John M. GIPSON	Yale University, Physics Department P.O. Box 6666 217 Prospect Street NEW HAVEN, CT 06520, USA
Shelly L. GLASHOW	Massachusetts Institute of Technology Department of Physics, 6-313 CAMBRIDGE, MA 02139, USA
George GOLLIN	University of Chicago Enrico Fermi Institute 5630 Ellis Avenue CHICAGO, IL 60637, USA
Steven GOTTLIEB	Fermi National Accelerator Center P.O. Box 500 BATAVIA, IL 60510, USA
Giuseppe GRELLA	Istituto di Fisica dell'Università Via Vernieri, 42 84100 SALERNO, Italy
Gregor HERTEN	Physikal. Inst. Rhein.-Westf. Tech. Hochschule Physikzentrum 5100 AACHEN, FRG

Thomas HIMEL	CERN EP Division 1211 GENEVA 23, Switzerland
Zvonimir HLOUSEK	Rudjer Bošković Institute Theoretical Physics Department Bijenička 54 P.O. Box 1016 41001 ZAGREB, Yugoslavia
Kenzo ISHIKAWA	DESY Notkestrasse 85 2000 HAMBURG 52, FRG
Wolfhard JANKE	Freie Universität Berlin Institut für Theoretische Physik Arnimalle 3 1000 BERLIN 33, FRG
G.C. JOSHI	University of Melbourne School of Physics PARKVILLE, Victoria 3052, Australia
Drasko JOVANOVICH	Fermi National Accelerator Laboratoy P.O. Box 500 BATAVIA, IL 60510, USA
Sinan KAPTANOĞLU	Middle East Technical University Physics Department ANKARA, Turkey
Frédéric KAPUSTA	Université Paris VI Laboratoire de Physique Nucléaire et de Hautes Energies 4 Place Jussieu, Tour 32 75230 PARIS CEDEX 05, France
Marek KARLINER	Tel-Aviv University Department of Physics and Astronomy Ramat-Aviv TEL-AVIV, Israel
Ehud KATZNELSON	Tel-Aviv University Physics Department Ramat-Aviv TEL-AVIV, Israel

PARTICIPANTS

Tetsuro KOBAYASHI	Tokyo Metropolitan University Department of Physics Setagaya-Ku TOKYO, Japan
Takahiro KUBOTA	Department of Physics, Univ. of Cambridge Cavendish Laboratory Madingley Road CAMBRIDGE CB3 0HE, UK
Ignazio LAZZIZZERA	Libera Università degli Studi di Trento Dipartimento di Fisica 38050 POVO (TRENTO), Italy
Richard LOVELESS	University of Wisconsin Department of Physics 1150 University Ave. MADISON, WI 53706, USA
Wolfgang LUCHA	Institut für Theoretische Physik der Universität Wien Boltzmanngasse 5 1090 WIEN, Austria
Erik Jan LUIT	DESY/F13 Notkestrasse 85 2000 HAMBURG 52, FRG
Usha MALLIK	Université de Paris Sud Laboratoire de l'Accélérateur Linéaire Bâtiment 200 91405 ORSAY, France
Paolo MARGOLUNGO	Istituto de Fisica "G. Galilei" Via Marzolo, 8 35100 PADOVA, Italy
Reinhard MECKBACH	Max-Planck-Institut für Physik Föhringer Ring 6 8000 MÜNCHEN 40, FRG
Stephen MIKHOV	Institute of Nuclear Research and Nuclear Energy Boulevard Lenin 72 SOFIA 1184, Bulgaria

Robert D.C. MILLER   University of Melbourne
                          School of Physics
                          PARKVILLE, Victoria 3052, Australia

Caroline MILSTENE   Tel-Aviv University
                          Department of Physics
                          Ramat-Aviv
                          TEL-AVIV, Israel

N.K. MONDAL   Tata Institute of Fundamental Research
                          National Centre of the Government of
                          India for Nuclear Science and Mathematics
                          Homi-Bhabha Road
                          BOMBAY 400005, India

David MORGAN   Rutherford and Appleton Laboratory
                          Theory Division
                          Chilton
                          DIDCOT OX11 OQX, UK

Emil MOTTOLA   Institute for Advanced Study
                          School of Natural Sciences
                          PRINCETON, NJ 08540, USA

Kalevi MURSULA   Helsinki University
                          Research Institute Theoretical Physics
                          Siltavuorenpenger 20C
                          00170 HELSINKI 17, Finland

René NACASCH   U.E.R. de Luminy
                          Département de Physique Théorique
                          70, Route Léon Lachamp
                          13288 MARSEILLE CEDEX 2, France

Carlo NOVERO   Istituto di Fisica Teorica dell'Università
                          Corso M.D'Azeglio 46
                          10125 TORINO, Italy

Theophil OHRNDORF   Universität Gesamthochschule Siegen
                          Fachbereich 7 - Physik
                          Postfach 21 02 09
                          5900 SIEGEN 21, FRG

Constantinos   Imperial College of Science and Technology
   PANAGIOTAKOPOULOS   The Blackett Lab., Theoretical Physics
                          Prince Consort Road
                          LONDON   SW7 2BZ, UK

PARTICIPANTS

Matej PAVSIC	University of Ljubljana Institut "Josef Stefan" P.O. Box 199 61001 LJUBLJANA, Yugoslavia
Mario PERNICI	Istituto di Fisica dell'Università Via Celoria 16 20133 MILANO, Italy
Alfred PETERSEN	University of Hamburg II. Institute für Experimental Physik Luruper Chaussee 2000 HAMBURG 52, Germany
Luc POGGIOLI	Université Paris VI Laboratoire de Physique Nucléaire et de Hautes Energies 4 Place Jussieu 75230 PARIS CEDEX 05, France
Krysztof REDLICH	University of Wroclaw Institute of Theoretical Physics Ul. Cybulskiego 36 50205 WROCLAW, Poland
Franco RIVA	Istituto di Fisica dell'Università Via Celoria 16 20133 MILANO, Italy
Thomas J. ROBERTS	University of Michigan Randall Laboratory of Physics ANN ARBOR, MI 48109, USA
Oswald ROEMER	DESY Notkestrasse 85 2000 HAMBURG 52, FRG
Graham ROSS	University of Oxford Theoretical Physics Department 1 Keble Road, Oxford, UK and Rutherford Appleton Laboratory Chilton, Didcot, Oxon, UK
Claudio SANTONI	Istituto Superiore di Sanità INFN Viale Regina Elena, 299 00161 ROMA, Italy

Gabriella SARTORELLI   Istituto di Fisica
                        Università di Bologna
                        Via Irnerio, 46
                        40126 BOLOGNA, Italy

Masa-Aki SATO   University of New York
                        Department of Physics
                        4, Washington Place
                        NEW YORK, NY 10003, USA

Dietmar SCHALL   University of Heidelberg
                        Institut für Theoretische Physik
                        Philosophenweg 16
                        6900 HEIDELBERG, FRG

Fidel SCHAPOSNIK   Université Pierre et Marie Curie
                        Laboratoire de Physique Théorique
                        et Hautes Energies
                        Tour 16 - 1er Etage
                        4, Place Jussieu
                        75230 PARIS CEDEX 05, France

Nathan SEIBERG   Weizmann Institute of Science
                        Department of Physics
                        P.O. Box 26
                        REHOVOT, Israel

Mordechai SPIEGELGLAS   Tel-Aviv University
                        Physics Department
                        Ramat-Aviv
                        TEL-AVIV, Israel

Mario SPINETTI   Laboratori Nazionali di Frascati
                        Istituto Nazionale di Fisica Nucleare
                        00044 FRASCATI, Italy

Pjotr STOPA   Max-Planck-Institut für Physik
                        Föhringer Ring 6
                        8000 MÜNCHEN 40, FRG

Morris SWARTZ   University of Chicago
                        Enrico Fermi Institute
                        5630 Ellis Avenue
                        CHICAGO, IL 60637, USA

PARTICIPANTS

Edward TELLER	University of California Lawrence Livermore Laboratory P.O. Box 808 LIVERMORE, CA 94550, USA
Raffaele TRIPICCIONE	Scuola Normale Superiore Istituto di Fisica Piazza dei Cavalieri 7 56100 PISA, Italy
Roberto URBANI	Istituto di Fisica dell'Università Viale Benedetto XV, No. 5 16132 GENOVA, Italy
Yosef VERBIN	Tel-Aviv University Mortimer and Raymond Sackler Institute of Advanced Studies RAMAT-AVIV, Israel
John WARD	MacQuarie University School of Mathematics and Physics NORTH RYDE, NSW 2113, Australia
Wolfgang WIENZIERL	Max-Planck-Institut für Physik Föhringer Ring 6 8000 MÜNCHEN, FRG
Eugene P. WIGNER	Princeton University Department of Physics P.O. Box 708 PRINCETON, NJ 08540, USA
Frank A. WILCZEK	University of California Institute for Theoretical Physics SANTA BARBARA, CA 93106, USA
Edward WITTEN	Princeton University Department of Physics P.O. Box 708 PRINCETON, NJ 08540, USA
Ullie WOLFF	Weizmann Institute of Science Department of Nuclear Physics P.O. Box 26 REHOVOT, Israel

Hidenaga YAMAGISHI   Princeton University
                      Physics Department
                      P.O. Box 708
                      PRINCETON, NJ 08540, USA

Hitoshi YAMAMOTO   Stanford Linear Accelerator Center
                      Bin 96
                      P.O. Box 4349
                      STANFORD, CA 94305, USA

INDEX

Abbott-Farhi model, 229-230
Anomalon, 786-788
Anti-SU(5), 9, 10
Antigoldstino, 628-630
Antimatter (see also Matter-antimatter asymmetry), 3-4
Antineutrino, total cross section measurement, 383-399
Antiphotino, 628-630
Asymptotic freedom of non-Abelian guage theories, 131-132, 136-137
Axions, 633-635

B mesons
  CP violation and, 501, 686-687
  decays
    charm in, 496-497
    CP violation in, 501
    inclusive, 488-493
    reconstructed exclusive, 499
  lifetime, 500
  production, 485-488
B quark, 471-506, 596
  decay, 493
    weak, 611-619
  experimental facilities, 471-474
  in left-handed weak doublet, 494-475
Baryon/photon number ratio, 246, 273-275, 279
Beauty production, 444-467
  reactions studied, 444-445
Beta-decay, double, 12
Big bang (see also Cosmology; Universe), 16, 251-260

Big bang (continued)
  density of neutron gas, 265-266
  magnetic monopoles and, 21-22, 243
  nucleosynthesis, 267-268
Big crunch, 285, 292
Birkhoff's theorem, 292-293
Black holes, 114, 287
  entropy of, 288-289
Bogomol'nyi bound, 182-183
Bohm-Aharonov effect, 25-26
Bohr orbits, 736-737
Bose-Fermi symmetry, 306-315
Bosons
  ETC, 203
  guage, see Higgs boson
  Higgs, see Higgs boson
  pseudo-Goldstone, see Pseudo-Goldstone bosons
  scalar, see Scalar bosons
  vector, see Vector bosons
  W, see W bosons
  Z, see Z bosons

C quark, 223
  decay, muons from, 614
Casimir effect, 354
Causally disconnected regions, 284-287, 301
Centauro events, 224-225
Charge quantization, monopoles and, 30-31
Charm particle lifetimes, 375-382
Charm production, 414-444, 468
  reactions studied, 414-415
Chemistry and physics, 769-780
Chromodyons, 99-101

Classical limit, nature of, 47–48
Coleman-Mandula theorem, 306, 308, 310–311, 359
Complementarity, 754–758, 761
Computing machines, 781–783
Confinement
  effects on non-Abelian magnetic change, 106–114
  strong coupling and, 151, 155
Conservation laws, exotic, 307
'Constant physics', points of, 142
Constants of nature, time-dependence of, 339, 344, 351
Continuum limit, 131–137
Contradictions in science, 758–759, 778
Cosmological constant, 290–291, 301, 369
  supersymmetry and, 314
Cosmology (*see also* Big bang; Universe)
  background radiation, 286–287
  closure, 260
  dynamic equations, 251–252
    relation to Hubble parameters, 252–253
    simple solutions, 253–255
  'flatness problem', 288, 302
  horizons, 284–285
  magnetic monopole production, 290
  missing mass, 271–272, 294–295, 297, 298
  neutrino decoupling, 264–265
CP violation, 279, 281, 282, 677–693
  detection in $p\bar{p}$ collider, 690–691
  guage theories and, 685–686
  in B decays, 501
  in neutral B-meson systems, 686–687
  in neutral D-meson systems, 686–687
  in neutral K-meson systems, 678–681
  matter-antimatter asymmetry, 683–684
CPT invariance, 692

D mesons
  cross sections, 421, 422, 429, 435
  D^+
    enhancement in $(K^-\pi^+\pi^+)$ spectrum, 423, 424
    observed with e, 425
    x_L distribution, 428
  D°
    CP violation, 686–687
    enhancement in $(K^-\pi^+)$ spectrum, 430, 431, 432
    p_T distribution, 434
    x_L distribution, 434
  $\bar{D}^\circ$
    decay, 440
    evidence for, 437, 438
    lifetimes, 380, 381
D quark, 684–685
De Broglie waves, 740
Diatomic molecule, 86–88, 101
  eigenstates, 87
  eigenvalues, 87
  Hamiltonian, 86
Dirac quantization condition, 26, 101
  for dyons, 32
  in classical physics, 33
  non-Abelian monopole, 51
  spin statistics puzzle, 39–42
Dirac string, 27–28, 97
  observability of, 107
  potential, 40
  singularity, 50, 74, 79
Dohkos-Tomaras monopole, 98–99
Double beta-decay, 12
Drell-Yan process, 729
Duality rotations, 31–32
Dyons, 31–33
  chromo-, 99–101
  Dirac quantization condition, 32
  effect of theta term on spectrum, 93–95
  from monopoles, 92–93
  spin statistics puzzle, 39–42

E-53A Fermilab experiment, 399–403

INDEX

E-531 Fermilab experiment, 375-382
  charm events, 379-80
  emulsion, 376-377
  mass distribution of tracks, 378
  run parameters, 379
  spectrometer, 376-379
E-616 Fermilab experiment, 383-399
  callibration runs, 386-388
  Cerenkov counter pressure curve, 389
  corrections to data sample, 394
  detector, 392, 393
  Dichronic beam line, 384-386
  ionization chamber response, 386-388
  measured k/π ratios, 390
  measured neutrino event energy, 395
  measured slope parameters, 396, 397
e^+e^- annihilation, see Electron-positron annihilation
e^+e^- machines, see Electron-positron machines
e p machines, see Lepton-hadron machines
electro-weak interaction models, 638-665
  determination of g_A and g_V, 662-665
  measurements of $\sin^2\theta_W$, 651-661
  tests of, 651-65
electron
  heavy, 636-638
    neutral, 626-628
  relativistic wave equation, 741-742, 768
  scalar, 628-631, 632
electron neutrino, mass, 12-13
electron-positron annihilation (see also Electron-positron machines)
  comparison with p$\bar{\text{p}}$ data, 696-720
  hadron production, 507-600
    working model, 508-510

E-616 Fermilab experiment (continued)
  hadronic final state, 529-532
    charged particle multiplicity, 529-31
    energy fractions, 531
  multijet events (see also Quarks), 559-582
    initial evidence for, 559-566
  muon pair production, see Muons
  particle composition, 546-550
  quark jets, see Quark jets
  single photon process, 507-508
  tau pair production, 641-5
  total cross-section, 510-29
    accuracy, 513
    final state radiation, 518-520
    first order QCD, 523-524
    initial state radiation, 515-518
    kinematic factors, 522
    low energy problem, 527-528
    measurement of, 511-513
    quark substructure, 527
    radiative corrections, 513-520
    second order QCD, 524
    vacuum polarization, 515-517
    weak interactions, 524-526
  two photon process, 508
Electron-positron machines (see also electron-positron annihilation), 158, 159, 160
  W and Z production, 185, 186
Electroproduction, QCD predictions, 165, 166
Entropy
  of sun, 288-9
  within horizon, 287-8
  within one 'curvature volume', 288
ETC bosons, 203
Event planarity in p$\bar{\text{p}}$ interactions, 713-715

F mesons, lifetime, 380
Fayet-Illiopoulos D term, 362

Fermilab, neutrino physics at, 374-408
  E-53A experiment, see E-53A Fermilab experiment
  E-531 experiment, see E-531 Fermilab experiment
  E-616 experiment, see E616 Fermilab experiment
Fermions, composite, in supersymmetric theories, 247
Fine structure constant, 746, 747-748
  time dependence of, 370-371

Galaxies
  black holes at centre, 15
  formation, 299-300
    neutrino mass and, 272-273
    massive neutrinos and, 14-16
Gamma decay, 585-586
  three gluon decay process, 583
  thrust distribution in, 583, 584
Gell-Mann Low function, 120
Generations, repetition of, 18-19
Georgi-Glashow model, 73, 78
  monopole masses, 81-82
    lower bound, 82-83
  non-singular monopoles, 77
  quantum monopoles, 84-93
GIM mechanism, 196, 197
Glashow-Weinberg-Salam theory, measurements of $\sin^2\theta_W$, 651-661
Glueball, 152, 586, 778
  difficulty of observation, 729
  mass, 156, 727
Gluino, 4-5, 209-210
Gluon(s), 147, 161
  bremsstrahlung, 541, 560-566
  coupling, 584-585, 586-588
    interpretation of, 579-582
    value of, 576-579
  experimental evidence, 223-224
  gamma decay, 585-586
  in standard (3,2,1) model, 161
  in upsilon decays, 478, 481
  jet fragmentation, 585
  massless, 147
  radiation model, 566-569

Gluon(s) (continued)
  spin, 575-576, 582-4
  technicolor, 200
  triple coupling, 171, 225-226
  vector nature, test of, 171
GNO monopole fields, 108, 112
  classification of monopoles, 48-53
  instability of, 62-68
  topological classification and, 62
Goldstino, 206, 207, 208, 628-630
Goldstone fermion, decoupling, 348-349
Grand unified theories, 21, 157-158
  hierarchy problem, 158
  instantons, 324, 364
  monopole masses, 81
  non-singular monopoles and, 77
  tests, 191-4
Gravitational constant, effect of time variation, 294-295
Gravitino, 208, 208-209
Gravity, 779-80
  animatter and, 684-685, 689-690
  K° meson and, 684-685
Guage bosons, see Higgs boson
Guage fields, 44-45
  on lattice, 120-122
Guage hierarchy problem, see Hierarchy problem
Guage invariance, 776
Guage theories, 42-47
  CP violation and, 685-686
  dynamics, 45
  group elements associated with paths, 46-47
  lattice (see also Wilson loops), 119-156
  non-Abelian, asymptotic freedom, 131-132, 136-137
  numerical studies, 119-156
  strong coupling and confinement, 151-155
  structure constants, 43
  temporal guage, 46
  transformations, 42-44
  two-dimensional, advantages of, 109

INDEX

Guage transformations, 42-44
  time-independent, 49
'Guiding field', 767
GUTs, see Grand unified
      theories

Hadron-hadron machines (see also
      Proton-proton
      interactions), 159, 160
  technicolor detection, 233-234
  W and Z production, 184
Hadrons, charmed, measurement of
      lifetimes, 375-382
Heavy flavour production, 409-470
  difficulty of detection in pp
      machines, 409
Heavy 'isotopes', 4-5
Heisenberg's matrix mechanics,
      737-738
Hierarchy problem, 158, 197-198,
      243, 415-324, 339-354
Higgs boson, 16, 179-182
  detection, 232
  expectation value, 316, 317
  in toponium decay, 178
  mass, 233
  multiplets, 235, 237, 240, 317
  pseudo-Goldstone boson and, 226
  scalar, 186-188, 213-215, 235
      mass, 197-198
Higgs dominant proton decay, 8
Higgs mechanism 72
  CP violation and, 685
  supersymmetric analogue, 208
      212-15
't Hooft-Polyakov monopole,
      78-80, 91-92
  transposition to SU(5), 98-99
Homotopy theory, 53-57, 58-61
Hypercolor, see Technicolor
Hyperpions, 631-633

Infinities in quantum machanics,
      744-746
Initial conditions, laws of
      nature and, 295-296, 766
Instantons
  grand unified, 324, 364
  SU(5), 366-367
'Isotopes', heavy, 4-5

J/ψ decays, 633-635
Jets
  gluon, 585
  quark, see Quark jets

Kaon masses, 691
Klein-Gordon equation, 741-742
K mesons
  CP violation, 678-681
  K° and gravity, 684-685
Kobayashi-Maskawa six quark
      model, 176-178, 227-228,
      617

Lambda b particle
  cross section estimates,
      461-467
  decay, 445-453, 457-461
  longitudinal momentum
      distribution, 454-457
  possible evidence for $\Sigma_b^\pm \to \Lambda_b^\circ \pi^\pm$,
      457, 459
Lambda c particle
  cross section estimates, 421,
      422
  decay, 441, 442
  experimental p_T distribution,
      418
  in $(pK^-\pi^+)$ spectrum, 417
  lifetime, 380
  longitudinal distribution of
      419, 420
'Large numbers problem', see
      Hierarchy problem
Lattice scale size, 153-154
Laws of nature, initial con-
      ditions and, 295-296, 766
Leading hadron effect, 697-701
Lepton-hadron machines, 159-160
  W and Z production, 184-185
Lepton number violation, 192
Leptons
  heavy
    neutral electron, 626-628
    new, 622-626
    tau, see Tau lepton
  scalar, 211
Lie algebras, 43, 58-60
Light scattering by light, 747
Lubkin classification of
      monopoles, 53-62

Magnetic charge
  Abelian, renormalization,
      101-106
  non-Abelian, effects of
      confinement on, 106-114
Magnetic monopole(s), 3, 21-117,
      748
  Abelian, 24-42
    force between distantly
        separated, 68
    magnetic charge
        renormalization, 101-106
    Witten effect, 93-95
  attractive force between, 69-71
  charge quantization and, 30-31
  classical limit, nature of,
      47-48
  combined electromagnetic and
      chromomagnetic, 97, 98,
      107
  Compton wavelengths, 103
  confinement and, 97
  cosmological production, 290
  coupling constants, 98
  Dirac quantization condition,
      27-33, 51
  Dirac string, see Dirac string
  Dokos-Tomaras, 98-99
  Dynamical (GNO) classification,
      48-53
  dyons from, 92-93
  field of two, 68-69
  formation in spontaneously
      broken guage field
      theories, 73-78
  fundamental, 102
  gedanken hoax, 25-26
  Hamiltonian for particle moving
      in field of, 33-39
  infinitesimal stability
      condition, 67
  interaction energy, 69-70
  isorotational excitations,
      84-93
  masses of, 21, 80-82
    Bogomol'nyi bound, 82-83
    in SU(5), 96
    Prasad-Sommerfield limit,
        83-84
    quantum correction, 90

Magnetic monopole(s) (continued)
  non-Abelian, 42-71
    force between distantly
        separated, 68-71
    stability analysis, 62
  nonsingular, condition for,
      75-77
  notational conventions, 24
  QED and, 31
  quantum, 84-93
    eigenstates, 89
    eigenvalues, 89
    geometrical degeneracy, 90-91
    internal-symmetry degeneracy,
        91
    'rotational spectrum', 88,91
    translational degeneracy, 90
  significance of absence, 22
  size of, in SU(5), 96
  stable fields in SO(3) and
      SU(2), 66
  SU(5), 95-101
  't Hooft-Polyakov, see 't
      Hooft-Polyakov monopole
  topological (Lubkin)
      classification, 53-62
  virtual, 103
  Witten effect, 93-95
  Wu-Yang construction, 28-30
Majorana neutrino, mass, 12
Manhattan Project, 761, 762
Many-body problem 784-785
Mass, missing, 271-272, 294-295,
      297, 298
Matrix mechanics, 737-738
Matter-animatter asymmetry,
      273-291, 683-684
  baryon/photon number ratio,
      273-275, 279
  CP violation, 279, 281, 282
  drift and decay, 278-279
    microscopic analysis, 279-282
  evidence for, 275-276
  theoretical requirements,
      277-278
  thermalization, 282-283
Meson-Meson interactions, 90
Mesons (see also Individual
      particles)
  charmed, measurement of
      lifetimes, 375-382

Mesons (continued)
    superheavy, 96
    Weinberg-Salam guage, 100
Migdal conjecture, 145-146
Monopoles, magnetic, see
    Magnetic monopoles
Monte Carlo simulations, 119-156
    algorithms, 122-124
    continuum limit, 131-137
    'experiments', 124-127
    phase transitions, 137-142
    renormalization group, 131-137
    SO(3), 137-140
    string tension, 127-131
    SU(2), 139-141
    variants on Wilson action, 137-142
Muons
    heavy, 635-636
    pair production, 641-651
        asymmetry measurements, 645-651
    scalar, 628-631

Neutral current
    new, 178-179
    right hand, 179
Neutrino(s), 11-13
    cosmic background, 297
    cosmologically stable, mass limit on, 266-267
    decoupling, 264-265
    double beta-decay, 12
    effect of flavors on ^{4}He bundance, 269
    electron, see Electron neutrino
    endpoint experiments, 12-13
    Majorana, see Majorana neutrino
    mass, galaxy formation and, 272-273
    masses, 11, 192
    oscillation (see also E-53A Fermilab experiment), 399-403
    permitted number of kinds, 298
    relic sea, 264, 265
    solar, 11-12
    stability, 269-270

Neutrino(s) (continued)
    tau, see Tau neutrino
    terrestrial oscillation experiments, 12
    total cross section measurement (see also E-616 Fermilab experiment), 383-399
Nuclear war, 18
Nucleosynthesis, 267-268

Onia, scalar, 212
Onium decay, 208
O'Raifeartaigh model, 318-319, 343

Parity violation, 678-679
Partons
    charged primary, 537-540
    fractional charge, 599
    spin, 537
Peccei-Quinn symmetry, 369, 370
Photino, 206, 207, 628-630
Photon
    hard photon effect, 595
    in standard (3,2,1) model, 161
    mass at low temperatures, 228
Photon/baryon number ratio, 273-275, 279
Physics, chemistry and, 769-780
Pion, massless, 149
Planck mass, 339, 349, 351
Poisson bracket, 738-739
Positron, discovery of, 743
pp machines, see Hadron-hadron machines; Proton-proton interactions
Prasad-Sommerfield limit, 83-84
Projective geometry, 736
Proton decay, 5-10, 161, 191-192
    anti-SU(5), 9, 10
    conventional scenario, 8-10
    $\Delta B = 2$ processes, 6-7
    $\Delta B = -\Delta L$ decay, 7-8
    Higgs Dominant, 8
    SU(5), 9-10
    total decay rate, 10
Proton lifetime, 6, 17-18, 348
Proton-proton interactions (see also Hadron-hadron machines), 409-470, 695-731

Proton-proton interactions (continued)
  charged particle multiplicity, 708-712
  charged/total energy ration, 712-713
  comparison with e^+e^- data, 696-720
  effective energy range, 723-726
  event planarity, 713-715
  experimental data, 696-720
  inclusive fractional momentum distribution, 702-708
  inclusive transverse momentum distribution, 708
  leading hadron effect, 697-701
  'wrong trigger' reactions, 415-416, 436, 439, 440, 441, 443
Pseudo-Goldstone boson, 200
  Higgs boson and, 226
  states, 202

QCD
  electroproduction predictions, 165, 166
  perturbative tests, 167-170
  strong interaction tests, 163-171
  technicolor properties inferred from 201
  upsilon spectroscopy and, 474-485
QED
  fundamental magnetic monopoles in, 101-102
  light scattered by light in, 747
  with cutoff, 745
  with magnetic monopoles, 31
Quantum chromodynamics, see QCD
Quantum electrodynamics, see QED
Quantum monopoles, see Magnetic monopoles
Quark(s)
  -antiquark pairs, see Partons
  baryon production, 553-558
  bottom, see B quark
  charm, see C quark
  down, 684-685

Quark(s) (continued)
  fragmentation, 540-559
    function, 552
    inclusive particle spectra, 543-546
    models of, 550-559
    particle composition, 546-550
    rapidity distributions, 540-543
  free, 4
  search for, 620-622
  gluon production, 541, 558
  jets, see Quark jet
  masses and mixing angles, 227-228
  nomenclature, 411
  pseudoscalar meson production, 553
  scalar, see Scalar quarks
  six quark model, see Kobayashi-Maskawa six quark model
  technicolor, see Techniquarks
  top, see T quark
  transitions, 411-412
  transverse momentum, 553
  types, 552
  vector meson production, 553
  with integer charge, 16
Quark jets, 532-559
  3-jets, identification of, 569-576
  4-jet event, 580-582
  angular distribution, 536-537
  charge correlations, 537-540
  multijets, 559-566
  planar events, 569-576
  sphericity, 534
  thrust, 534-535
  in 'qg' systems, 573-574
Quarkonia, 604

Red-shift, 258-259
Relativity theory, 734-735
Renormalization, 744-745
Renormalization group, 131-137

Scalar bosons, 186-188
  mass, 197
Scalar leptons, 211
Scalar onia, 212

INDEX

Scalar quarks, 211-212
  mass, 212
Schrödinger theory, 739-741, 767-768
  hydrogen molecular ion, 752
SFM spectrometer, 412-414
Six-quark model, see Kobayashi-Maskawa six quark model
SO(2), Monte Carlo experiments, 126
SO(3)
  Monte Carlo simulations, 137-140
  number of stable monopole fields, 66
Solar neutrinos, 11-12
Space shuttle, 783-784
Spin statistics puzzle, 39-42
Split-Field Magnet (SFM) spectrometer, 412-414
Spontaneous symmetry breakdown, 71-73
  magnetic monopole formation, 73-78
String guage, 61
String tension, 127-131, 145, 146, 153
Strong interaction tests of QCD, 163-171
Structure constants, 43
SU(2)
  Monte Carlo experiments, 124-127, 139-141
  number of stable monopole fields, 66
SU(2) x U(1) model, fermion sector tests, 171-172
SU(3), Monte Carlo simulations, 131
SU(3) x SU(2) x U(1) model, 16-17, 157, 158, 161-163
  fermion assignments, 162
  forbidden processes, 191
  infrared fixed points, 229
  scalar SU(2) x O(1) assignments, 162
  supersymmetric, 204-206
  symmetries, 188-189
  tests
    boson sector, 178-188
    fermion sector, 171-178

SU(5)
  experimental predictions, 233
  instantons as supersymmetry breaking mechanism, 366-367
  monopoles, 95-101
  Monte Carlo simulations, 138-140
  proton decay, 9-10
  self-breaking, 152
  supersymmetry with $Tr(-1)^F = 0$, 368-369
  tests, 192-193
  vacuum expectation values, 240-241
SU(7), 238-240, 242-243, 246-247
  magnetic monopoles in, 243
Supergravity, 209, 357
Superspace potential, 317-318
Supersymmetry, 204-207, 305-371
  Bose-Fermi symmetry, 306-315
  breaking, 206, 360, 366-367
    explicit, 357
  composite fermions, 247
  cosmological constant and, 314
  explicit breaking, 357
  extended, 231
  guage heirarchy problem, 315-324
  Higgs mechanism analogue, 209
  'ino sector, 207-210
  mass spectrum, 234
  Monte Carlo calculations, 363
  non-renormalization theorems, 322, 341
  non-zero temperature effects, 356, 361
  phenomenology, 207-210
  prediction of bosonic baryons, 235
  prediction of fermionic mesons, 235
  proton lifetime and, 348
  renormalizability, 362
  scalar sector, 211-215
  spontaneous breaking, 312-315, 340-341
  strangeness changing, 234-235
  unified models, 237-249
  with central charges, 358-359
Superweak interaction, 692

Symmetry breakdown, spontaneous, *see* Spontaneous symmetry breakdown

t quark, 172, 173-174, 187, 494
  mass, 223
  search for, 603-611
Tau lepton, 622
  pair production, 641-645
Tau neutrino, 172-173, 402-405, 406
TCP invariance, 678, 679, 681
Technicolor (*see also* Hyperpions), 198, 199-204, 212, 213
  extended, 201, 203, 248-249
    couplings, 203
    problems with, 248-249
  properties inferred from QCD, 201
  signs of, in pp colliders, 233-234
  theoretical difficulties, 234-235
Technihadrons, 200
Techniquarks, 200, 230
Temporal guage, 46, 85
  equations of motion, invariance under gauge transformations, 64
Three gluon coupling, 225-226
Time, fixed cosmic, 255-256
Time-of-flight (TOF) technique, 415
Time reversal violation, 678, 681-683
Topological charge, 53, 57, 58
  condition for nonsingular monopole, 75-77
  time-dependence, 78
Topological classification of monopoles, 53-62
Topological stability, 67
Toponium, 588, 607
  decay, 177-178, 208, 232
  mass, 174-176
$Tr(-1)^F$ (boson/fermion number difference), 325-338, 367
  calculation in Wess-Zumino model, 330-333

$Tr(-1)^F$ (boson/fermion number difference) (continued)
  lack of change with theory parameter changes, 327-330
  massless theories, 365
  modified, 367
  supersymmetric non-Abelian gauge theories, 334-338
  theories with guage group U(1), 338
Tumbling mechanism, 152

U(1) Monte Carlo experiments, 126
Ultraviolet divergences, 120, 238
Universe (*see also* Cosmology)
  age of, 294
  causally disconnected regions, 284-287, 301
  closed, 284-285
  global structure of, 255-258
    dynamics, 256-258
    fixed cosmic time, 255-256
  homogeneity of, 293-294
  isotropy of, 293-294
  physical history, 260-264
  radioactive age, 260
  stellar age, 260
  visible, mass of, 257-258
Upsilon particles
  hadronic cascades, 482-485
  leptonic pair widths, 476-478
  masses, 474-476
  spectroscopy, QCD and, 474-485
  $\Upsilon(1S)$ decays
    inclusive particles in, 481-482
    hadronic, 478-481
  $\Upsilon(4S)$ and decays, 485-488
    baryon production, 493
    electron production, 489-491
    kaon production, 492
    leptonic width, 504-505
    muon production, 492
    photon production, 489

Vacua, 72
Vacuum expectation value, 72
Vector bosons, spectrum, 161

INDEX

W bosons, 181
  in Abbott-Farhi model, 226-227
  in standard (3,2,1) model, 161
  mass, 184, 200
  search for, 182-186
Weinberg-Salam model, 95
  guage mesons, 100
  monopole masses, 81
  non-singular monopoles and, 77
Wess-Zumino model
  calculation of $TR(-1)^F$, 330-333
  massless, 309-310, 332, 361
White dwarfs, 269-270
Wilson action, variants on, 137-142
Wilson loops, 47, 107-108, 128, 134, 149

Wino, 210
Witten effect, 93-95

X particle, 352-354

Yang-Mills equations, 33, 49
  Lagrangian, 121-122
  sourceless, 45

Z bosons, 178, 179
  in standard (3,2,1) model, 161
  masses, 184, 675
  search for, 182-186
Zino, 210